THE PHILOSOPHY AND PRACTICE OF SCIENCE

This book is a novel synthesis of the philosophy and practice of science, covering its diverse theoretical, metaphysical, logical, philosophical, and practical elements. The process of science is generally taught in its empirical form: what science is, how it works, what it has achieved, and what it might achieve in the future. What is often absent is how to think deeply about science and how to apply its lessons in the pursuit of truth, in other words, knowing how to know. In this volume, David B. Teplow presents illustrative examples of science practice, history and philosophy of science, and sociological aspects of the scientific community, to address commonalities among these disciplines. In doing so, he challenges cherished beliefs and suggests to students, philosophers, and practicing scientists new, epistemically superior, ways of thinking about and doing science.

DAVID B. TEPLOW, Professor of Neurology (Emeritus) at UCLA, is an internationally recognized leader in efforts to understand and treat Alzheimer's disease. He has published more than 250 peer-reviewed articles, books, book chapters, and commentaries and served on numerous national and international scientific advisory boards. In addition to his basic science work, Dr. Teplow has had a life-long interest in general philosophy and the philosophy of science. He spent a year-long sabbatical at the University of Cambridge as a Fellow of Clare Hall and a visiting scholar in the Department of History and Philosophy of Science, where much of this book was written. This experience stimulated Dr. Teplow to create new courses at UCLA in the ethics of science, social aspects of scientific paradigm change, and the roles played by ignorance and failure in scientific progress.

THE PHILOSOPHY AND PRACTICE OF SCIENCE

DAVID B. TEPLOW

University of California, Los Angeles

CAMBRIDGE
UNIVERSITY PRESS

Shaftesbury Road, Cambridge CB2 8EA, United Kingdom

One Liberty Plaza, 20th Floor, New York, NY 10006, USA

477 Williamstown Road, Port Melbourne, VIC 3207, Australia

314–321, 3rd Floor, Plot 3, Splendor Forum, Jasola District Centre, New Delhi – 110025, India

103 Penang Road, #05–06/07, Visioncrest Commercial, Singapore 238467

Cambridge University Press is part of Cambridge University Press & Assessment,
a department of the University of Cambridge.

We share the University's mission to contribute to society through the pursuit of
education, learning and research at the highest international levels of excellence.

www.cambridge.org
Information on this title: www.cambridge.org/9781107044302

DOI: 10.1017/9781107360129

First published 2023

A catalogue record for this publication is available from the British Library

Library of Congress Cataloging-in-Publication Data
Names: Teplow, David B., author.
Title: The philosophy and practice of science / David B. Teplow,
University of California, Los Angeles.
Description: Cambridge, United Kingdom ; New York, NY : Cambridge
University Press, 2023. | Includes bibliographical references and index.
Identifiers: LCCN 2022062324 | ISBN 9781107044302 (hardback) |
ISBN 9781107360129 (ebook)
Subjects: LCSH: Science – Philosophy. | Science – Methodology. |
Science – Philosophy – History.
Classification: LCC Q175 .T464 2023 | DDC 501–dc23/eng20230512
LC record available at https://lccn.loc.gov/2022062324

ISBN 978-1-107-04430-2 Hardback

Some people have the ability to work independently and successfully to write a book of this type. I am not one of those people. Without my wife Arden's decades of understanding, patience, and editorial support, this book would never have been written. I lovingly dedicate this volume to her.

It ain't what you don't know that gets you into trouble. It's what you know for sure that just ain't so.

—*Mark Twain*

Contents

Foreword by Hasok Chang

The Philosophy and Practice of Science is a rare and remarkable kind of book –
a full-length treatment of the philosophy of science written by a highly respected
scientist. Why would we want a philosophy book written by a scientist, you may
wonder. In fact, there is a venerable tradition of such books, and David B. Teplow's
new offering is a significant work in that tradition.

Even though science and philosophy began to go their separate ways visibly by
the middle of nineteenth century, many renowned scientists continued to reflect
philosophically about science. Ernst Mach, Pierre Duhem, Henri Poincaré, and
Percy Bridgman are just a handful of examples among many. Their philosophical
books have helped many a practicing scientist reflect deeply and freshly about the
nature of their work: What it is that they are really trying to achieve in the hard graft
of daily scientific work, and how. They have also provided much-needed enlightenment to scholars in other fields and many interested members of the public. I
have every hope that Teplow's booking will do the same.

Times have changed, of course, since those scientists wrote their classic philosophical works a century ago. Back then, even scientists at the top of their game
seem to have had the sort of leisure that allowed serious philosophical reflections.
Bridgman's first philosophical book, *The Logic of Modern Physics*, was published
when he was 45, two decades before he won the Nobel Prize in Physics. He closed
down his high-pressure laboratory at Harvard during the summer and retreated to
his vacation home in New Hampshire, writing his philosophical books during those
summers. Today's top scientists have no such leisure, especially in the experimental sciences, with the need to finance and look after their labs and lab members
continually. David B. Teplow is a fine specimen of a modern scientist in that
regard; among other things, he has had the major responsibility of directing the
Biopolymer Laboratory and the Mary S. Easton Center for Alzheimer's Research,
both at UCLA. Yet he has managed to produce this major philosophical work, and

I know that it has only been possible because this book project has been a labor of love for more than a decade.

Another important difference between Teplow's milieu and that of the earlier scientist–philosophers is the fact that the philosophy of science has become established as its own specialist field in the second half of the twentieth century. We still have some philosophizing scientists ranging from Richard Dawkins to Roger Penrose, but their philosophical outputs are generally regarded with suspicion by professional philosophers of science. Being one of this strange band of people, I do understand why. We go through our own specialist training, we have our own exclusive professional associations and journals, and we give our own university degrees. We don't feel that scientists who lack this professional training and experience are likely to be able to jump in and say philosophical things that are worth paying attention to. But even if we philosophers of science have something uniquely worthwhile to say, it remains a fact that a clear majority of scientists and the general public have on the whole not been listening to it, which is partly because we are quite bad at expressing our insights in a way that anyone else can understand.

I am hopeful that some of the current new trends in the philosophy of science are beginning to make our work more relevant both to the general public and to the practicing scientists. But we also need people to help us by meeting us halfway. From the side of the practicing scientists, we need people who are willing to bring not only their obvious scientific expertise and experience but also a degree of respect for what we philosophers have been trying (and sometimes able) to achieve. Teplow is just such a scientist. In order to work up the knowledge and insights that went into this book, he has devoted an enormous amount of time and energy into philosophical study.

One of the most amazing moments in my academic life was when I was contacted, quite out of the blue, by David B. Teplow in late 2012, asking if I would host him as a visiting scholar in the Department of History and Philosophy of Science at the University of Cambridge. Back then, a decade ago, he had already written a detailed proposal for this book. Quite coincidentally, I had anonymously reviewed this proposal for a publisher, so when he contacted me, I already knew who he was and what a special project he had. My colleagues and I welcomed him enthusiastically, and so it was that a protein scientist from a major medical school came to spend his hard-earned sabbatical year in among philosophers and historians. During that 2013–14 academic year, we were also blessed with a visit by the neuroscientist and polymath Stuart Firestein, as well as the longer-term presence of Rick Welch, another philosophical scientist. Honoring and taking advantage of their presence, I started a discussion group called "Coffee with Scientists" bringing philosophers and historians of science into dialogue with practicing scientists, which is still running after a decade. All of that academic year we had stimulating discussions, at which we came to expect Teplow to jump in with his signature intervention: "But you're making an assumption!" That is a line that might be considered the philosopher's prerogative, and it was all the more refreshing coming from the scientist.

Books, meanwhile, have a habit of not getting done during a sabbatical year intended for their completion. Especially in writing a book like the present one you have in your hand, finding out one thing leads to the desire to learn five other things. Updating one important thing demands the adjustment of the whole structure. The process can be endless. Teplow continued adding to the manuscript and re-working it time and again. The outcome is a veritable intellectual feast, including a sweep of the whole history of scientific methodology in Chapter 4, going from ancient Egypt to today's data-driven research. This is something that professional philosophers these days dare not attempt, for their own safety but to the detriment of those who want to learn something. Teplow's focus in this historical survey is on how various scientific methods arose: Why were they necessary, in order to deal with what kind of problems? Throughout the book, there are scintillating insights that arise from a combination of the author's deep dive into esoteric philosophical and historical material, his direct experience of conducting and managing scientific research, and his refusal to accept conventional wisdom and hackneyed perspectives.

More than anything, this book makes something of a philosophical manual for working scientists, especially those who are in the early stages of their careers. In that sense, it continues in the tradition of the pathologist W. I. B. Beveridge's *The Art of Scientific Investigation* (1950) and the immunologist Peter Medawar's *Advice to a Young Scientist* (1979). I do hope that generations of students and researchers entering their scientific careers will discover this book and benefit from its abundant wisdom. I certainly wish that such a book would have been available to me when I was unsuccessfully trying to make my own way through scientific training, my efforts hampered by some misguided notions about what was really involved in the process of doing science.

The "process of science" is something that Teplow emphasizes throughout the book, explicitly and implicitly. In his discussions, science comes alive, as a process of learning and acting that is continuous with the process of intelligent life itself. Reading Teplow wonder about the process of science is a wonderful reminder of why the philosophy of science is needed, and why it needs to stay in close contact with science itself.

Hasok Chang
University of Cambridge

Foreword by Michael S. Wolfe

What is science? What does it mean to "do" science? How can science be done thoughtfully? These broad, fundamental, and challenging questions are taken on by David B. Teplow in the pages that follow. Motivated to write this book after a professional lifetime as a biomedical research scientist, Dr. Teplow, Professor Emeritus at the University of California in Los Angeles, sought to bring together the philosophical and the practical considerations of science into a single volume. In so doing, he hoped that philosophers of science would better understand and appreciate what practicing scientists actually do in the laboratory and that practicing scientists would better understand and appreciate the philosophical underpinnings of what they are trying to accomplish cumulatively through experimentation.

For too long, the philosophy and practice of science have been largely considered separately, as if the two resided in virtually nonoverlapping realms. Philosophers formulate ideas about the deeper meaning of the scientific enterprise as a whole – addressing problems such as how we can "know" if a hypothesis or theory is "true" – without considering how real experiments are actually designed and executed and the results analyzed, interpreted, and put into larger context. In parallel, practicing scientists forge ahead doing experiments to advance their specific fields of investigation, with little or no thought to how ingrained bias – in the individual and the scientific culture in which the individual is embedded – can influence experimental design, execution, analysis, interpretation, and perspective.

For philosophers of science, Professor Teplow imbues his text with many examples of how science is conducted in the laboratory, describing approaches and methods in sufficient detail that is free of jargon, so that one can appreciate the challenges of experimental design and how experimental results can lead to a better understanding of nature, despite limitations and caveats. For practicing scientists, he provides a rich historical perspective on the evolution of science and the philosophy of science over the course of some 5,000 years, from ancient Egypt to the present. In the process, reflective scientists should come to appreciate that what they so firmly believe to be true and obvious today may not be so in the future. For although scientists commonly view themselves as objective, they cannot escape the fact of being embedded in a particular moment in history. This does not mean that contemporary advances are doomed to become worthless in the future; rather, such discoveries are stepping stones in humankind's asymptotic approach to understanding our world and universe.

This book also provides practical real-world advice to young scientists (doctoral students, postdoctoral researchers, and early-stage investigators) on such issues as formulating a worthwhile hypothesis or question; the importance of experimental approach, design, rigor, and reproducibility; avoiding bias and preconceived notions in the interpretation of experimental results; and placing new findings into a broader context. Young scientists will also have to navigate the social, cultural, economic, and political milieu in which they find themselves, and in that regard, this book gives insight into the importance of conveying their findings as publications in reputable journals, interacting with colleagues and competitors in their field of investigation, and convincing funding agencies to support their research projects.

But perhaps the greatest service to young scientists rendered by Professor Teplow is instilling renewed meaning and value to the title of their degree: Doctor of Philosophy. In the modern scientific era, the Ph.D. degree is typically conferred for sophisticated technical abilities in the laboratory, employment of these abilities to address a scientific problem that was framed by the mentor, writing up the cumulative findings into a cohesive and coherent dissertation, and orally defending the dissertation before a doctoral committee. While preparing the written dissertation and defending it provide opportunities to think more broadly and place the work into a wider context, the newly minted Ph.D. scientist is too often closer to a glorified technician in a highly specialized field than to a scholar and a philosopher.

Professor Teplow forces the young scientist to stop and think about what they are actually trying to accomplish in the laboratory and ask themselves about the meaning and value of their work. Indeed, his advice includes this nugget: Don't just do something, sit there! He goes on to explain the value of sitting and thinking carefully and deeply about what one is trying to do and why, instead of just rushing ahead with more experiments. More broadly, he encourages young investigators to consider the wider meaning of the scientific enterprise and ask themselves why they believe they know what they think they know. Most profoundly, he asks what it even means to "know" something or to say something is "true." These are the deepest of philosophical questions, and the young scientist may think they are too abstract and not relevant. Yet how can one be a doctor (or teacher or scholar) of philosophy without giving these matters any thought?

Even for the established practicing scientist though, this book is highly worthwhile. I myself have been "doing science" for nearly 40 years and have been leading my own academic research laboratory for nearly 30 of those years, and I found Professor Teplow's book to be educational, enlightening, thought-provoking, and engaging. Reading this book reminded me to stop and think more deeply about the meaning and value of my own work, why I continue to "do science," and what I hope to accomplish. So, I encourage all those who aspire to be true Doctors of Philosophy or to reaffirm themselves as such: Don't just do something, sit there and read this profound and important book, thoughtfully.

Michael S. Wolfe, Ph.D.
University of Kansas

Preface

After more than half a century of studying, doing, and teaching science, I was shocked by the realization that *I* was in a lot of trouble. Not only did I find my fundamental knowledge of science was flawed, but I realized that many of my own life-long assumptions were wrong. Can you imagine my mortification when I found out that water, simple H_2O, may not boil at 100°C or that negative temperature $(-1K)$ exists?[1]

Can you imagine my shock that modern planeteria use Ptolemy's 2,000-year-old geocentric, not the correct heliocentric model of the heavens to project the movement of the planets and stars? Can you imagine my disorientation when I found out that time is not objective but actually determined periodically by the votes of committees of scientists around the world?

I'm a reasonably well-trained (UC Berkeley, Caltech) and knowledgeable scientist. I have been lucky to have been able to learn and work in diverse scientific fields, including biochemistry, bacteriology, immunology, virology, molecular biology, protein chemistry, spectroscopy, computational chemistry/physics, and neuroscience. I've "done science" as a faculty member at Harvard Medical School and UCLA, ostensibly excellent institutions. I've been involved in most aspects of peer review, including manuscripts, books, grant applications, faculty hiring and promotion, government panels, and private institutions. I myself have published more than 200 peer-reviewed papers. I've taught and mentored many undergraduates, graduate students, and postdoctoral fellows. How is it possible that I could be in so much trouble? And if I were, what about students and even other professors? What did I miss?

My answers to these and other questions comprise this book. They may not be *the* answers, anyone else's, or the common wisdom, but that is largely irrelevant because it will be the *experience* of reading and thinking about what you have read that provides answers, and these will be your own. "Experience" here means

---

[1] See Purcell [1] and Needham [2].

much more than simply learning new things, enjoying amusing anecdotes, becoming more familiar with the history and philosophy of science, gaining insight into how scientists select questions to answer and do so experimentally, or acquiring a sense of how science yields the truth about nature. You will do all of these, but far more importantly, you will be immersed in "the process of science."

The process of science, as I define the concept here, is a *behavior*. Its elements are theoretical, metaphysical, logical, philosophical, and practical, all at the same time. From an early age, we are taught the process of science in its empirical form: *what* science is, *how* it works empirically, *what* it has achieved, and *what* it may achieve in the future. What largely is absent is how to *think* about science and how to pursue truth, for if we cannot do the former it is unlikely we will be able to do the latter. We are taught the mechanics of research and mentored until we become independent investigators, after which we will do the same for our own students, which unfortunately means graduating students with a profound lack of recognition of factors implicit in their work that diminish scientific excellence and impede discovery. These factors include dogmatism, assumption, ignorance, lack of rigor, and most importantly, superficiality. Students are educated with an emphasis on "doing science" at the expense of "understanding science."

I argue that excellence in scientific practice requires inculcation not only in its empirical form, but also in its historical development, philosophy, and sociology, which the student and scientist should integrate into their design and performance of experiments, interpretation of results, dissemination of findings, and evaluation of the veracity and importance of their work. This is what I consider the "process of science."

The process of science *is* this book, and this book *is* the process of science. The manner in which it is written *is* scientific behavior, which means *thinking deeply and digging deeply*. As you read, you not only will be asking questions, but you will also be learning *how* and *why* to ask and answer them. You not only will learn about the history and practice of the scientific method, you will also question how it came about and whether it makes sense to believe the answers it produces. You will explore the method's rationality and find, contrary to what you may believe, that nonscientific factors (psychological and sociological) often play as much, or more, of a role in the process than does pure science.

The process of science is epistemological in nature. It seeks to question dogma and assumption. It reveals ignorance, especially where none was thought to exist. It is the antithesis of superficial, but it does provide the means to determine when one has "dug enough." In summary, and put pithily, *it is hoped that by reading this book aspiring and practicing scientists will know that the most important thing to know is to know how to know.*

This approach should be of particular value to graduate students and postdoctoral fellows learning science and the philosophy of science, faculty who have a responsibility to not only teach students facts, but to make them think for themselves, and scientifically literate nonscientists. Illustrative examples of science practice, history and philosophy of science, and sociological aspects of the scientific community are presented primarily in relation to the natural sciences, but the examples are relevant to scientific practice outside of them because they address commonalities among most disciplines. Readers also should note that the subjects discussed in this book are vast, which means that discussion of any topic must, perforce, be limited in length. However, citations throughout this book will help guide those interested in developing greater knowledge in specific areas to good sources of information.

Any information derived from others, including personal communication, quotations, web pages, books, peer-reviewed and non-peer-reviewed publications, presentations, and media sources, is cited. This is the ethos of science, should be practiced by all, and why 688 sources are cited. If one uses the insights, work, or opinions of others, these individuals should be acknowledged. Also, when "digging deeply," one always uncovers more than for what they were looking. Footnotes provide deeper insights into specific sections of text, clarify and further explain complex issues, introduce new ideas, stimulate thinking and further exploration of topics, correct common misconceptions, and reference sources of information. One need not read all 777 footnotes, but I encourage the reader to patiently take the time to do so because footnotes reveal interesting and surprising things not generally taught to students and often not known by even the most accomplished practicing scientists.

Acknowledgments

The production of this book was facilitated substantially by a sabbatical year in the Department of History and Philosophy of Science (HPS) at the University of Cambridge. I am deeply indebted to Professor Hasok Chang (HPS) for his guidance and constructive criticism of early drafts, but most importantly, for his friendship. I also thank the participants in the "Coffee with Scientists Reading Group" (HPS) for helpful suggestions and criticisms regarding the question of "what is science?" HRC Clark Rosensweig helped me in discussions concerning the age dependence of scientific acumen and intuition. I owe debts of gratitude to Tim Lewens, Shahar Avin, Stuart Firestein, Larry Laudan, Mary Domski, Lorraine Daston, Aphrodite Kapurniotu, Sherri Roush, Eric Scerri, Deborah Mayo, and Mike Wolfe for stimulating and informative discussions. George Benedek, Professor of Physics, Emeritus, MIT, deserves special mention for taking me under his wing as a new assistant professor at Harvard Medical School. His tutelage helped make me the scientist I am today. I acknowledge the many other students, teachers, colleagues, collaborators, and reviewers, too numerous to specify, that deserve praise for any success I may have enjoyed (but not for any failures, which were all my own fault). Finally, my long-suffering editors and colleagues at Cambridge University Press, Simon Capelin, Vince Higgs, and Sarah Armstrong, deserve thanks for their preternatural patience during the extraordinarily long gestation period of this book. A special "thank you" goes to my son, Gregory Teplow, for his insight into the development of intuition through historical study.

1

Introduction

Science is a primary force driving societal development. Science exists as a consequence of human curiosity about the natural world and of the practical needs of civilizations to survive and prosper. Knowledge (in Latin, "scientia") is what provided the means to manufacture the paper on which these words were printed, to fabricate the printing presses that did the imprinting, to extract the dyes that formed the letters, and to build the engines of the trucks that delivered the book to the bookstore. Science produced the medical knowledge necessary to prevent or cure illnesses that used to kill so many as infants or children. It enabled us to travel to the moon, explore distant planets, and study galaxies. Let's also not forget that without science there would be no Internet and no social media (whether this actually has been more of a benefit than a nuisance remains debatable).

Science is one of the few endeavors that provides those with insatiable curiosity a never-ending source of questions and answers. Curiosity may, in fact, be what all scientists have in common and the reason they became scientists. Scientists do want to discover new things about the natural world, things that will help humans exist in, and manipulate, the world to their advantage. However, most are committed to science for its own sake. To understand this better, let us consider young children asking the question "why?" This question is a natural expression of the curiosity of the child and a product of the evolutionary value of education for species survival. Parents, and others, often perceive the question as "cute." However, when the question is repeated incessantly, that cuteness can turn to annoyance. For the successful scientist, however, the "why" question initiates the search for greater understanding of our world. It drives the scientist to look deeper into questions and to search for ever greater elucidation of natural phenomena. With each new insight obtained, many other questions emerge. In short, scientists never are able to satisfy their desire to know why, which is not an annoyance, but a wonderful, challenging, fulfilling, and important endeavor that brings the great joy. Stuart Firestein put it this way:[1]

I am afraid that it is impossible to convey completely the excitement of discovery, of seeing the result of an experiment and knowing that you know something new, something fundamental, and that for this moment at least, only you, in the entire world, knows it.

---

[1] Quotation from Firestein [3], p. 160.

In a *Science* article published in 2017 [4], scientists offered their opinions about why science is important to them. These scientists were from diverse fields and emphasized the importance of science for, among other things, sustainability and conservation, human health, truth, shared history, shared future, and discovery and wonder. Samantha VanWees offered a particularly personal and insightful comment about the question "why?"

Thanks to public education and television, I fell in love with science and research at a young age. I've studied why bread rises, why meat browns, why canned peas soften, and why food tastes so darn bad on airplanes. We like to think of science as facts and logic, but really science is about finding out just how much we don't know. We must continue investing in science so children know the importance of asking "why?" Science and wonder are all around us, all the time. You just have to look for it.

One may gain satisfaction in knowing why and then moving on. The question has been asked and answered. "I have learned the answer and thus I am more knowledgeable." However, each quantum of knowledge raises more questions than it answers. In this way, the scientist, over years and decades, may increasingly appreciate the surprising breadth of ignorance they exhibit. In essence, and paradoxically, the more one knows the more one realizes how ignorant they are. This may, at first, appear unfortunate. I argue that it is not. Instead, it is a reflection of the success of the effort to reveal the truths of the physical world, and in doing so, simultaneously reveal the world's complexity. It should be a humbling experience, and this is critically important for reminding the scientist of how difficult and tenuous the search for truth is. Darren Saunders, University of New South Wales,[2] expressed this notion quite nicely when he said:

So many questions without answers. So many experts with differing views. A brutal realisation that things don't work the way we always thought. Seeing infinite shades of grey instead of that comforting black and white world. Sound familiar? Welcome to the mind of a scientist.

Science is a field of endeavor that is endlessly fascinating, stimulating, and enriching. Science also can be incredibly difficult, frustrating, and exhausting. I have loved science since the time, as a child, that I first visited my parents' clinical laboratory. Hearing the whining of centrifuges, seeing donut-shaped, pink erythrocytes and purple and white polymorphonuclear leukocytes in a microscope, or looking at clear solutions in a water bath suddenly become turbid, were magical experiences. For the next half-century, from public school to the university to graduate school to postdoctoral fellowships and through the professorial ranks, science has remained a love. In fact, doing science has been and remains more of an avocation than a vocation. Most scientists feel very lucky to be "doing science" rather than having a "real job."

What does it mean to "do science?" Doing science is not a job at which you punch the clock when you arrive and when you depart. It is not a compartmentalized component of your existence. Science can be, and often is, a way of life. In this sense, it has stages much like birth, childhood, adolescence, adulthood, old age, and death. Doing science requires parenting (mentoring), schooling, learning to function in a community, developing a sense

---

[2] See "Uncertain? Many questions but no clear answers? Welcome to the mind of a scientist." https://theconversation.com/uncert ain-many-questions-but-no-clear-answers-welcome-to-the-mind-of-a-scientist-134388

Table 1.1 *What if? This table uses the well-known fishing metaphor to illustrate, in science education and beyond, the value of integrating the history and philosophy of science with the practice of science.*

If you give a person a fish	→	they can have *a* meal
If you teach a person how to fish	→	they can have *many* meals
If you provide a context for fishing	→	they can find out *why* one would fish in the first place, evaluate the desirability of fishing, learn about successful and unsuccessful fishing methods utilized from antiquity to the present, determine if fishing might be damaging to the environment, decide if one was or was not concerned about such damage and why or why not, appreciate the potential emotional rewards or frustrations of the activity, consider whether teaching one's progeny how to fish would be desirable, develop awareness of other methods of food acquisition, evaluate the moral question of whether one should kill and consume fish, etc.

of one's own individuality, making mistakes and overcoming them, establishing one's own niche in the scientific world, and finally, raising your own scientific children.

The *gestalt* of the scientific enterprise is rarely, if ever, considered explicitly in science education. Yet this consideration is critical if new generations of scientists are to acquire a holistic perspective of their own field. A holistic treatment of the scientific enterprise provides the practical and philosophical knowledge to enable a person to appreciate and to "do" science better. This is not a goal relevant only to a small community of cloistered practitioners of the philosophy of science. If societies are to thrive and evolve, it is incumbent upon them to produce the most capable scientists.

In much the same way as strategic planning is a prerequisite and a foundation for the execution of tactics that will achieve a desired outcome, an appreciation of the *gestalt* of science is a prerequisite for optimal experimental planning and execution. In short, to be a successful scientist, one must know substantially more than a series of useful facts like Avogadro's number, Michaelis-Menten kinetics, electron orbital geometry, or nucleosome structure. It is not enough to be a bright, energetic, objective participant in paradigmatic science. One must dig deeper and look more broadly,[3] especially into the historical and philosophical principles providing the foundation for science practice. Table 1.1 provides a conceptual illustration of the value of this approach. Table 1.2 lists achievements fostered by the integration of science history, philosophy, and practice.

The thesis of this volume is that if one is to "do science" most effectively, and to derive maximal satisfaction from the endeavor, one cannot simply execute the technical processes

---

[3] It is interesting that the terminal degree in science is a Ph.D., which designates one as a *Doctor of Philosophy*, yet it is rare that graduate programs have any requirements for education in philosophy itself. This has led some to question whether the scientific Ph.D. represent anything more than a technical degree (see foreword by Miller in Gauch [5]).

Table 1.2 *Why integrate science history, philosophy, and practice?*

To understand the philosophical and historical reasons science is done as it is	To understand and use the scientific method
To create a foundation for truth	To find truth
To create a scientific experiment or project	To appreciate the loveliness of nature
To formulate a hypothesis	To evaluate hypotheses
To understand	To explain
To improve analytical and theoretical thinking	To justify public policy decisions
To apply logic in inferential processes	To insure assumptions are reasonable
To explore assumptions intrinsic to knowledge creation	To integrate probabilism into hypothesis creation
To understand different methods of knowledge creation	To understand the limits of your understanding
To chart new directions for science	To solve intractable problems
To enjoy thinking deeply	To know "why"

that comprise experimental science – *one must "understand" science*. A person who is expert at such processes certainly is a gifted technician, and science cannot be accomplished without such expertise, but a scientist is more than a person *doing* experiments. The scientist is a thinker, not just *in addition* to manipulating scientific instruments and materials, but first and foremost. Chemical robots that perform high throughput chemical reactions are superb at this task, but they are not scientists in the classical sense.

As with any book on a particular subject, one generally starts by defining it. Chapter 2, "Defining Science," considers this question. It turns out that it is difficult, if not impossible, to reach consensus on what science is, but the attempt to do so introduces the reader to a general theme of this book, when one "digs deep," one may be surprised that their knowledge base is not as firm as they thought, if at all. One has had the rug pulled out from under them.[4] Digging deeply and holistically is one approach for minimizing, but not eliminating, this risk.

If one can't define science, then distinguishing it from nonscience also appears hopeless, yet this distinction has become critical in this political, internet age. With the ability to disseminate information with single key click, we are now experiencing not just an "information age," one in which the amount of credible, logically supportable, fact-based information has exploded, but an age in which, as in Orwell's book *1984*, language has been transmogrified into Newspeak, which facilitates and encourages contradictory expressions such as "War is peace," "freedom is slavery," and "ignorance is strength." In 2022, we have

---

[4] I can think of no better vignette for this metaphor than the following. "The rabbinical student is about to leave for America. When he asks his mentor for advice, the rabbi offers an adage that, he tells the student, will guide him for the rest of his life. 'Always remember,' the rabbi said sagely, 'life is like a fountain.' Deeply impressed by his teacher's wisdom, the student departs for a successful career in America. Thirty years later, he learns that the rabbi is dying, so he returns for a final visit. 'Rabbi,' he says, 'I have one question. For 30 years, whenever I was sad or confused, I thought about the phrase you passed on to me, and it has helped me through many difficult times. But to be perfectly frank, I have never understood the full meaning of it. Now that you are about to enter the realm of truth, tell me, dear rabbi, why is life like a fountain?' Wearily, the old man replied, 'All right, so it's not like a fountain.'"

shrill statements such as *"the truth of science is false."* What are we to believe, and why? These questions are among those addressed in Chapter 1.

Chapter 3, "Learning Science," focuses on concepts of science outside the purview of rote learning, concepts that are more metaphysical and sociological in nature. I discuss how history has shown that facts may be fallacies and fallacies may be facts ... depending on historical context. I then discuss how personalities, not just scientific knowledge, can drive science forward. Personality traits such as courage, unwavering commitment, immunity to criticism, farsightedness, intuition, perspective, and yes, egotism, all are associated with great leaps in science, from before Galileo and on to Isaac Newton and Gottfied Leibniz, Niels Bohr and Albert Einstein, Max Delbrück and Seymour Benzer, and more recently, Stanley Prusiner.

One would think that knowledge drives science forward, but one would not be entirely correct. It is not knowledge *per se*, but rather ignorance, that drives scientific advancement. If one *already* knows something, i.e., one possesses knowledge, there would be no reason to search for more – unless this knowledge raised questions or issues about which one was ignorant. It is the recognition of ignorance that propels us forward. "I don't know how this works. Let's create a hypothesis why and then test it experimentally to see if it explains." What happens, however, if we don't know what we don't know? How are we to reveal questions about a universe that exists in all its complexity without knowing the questions that could be asked? In Chapter 3, we discuss this and learn how to be *more* ignorant.

When we "create a hypothesis why, and test it experimentally," it is highly likely we will do so using the "scientific method." It may seem to some that this method is a product of modern science, but surprisingly, elements of it can be found in antiquity, in Egypt 5,000 years ago, and in the writings of Aristotle in the fourth century BC. How the current scientific method came into being is the subject of Chapter 4, "Development of the Scientific Method: From Papyrus to Petaflops." Here, I trace what I believe are the key moments and key people that contributed significantly to this process. I move from the time of Ptolemy to the great leaps made in the Islamic world during the tenth and eleventh centuries, to the scientific revolution of the sixteenth and seventeenth centuries, and on to what I call the "probabilification" of the method in the early twentieth century. The scientific method has been the mainstay of science throughout history, but does it remain so in the twenty-first century? Some think not, arguing that there *is* no scientific method or "anything goes," and with the development of powerful computers and algorithms, a new scientific method, "data-driven science," may signal the end of the scientific method as we have known it.

Putting aside this eventuality for the moment, Chapter 5, "Science in Practice," discusses the process of science, including how scientists choose projects, experiment, observe and interpret, and infer. Inference has always been the bread and butter of science. We see something happening consistently, and we generalize this observation into a law, a hypothesis, or a theory. We do this through abduction, induction, and deduction, three modes of inference. But how does one actually conceive of a theory or select one from among many candidates? Once determined, how does one know that their theory holds water? How are theories tested? These are among the issues discussed.

When one develops a theory, one does so believing it is a true reflection of processes occurring in the natural world. Is this belief warranted? *Does* science actually explain the real world or does it just explain a world we imagine? This question is at the center of debate among those who answer "yes" (realists) and those who answer "no" (anti-realists). But wait, what do people mean when they use the term "realism?" Maybe in one context our theories *are* real, but in others they are not. Could our sense of realism be somewhere between the poles? Maybe realism is neither solely empirical nor theoretical, i.e., maybe it is metaphysical in nature. Even if we *are* realists, does knowledge = understanding?

Knowledge has been characterized as "justified true belief," which in many cases comes through statistics. Statistics is an obligatory part of training in science because it gives us the tools to determine if our observations are or are not "significant," i.e., whether we are justified in believing them. Significance in this context usually means $p < 0.05$, which is a value almost universally accepted in the biomedical sciences.[5] But why? As we dig a little deeper into the question, we find that the convention of considering an observation with a $p < 0.05$ significant was *arbitrary*. One could just as well have chosen $p < 0.01$ or $p < 0.001$. What is more disturbing is how confidence intervals are (mis)used and so lead us to make unjustified conclusions. This is a particularly large problem in clinical trials and one that is responsible, to a significant degree, for their low replicability (10%!?). All of these issues beg the question of "how do we know something is true?" a question addressed in the last section of Chapter 5.

We have discussed practical, theoretical, historical, and philosophical aspects of science. Each can be said to comprise elements with certain levels of predictability. For example, we have the practical skills necessary to determine the melting and boiling points of chemicals, which are constants. We construct theories guided by logic, laws of nature, and observations of our own or of those of others. We consult the written historical record to understand the genesis of our world and the things in it, a record we can always come back to knowing its elements will remain relatively constant.[6] We use the tools of logic to frame philosophical arguments. The *ways* we go about doing these things, although not rigidly codified, do lead to the expectation that two people ostensibly doing the same thing will produce the same result. However, as they say in cheesy[7] commercials for useless gimmicks on day-time television in the United States (all for $19.95 [shipping not included]), "But wait! There's more!" The "more" is illogic, irrationality, impulsiveness, dogmatism, competitiveness, the quest for fame, hubris, career advancement, political or commercial agendas, etc. – in a word, "sociology." Few may appreciate the importance and impact of sociology on science practice. It is immense and unpredictable and is considered in Chapter 6, "Science as a Social Endeavor." The Gershwin song, "Let's call the whole thing off," provides a lyrical example of the unpredictability of human (including scientist) behavior.

---

[5] The *p*-value is the likelihood a given experimental result is due to chance.

[6] Divergent historical reports may exist and interpretations of those reports may differ, but the root literature remains constant even as new information is added. We don't rewrite Newton's *Principia* or purge the historical record of things we don't like (unless we live in totalitarian societies in which the government determines to what information people will have access and the level of veracity of that information).

[7] For those not familiar with this expression, definitions include "bad quality or in poor taste, tacky, tasteless."

You say eether and I say eyether
You say neether and I say nyther
Eether, eyether, neether, nyther
    Let's call the whole thing off!
You like potato and I like potahto
You like tomato and I like tomahto
Potato, potahto, tomato, tomahto!
    Let's call the whole thing off!

2

Defining Science

> Science does not rest upon a solid bedrock. The bold structure of its theories rises, as it were, above a swamp. It is like a building erected on piles. The piles are driven down from above into the swamp, but not down to any natural or given 'base': and if we stop driving the piles deeper, it is not because we have reached firm ground. We simply stop when we are satisfied that the piles are firm enough to carry the structure, at least for the time being.[1]

It is natural to begin the discussion of a subject with an explanation of what the subject is. This provides a scaffold upon which details about the subject can be placed, and importantly, integrated into an organized whole. It is likely that the reader has experienced this process in grade school and beyond as they received formal education in English, history, art, biology, and other subjects. This approach is reasonable and useful. When we begin to learn about a new subject, we are largely ignorant about it and thus we count on our teachers to educate us. We assume that the facts they teach us are, in fact, facts. Some readers may trust that my exposition of science, in this section and throughout the text, faithfully represents what is true. They should not! What they *should* do is carefully observe the approach taken to produce the exposition and then determine for themselves whether to believe what has been written.[2]

Descartes[3] may have said it best when opining about his own education:

Several years have now lapsed since I first became aware that I had accepted, even from my youth, many false opinions for true, and consequently that what I afterward based on such principles, was highly doubtful; and from that time I became convinced of the necessity of undertaking once in my life to rid myself of all the opinions I had adopted, and of commencing anew the work of building from the foundation, if I desired to establish a firm and abiding superstructure in the sciences.

We want to know what science is, but before we can be confident in what we learn we must have reason to believe that what we learn is true. The study of "how we know what

---

[1] Quotation from Popper [6], p. 94.
[2] As argued in the *Preface*, this approach itself is the most important thing to be learned from reading this book – more important than any facts revealed therein.
[3] See Descartes, *Meditations on a First Philosophy*, Meditation I, p. 1.

we know" is epistemology and it is the bedrock upon which the entire scientific enterprise rests. The Oxford English Dictionary (OED) defines epistemology as "The theory of knowledge, especially with regard to its methods, validity, and scope, and the distinction between justified belief and opinion."[4] Let us now apply the principles of epistemology to the original question posed, namely "what is science?" Our methods will include an examination of the etymology of the term and its history of use, ascertainment of the opinions of prominent scientists and philosophers of science, and consideration of why science is viewed by many as something "special," that is, especially authoritative and true, relative to other pursuits, for example, art, law, and religion.

Our epistemological quest begins with the Greek philosopher Aristotle (384–322 BC), and interestingly, with the word root of epistemological, epistēmē (Greek: $\epsilon\pi\iota\sigma\tau\eta\mu\eta$). In Aristotle's era, epistēmē meant knowledge. The word he used for this type of knowledge was *scientia*, which is the root of the Latin word *science*. Science thus was *a particular type of knowledge*, knowledge gained through "demonstration," which for Aristotle meant "a deduction that produces knowledge." Surprisingly, the first reported use of this term in the English language did not occur until the seventeenth century [8], when the term was borrowed from the French. In the Early Modern Era, the term is found in the 1620 publication *Novum Organum*, "New Method," by the English philosopher Sir Francis Bacon, 1st Viscount St Alban.[5] Bacon is considered by many to be the father of the scientific method (based on his Baconian method [9]) and of empiricism, the notion that "sense experience is the ultimate source of all our concepts and knowledge" [10]. A key tenet is knowledge must be obtained through experience (e.g., experimentation). Experimentation must be conducted in a rigorous manner, one that allows appropriate inferences to be made from results. Rigor was integral to the Baconian method (later the scientific method), but many at that time did not apply sufficient rigor to their studies. Bacon bemoans this fact in saying:

This science therefore (as I understand it) I may justly report as deficient: for I see sometimes the profounder sort of wits, in handling some particular argument, will now and then draw a bucket of water out of this well for their present use: but the spring-head thereof seemeth to me not to have been visited; being of so excellent use both for the disclosing of nature and the abridgment of art.[6]

A lack of scientific rigor remains a serious problem in science today, four centuries after Bacon, and for this reason, I emphasize here the obligatory nature of understanding and practicing scientific rigor in all aspects of the scientific process, including the practical, theoretical, and philosophical.

The meaning of science, that is, as "demonstrated knowledge", has evolved since Aristotle's time. Language is malleable [11] and new words are created to represent novel things or ideas. Existing words may become archaic as they are supplanted by newer words or

---

[4] Oxford English Dictionary, 3rd edition, Oxford University Press, 2014 www.oed.com/view/Entry/63546? redirectedFrom=epistemology#eid. We will not discuss here the interesting question of whether or not we should accept this definition, as definitions are not determined dispassionately through the application of some universally agreed upon algorithm but rather, and quite often, following deliberation among members of a committee of linguists appointed through a selection process (a process that itself could be biased) [7].

[5] Oxford English Dictionary, 3rd edition, Oxford University Press 2014 www.oed.com/view/Entry/172672? redirectedFrom=science#eid.

[6] See Bacon [9].

their original meanings change. From the time of the ancient Greeks through the Middle Ages and into the early modern period, scientific investigation was conducted by philosophers. The subject of science was subsumed under the broad rubric of philosophy and those practicing "science" were known as natural philosophers.

This changed in the age of Sir Isaac Newton when, instead of referring to science as natural philosophy, he used the term "experimental philosophy" in his *Principia* [12]. It was not until the eighteenth century that the word science began to be used in connection with those branches of philosophy dealing with the natural world [8], and it was not until the late nineteenth century that the Baconian meaning of the word, and its meaning in common usage, were essentially the same.[7]

The supplanting of the concept *natural philosophy* by that of *science* occurred as a result of the massive expansion of knowledge about the natural world that began during the scientific revolution of the sixteenth and seventeenth centuries, and the development of novel methods to obtain this knowledge. Science could no longer be accommodated within the bailiwick of the natural philosopher because its scope and knowledge content were too broad. Most importantly, science could no longer be characterized solely by its historical meaning *per se*, that is, association with passive observation of the natural world, or its use of pure thought and formal logic. Science had become a *physical activity*, something that scientists did with their hands and with an expanding array of instruments, which allowed them not only to observe how the world worked but to manipulate it. How then should we now define science?

The OED provides 10 definitions of "science", including 7 sub-definitions [14]. Some are archaic, but many are not, and this diversity of meanings reflects varying perspectives on science itself. This means it may be quite difficult, or even impossible theoretically, to establish the accuracy or truth[8] of anything in the absolute, including definitions. However, in practice, one useful definition would be #4b, which states science is

[a] branch of study that deals with a connected body of demonstrated truths or with observed facts systematically classified and more or less comprehended by general laws, and incorporating trustworthy methods (now esp. those involving the scientific method and which incorporate falsifiable hypotheses) for the discovery of new truth in its own domain.

This is how lexicographers define science, but how do scientists and philosophers do so? Their definitions are varied (see [6, 15–19]), but in their examination one begins to appreciate the breadth and richness of definitions of science.

---

[7] Historically, Pasnau states that English starts to use the latinate word 'science' beginning in the fourteenth century [13]. Ross argues that, from ≈1800 to 1850, the words philosophy and science were in fact synonymous in common usage [8].

[8] Throughout this volume, I will distinguish the words "Truth" (uppercase T) and "truth" (lowercase t). The former term refers to absolute truth, i.e., knowledge that is unassailably correct, universal, immutable, and unaffected by the passage of time or the acquisition of new knowledge. The latter term refers to knowledge obtained within a particular system of knowledge acquisition and at a particular time. It provides for the possibility that a current definition may need to be modified or eliminated in the future should new knowledge be obtained. Karl Popper addressed this issue eloquently in his book *The Logic of Scientific Discovery* [6], where he said, *"[S]cience is not in the truth business per se, at least not the 'Truth' business, although it may be in the 'truth' business."* The reader must, above all else, understand this difference, and the basis for it, namely absolutism vs. relativism.

According to Nagel [17],

... [science] is the desire for explanations which are at once systematic and controllable by factual evidence that generate science; and it is the organization and classification of knowledge on the basis of explanatory principles that is the distinctive goal of the sciences.

A more detailed definition was provided by Weaver [15], who said:

Science clearly is a way of solving problems-not all problems, but a large class of important and practical ones. The problems with which science can deal are those in which the predominant factors are subject to the basic laws of logic, and are for the most part measurable. Science is a way of organizing reproducible knowledge about such problems; of focusing and disciplining imagination; of weighing evidence; of deciding what is relevant and what is not; of impartially testing hypotheses; of ruthlessly discarding data that prove to be inaccurate or inadequate; of finding, interpreting, and facing facts, and of making the facts of nature the servants of man.

Weaver's quotation emphasizes the fact that science comprises not only intangible aspects related to theory and logic, but also the practical aspects of how the enterprise is executed physically and how the scientific product is used by society. The search for truth is a fundamental component of science.[9]

Bas Van Fraassen, a prominent philosopher of science, has discussed a number of views on science.[10]

[S]cience aims to find a true description of unobservable processes that explain the observable ones, and also of what are possible states of affairs, not just of what is actual.

Science aims to give us, in its theories, a literally true[11] story of what the world is like.

[Science aims] to discover facts about the world ... about the regularities in the observable part of the world.

... science gives us a picture of the world as a net of interconnected events, related to each other in a complex but orderly way.

These views of science share a common feature, the idea of *systematized knowledge*. One sees this in the use of the terms *organization, classification, regularities, net, orderly*, and *interconnected*. Thus, science is being defined less by its methods and more by how it is *used*, viz., to yield knowledge. This view accords with that of the philosopher Ludwig Wittgenstein, who, in his study of the philosophy of language, addressed the question of what we mean when we use a particular word.[12]

For a large class of cases – though not for all – in which we employ the word "meaning" it can be defined thus: the meaning of a word is its use in the language.

---

[9] We address more fully how truth is uncovered, and what it actually may or may not be, in Section 5.7.8.

[10] See Van Fraassen [18].

[11] Van Fraassen's use of the term "truth" in this context may be construed to mean that science gives us a *realistic* view of nature. Whether this is itself true, or whether our view of science is simply some type of intellectual construct resting upon the results of visualizing nature with our own eyes or with an array of instruments, remains the subject of debate between "realists" and "anti-realists" (see Section 5.7).

[12] See paragraph 43 in Wittgenstein, *Tractatus Logico-Philosophicus* [20].

One thus might say that science is systematization of knowledge, or as propounded by Hoyningen–Huene, "systematicity" [21]. This concept came about as Hoyningen–Huene, like many before him, was searching for a general answer to the question "what is science?".[13] And, like others before him, he understood the difficulty involved, *if the question were posed in this way*. However, by asking instead, "how does scientific knowledge, in general, differ from other knowledge types?" Hoyningen–Huene was able to come up with an answer [21].

Scientific knowledge differs from other kinds of knowledge, especially from everyday knowledge, by its higher degree of *systematicity*.

To explicate systematicity, Hoyningen–Huene starts by likening it to an "umbrella," under which the concept could be described and understood in different scientific contexts. If we use biological evolution as metaphor for systematicity, we see this concept as a network emanating from an all-encompassing primordial precursor and forming a tree-like structure characterized by ever more complex branching (Fig. 2.1). One might interpret this as an example of the unity of science. Here, though, perspective comes into play. From Hoyningen–Huene's perspective, the concept of systematicity was *not* to be a means of demonstration of scientific unity. Instead, it was to be an abstraction that allowed one to recognize and explore commonalities among disparate scientific enterprises, but not from a strictly operational/methodological perspective. Hoyningen–Huene admits this conception of systematicity is nebulous, but at the same time notes that every new branch provides a more concrete *de facto* definition of systematicity because its scope narrows in the evolution from major branches of science (physics, biology, chemistry, etc.) to scientific sub-disciplines (e.g., developmental biology, molecular biology, biological clocks, cellular organelles) and their distinct practices, methods, strategies, etc.

The umbrella of systematicity is related to its component elements through what is called "family resemblance relationships" [21, 22]. These relationships occur among nine "interconnected dimensions" that comprise descriptions, explanations, predictions, the defense of knowledge claims, critical discourse, epistemic connectedness, an ideal of completeness, knowledge generation, and the representation of knowledge.[14] If one considers each dimension to be an element of a weighted average, it is easy to see that by changing the weights of each element one can create an enormous number of unique systematicities. This enables the application of different systematicities to disparate scientific disciplines and their sub-fields. Each then would have a distinct form of systematicity, but a form tailored from some or all of the same elements comprising the general concept.

Hoyningen–Huene's arguments do not involve the question of the demarcation of science from "pseudo-" and "non-science." His goal is not to offer a normative argument for what is, and isn't, science. Instead, he discusses another type of demarcation, that between the "special nature" of scientific knowledge and what he terms "everyday knowledge," the

---

[13] For Hoyningen–Huene, science was more akin to the German notion of *wissenschaft*, the direct translation of which is "knowledge making." There is no word equivalent in English, but the scope of *wissenschaft* comprises not only science qua natural science, but social science and the humanities as well.

[14] Space limitations preclude deeper discussion of these terms, but one can be found in Chapter 3 of Hoyningen–Huene [22].

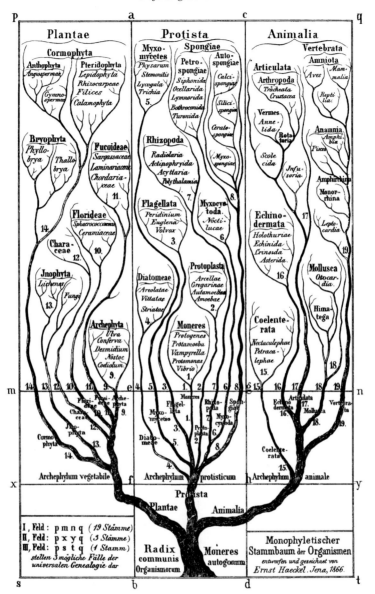

Figure 2.1 Haekel's evolutionary tree.

kind with which most are familiar in the contexts of their daily lives. Hoyningen–Huene does not explicitly define everyday knowledge, but, in his quoting Einstein, its meaning crystallizes.

The whole of science is nothing more than a refinement of everyday thinking.

Another Nobel prize-winning physicist, Richard Feynman, provided a more irreverent take on "science" in a talk he gave to the National Science Teachers Association in 1966.

The talk was entitled "What is science?" but Feynman demurred with respect to providing an absolute definition. Instead, he said:

Science is the belief in the ignorance of experts.

Feynman's sense of science might be rephrased *"I don't believe you. Where's the proof!"* Thus science, at its core, is an epistemological process applied to *ostensible* knowledge. Scientists, and especially students of science, should never passively accept facts provided to them but instead determine for themselves whether what they are being told or what they read is true or not (in whatever way they define truth).

It has been suggested that when confronted with a situation in which determining what something *is* is difficult or impossible, one might gain insight by determining what something *is not*. An example of such a suggestion occurred in 1968 in an episode of the iconic television show *Star Trek* [23]. While on patrol deep in space on the starship Enterprise, Captain James T. Kirk and Science Office Mr. Spock had encountered an unknown and potentially life-threatening phenomenon, a "dark area" in space. This led to the following conversation.[15]

KIRK: Spock, give me an update on the dark area ahead.

SPOCK: No analysis due to insufficient information.

KIRK: No speculation, no information, nothing. I've asked you three times for information on that, and you've been unable to supply it. Insufficient data is not sufficient, Mister Spock. You're the science officer. You're supposed to have sufficient data all the time.

SPOCK: I am well aware of that, Captain, but the computers contain nothing on this phenomenon. It is beyond our experience, and the new information is not yet significant.

KIRK: If you can't tell me what it is, let's use reverse logic. Perhaps it'll help if you tell me what it isn't.

The philosopher Thomas Nickles (a nonfictional person!) has termed this an "exercise in exclusion" [24], that is, we exclude everything we don't consider science and what's left *is* science. Samuel Schindler echoed Nickles when he said explicitly "...demarcation can be understood broadly as the demarcation of science against all intellectual enterprises which are not science – for example, philosophy, religion, logic, mathematics, and engineering" [25]. This problem of distinguishing science from nonscience has existed since the time of Aristotle (384–322 BC) [24] and has been referred to as the "demarcation problem" [6, 26]. Demarcation requires establishment of boundaries separating different entities, for example, science, nonscience, and pseudoscience. Demarcation is a problem because there is no general consensus, at least among philosophers, on how to determine where these boundaries should be [27], no doubt, as discussed above, because to distinguish one thing from another one must know what each is. Nevertheless, the failure to answer the question in the last 2,400 years, although it would be edifying to do so, has significant effects on society.

---

[15] From "The Immunity Syndrome," Stardate: 4307.1, original air date: January 19, 1968.

However one defines science, societal development and functioning depend on it. It can be argued that without science, society would comprise high-functioning primates displaying instinctual behavior, having and using empirical knowledge and some form of communication, and expressing themselves artistically and musically, but remaining in the stone age technologically. Science has provided us the means to understand the natural world and to use this understanding to our advantage (and, unfortunately, often to our disadvantage). Harnessing electricity enabled us to create light, radio waves, radar, global positioning systems, computers, telephones, motors, batteries, and myriad other devices. Chemical knowledge has enabled production of medicines, fuels, clothing, materials (metal composites, plastics, carbon fiber jet planes), and weapons (bombs, missiles, and nerve gas). Exploration of magnetism resulted in creation of electric cars, generators, and magnetic resonance imaging systems. Scientific studies of living organisms, from those as small and simple as viruses to those as large and complex as humans, have revolutionized our understanding of development, physiology, and disease. Scourges of history, including smallpox, polio, tuberculosis, syphilis, AIDS, and bubonic plague have either been eliminated or can be managed effectively with medication. No one can deny these obvious societal benefits of science. These successes, and their impact on life on Earth, have given science special importance with respect to societal issues, for which science often is the final adjudicator of fact. This is because the knowledge provided by science has, since the scientific revolution (c. 1550–1700 AD), been considered to be the truth.

Demarcating truth from mere opinion is an even more fundamental demarcation problem, one discussed by Parmenides (c. 515–450 BC) two centuries before Aristotle's consideration of science and nonscience [28]. Here, I use the term *opinion* in the modern sense of Parmenides' *doxa*,[16] that is, to connote a verbalization of a common belief or public opinion, the epistemological foundations of which may be unclear, irrational, or nonfactual. Expressing one's opinion can be a valuable social exercise when it informs and encourages congenial or constructive interactions among people. In politics, it is obligatory. Expressing oneself also has positive psychological effects on the expresser (although maybe not on the listener!).

The interchange of opinions often augments creativity. However, opinions are not truths *per se*, unless their epistemological foundations are considered solid.[17] The fact of a book is irrelevant to the question of whether I or another likes or dislikes it. There is no right answer to this question as there are no right answers to the questions of whether one likes or dislikes caviar, opera, Pilates, Picasso, traveling, the New York Yankees baseball team,[18] or even other people. By contrast, there *are* right, that is, true, answers in science because

---

[16] In the fifth century BC, Parmenides used the term *doxa* to describe the *perception* of reality of the physical world, which could be false and misleading [29]. This distinction is directly related to our earlier distinction of "Truth" from "truth" (see Footnote 8), i.e., something absolute and unchangeable rather than something subject to reinterpretation, change, or discussion.

[17] I will not address the question of how "solid" may be defined. Instead, I will stipulate, in this context, that solid means unassailable.

[18] This may not be a good example because, based upon decades of careful observation of baseball fans in New England, there does appear to be broad dislike.

these answers have factual[19] bases. It is a fact that we are born.[20] It is a fact that the sun is hot. These facts are indisputable scientifically because they are empirically true. Anyone expressing the opinion that these facts are untrue would be wrong, but they certainly would be free to express that opinion.

Questions of truth or opinion, scientific or nonscientific, or right or wrong histori- cally have been the province of philosophers. Whether such questions were discussed in the Agora in Athens by Socrates, by natural philosophers centuries later, or by scientists today, they not only have intrinsic historical, educational, and academic value, they have tremendous practical importance. We saw above examples of how science has contributed unequivocally to the health and advancement of society. However, science also plays a vital role in ostensibly nonscientific milieus in which its special importance as a "gold standard" of knowledge is used to inform decisions with local, national, and global impact, for example, environmental policy, education, and public health.

In such cases, stakeholders must know what is scientifically true (and not) if they are to make informed decisions. One of the best examples of this was the decision Federal District Court Judge William Overton was required to make in a monumentally important court case, "McLean v. Arkansas Board of Education."[21] The court had to determine whether "creation science" was indeed "science" and thus should be taught alongside evolution in the public schools. To do so, the court first had to define "science" and then determine if creationism shared science's precepts. After considering the evidence presented by sci- entists, philosophers of science, creationists, and others, Judge Overton provided a legal definition. Not surprisingly, it incorporates many of the ideas we have discussed above. The definition may be more abstract and less methodological than others we have discussed (e.g., systematicity is not mentioned), but its elements do reflect the canons of science in practice.

1. It is guided by natural law;
2. It has to be explanatory by reference to natural law;
3. It is testable against the empirical world;
4. Its conclusions are tentative, i.e., are not necessarily the final word; and
5. It is falsifiable.

We have seen that many definitions of science have been proffered, each of which may have merit but none of which can define science in the absolute. The scope and complexity of science preclude this. One might think this puts the idea of science on a weak foot- ing. After all, how can we practice something if we don't even know what it is? Answer: many forms of science are practiced and this diversity not only is a good thing, but it is one of the most important ways quantum leaps science are made. Paradoxically, there has

---

[19] The astute reader may now ask "what is a fact?" or what do you mean by "factual?" I will not attempt to answer these superb and difficult philosophical questions here. Instead, I refer the reader to two excellent sources of information on the topic [30, 31].

[20] T.S. Eliot also considered this a fact, along with two others, when he penned "*Birth, and copulation, and death. That's all the facts when you come to brass tacks*" [32].

[21] See McLean v. Arkansas Board of Education, 529 F. Supp. 1255 (E.D. Ark 1982).

been a consensus about the lack of consensus[22] on science [33]. Allan Franklin [34] and Andrew Pickering [35], in talking about experimental science, have suggested this is a strength.

'...the lack of a consensus [i]s a strength, rather than ...a weakness .... The absence of an accepted framework ...can only encourage further study. As Pickering might put it, 'We have accepted the lack of a single framework because we can ply our trade more profitably in a world in which no such framework exists.'

By this criterion, and as shown throughout this volume, modern science is in a very strong position indeed!

---

[22] In Chapter 6.2, we will find that one reason for this lack of consensus is that defining science involves psychological and social factors that themselves are nonscientific.

3

Learning Science

If you don't know your history, then you don't know anything; you
would be like a leaf that does not know it is part of a tree.
—Pohnpeian proverb

One can comprehend science only by taking into account its history
and its institutions, its social aspects as well as its cognitive ones; and
... [merging] sociological with philosophical viewpoints.
—Henry H. Bauer[1]

Now that we have seen that no consensus exists as to what science is, or even what it isn't,
let's "learn science!" One might reasonably ask how one could learn something for which
there is no categorical definition. The answer to this conundrum comes in how we define
the term "learn." In the context of this volume, "learning" is about more than simple information accumulation and prescriptive knowledge. A computer can accumulate and process
information, but it is not sentient in the same sense as humans are (although advances in
computer hardware and artificial intelligence software are making this distinction increasingly blurred). A computer has no perspective nor does it appreciate the scientific gestalt. It
is the epitome of a "black box," into which certain inputs are made and predetermined outputs are observed. What separates the scientist from the computer, or robot scientist [37], is
the often idiosyncratic and unpredictable manner in which the scientist uses knowledge to
develop novel conceptions of nature, to have epiphanic episodes displaying human genius,
and maybe most importantly, to make incredible mistakes that lead to even more incredible
discoveries. "Learning science" in our context thus means, a là Louis Pasteur, preparing the
mind for discovery.[2]

Learning science is not simply learning science *per se*, it is, more importantly, about
learning how to learn science. This pedagogic process, if it is to be most effective, i.e., to
produce the most thoughtful, creative, and successful scientists, must involve more than
merely learning how to take an infrared spectrum, culture a bacterium, synthesize a chemical compound, detect neutrinos, or do carbon dating. It must provide the budding scientist

---

[1] Taken from Bauer [36], p. 51.
[2] During his inaugural address as newly appointed Professor and Dean (September 1854) at the opening of the new Faculté
des Sciences at Lille (December 7, 1854), Pasteur said, *"Dans les champs de l'observation le hasard ne favorise que les
esprits préparés."* ["In the fields of observation, chance favors only the prepared mind."]

with an *intuition* of science that will allow the scientist to identify important questions and to determine or develop the tools necessary to answer them. The intuition of science about which I speak has been termed "tacit knowledge" [38]. It is knowledge that cannot be directly articulated, e.g., as in knowing how to ride a bike. One can be instructed how to do so, but one must actually ride a bike to learn how to ride a bike. Relative to this metaphor, learning science is equivalent to bicycle training wheels. It provides support for the development of scientific intuition, and it comprises a broad subject area that includes knowledge not only in basic and specialized areas of natural science but also of the history, philosophy, and sociology of science, all of which help us to optimize our own scientific practice and facilitate discovery. It does so through the provision of concrete, practical examples of *how* science was and is done, *why* it was or is done in a particular manner, and how societal factors within and outside of the scientific community affected and affect not only the questions addressed but also the interpretation of the observations and experiments designed to answer them.

3.1 To Learn Science, Learn the Humanities!

Knowledge of the history and philosophy of science are two of the most important aspects of learning science. Why? What benefits do they have for the student or practicing scientist? One method of answering this broad question is to break it down into narrower parts and then relate each answer to the process of science. For example, with respect to the philosophy of science (PoS), we might ask:

1. What *is* PoS?
2. What are the "big questions" in PoS?
3. Is PoS exclusively a theoretical exercise or does it have practical value?
4. How do we avoid bias in our thinking? How can one advance in knowledge if one already believes they have the answer?

My approach to answering questions such as these is idiosyncratic. I argue that there are no absolute answers to these questions. I will define, with respect to this book only, what I consider the study of the PoS to be (an opinion with which many are sure to disagree). The boundaries I place on PoS are done to focus the reader's attention on the parts of PoS that are most relevant and significant with respect to understanding its role in the process of science. Combining these parts with relevant topics in the history of science (HoS) has tremendous value by providing a sense of the *gestalt* of science from the time of the ancient Egyptians to the present. One of the most important areas of study within the PoS is epistemology. Simply put, we must understand "how we know what we know." This is *not* obvious when one begins to critically examine the history of scientific discovery and the bases upon which our current pedagogical approaches sit.

In his book *Inventing Temperature*, Hasok Chang[3] presents his view of the importance of study of HoS and PoS for acquisition and protection of knowledge about nature. He

---

[3] See Chang [39], pp. 3–4.

terms this view "complementary science" because it augments knowledge obtained solely through empirical (experimental) means.

This book ... aspires to be a showcase of what I call "complementary science," which contributes to scientific knowledge through historical and philosophical investigations. Complementary science asks scientific questions that are excluded from current specialist science. It begins by re-examining the obvious, by asking why we accept the basic truths of science that have become educated common sense. Because many things are protected from questioning and criticism in specialist science, its demonstrated effectiveness is also unavoidably accompanied by a degree of dogmatism and a narrowness of focus that can actually result in a loss of knowledge. History and philosophy of science in its "complementary" mode can ameliorate this situation. ...

The progress of science, like that of societies in general, is characterized by greater and greater specialization. We know more and more, and we do more and more, which requires us to specialize more and more. No longer can polymaths and jacks-of-all-trades with expertise in a broad range of areas exist. We are, and have increasingly been, in the age of specialization. If one searches for subdisciplines of science, defined as "ologies," e.g., beginning with acarology (the study of mites and ticks) and ending with zythology (the study of beer), one will find ~700. Humankind progressed from a perspective of marvel and fear about a unitary entity, the "natural world," to a recognition of the manifold parts comprising it and how each of these parts works mechanistically. Like the parable of the blind men describing an elephant,[4] scientists have been remarkably successful in describing and understanding smaller and smaller parts of nature, but in doing so, they also are danger of losing their ability to "see the elephant." Seeing the whole is critical if scientists are to maximize the value of their work, and by "value," we mean not only with respect to purely scientific knowledge but also with respect to the larger impact of the results of scientific investigations on the world's health and well-being. Examination of the HoS helps us to see the whole as it was in the past and how it may be in the future, a vision that helps guide us in our scientific work.

Curiosity stimulates us to learn how the pieces of the natural world fit together and work. How do we do this? When we study the history and philosophy of science, we learn *how* and *why* science came to be practiced the way it is today (Table 3.1). This provides a starting point for our own investigations, but even more importantly, it gives us the means to consider the *hows* and *whys* that are implicit in our own determination of appropriate scientific questions to ask and how to answer them properly. George Santayana said "Those who cannot remember the past are condemned to repeat it."[5] These words are oft quoted to warn of the negative consequences of not remembering the past. However, in the context in which George Santayana wrote these words, that of *progress*, the sentence also should be considered a valuable and positive prescription for optimizing one's scientific practice. From this perspective, we study the history and PoS not to avoid some negative outcome

---

[4] The parable is from the Paṭhamanānātitthiya Sutta, part of the Udāna, or "Inspired Utterances of the Buddha," which belongs to the Sutta Piṭaka of the Pāli Canon. See Ireland [40], p. 86.

[5] Spanish-born American philosopher (1863–1952). See [43], p. 284.

Table 3.1 *What we learn from the study of the history and philosophy of science. "Learning" in these contexts focuses primarily on understanding and discussing the questions themselves as opposed to seeking textbook answers. It teaches one how to think about the questions and to understand their depth and complexity. We learn how to navigate the world of science without a map, a process that is, in essence, the journey of discovery.*

What characterizes the scientific method?[a]
What constitutes critical thinking about empirical statements?[a]
What is a scientific explanation?[a]
Why and how are theories developed?
What role do value judgements play in the work of scientists?[a]
What are science's strengths and limitations?
What is the role of science in guiding public policy?
How does society affect science?
What is the nature of science and why is it done?[b]
Why and how were certain instruments developed?
What are the concepts and methods of science?[b]
Why and how were certain techniques developed?
How was knowledge disseminated?
Does science comprise autonomous disciplines or are disciplines integrated into a single discovery process?[b]
What are the structures of scientific disciplines?[a]
How does knowledge of the past provide the means to envision the future?

[a] Adapted from Ennis [41].
[b] Adapted from Table 1 of Matthews [42].

but to enable the achievement of extraordinary scientific goals. We learn from the experience and practice of those that came before how different approaches to science succeeded, or did not succeed. We learn how different personalities and scientific practice styles contributed to successes and failures. We learn how scientists, and scientific thought, affected society as a whole, and vice versa. We learn about incremental and quantum advances in scientific knowledge. We learn about serendipity in scientific discovery.

In the eighteenth century, Robert Boyle was conspicuously aware of serendipity, failure, and the value of both when he opined:

...in experiments it not unfrequently happens, that even when we find not what we seek, we find something as well worth seeking as what we missed and ... unexpected accidents, that defeat our endeavours, sometimes cast us upon new discoveries of much greater advantage, than the wonted and expected success of the attempted experiments would have proved to us.[6]

---

[6] See Boyle [44], p. 353.

Those embarking on a scientific career rarely are provided an educational experience that includes the history, philosophy, or sociology of science.[7] These topics usually are addressed *en passant* in lectures and readings associated with specific domains of science as, for example, in physics classes in which Newton and Einstein are discussed in the contexts of mechanics, optics, gravity, and relativity. We learn what Newtonian mechanics is, how light is refracted by a prism, the role of a postulated gravitational force in mediating the structure of the solar system, and the physical consequences of movements of massive objects near the speed of light. Darwin's theory of evolution is a core component of biology classes. If we're lucky, we would learn the nationalities and dates of birth and death of the scientists who made such discoveries. This type of education is equivalent to our being given a picture of Sir Edmund Hillary on the top of Mt. Everest (after he became the first to do so) and then being expected to duplicate his achievement ourselves. Why would he want to climb this mountain in the first place and how would we do it? The picture of Hillary does not teach us. We need to know more. Quoting Henry Bauer once more:

Through learning textbook science, one is misled about the nature of scientific activity by learning only about relatively successful science, the things that have remained within science up to the present. ... [O]ne hardly ever encounters the phenomenon of unsuccessful science. ...[8] We see nothing in it of the trial and error, backing and filling, dismantling and rearranging that actually took place in the past, be that centuries ago or just a few years ago. Only when we read the actual accounts written by early students of nature do we begin to realize how many errors and false starts there were that left no traces in modern scientific texts. One can give excellent, objective, rational grounds now for the science in the textbooks, but that does not mean that it was actually assembled in an impartial, rational, steady manner.[9]

We thus read history and philosophy of science to learn how to address questions experimentally, deal with failure and success, and interact with the society of scientists and with society at large. What may be the most important and neglected reason is to ensure that our knowledge base is as firm as we were thought it was. Returning to complementary science for the moment, we see it provides us a process through which we can examine the history and philosophy of science to understand how science *actually* was done in the past, and even more importantly, whether we should consider the knowledge thus produced as *a priori* truth [39]. Not only may we be surprised that this process reveals weaknesses in our thinking, or outright untruths, but that it also can produce new scientific knowledge. In this last sense, complementary science is "history and philosophy of science as a continuation of science by other means,"[10] i.e., it is itself a type of science practice. It encourages and facilitates the recognition by scientists of the uncertainties and inconclusiveness inherent

---

[7] An example is the five core concepts that biology educators have agreed, by consensus, that undergraduates should grasp before graduation, viz. evolution; pathways and transformations of energy and matter; information flow, exchange, and storage; structure and function; and systems [45]. Interestingly, in the same report, when undergraduates were asked how biology education could be improved, one of their responses was "Biology majors should take a history/philosophy of science or a science and society course."

[8] Taken from Bauer [36], p. 11.

[9] Taken from Bauer [36], p. 36.

[10] Quotation from Chang [39], p. 235.

in their own work, a recognition that can lead to the design and execution of new experiments testing accepted hypotheses as well as the formulation of entirely novel hypotheses. It also can recover forgotten ideas, ideas that may have been rejected at the time on the basis of prevailing, but incorrect, ideas or theories, or because the means to further explore these ideas were not available. Complementary science thus can provide a means to escape the dogmatism that can stifle the advancement of science, and as discussed by Michela Massimi in a lecture entitled "Why Philosophy of Science Matters to Science" [46], it is essential for generating new knowledge.

Genuine new knowledge is produced through ... critical engagement between philosophy and science. ... Investigating the nature of scientific confirmation, procedures for checking datasets consistency, inferences used in parameters calibration, methods adopted for model selection, the reliability of computer simulations to retrieve particular phenomena, these all fall within the remit of philosophers of science – and they are (or should be) an integral part of what a well-rounded scientific inquiry ought to look like.

One cannot study everything, and as we will see throughout this book, temporal boundaries must be drawn between theory and practice if any experimentation is ever to be done.[11] One also must set boundaries with respect to truth. We can never obtain it, but how close must we come to it to use it as a basis for our theories and experiments? How do we even know how close we are to it? Science is in the progress business, not in the perfection business. It is unlikely that a perfectionist will be able to accomplish anything (other than in the field of mathematics). It is equally unlikely that anything of value could be accomplished in the absence of any attempts to do things as correctly as possible. Our knowledge of Newtonian mechanics enabled us to get to the moon and beyond, yet scientists know that Newtonian mechanics is flawed. It is clear from the incredible success of science that its progress does not depend on perfect. "Good enough" often *is* good enough. The problem is that good enough is *not* good enough for learning science and becoming a successful scientist. The scientist must understand *explicitly* the strengths and weaknesses of the theoretical and factual foundations of their work and determine for themselves whether they are comfortable constructing a scientific edifice upon them. This begs a new question, *"Can I be comfortable with the foundation I use?"* We know that facts and theories often change with time. If so, we may be building on quicksand.

3.2 What We Thought We Knew but Didn't

... it is important to remain alert to the judgments we do have about epistemic values operative in past science, because they will affect our judgments about present science. What we celebrate and condemn in past science, however implicitly or subtly, cannot remain safely separate from how we deal with present science.

—*Hasok Chang*[12]

---

[11] Of course, if one is not interested in actually *doing* science, one always could become a philosopher of science!
[12] Taken from Chang [47], p. 27.

Water boils at 100°C (212°F). This is a fact that we learn quite early in our science education, and it's an important one. It is a standard for the calibration of thermometers, as it has been since 1777 when the Royal Society of London, by fiat, assigned this temperature to the phenomenon [48]. By using this temperature as a reference point, one can determine accurately the boiling points of other liquids. These data are important for a variety of reasons. Boiling points (bp) are intrinsic properties of liquids that assist in their characterization. Measuring the boiling point of an ostensibly pure liquid allows us to determine its level of purity. This in turn provides the information necessary to distill one component from a multi-component liquid. Distillation is a process in which the temperature of liquid within a vessel is slowly increased to the points where various of its components become gases. If these boiling points are sufficiently different, then one can condense the vapors emanating from this vessel at specific temperatures by conducting them through a cooling coil, thus obtaining pure liquids.[13] The distillation of alcohol (ethanol) in the manufacturing of bourbon whiskey or Scotch whisky is a common example of this process. One must slowly raise the temperature of the liquid from fermentation so that compounds with lower boiling points (e.g., methanol [66°C]) can be discarded, ethanol collected at its boiling point (78.4°C), and its collection terminated before compounds with higher boiling points condense (e.g., isopropyl alcohol [rubbing alcohol] 82.6°).

This highlights how important it is to know that water boils at 100°C. The problem is that water often does *not* boil at 100°C! I didn't learn this until *after* I had become a Professor Emeritus and started writing this book. I spent my entire scientific career believing what I had been taught in elementary school and beyond. Had I taken the time to question my assumptions, and training, and asked "how do I know this?" I would have studied the history of the determination of water's boiling point and discovered that only under some conditions it is exactly 100°C.[14] In *fact*,[15] boiling points for water at atmospheric pressure have been reported to be as low as 91°C [49] and as high as 270°C [50].

Equally disturbing is the fact that I always assumed I knew what time it was, as long as I could count to 12 and knew where the "little hand" and the "big hand" were.[16] If I wasn't sure my watch or clock was accurate, I could call "time" at "popcorn" (767-2676; in the 1950s no area code was needed) and a recorded phone message would tell me.[17] This mode of telling time kept me from being late to school and enabled myriad other temporal features in my life. The problem was, and is, there is no such thing as the "correct time," at least not in an absolute sense.

---

[13] We are, for the moment, neglecting the fact that the boiling points of various substances in water, and of water itself, change depending on the precise composition of the starting mixture.

[14] An informative and amusing historical and experimental account of "The Myth of the Boiling Point" may be found here: www.sites.hps.cam.ac.uk/boiling/index.htm.

[15] I hope the reader understands that "facts" may not always be facts!

[16] I was raised in an era without digital clocks, and this is how we were taught to tell time.

[17] Now we can click on a link, www.time.gov/, and be taken to a page run by the U.S. Naval Observatory (USNO) and by the Time and Frequency Division of the National Institute of Standards and Technology (NIST), which maintains the standard for frequency and time interval for the United States, provides official time to the United States, and carries out a broad program of research and service activities in time and frequency metrology, which will tell you the time with 1-s precision.

Time is quite a complex concept.[18] Depending on the purpose for which time is being used, e.g., commercial, scientific, or for global navigation satellite systems, different "times" may exist. For example, International Atomic Time (TAI; from the French *temps atomique international*) is determined by the International Bureau of Weights and Measures (French: *Bureau international des poids et mesures* [BIPM])[19] in Sevrès, France, by acquiring time information from ≈70 national laboratories running a total of ≈420 atomic clocks. These data are subjected to a weighted averaging process that produces the TAI, a process that includes the periodic elimination or addition of particular atomic clocks, depending on their prior performance. Unfortunately, the irregular rotation of the Earth is not accounted for in the TAI and thus the time at which the sun crosses the meridian at Greenwich, England, may not be exactly (within 0.9 s) noon, as it must be for many terrestrial timing purposes. If this is the case, the TAI then is corrected by adding or suppressing "leap seconds," which produces another type of time "Coordinated Universal Time" (UTC). "But wait, there's more," i.e., another kind of time, terrestrial time (TT), which astronomers use to avoid problems with Earth's irregular rotation. TT runs 32.184 s ahead of TAI! So, ... what time is it? I still don't know.

If the reader found the paragraph above a bit complicated, as the author did when it was written, then the point has been made – time may be the most fundamental and important factor in our lives, but it is difficult to determine. However, before we descend into the chaos of an atemporal world, let's dig a little deeper. Maybe for some purposes, the time we get from "popcorn" or NIST is sufficiently accurate and precise. Indeed, since my school teachers did not have atomic clocks, the time appearing on the wall clock in the classroom *was* sufficiently accurate and precise to determine if I was in class on time and when class was over. This is not the case for many other purposes, including international time distribution services, delivering precise carrier wave frequencies for television broadcasts, operation of global positioning systems (GPS), radioastronomy, internet function, quantum computing, and sensor development. Businesses also need very accurate clocks to guarantee the time and date stamps of high frequency transactions with microsecond accuracy [51]. How is this done?

Historically, time intervals (astronomical time) were determined by measuring celestial movements of various kinds. For example, solar time defines 24 h as the time it takes for the Earth to rotate once about its axis and 1 year as the time it takes for the Earth to complete one orbit around the sun. The problem is that the Earth's rotation rate is gradually slowing, which means that 24 h gets longer and our trusty astronomical seconds are lengthening. In 1967, this problem was obviated by the introduction of atomic clocks, which use atomic transition frequencies (the frequencies of radiation emitted from electronic movement between energy levels) to define time. For example, the International Astronomical Union defines the second to be exactly "the duration of 9,192,631,770 periods of the radiation corresponding to the transition between the two hyperfine levels of the ground state

---

[18] For an amusing and enlightening treatment of the concept of time, see *The Order of Time* by Carlo Rovelli.

[19] The BIPM is overseen by the International Committee for Weights and Measures (French: *Comité international des poids et mesures*, CIPM), which in turn is overseen by the General Conference on Weights and Measures (French: Conférence générale des poids et mesures, CGPM), the body that is responsible for defining terms in the SI.

of the caesium-133 atom" (at 0 K). Today, the best cesium clocks have accuracies of 1.6 parts in 10^{16}. However, optical lattice clocks now being perfected [52] would define the SI[20] second (s) as 4.29×10^{14} oscillation/second. These clocks would have an error rate of ~1.4 in 10^{18}, > 100-fold lower.[21] Translated into more readily appreciated magnitudes, these new clocks "are so accurate they would lose less than a second over the lifetime of the universe, or 13.8 billion years."[22]

The boiling point of water and knowing what time it is are just two of many examples of "what we thought we knew but didn't." We recognized our ignorance only when we questioned are assumptions. I cannot emphasize enough how important it is for a scientist to do this, yet seldom is this philosophical principle taught to science students. "Digging deep" is a core philosophical tenet of the approach to learning and doing science espoused in this book. It is reflected, in practice, in the detail with which many topics in the book are addressed and, especially, within footnotes. If the reader is to develop a sense of the scientific process, which takes substantial effort and dedication, and experience the joy of learning, especially of the interesting, unexpected, and stimulating, then digging deeply into the "text" and "footnotes" of knowledge extant certainly will help them to achieve this goal.[23] Stuart Firestein, in his book "Ignorance," discussed how important "digging" had been to him in his work on olfaction:

I had not imagined how much there could be to know . . ., how many deep questions there could be, if you just refused to be satisfied with a superficial or cursory exploration. This was an adventure.[24]

3.3 Personality Drives Scientific Progress

Thus far, we have seen how the HoS can prepare us to be knowledgeable, critical, thoughtful, technologically savvy, and socially aware practitioners of what one may call "the art of science" – as opposed to being human automatons following prescribed paths of experimentation and analysis. The history of scientific discovery also illuminates the personalities of great scientists, revealing characteristics that helped them to produce the quantum leaps in knowledge for which they were famous. In the next sections, we discuss the role of intuition in scientific discovery, followed by an analysis of the scientific development and personality traits of two historically great scientists, Max Delbrück, whose worked helped spawn three different fields, biophysics, bacterial genetics, and molecular biology, and

---

[20] SI is the abbreviation of the French "Système international" (d'unités) (The International System of Units), administered by The International Bureau of Weights and Measures, which defines the second, s, as well as other quantities in the metric system. The s is the foundational time unit for the natural sciences and engineering.

[21] These clocks are *so* precise they have enabled the development of an entirely new and highly sensitive method for detecting gravitational waves, the "ripples in the fabric of space-time" [53].

[22] Quotation from https://phys.org/news/2015-06-atomic-clock.html.

[23] A shock to physical chemists – and their textbooks and students – was the recent finding that chemical bonds do *not* always become stronger as the electronegativity difference between their participating atoms increases [54]. Catharine Esterhuysen (Professor, Department of Chemistry and Polymer Science, Stellenbosch University, South Africa), in commenting on this surprising finding, said "This is an intriguing reminder that in science we should always be careful of giving a simple answer. As the [researchers] point out, if one really wants to obtain a full understanding of a bonding mechanism, it must be studied at a *deep and fundamental level*."

[24] Taken from Firestein [3], p. 155.

Leroy E. Hood, whose preternatural vision led to the development of the field of biotechnology, which took the embryonic molecular biology of Delbrück to the next level, and in doing so, revolutionized it and enabled the sequencing of the human genome.

3.3.1 Scientific Intuition

Intuition, as defined in the dictionary,[25] is a kind of understanding that does not evolve from the conscious, systematic consideration of knowledge extant. It is not the type of deductive reasoning practiced by Sir Arthur Conan Doyle's great detective, Sherlock Holmes. It also is not something that can be acquired through didactics of the usual sort. Intuition is a more ethereal quality. When we intuit, *we just know or sense* something. In this way, intuition occupies the same type of neurobehavioral niche in which one finds curiosity, creativity, and imagination, three additional qualities necessary for scientific excellence.

The importance of intuition for scientific discovery suggests that it would be helpful, and enlightening, to understand how one acquires this quality. This in turn, would be relevant to a consideration of the origins of curiosity and creativity. How is that we just know? I argue that, in fact, we don't (at least not initially). Intuition comes from learning science. But wait, I just said that intuition "is *not* something that can be acquired through didactics . . ..". How can both statements be true? The answer, as we will find over and over again in our consideration of science, is that the natural world is incredibly complex, the result of which is that both statements are partially true *and* partially false.

A useful approach for considering this apparent conundrum is to delve more deeply into the neuroanatomical and neurophysiological bases for behavior, including that behavior defined as intuition. The reader should note that, in doing so, intuition now has firmly been placed into the realm of natural science – it results from the complex connections (neuroanatomical) and electrical activity (neurophysiological) within specific systems of neurons in the brain. Intuition now must be considered a tangible entity, not a metaphysical or spiritual one.

Evolution of the human species has produced an organism in which embryonic and fetal development is grossly identical in all. Normal development produces a bipedal animal with two arms, a heart, a thorax, a cranium, a brain, and all the other anatomical structures that define the humans species. These characteristics are immutable (again, considering only normal development). Gross brain development and anatomy also are preprogrammed. Information flow within the brain occurs through connections among neurons within and among different brains regions. The result is behavior.

The hard wiring of the human brain results in a species that possesses certain intrinsic traits. Among these traits *are* intuition, curiosity, imagination, and creativity. However, although each person does possess these traits in the absolute sense, the relative "strength,"

---

[25] 1a. The act or faculty of knowing or sensing without the use of rational processes; immediate cognition. See Synonyms at reason. 1b. Knowledge gained by the use of this faculty; a perceptive insight. 2. A sense of something not evident or deducible; an impression. (Taken from The American Heritage® Dictionary of the English Language, 4th edition, © 2010, Houghton Mifflin Harcourt Publishing Company.)

if you will, of traits can vary substantially. Some people are highly intuitive. Others are not. Some are very intelligent. Others are not. Does this mean that we all are limited by our innate abilities? The answer is "no." We start with certain innate capabilities, but we augment and mold these capabilities as we develop and live, which brings us once again back to intuition. We can't teach it, but we can acquire it. We develop intuition through experience. Our experiences are imprinted physically in the organization of networks of synapses in our brains. Intuition arises as an emergent principle of these tangible neuronal structures and networks, both of which are highly plastic and are modifiable through experience.

When we study the history, philosophy, and sociology of science, we physically change our brain structure. It is difficult to determine precisely what structural changes are induced through study or the effects of these changes on brain electrophysiology (although with modern electroencephalographic and functional MRI techniques, we can grossly monitor and observe brain activity contemporaneously). What we do know is that scientific intuition develops over time, and requires time. For most, in fact, the development of an intuition about a subject is a given consequence of that individual's knowledge of, and experience in, that subject.

We discussed above how modern science education often tells us about milestones in scientific history – the *what* – but not the intangible elements allowing these milestones to be reached – the *how*. The progress of science depends less on what *is* done and more on what *might* be done, less on what *is* known and more on what *could* be known, less on what *other* scientists think and more on what *you* think. Imagination, intuition, vision, open-mindedness, objectivity, eclecticism, and persistence are personality traits critical to scientific progress. There are no books that provide formulas for learning these traits. Each of us is endowed innately, but to varying degrees, with abilities in each. However, we can enhance our skills in particular traits by emulating those scientists who displayed them in abundance. In this section, we consider the monumental achievements of scientists who are exemplars of *how* science *was*, *might*, and *could* be done and how their achievements reflected their own unique backgrounds and personalities.

3.3.2 From the Descriptive to the Quantitative: Max Delbrück Invents Biophysics, Bacterial Genetics, and Molecular Biology

Max Ludwig Henning Delbrück was born in Berlin in 1906. He enjoyed the advantages of modest affluence in a neighborhood of Berlin in which many distinguished academic families resided, including that of future Nobel prize winning physicist Max Planck. The Delbrück "clan" was quite large[26] and well known. Delbrück's father, Hans Delbrück, was a Professor of History at Berlin University and the editor of a monthly journal, *Preussische Jahrbücher* (Prussian Yearbooks). Professors of German Literature, Chemistry (and a foreign member of the Royal Society), and Theology (and Director of the Prussian State Library), the President of the Imperial Supreme Court, and a Minister of State also were

---

[26] Members actually were numbered according to their family trees. Max was 2.517, the seventh child of the eldest child of the fifth child of Gottlieb Delbrück (1777–1842) who was the next younger brother of Johann Friedrich, therefore number 2.

relatives. Delbrück developed an interest in astronomy during his *gymnasium*[27] education. One reason he cultivated this interest was as "a means of finding and establishing his own identity ...."[28] As Delbrück himself wrote, "the name 'Delbrück' was quite well known; too well, in fact, for the comfort of a youngster, who felt ambiguous about being recognized and identified immediately as a member of the clan, but not for himself."[29] Just as Delbrück sought to establish his own identity for psychological and sociological reasons, a scientist desiring to make their mark within the community of scientists depends, in part, on their ability to distinguish their research from those of others – to make a field their own.

Students of science would be well served by emulating aspects of Delbrück's early education. One was taught to Delbrück by Karl Friedrich Bonhoeffer,[30] a neighbor and physical chemist, who helped young Max to appreciate the difference between "vague talk and real insight." This was an appreciation that many would face in the future when Delbrück would roundly criticize what he considered their "vague talk." Bonhoeffer also inculcated in Delbrück the value of science's quest for new knowledge. The truth of the natural world is what such knowledge provides scientists and it is sacrosanct. Delbrück has said that as a child, he felt *"the truth shall make you free."*[31] Ironically, this turned out to be the motto of Caltech, where Delbrück was a professor from 1947 to 1977. At Caltech, in 1978, Delbrück delivered the commencement address and discussed how important this motto had been to him.

When I first saw the motto, ... it thrilled me. It made a strong impact on me. Why the emotional response? Perhaps because in science the name of the whole game is truthfulness. If you cheat in science, you are simply missing the point.[32]

Another important aspect of Delbrück's approach to learning was to "go to the source." As valedictorian of his *gymnasium* class, Delbrück delivered a lecture on Kepler. To prepare the lecture, Delbrück examined the actual books in which Kepler's work had been published 300 years earlier. Delbrück recalled *"to see and handle these old books ...was a tremendous experience,*[33] *especially when I found in one of them some of the speculations about the celestial harmonies expressed in terms of musical notes. It showed that Kepler literally thought in terms of a 'heavenly dingdong' ...rather than in terms of abstract mathematics."*[34]

What we are taught usually is a distillation of what exists or was done before. We do not see knowledge in context. We do not see the solution from which the distillate was produced, yet it is this solution from which the knowledge our teachers seek to instill in us

---

[27] Not to be confused with the athletic facility. *Gymnasiasten* are equivalent to college preparatory schools that students attend for 8–9 years prior to entering college.

[28] Quotation from Hayes [55], p. 70.

[29] From Delbrück's unpublished personal biographical notes.

[30] Bonhoeffer's children, Klaus and Dietrich, were executed in 1945, along with Planck's son Erwin, after being convicted of plotting the assassination of Adolph Hitler.

[31] Taken from Delbrück [56].

[32] *Ibid.*

[33] The author also was fortunate to be able to experience the joy of reading the original text reporting a great scientific discovery. In this case, it was Le Châtelier's original 1884 article, *"Sur un énoncé général des lois des équilibres chimiques,"* in the journal *Comptes Rendu*, explaining the principle of chemical equilibrium that would later bear his name.

[34] *Ibid.*

is obtained. How are we to distill new knowledge if we do not know from what mixture of information we are to produce it? How are we to know if what we are taught is accurate? A deeper examination of the facts upon which we base our knowledge and beliefs often reveals uncertainty in that knowledge, and sometime outright error. Equally important is the stimulatory effect that an appreciation of the history of discoveries has on one's own conception of novel ideas from a knowledge base long forgotten or ignored entirely.

From 1924 to 1929, Delbrück led a peripatetic academic life, first studying astrophysics at the University of Tübingen and then moving to the Universities of Berlin, Bonn, Tübingen, and finally Göttingen. He had failed in his first project, to develop a theory or novae, because of his inability to read the key papers in the field, which were written in English, a language that Delbrück did not know. He instead began working in the exciting new field of quantum mechanics (QM), which had been developed by Max Born, Werner Heisenberg, and Pasqual Jordan in Göttingen. Wolfgang Pauli, an assistant of Born, also was there, as were Edward Teller and Robert Oppenheimer. Born, Heisenberg, and Pauli all later would win Nobel prizes in physics. Delbrück used QM to study bond energies in a simple molecule, Li_2, and received his Ph.D. degree in 1930 (after flunking his first oral examination the year before!).

At the University of Göttingen, Delbrück was surrounded by highly accomplished academics who could both guide and stimulate him, as he had been during his childhood. Delbrück's accomplishments in Göttingen led to his receipt of a fellowship to study at the University of Bristol in England with John E. Lennard-Jones, Professor of Theoretical Physics, developer of the Lennard-Jones potential, which models the interaction energy between a pair of neutral atoms or molecules and is a foundational element in molecular dynamics simulations. Delbrück had among his friends Cecil Powell, Gerhard Herzberg, Patrick Blackett (later to become President of the Royal Society), and Paul Dirac, all of whom later won Nobel prizes.

The student of science should not underestimate the value of environment, and changing environments, in supporting their development. Even if one cannot walk down a hallway to talk to an esteemed professor or colleague, as could Delbrück, one can, through their own initiative, establish and maintain interactions with "virtual" mentors and peers in a scientific community. Attendance at scientific meetings, either in person or through teleconferencing, is a must because this is where real time, face-to-face scientific intercourse is possible. As one advances in their career, they likely will find that interactions with colleagues at meetings stimulated them more, and fostered more new ideas and thinking, than attendance at the actual talks and poster sessions. Such interactions help the scientist know what is at the cutting edge of their field and of other fields. This allows the scientist to compete in crowded, competitive fields, but it also tells one what questions remain to be addressed, questions that the scientist can make their own and use to establish their own unique identities.

Delbrück also spent a spring and summer in Copenhagen with Niels Bohr, who had won the Nobel prize in Physics nine years earlier. This interaction made a profound impact on Delbrück, whose "... *mind and style had been formed by Niels Bohr, the physicist, philosopher, poet and incessant Socratic questioner . . . .*"[35] Delbrück emulated the Copenhagen group's practice of scientific openness, saying "*. . . the first principle had to be openness. That you tell each other what you are doing and thinking. And that you don't care who – has the priority.*"[36] However, Bohr's greatest impact was his stimulating Delbrück to switch from quantum physics to biology, as Delbrück later recounted quite precisely:

The transition from physics to biology was catalyzed by a lecture of NIELS BOHR's [Delbrück's capitalization] at 10:00 a.m. on August 15, 1932. . . . the ultimate effect was to change the course of my life.[37]

What had Bohr said that was so impactful? He suggested applying the principles of quantum physics, in particular the idea of complementarity, to understanding life itself.

. . . it might perhaps be of interest . . . to enter on the problem of what significance the results reached in the limited domain of physics may have for our views on the position of living organisms in the realm of natural science. Notwithstanding the subtle character of the riddles of life, this problem has presented itself at every stage of science, since any scientific explanation necessarily must consist in reducing the description of more complex phenomena to that of simpler ones. . . . The spatial continuity of light propagation, on one hand, and the atomicity of the light effects, on the other hand, must, therefore, be considered as <u>complementary</u> aspects of one reality, in the sense that each expresses an important feature of the phenomena of light, which, although irreconcilable from a mechanical point of view, can never be in direct contradiction. . . .[38]

As with the complementarity of the particle and wave descriptions of light, Bohr suggested that, in understanding life, "*. . . teleological (organismal level) and mechanistic (molecular level) descriptions are mutually exclusive yet jointly necessary.*"[39] Put another way, the behavior of an organism cannot be gleaned from the examination of its component parts in isolation and the nature of these parts cannot be gleaned from the observation of organismal behavior. This idea immediately resonated[40] with Delbrück and provided an impetus for him, in essence, to study the physics of biological systems. Fisher and Lipson[41] described the manner in which he did so:

Max began his search for complementarity in biology by analyzing how ionizing radiation influenced the genetic material. Genes were stable elements; as a special feature, they could be shifted to a different form, which again was stable. Could the new quantum mechanics explain this, or did biology run into a paradox right here?

To answer this question, Delbrück collaborated with the geneticist Nikolai Timofeev-Ressovsky, who directed the project, and a radiation physicist, Karl Günther Zimmer, to study mutations in the fruit fly (*Drosophila melanogaster*) produced by ionizing radiation.

---

[35] Quotation from Judson [57], p. 50.
[36] Max Delbrück, as quoted in Judson [57], p. 61. It is unfortunate that science, by in large, does not practice this dictum *because* scientists need priority to advance in their careers.
[37] From Delbrück [58].
[38] Quotation from [59].
[39] Quotation from McKaughan [60].
[40] Pun intended.
[41] Quotation from Fischer [61].

Zimmer varied a number of factors including radiation dose, wavelength, and duration. Timofeev-Ressovsky determined the effects of irradiation on fly phenotype. Delbrück developed a quantum mechanical theory of gene mutation. The theory was consistent with a quantum mechanical mechanism of genes and inheritance in which a baseline genetic state was associated with a stable, heritable phenotype, but this state could be mutated to produce a second stable, heritable phenotype. Timofeeff-Ressovsky, Zimmer, and Delbrück published the work, *"Über die Natur der Genmutation und der Genstruktur" ("The Nature of Genetic Mutations and the Structure of the Gene")*, in German, in 1935 in a journal[42] that, according to Delbrück, gave the paper *"a funeral first class."* Only limited dissemination and recognition of the work occurred, and only because the Nobel prize winning physicist Erwin Schrödinger enthusiastically supported its conclusions and discussed the work in his lectures, and subsequently, in 1944, in his book *What is Life? The Physical Aspect of the Living Cell* [62], where he asked a provocative question, *"How can the events in space and time which take place within the spatial boundary of a living organism be accounted for by physics and chemistry?"* Timofeeff-Ressovsky, Zimmer, and Delbrück showed one way in which this could be done, and in doing so, revealed, at a time before anyone had a physical notion of what genes were, that they were *stable physical entities with specific sizes*, as opposed to properties produced through some emergent principle of cells. In their paper [63], later referred to as the "three-man work" or "three-man paper" (3MP) (German: *Dreimännerwrk, Driemännerarbeit*), they summarized this idea as follows:

...we view the gene as an assemblage of atoms within which a mutation can proceed as a rearrangement of atoms or dissociation of bonds.

The 3MP was a landmark in the history of molecular biology[43] and biophysics, one that led to physiological and chemical studies that would reveal that genes were linear sequences of deoxyribose nucleotides, and eventually, to Watson and Crick's determination of the structure of DNA and the sequencing of the human genome (see Section 3.3.3).

Delbrück's work on the 3MP was done in Berlin, where he was "a consultant on theoretical physics" in the group of Lise Meitner, who, with Otto Hahn and Otto Robert Frisch, discovered the fission of heavy atomic nuclei.[44] Although Delbrück did publish seminal papers on γ-ray scattering (later named "Delbrück Scattering" by Hans Bethe) and on the application of QM to statistical mechanics, his primary interests continued to lie in the application of the principles of physics to biological systems. Of course, one actually must choose "a system" to study and Delbrück was intrigued by viruses. In 1937, writing to

---

[42] The paper appeared in the journal *Nachrichten von der Gesellschaft der Wissenschaften zu Göttingen: Mathematisch-Physische Klasse* (Treatises of the Society of Sciences in Göttingen: Mathematical-Physical Class), which was published by the University of Göttingen and comprised only three issues in its lifetime. An English translation of the paper was not available until 2011, 76 years after its initial publication.

[43] The term "molecular biology" was first used by Dr. Warren Weaver, Director of the Natural Sciences Division of the Rockefeller Foundation, when he described a program to fund research into "sub-cellular biology" and the "biology of molecules" [64]. This program would involve physics, as well as mathematics and chemistry [61].

[44] This discovery led to Hahn being awarded the Nobel Prize in Chemistry in 1944. Meitner and Frisch deserved to be co-recipients, but for reasons not revealed until more than 50 years later, the committee did not do so. The discussion can be found at Crawford E, Sime RL, Walker M: A Nobel Tale of Postwar Injustice. *Physics Today* 1997, **50**(9):26–32 https://doi.org/10.1063/1.881933.

Niels Bohr, Delbrück rationalized this interest in a draft document addressing the "riddle of life."

We inquire into the relevance ... of virus research for a general assessment of the phenomena peculiar to life. ... viruses are things whose atomic constitution is as well defined as that of the large molecules of organic chemistry. ... Therefore we will view viruses as molecules ... [and] look upon the replication of viruses as a particular form of a primitive replication of genes .... Such a view would mean a great simplification of the question of the origin of the many highly complicated and specific molecules found in every organism in varying quantities and indispensable for carrying out its most elementary metabolism.

Delbrück initiated his studies of viruses in 1937 when he traveled to the California Institute of Technology (Caltech) in Pasadena, California, to study "Theoretical Genetics."[45] There, Emory Ellis introduced Delbrück to viruses of bacteria, bacteriophages (phages). The simplicity of this biological system and the ease and rapidity with which experiments could be done impressed Delbrück.

I was absolutely overwhelmed that that there were such simple procedures with which you could visualize virus particles. I mean, you could put them on a plate with a lawn of bacteria, and the next morning every virus particle would have eaten a macroscopic one-millimeter hole ("plaque") in the lawn. You could hold up the plate and count the plaques. This seemed to me just beyond my wildest dreams of doing simple experiments on something like atoms in biology.[46] [Author underlining.]

Delbrück now had a simple biological model system, phage ("a gadget of physics" according to Delbrück), for elucidating fundamental principles of genetics and "life." Phage was to genetics what the hydrogen atom had been to quantum physics, Delbrück's original area of expertise. However, in contrast to hydrogen, very little was known about phage in 1937. For example, no distinction could be made among "phage," "protein," or "gene." Phage could propagate within a cell, but how this was done was not understood and thus the process could only be described descriptively. Delbrück sought to bring the quantitative rigor of physics to phage to tease apart the kinetics of replication and the individual steps involved.

Phage replication was known to consist of three phases: (1) adsorption of a phage to a bacterium; (2) phage multiplication ("latent period"); and (3) release of large numbers of phage progeny (the "burst" phase). By simply "counting plaques," Ellis and Delbrück [66] were able to quantify phage concentration, the percentage of phage binding to bacteria, plating efficiency, the number of cycles of adsorption, multiplication, and release necessary to create a plaque, and burst size (number of phage released after multiplication). Delbrück did so in the same manner as he had done in the 3MP studies, by viewing phage replication as a physical system for which the tools of physics (mathematics, statistics) could be used to describe its behavior.

Once the phage system has been standardized, Delbrück sought to answer a much bigger question of the time, "how do bacteria become resistant to phage infection?" Resistance had

---

[45] This was the title of the fellowship the Rockefeller Foundation awarded to Delbrück to come to the United States.
[46] Quotation from Harding [65].

been observed when bacterial colonies eventually grew inside the clear plaques in which, presumably, all the bacteria had been lysed. What was the explanation for this? Viruses were not known to have genes at the time, so did phage infection cause the bacteria to *acquire* immunity or was *mutation* the operative principle? Salvador Luria, who Delbrück had first met in 1940, sought to answer this question by comparing the numbers of colonies arising in many independent experiments in which only small numbers of bacteria were used. Delbrück provided the mathematical theory to do so. If a preexisting cellular element was responsible for resistance, the same number of resistant colonies, on average, would grow in each experiment, but if resistance was due to mutation, then significant *fluctuations* in the number of colonies would be observed. In addition, such fluctuations would be larger if the bacteria used in the experiments were allowed to grow for different amounts of time before their use. Luria and Delbrück published their results in 1943 in a paper entitled *"Mutations of bacteria from virus sensitivity to virus resistance"* [67]. Their conclusions?

... the experiments show clearly that the resistant bacteria appear in similar cultures not as random samples but in groups of varying sizes, indicating a correlating cause for such grouping, and that the assumption of genetic relatedness of the bacteria of such groups offers the simplest explanation for them. [Thus] ... the resistance to virus is due to a heritable change of the bacterial cell which occurs independently of the action of the virus.

Luria and Delbrück, along with Hershey, receiving the Nobel Prize in Physiology or Medicine in 1969 for "... for their discoveries concerning the replication mechanism and the genetic structure of viruses." Professor Sven Gad, Member of the Staff of Professors of the Royal Caroline Institute, gave the presentation speech [68], saying, "The honour in the first place goes to Delbrück who transformed bacteriophage research from vague empiricism to an exact science. He analyzed and defined the conditions for precise measurement of the biological effects. Together with Luria he elaborated the quantitative methods and established the statistical criteria for evaluation which made the subsequent penetrating studies possible."

Delbrück had been instrumental in creating the new field of bacterial genetics, and along with it, the more general field of molecular biology. As Theodore Puck said, Delbrück's "passionate rejection of vagueness in the building and testing of conceptual models has helped to change radically the entire philosophy of biological research."[47]

Passion, curiosity, persistence, intelligence, vision, collegiality, cooperation, and rigor all were traits exhibited by Delbrück that helped him succeed. Two traits that may have been equally or more important were confidence in one's ideas and attacking scientific problems in one field with the techniques and perspectives from another. When Delbrück arrived at Caltech it was as a member of the Division of Biology, yet at that time he had essentially no experimental training. He was a theorist. When Delbrück was invited by the famous geneticist Alfred Sturtevant to give a departmental seminar, ostensibly on gene mutations, Norman Horowitz, then a graduate student in embryology, recounted that "this 'strange creature' ... opened his presentation by saying 'Let us imagine a cell as a

---

[47] Quotation from Kendrew [69], p. 143.

homogeneous sphere,' at which point he was greeted by bursts of laughter. At the end of the seminar, Sturtevant thanked Max, assuring him that now they knew everything they wanted to know. No questions were asked following the talk; the gap between theory and experiment was too wide."[48] Delbrück was confident enough in his own ideas that he was not dissuaded by the very dismissive reception of his ideas by the world leaders in cell biology and genetics at Caltech. They did not consider Delbrück's ideas with open minds. Instead, because of their intrinsic biases and dogmatic perspectives on biology, they prevented themselves from learning of a new biological paradigm, molecular biology, that Delbrück helped create and that revolutionized biology.

Ironically, another revolution in biology also was occurring at Caltech late in Delbrück's tenure there. Its leader was Leroy E. Hood, and it overthrew the prevailing dogma that biology was about viruses (phages), bacteria, archaea, and eukaryotes, about their physiology, about their morphology, or about their genetics. In fact, it eschewed studies of life at all! Instead, it argued that advances in biology must be driven by technology. It rejected the laboratory and embraced the machine shop, the electrical engineer, and the computer scientist. In doing so, it created so called "gene machines" and a new field, biotechnology.

3.3.3 How Do We Get to the Next Level? Lee Hood Invents Biotechnology

Discovery consists of seeing what everybody has seen and thinking what nobody else has thought.
—Albert Szent-Györgi

New directions in science are launched by new tools much more often than by new concepts. The effect of a concept-driven revolution is to explain old things in new ways. The effect of a tool-driven revolution is to discover new things that have to be explained.
—Freeman Dyson, physicist and futurist

Leroy E. Hood is a unique individual and scientist. He has indeed "thought what nobody else has thought" and developed the tools that have enabled the revolutions in molecular biology leading to the successful sequencing of all 3 billion bases of the human genome. Hood was born in 1938 in Missoula, Montana, and spent his entire childhood living in "Big Sky Country." It may have been the vastness and grandeur of the five mountain ranges surrounding Missoula[49] that facilitated his uncanny ability to see the "big questions" of science. It may have been his camping alone in those mountains while still in elementary school that taught him self-reliance and confidence. It may have been the quarterbacking of his high school football team to an undefeated season that taught him how to focus on a goal and overcome all obstacles to reach it.[50] It may have been his parents' expectations of

---

[48] Quotations from Fischer [61], p. 112.

[49] These are the Bitterroot Mountains, Sapphire Range, Garnet Range, Rattlesnake Mountains, and the Reservation Divide.

[50] Stephen Friend, physician-scientist at the Fred Hutchinson Cancer Research Center in Seattle, characterized Hood thusly: I had been around brilliant people. The attribute that stood out with Lee Hood was the ability to be completely consumed by a problem and not see any barriers. The closet thing I know to it, in sports, is the ability to be 'in the zone.' You think of Tiger Woods at his peak. ...If you want to know why Lee is successful, it comes down to total faith, not an atom, not a molecule of hesitation. ...I have not ever met a scientist who ever had that kind of force field that Lee has about him. Quotation from Timmerman [70], p. 269.

excelling academically that prepared him for a life in science, a life dedicated to learning and discovery.

Nature, literally and figuratively, and nurture certainly operated synergistically in Hood's development as a person and a scientist. One place Hood found nurturing was in the laboratory of William Dreyer at Caltech. Of Dreyer, Lee later would say "Bill Dreyer was a remarkable mentor. I learned from him to think conceptually. He was always willing to explore any problem. He was incredibly creative. Bill also had a deep interest in technology. ... Bill gave me the two dictums that have guided my subsequent scientific career. First, 'Always practice biology at the leading-edge.' It is more fun – always exciting and challenging. Second, 'If you really want to change biology, develop a new technology for pushing back the frontiers of biological knowledge.' " [71] This advice is typical of fields like physics and chemistry, but it certainly was not a part of practice in the biology communities of the 1980s.

Hood has an unusual ability to recognize the big questions in biology,[51] to invent new technologies required to answer them, to see where these answers may lead far into the future, and to understand their impact on society itself. Of course, as with most things, intelligence and hard work (until recently he generally worked an 84 (12/7) hour week) have been key contributors to his success.

Hood "grew up" at Caltech, where he received both his undergraduate and graduate degrees, with a brief hiatus in between to obtain his M.D. degree at Johns Hopkins Medical School. Following postdoctoral study at the NIH, Hood joined the Department of Biology at Caltech in 1970 as Assistant Professor, becoming Full Professor only five years later and then Chair of the department five years after that. Hood's initial studies focused on antibody diversity, i.e., the mechanism through which the immune system could make the specific antibodies needed to fight the myriad of bacterial and viral invaders humans encounter. This was a big question at the time because its answer would elucidate the most fundamental aspects of humoral immune responses (responses producing antibodies).[52] In addition, and even more fundamental, would be an understanding of the genes responsible for antibody production. How was it that a limited number of genes encoding proteins (at that time thought to be $\sim$100,000 but now known to be closer to $\sim$20,000 [72, 73]) could produce an apparently unlimited number of different antibodies?

The "central dogma of molecular biology" at the time was DNA→RNA→protein. Each protein is translated from a single messenger RNA (mRNA) that is transcribed from one gene, i.e., "one gene-one polypeptide." However, the provocative suggestion had been made by Dreyer that antibody production violated this principle because it appeared that two genes combined into one to produce an antibody. If the dogma of molecular biology was true, immunologists were faced with quite a paradox. In a famous paper published in 1965 in the *Proceedings of the National Academy of Sciences* entitled "The molecular

---

[51] Hood himself acknowledges that this is difficult to teach. "What makes a really good scientist great is the intuition to choose your path to optimize the yield of what you can learn, and move in directions that are most fruitful. It's not something easily taught."

[52] Other types of immune responses are cellular in nature, e.g., killer T lymphocytes are required to rid the body of virus-infected cells.

basis of antibody formation: A paradox," Dreyer and J. Claude Bennett discussed this paradox and provided a simple but revolutionary solution to it, *viz.* "*two* genes make one polypeptide." The dogma of molecular biology was wrong! Quoting from that paper [74]:

> The paradox results from the observation that one end of the [antibody] light chain behaves as if it were made by the genetic code contained in any one of more than 1000 genes, while the other end of the L chains can be shown to be the product of a single gene.... It appears therefore that immunologically competent cells have evolved a pattern of somatic genetic behavior which is radically different from anything normally found in modern molecular genetics [Author underlining].

The experimental basis for Dreyer and Bennett's theory had been studies of the structures of different antibodies. The sensitivities of the methods of protein structure analysis required large quantities of proteins. The only sources were tumors, known as myelomas, that secreted large quantities of a single type of antibody. Antibodies produced in mice and humans by normal cells were not present in sufficient amounts to study, nor were the vast majority of other proteins. Instead of accepting the fact, according to Hood,[53] that "*... a natural barrier was the fact many interesting proteins were available in very low quantities,*" he sought to create a protein sequenator (an instrument that reads the sequence of amino acids that comprise a protein) that was $100\times$ more sensitive than any in existence. Hood's scientific philosophy was simple:

> ...there is an intimate interrelationship where biology dictates the choice of technology and the technology in turn opens up new frontiers in biology (Fig. 3.1). Hopefully, this is an iterative cycle as new technologies emerge. Clearly this approach integrates technology with biology.[54]

The development of an improved protein sequenator (Fig. 3.2), according to Timmerman[55] "*...was in some respects to biology what the semiconductor was to the computer revolution.*" It enabled theretofore unimagined advances in understanding the structures of bioactive peptides and proteins and opened up entirely new fields of biology (Table 3.2). It also provided the information necessary to use reverse genetics to synthesize DNA probes for cloning the genes encoding the proteins. Machines to automate DNA synthesis and sequencing were, in fact, the next two machines created. These were the second and third pieces of the puzzle. The fourth was of an automated peptide synthesizer, which would enable researchers to tailor-make peptides that were unavailable or too costly to get from natural sources. These peptides could be used to immunize humans against specific diseases or to make antibodies for studying cell biology. They also allowed protein chemists to make defined changes in peptide structure to understand how proteins worked. These four instruments revolutionized molecular biology.[56] The instrument many consider the most important was the DNA sequencer, which made the human genome project possible (the effort to completely sequence the $\sim$6 billion nucleotides comprising the diploid DNA

---

[53] Taken from Hood [71].

[54] *Ibid.* Hood later became the William Gates III Professor, Chairman, & Founder, Department of Molecular Biotechnology, University of Washington. Molecular Biotechnology now had become its own scientific subdiscipline.

[55] Taken from Timmerman [70], p. 105.

[56] As an example, Luke Timmerman, writing in the Seattle Times in 1992, said about the DNA synthesizer: "A gene synthesis that took 20 chemists five years to build can now be done by a technician using part of one day."

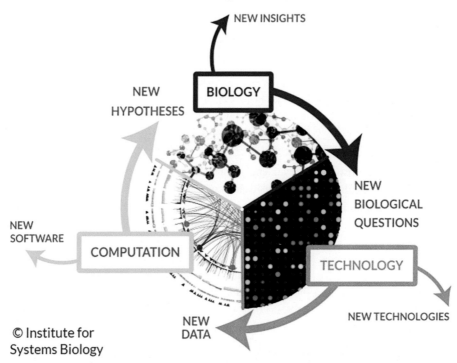

NEW INSIGHTS

NEW
HYPOTHESES

BIOLOGY

NEW
BIOLOGICAL
QUESTIONS

NEW
SOFTWARE

COMPUTATION

TECHNOLOGY

© Institute for
Systems Biology

NEW
DATA

NEW TECHNOLOGIES

Figure 3.1 The cycle of molecular biotechnology. "Biology" poses "new questions," some
of which are unanswerable in the absence of improved "technology." Once developed, new
technologies yield "new data" (huge data sets in the case of the human genome project) that
are analyzed "computationally" using multivariate analysis and artificial intelligence algorithms. These analyses may produce "new hypotheses" for biologists to test. The cycle of
new concepts→new technologies→new software continues *ad infinitum*. (Figure provided
by Lee Hood, with permission.)

from the 23 human chromosomes) and led to the creation of the new field of genomics.
Timmerman considered the invention of the DNA sequencer to be *"comparable to that of
the printing press...."*[57] Surprisingly, a US District Court judge, not a scientist, may have
summarized the scientific and societal impact of the DNA sequencer best when he wrote:

This case involves what was arguably one of the most important advances in biology in the 20th
century: the automatic DNA sequencer. The automation of DNA sequencing substantially reduced
the time, skill, and money necessary for DNA sequences to be decoded, and consequently enabled
such advances as mapping the human genome.[58]

In an interview for the *New York Times* in 1981, Hood estimated, with then current methods
for DNA sequencing, that it would take 95 years to sequence the human genome. However,
he also said *"The pace of recombinant DNA research continues to accelerate beyond most*

[57] Quotation from Timmerman [70], p. 163.
[58] See [75].

(a) Prototype Caltech gas–liquid solid phase
protein sequenator (c. 1980).

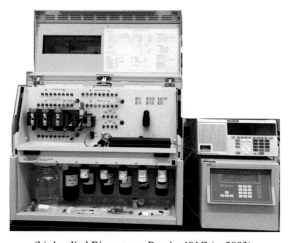

(b) Applied Biosystems Procise491C (c. 2003).

Figure 3.2 Shown are the original breadboard protein sequenator and its commercial version. The prototype instrument required daily liquid N_2, two noisy vacuum pumps, an off-site HPLC system to identify the amino acids (not shown), and a control unit (blue box at center of (a)), requiring actuation of a set of toggle switches to program reagent, solvent, and N_2 flow. The commercial instrument required no liquid N_2 or vacuum pumps. Amino acids were ported directly to an on-board HPLC for analysis (right side of (b)). Instrument programming and control were done with an integrated PC (not shown).

expectations. Consequently almost any predicted development is likely to occur far sooner than anyone could conservatively foresee." In 1990, nine years later, the human genome project began. It succeeded in sequencing the entire human genome $6\times$ faster than the earlier estimate, in 13 years, but at a cost of $\sim$\$1 billion. Continued advances in technology led to the sequencing a human genome in 2013 in only $\sim$1–2 days at a cost \$3,000–5,000, and in 2017, Francis DeSouza, the CEO of Illumina, a leading DNA sequencing company,

Table 3.2 *Integrating biology and technology. New areas of biology and medicine opened up by the gas–liquid solid phase protein sequenator.*

Protein sequences	Biological opportunities
Platelet-derived growth factor	An understanding of how oncogenes cause cancer
Erythropoietin	Stimulation of red cell production for cancer patients (biotechnology's first $1 billion drug)
Prion protein	First nucleic acid–free infectious disease agent (Mad Cow disease, kuru, Creutzfeldt–Jakob disease)
Interferons	Key biotechnology drugs for certain cancers and triggers of antiviral immune reactions
Acetylcholine receptors	Opened the way for understanding mechanisms of neural receptor action

predicted that in the future, the entire process would be done within 24 h and cost only $100.[59] It now can. Hood's vision had been on the mark and he didn't stop there. In 1996, he predicted that within 10 years everyone would carry a digital card that would contain their genome sequence and their entire medical history. This now is possible and will be implemented in the future.

Now imagine you had just succeeded in filling in a spreadsheet consisting of two columns, the first a series of numbers from 1 to 3 billion, and the second a series of letters from a four-letter alphabet, A, C, G, and T. This is what the human genome project yielded. Now what? We know that different combinations of three letters encode particular amino acids. We also know that most of the human genome does not encode proteins. What do the rest of the letters do? Some letter sequences are sites to which DNA-binding proteins attach to perform functions like DNA replication and repair or turning genes on and off. Others encode rRNA, a type of RNA built into ribosomes, the cellular machines that convert mRNA into protein. How does one convert one-dimensional data into an explanation of how cells and organisms are made and function? How does one pinpoint mutations in genes that may be related to disease, and then treat those diseases? The answer, according to Hood, was to *"...bring people from hard-core computer science, applied math, physics, chemistry, and engineering into this whole process."* This vision was what Hood sought to actualize in his creation of the Department of Molecular Biotechnology at the University of Washington [76] in 1992, a department with the goal of bringing "...together scientists in mathematics, chemistry, engineering and computer science with those in biology and medicine" for the purpose of "exploring problems at the frontiers of biology and medicine."[60] Hood summarized this type of endeavor thusly:[61]

[59] From www.sandiegouniontribune.com/business/biotech/sd-me-illumina-novaseq-20170109-story.html.
[60] The department later was consolidated with the Department of Genetics to form yet another new department and discipline, the Department of Genome Sciences.
[61] Quotation from Timmerman [70], p. 278.

The future of biology is the analysis of complex networks and systems, such as the ensemble of 100,000 genes [later discovered to be closer to 20,000] that are the blueprint for human beings, or the interactive assemblies of protein species that regulate transcription and development. To analyze and understand these complex networks and systems, one must identify the component elements, establish their interconnectivity, model their behavior computationally, and finally test the models against biological reality.

The future to which Hood referred became the present in 2000 when Hood created yet another new scientific discipline, "systems biology" [77], and an institution, the Institute for Systems Biology, where the new scientific subdiscipline would be studied [78]. The goal was characteristically ambitious and far sighted, *viz.,* "to make profound breakthroughs in human health, leveraging the revolutionary potential of systems biology."

Lee Hood is a visionary, one who through his interest in big questions and big science was instrumental[62] in creating the fields of molecular biotechnology and genomics. His recognition of the need for interdisciplinarity to solve complex biological and medical problems led him to create another new field, systems biology. These endeavors, in turn, now have led to what some have called the fourth paradigm of science, "data-driven science." Data-driven science ("data exploration," "data-intensive science," or "eScience") is the successor to paradigms of scientific practice that centered upon: (I) strict empiricism, which characterized science up to the scientific revolution beginning in the 16th century; (II) theory and model construction (Kepler's Laws, Newton's Laws of Motion, and Maxwell's equations); and (III) *in silico* methods to simulate natural phenomena too complex to study experimentally. Data-driven science requires no *ab initio* hypothesis to be tested. Instead, it uses sophisticated statistical and computational methods to interrogate huge data sets and derive hypotheses from the data themselves.[63]

How did Hood accomplish all that he did? In discussing Hood's legacy, Timmerman may have answered the question best when he said:

He wasn't afraid to step out of his comfort zone, crossing disciplines, enlisting smart people of all stripes to help tackle big problems in biology. . . . He appreciated thoughtful dialogue from people in humanities. . . . [and when] science opened up new ethical questions, he welcomed the input of science historians and ethicists. Some scientists could be arrogantly dismissive of ethicists looking over their shoulders, but Hood never claimed that scientists had all the answers.[64]

I was a postdoctoral fellow in Lee Hood's laboratory at Caltech during those exciting years when the so-called "gene machines" were being created and used to answer big questions in biology. I worked on protein sequencing using the breadboard instrument (Fig. 3.2a). In my estimation, what did, and continues to, characterize Lee well was his incredible drive to "get to the next level" *before* the current level had even been achieved! One may not able to look into the future as far, and with the clarity of vision, that Hood does, but one can learn not be contented with simply asking a scientific question. One can require of oneself that any question must be followed by as deep, imaginative, open-minded, and critical a

---

[62] Pun intended.
[63] We discuss data-driven science in more detail in Section 4.8.
[64] Quotation from Timmerman [70], pp. 382–383.

Table 3.3 *Creativity and character. Attributes associated with creativity (positive) and lack of creativity (negative). Creative and noncreative persons may share selected attributes from each other's lists . The list is not comprehensive and does not address the question of "how creative" a person may be. An individual's creativity may vary from "preternaturally creative" to "not creative," depending on how one defines these terms.*

Positive	Negative
Nonconforming	Questions rules and authority
Impulsive	Impulsive
Independent	Stubborn
Energetic	Hyperactive
Thorough	Careless and disorganized
Emotional	Forgetful
Original	Egotistical
Risk taking	Indifference to common conventions
Capacity for fantasy	Rebellious and argumentative
Emotional maturity	Tendency to be emotional
Attracted to complexity and ambiguity	Low interest in details
Artistic	Neurotic
Need for "alone time"	Absent-minded
Less driven by external recognition	
Curious	
Open-minded	
Perceptive	
Ethical	
Persevering	
Awareness of creativeness	
Self-confident	
Courage to actualize one's abilities	
Sense of humor	

Gedankenexperiment as one can muster. In doing so, it is possible to recognize initially unconceived possibilities and confounding factors that might affect the quality and rate of progress of one's experiments. A *Gedankenexperiment* often also reveals new ideas and concepts as its possible outcomes are considered.

No single person has done more to create the genomics era than Leroy Hood.[65]

Delbrück and Hood had many behavioral traits that contributed to their successes and led to quantum leaps in scientific progress. "Gifted" children have many of these traits as well, traits that are maintained into adulthood. Psychological and behavioral studies have shown that these children, and those not considered gifted, display distinct sets of personality traits (Table 3.3). Each group many not express all traits common to that group and some display

---

[65] Quotation of Michael E. Phelps, Professor and Chair, Molecular and Medical Pharmacology, UCLA, and co-inventor of the positron emission tomography (PET) scanner.

selected traits from the other group. For example, impulsivity appears on both lists and subtle differences in traits are distinguished (e.g., energy vs. hyperactivity and emotional maturity vs. emotional). Artistic, among other things, may include musical, as for Nobel Prize winners Richard Feynman (who played a Brazilian "frigideira," a metal percussion instrument based on a frying pan, and bongo and conga drums) and Albert Einstein (who played violin).

3.4 Knowledge Also Is Important ...

One can possess all the positive personality traits of Delbrück and Hood, but without some foundational of knowledge, one's scientific work would be limited to unproven intuitions about nature. We experience life and nature from birth until death. We know, for example, that the sun rises and sets each day, winters are cold (especially as we move away from the equator) and summers are hot (especially as we approach the equator), we may become ill and then get better (or not), we procreate, we change physically and mentally with aging, we have emotions and physical sensations, and we will die. This knowledge requires no formal education. It comes simply from living in and have direct experience with the world. This is empirical knowledge.

Empiricist philosophers of science in the early twentieth century argued that empirical knowledge was "all there is." One experiences the world through their senses. What one cannot experience does not exist but rather is an intellectual fabrication. This view is now archaic. It could not be defended because of the overwhelming mass of evidence accumulated with time that showed that the "world," in a very practical and obvious sense, includes not only what we can sense but also what we cannot. In fact, what we cannot sense may be the largest component of our world.

How is it that the "invisible" world is made visible? How do we learn about this world? The answer is the practice of science, and through it, science education. Science education seeks to reveal to the student the "whats" and "whys" of science. For example, what is the sun and why does it rise and set? Science education should provide us answers to key questions (Table 3.4[66]) and help us to avoid common misconceptions (Table 3.5[67]). However, we generally are not taught *how* to think about science, but rather simply to learn and appreciate the "whys." *We memorize, but we rarely ponder.* We learn simple concepts of general science in grade school, and we supplement these concepts with more sophisticated knowledge of physics, chemistry, and biology as we move through high school. We learn fundamental equations, e.g., $f = ma$, simple chemical equilibrium relationships, e.g., ($HCl \rightleftharpoons H^+ + Cl^-$), and basic cell biology and genetics (cell division, meiosis, mitosis, sexual reproduction). In these stages of education, everything is relatively simple and straightforward. As undergraduates, we delve deeper into the concepts, laws, and equations of science, and we use these concepts experimentally in laboratory classes. It is only in

---

[66] Adapted from Fig. 2 of Clough [79], p. 57.
[67] Adapted from Table 1 of Clough [80], p. 41.

Table 3.4 *Nature of science questions worth addressing in science education.*

In what sense is scientific knowledge tentative? In what sense is it durable?

To what extent is scientific knowledge based on and/or derived from observations of the natural world? In what ways is it based on reasons other than observational evidence?

To what extent are scientists and scientific knowledge subjective? To what extent can they be made less subjective?

To what extent is scientific knowledge socially and culturally embedded? In what sense does scientific knowledge transcend particular cultures?

In what sense is scientific knowledge invented? In what sense is it discovered?

How does the notion of a scientific method distort how scientists actually work? In what sense are particular aspects of scientists' work guided by protocols?

In what sense are scientific laws and theories different types of knowledge? How are they related to one another?

How are observations and inferences different? In what sense is an observation an inference?

How is the private work of scientists similar to and different from what is publicly shared in scientific papers?

graduate school that we are encouraged to begin thinking for ourselves, but under the tutelage of our mentors. We are given a project question/hypothesis and we strive to answer/test it. We take advanced classes that delve still deeper into our specific area(s) of interest and our knowledge is assessed in our written and oral examinations. As postdoctoral fellows, we almost always come to a laboratory to work on a question that already has been asked. We do not formulate the question ourselves. This is the time we must demonstrate our productivity and creativity to move onto the next career level, e.g., Assistant Professor.[68]

Each of these stages of science education is, and must be, tailored to the ages of the students being instructed. I argue at the outset that there is no correct way of teaching science. Instead, different approaches yield different benefits, and detriments, vis-à-vis the student. However, what teaching approaches at all these levels have in common is a relative lack of examination of how the science being taught actually came to be known, i.e., an in-depth epistemological – not solely historical – analysis of the development of scientific knowledge. The student tends to be given the "bottom line" or "take home message," but not the often tortuous, contentious, confusing, contradictory, and frustrating paths and dead ends that are part and parcel of scientific discoveries. In fact, it is a thorough understanding of what one might call "pathways of discovery" that the student comes to appreciate what Chang has characterized as "scientists [getting] down-and-dirty in the complexities of nature."[69] In some ways, this is the social equivalent of contrasting the ideal world that youngsters are taught with the real world in which they live, and in which they will face a multitude of non-ideal life experiences. Science *can* be messy, as Chang suggests, and

---

[68] Whether one decides to pursue an academic career or work at a commercial firm (biotechnology, chemistry, pharmaceuticals, etc.), one's attractiveness as a candidate depends in large part on their publication record as a graduate student and postdoctoral fellow.

[69] See Chang [47], p. 115.

Table 3.5 *Common misconceptions about the nature of science.*

Disagreements regarding competing scientific explanations for natural phenomena are resolved through polling scientists on their view of the best explanation.

Science and those who do science can and should be free from emotions and bias. Scientific ideas arise directly from data.

Data supporting a contentious scientific idea demand that doubting scientists drop their objections to the idea.

Data that are at odds with a prevailing science idea should result in the rejection of that idea.

Science, when well done, produces true knowledge.

Scientific knowledge falling short of that status is unreliable.

While creativity and inventiveness assist scientists in setting up their research, the resulting science ideas are discovered, much like finding something.

Scientific models are exact copies of reality.

Science is equated with technology, and all science research is thought or expected to be in some way directed at solving societal problems.

Science research follows a step-by-step scientific method and carefully adhering to this systematic method accounts for the success of science.

The status of, and relationship between, scientific laws and theories is misunderstood.

Methodological naturalism is equated with philosophical materialism.

one better know how to recognize the mess and to clean it up if one is to make valuable scientific discoveries.

Lee Hood may have opined the most succinctly about knowledge acquisition (memorizing) and pondering when he stated what he felt were the two most important things one should learn in graduate school.

[The first is] to learn how to read the literature and realize that not everything you read is true. The second is to learn how to think about experiments and designing experiments – thinking in a bigger context about what the major problems are.[70]

Scientists must have as broad a knowledge base as possible if they are to stay at the forefront of their field, be stimulated to develop new ideas and hypotheses, and understand what types of research are most interesting, novel, or popular so that the projects they propose in grant applications are most likely to be funded.[71] Knowledge is acquired primarily through reading and is augmented through verbal means in classes, scientific meetings, seminars, colloquia, and discussions with teachers and colleagues. We need not discuss these mechanisms further as they need no explanation. What is important is what we mean when we

---

[70] Quotation attributed to Lee Hood.

[71] This last point is pragmatic in nature. It should not be read as suggesting that all proposed research projects should reflect prevailing knowledge and norms. However, it is necessary to fund one's research efforts and most grant review boards are more likely to fund ideas with which they are familiar. Pursuing research areas that are lower in risk but likely to produce publishable data should be combined with higher risk, higher reward projects so that a laboratory can support itself and also have the potential to produce quantum leaps in understanding.

say "knowledge." This is an epistemological question that explores how we define knowledge and why, convert experimental information into new knowledge, and disseminate the knowledge we produce.

3.5 ...But Ignorance May Be More Important!

The importance of ignorance in science cannot be overstated. It's important for the simple reason that if we already knew everything about nature, there would be no reason to study it. There *is* a symmetry here – ignorance↔knowledge. What drives science are questions, and these questions are asked out of ignorance. Of course, if we possess no knowledge to begin with, we can't be ignorant about anything. The role of ignorance in knowledge creation has been discussed as far back in history as Confucius. Table 3.6 presents his view, as well as those of other important historical figures.

Lewis Thomas, famous for his conversion of the intricacies of science into interesting, stimulating, informative books[72] understandable by nonscientists, addressed the value of ignorance in his book *The Medusa and the Snail*.

The only solid piece of scientific truth about which I feel totally confident is that we are profoundly ignorant about nature. Indeed, I regard this as the major discovery of the past hundred years of biology. It is, in its way, an illuminating piece of news. It would have amazed the brightest minds of the eighteenth-century Enlightenment to be told by any of us how little we know, and how bewildering seems the way ahead. It is this sudden confrontation with the depth and scope of ignorance that represents the most significant contribution of twentieth-century science to the human intellect. We are, at last, facing up to it. In earlier times, we either pretended to understand how things worked or ignored the problem, or simply made up stories to fill the gaps. Now that we have begun exploring in earnest, doing serious science, we are getting glimpses of how huge the questions are, and how far from being answered.[73]

Table 3.6 *Views of ignorance from the sixth century BC to the twentieth century.*

"Real knowledge is to know the extent of one's ignorance." Confucius (551–479 BC)
"...although I do not suppose that either of us knows anything really beautiful and good, I am better off than he is – for he knows nothing, and thinks he knows. I neither know nor think I know." Socrates (c. 470–399 BC) from Plato's *Apology*
"To know that we know what we know, and to know that we do not know what we do not know, that is true knowledge." Nicolaus Copernicus (1473–1543)
"Education is a progressive discovery of our own ignorance." Will Durant (1885–1981)

---

[72] These included *The Lives of a Cell: Notes of a Biology Watcher*, which won the 1974 annual National Book Award in both the Arts and Letters and The Sciences categories, *The Medusa and the Snail* (which won the 1981 National Book Award in science), and *Late Night Thoughts on Listening to Mahler's Ninth Symphony*.
[73] Quotation from Thomas [81].

This certainly was high praise for ignorance. But why? Because knowing what we don't know is the prerequisite for eliminating our own ignorance. As Merton has opined,[74] *"The specification of ignorance amounts to problem-finding as a prelude to problem-solving."*

In his book *Ignorance* [3], Stuart Firestein expressed this idea thusly:

. . . scientists don't concentrate on what they know, which is considerable but also miniscule, but rather on what they don't know. The one big fact is that science traffics in ignorance, cultivates it, and is driven by it. Mucking about in the unknown is an adventure: doing it for a living is something most scientists consider a privilege.[75]

Ignorance does indeed drive science, but can it also retard scientific progress? Not surprisingly, considering the contradictions and paradoxes discussed in this book, the answer is "yes" *and* "yes," as we shall see. First, we must disabuse ourselves of the common notion that ignorance is an intrinsically bad thing. It isn't. Ignorance is simply the antonym of knowledge, meaning "that which we do not know." However, defining ignorance is not that simple, for if we define it as the antonym of knowledge, then we need to define knowledge itself, and this is quite difficult. In fact, an entire field of study, epistemology, deals with this very question, but a consensus answer has not been obtained. Ignorance is not one thing, and it has many levels that are associated with passive and active types of "not knowing."

In practice, whether ignorance is good or bad, benign or damaging, or unnecessary or necessary depends on context. For example, if one happens to be standing on the railing of the Golden Gate bridge in San Francisco, it is necessary to know that jumping off will kill you, i.e., if you don't want to commit suicide. However, it is unnecessary that you know the magnitude of the acceleration of gravity (≈ 9.8 m/s^2) to understand that you will fall.

The amount of knowledge extant is huge. As an example, in May 2019, Wikipedia estimated that its English site contained 5,854,191 articles.[76] Google estimates that there are $\sim$130 million books in existence.[77] It is thus impossible, at least for humans as they now exist, to acquire all this knowledge. In fact, it is also impossible for anyone to possess much more than an infinitesimal percentage of all knowledge. All of us thus are quite ignorant. Good science requires humility in the face of this fact. Great science requires even greater humility and the greatest scientists often have been the most modest. William Harvey, the discoverer of the human circulatory system, opined:

True philosophers[78] who are burning with love for truth and learning never see themselves . . . as wise men, brim-full of knowledge . . . . For most of them would admit that even the very greatest number of things of which we know is only equal to, the very smallest fraction of things of which we are ignorant.[79]

---

[74] See Merton [82], p. 10.
[75] Quotation from Firestein [3], p. 15.
[76] See https://en.wikipedia.org/wiki/Wikipedia:Size_of_Wikipedia#Graphs_of_size_and_growth_rate.
[77] See http://booksearch.blogspot.com/2010/08/books-of-world-stand-up-and-be-counted.html.
[78] Harvey is referring to "natural philosophers" because scientists in the seventeenth century were referred to as "philosophers" or "natural philosophers." The term "scientist," as we know it today, was not used until the late nineteenth century.
[79] From Harvey [83], p. 7.

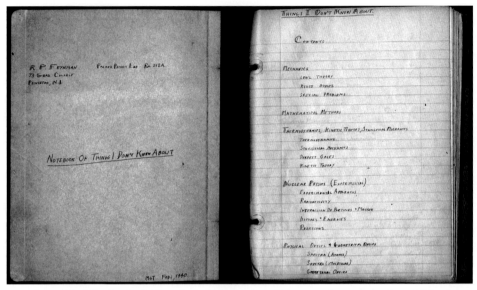

Figure 3.3 "Notebook of things I don't know about." A copy of two pages from Richard Feynman's notebook when he was a graduate student at Princeton. (Image from Feynman [84], p. 45.)

The Nobel prize–winning physicist, Richard Feynman, was another avatar for the importance of recognizing and using one's ignorance to one's advantage in science. Feynman not only embraced this concept philosophically, he embraced it practically by keeping a written record of things he did not know (see Fig. 3.3).

Sir Isaac Newton, one of the greatest scientists in history, was acutely aware of his ignorance and exceptionally modest in his own self-assessment. According to Brewster [85]:

The modesty of Sir Isaac Newton, in reference to his great discoveries, was not founded on any indifference to the fame which they conferred, or upon any erroneous judgment of their importance to science. The whole of his life proves, that he knew his place as a philosopher .... His modesty arose from the depth and extent of his knowledge, which showed him what a small portion of nature he had been able to examine, and how much remained to be explored in the same field in which he had himself laboured. In the magnitude of the comparison he recognised his own littleness; and a short time before his death he uttered this memorable sentiment: 'I do not know what I may appear to the world, but to myself I seem to have been only like a boy playing on the sea shore, and diverting myself in now and then finding a smoother pebble or a prettier shell than ordinary, whilst the great ocean of truth lay all undiscovered before me.[80]

Newton's self-deprecating metaphoric depiction of his scientific life contains the word "playing," and indeed many scientists consider their "work" actually to *be* play because of the joy experienced in discovery. The joyful aspects engendered by ignorance were the subject of a book, *The Pleasures of Ignorance* [86], written in 1921 by Robert (Y.Y.)

---

[80] Quotation from Brewster [85], p. 407.

Lynd,[81] a writer, essayist, and editor who was known for his urbane weekly column in the *New Statesman*.[82] In Chapter 1, Lynd celebrated the simple beauties and wonder of nature. He wrote:

It is impossible to take a walk in the country without being amazed at the vast continent of one's own ignorance. . . . This ignorance, however, is not altogether miserable. Out of it we get the constant pleasure of discovery. . . . [T]he happiness of the naturalist depends in some measure upon his ignorance, which still leaves him new worlds of this kind to discover. . . . The great pleasure of ignorance is, after all, the pleasure of asking questions. The man who has lost this pleasure or exchanged it for the pleasure of dogma, which is the pleasure of answering, is already beginning to stiffen. . . . We forget that Socrates was famed for wisdom not because he was omniscient but because he realized at the age of seventy that he still knew nothing.

3.5.1 Knowing and Not Knowing What You Know and Don't Know

Ignorance and knowledge are inextricably linked because *ignorance* implies a lack of *knowledge* and *knowledge* implies a lack of *ignorance*. One concept cannot exist without the other. "Unknown" and "known" have an equivalent relationship, one that largely determines how science will be done, as science requires that one be aware of what they know and don't know, things they don't know they don't know, things they don't know they know, and finally, things that simply are unknowable. Unless one understands precisely what the limits of their ignorance and knowledge may be, they will not be in a position to optimally determine fertile and significant areas of study. These five general knowledge states are discussed below and schematized in Fig. 3.4.[83]

Known knowns are knowns we know we know. Examples would include our names, the alphabet, Avogadro's number, and the existence of gravity.

Known unknowns are unknowns we know we don't know. Examples might include the value of π to 37 places, the weight of an electron, the winner of an upcoming horse race, or Latin. We do have a superficial sense of what these things are otherwise we could not know we didn't know, but we consider this sense to be inadequate in depth, scope, or accuracy to be considered "knowledge." This type of ignorance has been referred to as "simple ignorance" or "Socratic ignorance," as opposed to "double ignorance," not being aware of one's ignorance while thinking that one knows.

Unknown unknowns are unknowns we don't know we don't know. There does exist the possibility that we may come to know these unknowns, in which case they would become known unknowns. This phrase is often used in project planning to denote "future circumstances, events, or outcomes that are impossible to predict, plan for, or even to know where

---

[81] Y.Y. was Lynd's pseudonym, the two letter Y's being interpreted as "wise."

[82] A British magazine founded in 1913 and still in existence today, is "the leading progressive political and cultural magazine in the United Kingdom."

[83] The first three states were made famous by Donald Rumsfeld, then United States Secretary of Defense, who used these expression in a news briefing at the Department of Defense on February 12, 2002 regarding Iraq's possession of weapons of mass destruction [87].

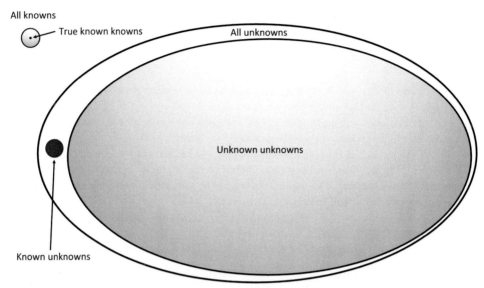

Figure 3.4 Diagram illustrating the relationships among four general knowledge classes. The set of "all unknowns," which includes known unknowns, unknown unknowns, and unknowable unknowns, is shown in pale blue. Subsets of this set are unknown, but knowable, unknowns (orange–brown), and known unknowns (red). The set determined by elimination of the sets of unknown unknowns and known unknowns is unknowable unknowns. The small blue circle at the upper left represent all knowns, whether they are, in fact, true or not. The dot inside the circle represents the set of true known knowns. The remainder of the blue area represents false known knowns, i.e., things we falsely believe we know.

or when to look for them."[84, 85] These unforeseen problems abound, not only in economics, but in almost all human endeavors. In cybersecurity, for example, threats from unknown malware are particularly dangerous because we don't know they even exist. Unknown unknowns are particularly important and frequent in scientific research. They are found in genetics, where substantial ignorance exists with respect to elucidating the genetic bases for particular phenotypes [88]. This type of ignorance is what geneticists term "missing heritability." It is possible to estimate phenotypic variability in a population by statistical analysis of phenotype frequencies. This provides an estimate of how much heritability is "missing." However, these methods cannot predict whether rare genetic variants may

---

[84] See www.businessdictionary.com/definition/unknown-unknowns-UNK.html.

[85] The question of how we use known knowns or unknown unknowns in scientific research gives rise to an interesting paradox – Meno's Paradox, which is taken from Plato's dialogue *Meno:80d5-e5*. "*Meno*: And in what way, Socrates, will you look for that of which you do not know at all what it is? What sort of thing from among those things which you do not know will you posit to look into? Or indeed, even if you come across it as much as you like, how will you know that this is the thing which you did not know?
Socrates: I know what you mean, Meno. Do you see what a contentious argument you are conjuring up, that it is not possible for a man to inquire either into the things he knows or into the things he does not know? For he would not look for the things which he knows – for he knows them, and has no need for such inquiry – nor what he does not know – for he does not know what he is looking for."

exist – an unknowable unknown. Unknown unknown variables often affect experimental results and reproducibility, creating vexing problems in interpretation and hypothesis formulation.

It has been said that the ability to know that you don't know what you don't know is a mark of intelligence. It certainly seems so based on the humility of Newton and other geniuses, as discussed above, but clinical studies do support this contention.

Unknowable unknowns, unlike unknown unknowns, are unknowns that cannot now, or possibly ever, be known. Unknowable unknowns have been defined as "uncertainties that result from the inherent unpredictability of the system being analyzed,"[86] the key word being "inherent." If there were not an inherent unpredictably in the system, the unknown unknowns might eventually become known. Unknowable unknowns also may arise for practical reasons. A classic example is the attempt to establish a limit of acceptable radiation dose, i.e., a dose that does not increase the rate of spontaneous mutations. The problem here is that the lower the dose, the lower the mutation rate and the larger the number of subjects must be to establish statistical significance. Alvin Weinberg, a Nobel prize–winning physicist, has calculated to experimentally demonstrate, in mice, that a 150 millirem yearly limit for x-rays[87] is indeed safe at the 95% confidence level would require 8 billion animals [91]. Of course, if one only required a 60% confidence level, that number would drop to only 195,000,000 animals! The point Weinberg makes is *"to measure an effect at extremely low levels usually requires impossibly large protocols. Moreover, no matter how large the experiment, even if no effect is observed, one can still only say there is a certain probability that in fact there is no effect. One can never, with any finite experiment, prove that any environmental factor is totally harmless."* The unknown (safe dose) is thus unknowable.

A second example of an unknowable unknown comes from the field of artificial intelligence/machine learning [92]. This unknown is the answer to the question "why did a self-modifying algorithm make the decisions it made?" Geer argues that the question currently is unanswerable, although it may be in the future, so that for now the operation of the algorithm must be considered a "black box." J.B.S. Haldane addressed what is the black box of the universe when he said, *"Now, my own suspicion is that the universe is not only queerer than we suppose, but queerer than we can suppose."*[88] It is hard to think of anything more exciting than finding out!

Unknown knowns are knowns we know, but don't realize we know. A typical example would be something that is brought up in conversation that induces us to say "I knew that" because up until that point we did not realize we possessed that knowledge. One might term this an "amnestic unknown known." Zizek [94] addresses unknown knowns in the realm of public policy. He has likened them to the "Freudian unconscious"[89] and

---

[86] Quotation from Johnson [89].

[87] The International Commission on Radiological Protection (ICRP) actually has established a significantly lower limit, 100 millirem [90], which would mean ever larger numbers of animals would be necessary.

[88] Quotation from Haldane [93], p. 286.

[89] This term has been described as "mental activity that is not mediated by conscious awareness" as well as "the registration and acquisition of more information than can be experienced through conscious thoughts." See https://en.wikipedia.org/wiki/Unconscious_mind#Contemporary_cognitive_psychology.

see them as *"unconscious beliefs and prejudices that determine how we perceive reality and intervene in it."* Natali has examined how knowledge and ignorance have affected companies and individuals involved in the global gas market. He identifies two types of unknown knowns, knowledge that is readily available but not accessed and *"information that we are perfectly conscious of, but that we have not connected in a causal way."* In an amusing blog, Tim Kastelle, a professor at the University of Queensland Business School, provided an example of the former type of unknown known – the line at Starbucks.[90] He wrote *"... there are ... things that we know are true, that seem to be unknown because no one ever acts on them. Here's an example. I took this picture a couple of minutes ago at 10:39 am [photograph of a long line of people]. ... That's the line of people that left their desks at exactly 10:30 to get their morning coffee. If you had left your desk at, say, 10:25 instead, the number of people in line was 0. At some level, everyone knows this. But still, everyone still leaves their desk at 10:30, not 10:25. It's an Unknown Known."*

The five types of ignorance may be considered logical semantic constructs if we grant that "unknowns" and "knowns" are indeed knowns and unknowns, i.e., that the truth and falsity accorded to each term are, in fact, accurate. However, this accuracy is being assaulted today because of societal movements from the logical to the psychological, truth to belief, scientific to political, unbiased to biased, attentive to inattentive, and other such contrasting states. This movement from the epistemic to the doxastic substantially increases the complexity of our examination of ignorance, which is necessary because of the vagaries of human behavior and its effects... If [the] search for truth is to be successful, [we] must be able to distinguish truth (actual knowledge) from purported truth (actual ignorance).

3.5.2 *Ignorance Is Complex*

The information age has provided us with an immense amount of knowledge – knowledge that we use in innumerable ways. The most important issue that this wealth of information raises is whether the knowledge we encounter actually is knowledge or is it a fabrication, untruth, uncertainty, or distortion? Is public opinion or common belief being preached as truth or "alternate truth" (of which there is none, by definition). Assuming that even some of our information is knowledge, are we using this knowledge wisely and appropriately? Are we missing something that could have serious ramifications for ourselves or our society? Are we opening up a Pandora's Box of problems?

When the concept of ignorance is examined in depth, we find that it comprises much more than a simple lack of knowledge. Do we exhibit the ignorance of a child, one who must learn everything about the world? Do we choose to ignore certain things we don't know about? Are we precluded from learning? Do we avoid knowledge to prevent our entrenched beliefs from being questioned? Are we ignorant because the type of knowledge we seek is unattainable? These and other questions are being addressed with

---

[90] The photograph was not actually taken at a Starbucks, but the meme is useful here for the purpose of visualization.

increasing urgency within an entirely new field, "ignorance studies" [95].[91] Ignorance studies are relevant to a remarkably large number of fields, including art, anthropology, business,[92] sociology, psychology, history, politics, economics, philosophy, the law, and most importantly, science and medicine.

The urgency of ignorance studies has been argued by Smithson [99], who points out that the pace of "technological and intellectual achievement" has made many truths what I would call "wasting assets," meaning that as time passes, the veracity of many scientific truths decreases due to new discoveries, corrections, contradictions, exceptions, etc. This temporality of truth has been accompanied by the explosion of knowledge and the necessary consequence thereof, *viz.*, an explosion of ignorance. Blaise Pascal (1623–1662), a famous mathematician, is often quoted as saying *"Knowledge is like a sphere, the greater its volume, the larger its contact with the unknown."*[93] If one considers the radius of the sphere of an explosion to be the measure of the knowledge it represents, then as knowledge increases linearly the surface in contact with ignorance increases exponentially. This is what we now are experiencing in the world, in general, and in science in particular.

Let us now address the complexity of ignorance. This has been illustrated by Smithson [100] through the development of a "taxonomy of ignorance" (Fig. 3.5)[94] and by others through various subcategorizations. Smithson considers ignorance to be "the absence or the distortion of 'true' knowledge."[95] He distinguishes, in essence, two states of ignorance, an active state of "ignoring" and a passive state of "being ignorant," with each state being subcategorized. The former state considers that about which we are ignorant to be irrelevant. The latter is an erroneous cognitive state and is the one most commonly associated with ignorance. Smithson refers to this as error, a usage originating with Socrates, that describes someone who thinks they know something when they don't. This type of ignorance also has been termed "meta-ignorance" [100]. Socrates used "ignorance" to describe those that knew they did not know something. The active and passive types of ignorance defined by

---

[91] A masterful compendium of scholarly works on ignorance may be found in the *Routledge International Handbook of Ignorance Studies* [96]. This volume's cover image contains the warning *"Not ignorance, but ignorance of ignorance, is the death of knowledge."* An early effort to establish the importance and value of ignorance to scientific progress was *The Encylopaedia of Ignorance*, published in 1977 [97]. The editors opined *"Compared to the pond of knowledge, our ignorance remains atlantic [sic]. Indeed the horizon of the unknown recedes as we approach it. The usual encyclopaedia states what we know. This one contains papers on what we do not know, on matters which lie on the edge of knowledge. ... Clearly, before any problem can be solved, it has to be articulated."* Fifty-two of the most prominent scientists in the world contributed essays about their ignorance, many of whom were Nobel Prize winners. Interestingly, although not entirely unexpected, the more eminent a scientist a contributor was, the more likely they were to contribute to the volume, likely because they understood better than most their own profound ignorance.

[92] To be successful, businesses must be innovative, but to be innovative, they must actively seek out ignorance. Worried that businesses instead were shifting their balance from "spread your wings" to "stick to your knitting," thus stifling creativity, the editor of the *Journal of Product Innovation Management*, in an editorial [98], encouraged authors to address issues of ignorance by saying the journal "will gladly carry the story of the functions of ignorance alongside its presentation of the lessons of the learned."

[93] This quotation has been attributed to Pascal, but digging deeply into the books and manuscripts extant, neither I nor reference librarians have found the actual quote in any of Pascal's written work. This is another example of the value and necessity of not assuming anything. Pascal did, in fact, use the sphere as a metaphor, but not for knowledge, for God. Pascal actually said in the Pensées "Dieu est une sphère infinie, dont le centre est partout et la circonférence nulle part." Translation: God is an infinite sphere whose center is everywhere and circumference is nowhere.

[94] For a detailed explanation of the elements of the figure, the reader is referred to Chapter 1 of Smithson [100].

[95] Quotation from Smithson [99], p. 136.

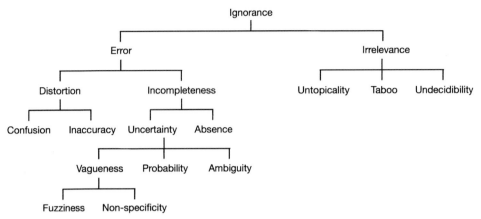

Figure 3.5 Taxonomy of ignorance. (Credit: Reprinted by permission from Springer Nature [102 Smithson], ©1989.)

Smithson are but two of a number of subcategories that have been created to simplify the concept of ignorance.

In an intriguing article entitled "What ignorance really is. Examining the foundations of epistemology of ignorance," Nadja El Kassar discusses the current state of ignorance studies [101]. She categorizes ignorance as: (1) lack of knowledge/true belief; (2) actively upheld false outlooks; and (3) substantive epistemic practices. The first category is the standard conception of ignorance, the "standard view," *viz.* we simply don't know something – call it *"p"* [102, 103]. The second category, also referred to as "pernicious ignorance," refers to the active maintenance of false beliefs. People who deny the existence of global warming exhibit pernicious ignorance. Active ignorance also includes *pre facto* rejection of new or controversial hypotheses, which may occur because of one's arrogance, laziness (to actually study the question dispassionately), defense of one's own cherished hypothesis, dogmatism, or closed-mindedness [104]. Rejection of the prion hypothesis in the 1980s and beyond is a good example of this type active ignorance (see the discussion of this issue in Section 3.5.2). It is important to note that this type of ignorance can exist at multiple levels. It begins at the level of the individual, but as individuals comprise social and political structures, it then extends to these as well [105, 106]. Of course, this extension also works in the opposite direction, and this is the focus of Kassar's third category, substantive epistemic practices. Here, ignorance does not exist passively, as for the first category, it is something that is practiced as a result of active inculcation in the individual by societal or political groups. These groups decide about what one should be ignorant, and through training, pressure, expectation, peer pressure, etc., and one's implicit consent or obeisance, make this standard practice.

A new field, "agnotology," has emerged that studies this type of ignorance.[96] Examples of areas studied by angnotologists include the perpetuating, by the tobacco industry and the

---

[96] For an excellent treatment of this subject by its founder, Robert N. Proctor, see [107].

physicians they paid to advertise their products, of public ignorance of the direct relationship between smoking and lung cancer, the role of the media in the propagation of false or misleading information, the explosion of ignorance masquerading as knowledge through the Internet, and the dissemination of inaccurate, misleading, incorrect, unverified, or fabricated scientific information – a particularly insidious and dangerous thing for science and the public.

A "New View" (conception) of ignorance has challenged the standard view that ignorance is simply lack of knowledge. The New View argues, in essence, that psychological and doxastic factors contribute to one's disbelieving, or suspending belief in, p, though p is, in fact, true [108, 109]. Peels has defined four subcategories of the New View of ignorance, disbelieving, suspending, conditional disbelieving, and conditional suspending [108, 109]. For the first two subcategories, one has considered p, whereas for the latter two, one has not considered p but would disbelieve or suspend judgement in it if they did. These behaviors are particularly dangerous with respect to science's search for truth.

DeNicola [110] has proposed a type of active ignorance he terms *nescience*, "which designates what we or others have determined we are not to know." He specifies five categories of nescience, each of which seeks to achieve a different goal through premeditated ignorance.

1. *Rational ignorance* corresponds to Smithson's category of "irrelevance." We choose not to know, e.g., what is in the fine print of a long legal document, all the words in the Oxford English Dictionary,[97] or the gross domestic product of Ecuador. This term originated in economics and its "key element is that the individual makes a judgment about learning based on the perceived costs of learning in relation to the perceived benefits of knowing."[98]
2. *Strategic (or tactical) ignorance* protects us from something. It is "ignorance is bliss." It may be refusing to learn the ending of a book before reading it or to know whether one's client (if you're a defense attorney) did or did not commit the crime. In this latter context, it is avoidance of the responsibility of knowing. Strategic ignorance is essential in every aspect of scientific practice to avoid bias. A particularly important example is double blinding in drug trials. Here, neither the individual administering a putative treatment nor the person receiving it knows what was administered. Further blinding involves ensuring those evaluating the clinical effects of the treatment or analyzing the resulting data also are not aware of what was administered to whom. This approach ensures that investigator bias and placebo/nocebo effects do not compromise the study. The routine employment of strategic ignorance during experimentation in the basic sciences provides similar benefits. Haas and Vogt use the term "preferred ignorance" to describe the type of premeditated ignorance that scientists and clinicians practice in blinded studies.[99]

---

[97] Unless you are Ammon Shea, who read the entire 21,730 pages of the *Oxford English Dictionary* in a year [111].
[98] Quotation from DeNicola [110], p. 80.
[99] See Gross [112], p. 17.

3. *Willful ignorance* is another ignorance is bliss category that "connects *being ignorant of something* with *ignoring that thing*." DeNicola uses the example of a wife not wanting to acknowledge that her husband is unfaithful. She self-deceives herself because she may feel that *"[l]earning the truth can be difficult; embracing it can be even more difficult. Better not to know."*[100] Another example is that of a scientist who may choose not to read papers that could contradict their own hypotheses or experimental findings, were written by a researcher for whom the scientist has low regard, or could bias the conception, execution, analysis, and interpretation of their experiments. The latter is another example of the use of premeditated ignorance to avoid bias and is a valuable use of ignorance. By contrast, willful ignorance also is a part of close-mindedness.

4. *Privacy and secrecy* is a type of enforced ignorance, one that is imposed on others.

5. *Forbidden knowledge (Smithson's "taboo")* may be considered required ignorance. Such ignorance is produced by fiat, usually by religious or political institutions. It is particularly relevant to science in America's current political climate. DeNicola defines it as "systematic suppression of scientific investigation, as by the ancient Church or the Trump administration, or by scientists themselves ...." Enforcement of such taboos in science often takes the form of funding restrictions. We (e.g., the NIH or NSF) will pay for you to study *this* but not *that*. Peer pressure also functions to enforce taboos by making clear that research on certain topics will be met with censure and damage to one's scientific reputation. An example of this type of enforcement were scholarly studies of sexuality. Such studies were taboo until Kinsey, and Masters and Johnson, had the courage to publish their seminal works on human sexuality in the mid-twentieth century [113–116]. The continued maturation and enlightenment of society have resulted in the elimination of many taboos, enabling significant advancements in knowledge. Whether this has been a good thing or a bad thing, we will leave for philosophers, ethicists, and theologians to determine, as they now are doing with respect to human cloning.

Thus far we have discussed a variety of types of ignorance that have roots in semantic logic, psychology, politics, religion, personal freedom, and science practice. The assumption intrinsic to this discussion has been that valid reasons exist for their practice, though what we mean by valid has not been specified. The last category of ignorance is *double ignorance*. It is a subset of known knowns that represents things we *think* we know that we actually don't. Socrates considered those exhibiting this characteristic to be "in error," a term used by Smithson in his taxonomy of ignorance (see Fig. 3.5). The term meta-ignorance also has been used to describe this type of ignorance. This is a very large subset, one that is increasing exponentially in this information age, especially through the Internet and other electronic means, where the veracity of information is questionable. This is by far the most dangerous knowledge state for a scientist. It blocks access to the truth, misleads the scientist and others, and results in wasted time and resources. It also creates controversy where there should be none and can have disastrous public policy implications, as e.g., after

---

[100] Quotation from DeNicola [110], p. 84.

the Trump administration based its policies on global warming, biomedical research, and ecology, to name but a few, on known knowns that actually weren't.

3.5.3 Learning to Be Ignorant

Ignorance is important and complex. How do we learn it? Science students are educated by acquiring knowledge. This knowledge may be theoretical or practical in nature, as, for example, in lecture and laboratory classes in physics or chemistry. It also may have historical, political, sociological, and even philosophical, components. What it has not had is instruction in ignorance. Rene Descartes, after his scientific training was completed, found that rather than acquiring knowledge during his education he had instead acquired ignorance. In 1637, he wrote:

From my childhood, I have been familiar with letters; and as I was given to believe that by their help a clear and certain knowledge of all that is useful in life might be acquired, I was ardently desirous of instruction. But as soon as I had finished the entire course of study, at the close of which it is customary to be admitted into the order of the learned, I completely changed my opinion. For I found myself involved in so many doubts and errors, that I was convinced I had advanced no farther in all my attempts at learning, than the discovery at every turn of my own ignorance. And yet I was studying in one of the most celebrated Schools in Europe, in which I thought there must be learned men, if such were anywhere to be found. I had been taught all that others learned there; and not contented with the sciences actually taught us, I had, in addition, read all the books that had fallen into my hands, treating of such branches as are esteemed the most curious and rare. I knew the judgment which others had formed of me; and I did not find that I was considered inferior to my fellows, although there were among them some who were already marked out to fill the places of our instructors. And, in fine, our age appeared to me as flourishing, and as fertile in powerful minds as any preceding one. I was thus led to take the liberty of judging of all other men by myself, and of concluding that there was no science in existence that was of such a nature as I had previously been given to believe.[101]

Three hundred years later, Ravetz [118] reiterated Descartes's lament when he said:

Ever since Descartes, the awareness of ignorance has been inhibited in our intellectual culture. Consider the teaching of science and mathematics, where over centuries, students have been stuffed like geese, full of facts accepted as important and true at the moment of teaching. The curriculum is always crowded; there is always too much to teach; there is never time for reflection, for perspective, for the cultivation of awareness. Examinations in history or literature normally include questions like, "Critically evaluate ...."; but such questions will hardly ever be found in science [education].

Dr. Lloyd H. Smith, Associate Dean for Admissions and Special Projects at the University of California at San Francisco, was even blunter in his assessment of medical education.[102] He said *"Our students should be freed from the stupefying excesses of fact engorgement, which threaten to convert them into floppy disks encoded for our present ignorance."*

The lack of awareness of the value, and necessity, of an intimate knowledge of ignorance is especially important in medicine, where *"... physicians daily confront mountains*

---

[101] From Descartes [117], p. 151.
[102] Quotation taken from Burrow [119].

Table 3.7 *Goals and methods of the curriculum on medical education at the University of Arizona College of Medicine.*

Gain understanding of the shifting domains of ignorance, uncertainty, and the unknown: philosophical and psychological foundations and approaches to learning, questioning, and creating "knowledge"; history and development of ideas and methods in basic science and medicine.

Improve skills to recognize and deal productively with ignorance, uncertainty, and the unknown: question critically and creatively focus on raising, listening to, analyzing, prioritizing, and answering questions from different points of view.

Reinforce positive attitudes and values of curiosity, optimism, humility, self-confidence and skepticism.

Questions and questioning exercises.

Weekly ignorance logs.

Adapted from Fig. 1, p. 8, Witte [125].

of confusing and conflicting information, entrenched dogma, and half-way technologies and are often forced to make rapid decisions in this context while lives hang in the balance. The ultimate victims of medicine's collective ignorance ... are the patients."[103] There is a vast amount of knowledge available to clinicians – and scientists – and this amount continues to increase. For this reason, medicine and the basic sciences have used informatics[104] to extract the most significant, helpful, or informative bits of knowledge relevant in their domains of study. Training in informatics is essential for future doctors and scientists, but Witte *et al.* have argued compellingly that "ignoramics – the art and science of ... recognizing and 'managing' unanswered questions, unquestioned answers, and unmanageable questioners and analyzing the indecisions implicit in everyday medical practice and laboratory research" – also is essential [120, 121].

A pioneering program in ignoramics was the "Curriculum on Medical Ignorance" (CMI), introduced at the University of Arizona College of Medicine in 1985. As shown in Table 3.7, the CMI was a comprehensive program to make medical students aware of the importance of recognizing, understanding, and dealing with the ignorance and uncertainty pervasive in the practice of medicine. The program included philosophical and psychological components, examination of the history and use of the scientific method (with its intrinsic limitations and uncertainties), dispassionate skepticism, open-mindedness, and creative thinking exercises. Students also were required to keep a log of their own ignorance, much like Feynman used to do (see Fig. 3.3). In addition, the students received instruction from "Visiting Professors of Medical Ignorance." This program was emulated a few years later by Peden and Keniston at the University of Wisconsin [122, 123]. Their program taught ignorance to undergraduate psychology students. Student were instructed on how to embrace their ignorance and use it to conceive of new questions for study and to critically evaluate the knowledge they ostensibly were acquiring in standard coursework.

---

[103] Quotation from Witte [120], p. 897.

[104] This term has been used since the mid-twentieth century to describe information processing by computers (computer science), but it is now applied much more widely. In the context of medicine and science, it describes a range of activities including information acquisition, archiving, processing, analysis, and dissemination.

According to Peden,[105] students *"... really believe the textbook, that this is it – if they learn that, they know everything there is to know about a subject."* However, as the class proceeds, students get the *"sense of what is not in the text [i.e., their ignorance]. They come to realize that a text can't do everything. It is not a compendium of all that is known .... It is a particular view and selection of material."* The students learn to appreciate the value of critical thinking, and most importantly, skepticism. These qualities are prerequisites for the advancement of science and medicine because they free the student, scientist, or physician from the bonds of dogma and conventional wisdom that block the exercise of creativity and the possibility of discovery.

Few have access to curricula in ignoramics. However, because ignoramics eschews *a priori* knowledge *per se*, someone who is determined to learn it need only recognize and practice certain behaviors to acquire knowledge of it. A good place to start is the study and emulation of Socratic ignorance and the Socratic method. Practicing Socratic ignorance is straightforward. You make yourself aware of what you don't know. However, this may be difficult for students, fellows, and even faculty, who may want to impress others, especially mentors and colleagues, with how much they know and how little they don't. Students, in particular, may not want to appear stupid or ask "stupid questions." Interestingly, one finds that the most knowledgeable people are those that appear to be asking the "stupidest questions" over and over and over again! Those that don't ask remain ignorant. This was a lesson I learned when I was shocked that a famous Caltech biologist, and member of the National Academy of Sciences, asked a seminar speaker such a simple and apparently unimportant question as "what buffer was used in the experiment." I had expected the question to have greater profundity, but after my shock I had an epiphany – maybe this scientist became as accomplished as they did *because* they weren't afraid to ask simple, "stupid" questions. An insatiable need to ask questions, i.e., to be unafraid to exhibit one's ignorance, is a key element in the advancement of science. Schwartz has termed this "productive stupidity" [126], although a more accurate term would be "productive ignorance" because the concept has nothing to do with intelligence, only a lack of knowledge.

Questions are the remedy to Socratic ignorance. What about the Socratic method? Here, the quality of the questions must change from simple information acquisition to what one might call "probing skepticism." One becomes the devil's advocate, beginning by rejecting assumptions, givens, dogma, established facts, what you read, what you were taught, and what your parents or mentors told you. In short, believe nothing and assume nothing, and then seek to understand the epistemological bases of the beliefs and "facts" you encounter. Do so by seeking to uncover the assumptions, including assumptions of fact, that actually underpin so-called knowledge. In essence, one must become ruthlessly critical of one's own knowledge and that of others. This tacit assumption of ignorance appears to be characteristic of the most creative among us. Kerwin [127] has suggested that *"highly refined capacities to 'not know' – such as abilities to suppress preconceptions, embrace serendipity, selectively 'forget' irrelevancies, see and create new patterns of possibility – distinguish highly creative individuals from less inventive types."*

---

[105] Quotation from Stocking [124].

4

Development of the Scientific Method

From Papyrus to Petaflops

We discussed in Chapter 2 the fact that society has considered science to be "special," i.e., especially authoritative and true. Why this is so is implicit in the various definitions of science that have been propounded over the centuries, all of which involve the notion that science is empirical knowledge, i.e., knowledge obtained by observation. However, observation *per se* is not science. It is only when observation of the natural world is incorporated into a logical framework for scientific discovery, i.e., systematized, that observation can be termed "scientific." The application of this framework in practice is the "scientific method," an idealized version of which is illustrated in Fig. 4.1.

Curiosity drives science and the need to provide reliable answers to the questions resulting therefrom drove development of the scientific method. The ancients saw the moon, the sun, and the stars move through the sky, which led to fundamental questions such as what are these celestial objects and why and how do they move? At that time, no one could provide an explanation, other than mystical. Today, science is information rich, and anyone interested in a particular question first would access this information to determine if the answer might already be known. If it were not, or if an understanding of a phenomenon were considered incomplete, this could justify new investigations of the phenomenon.[1] The scientist then uses their ingenuity, creativity, intelligence, and intuition, among other qualities, to formulate a hypothesis that might provide the answer. One is free to communicate this hypothesis to whomever one wishes, but as we discussed earlier, without evidence of the truth of the hypothesis, it remains nothing more than an opinion. Converting this opinion into a scientific fact is done by experimentation and *it is the logical framework of the scientific method that allows scientists to do so.*[2]

We use experimental systems to test predictions of hypotheses. When the experimental data are analyzed, we determine if our predictions were true. If so, we have generated empirical support for the hypothesis. If not, we have done the opposite. However, many times (maybe most times!), we obtain results that do not allow us to confidently determine truth or falsity. In these cases, we may modify the hypothesis to explain deviations from predicted behavior and repeat the experimental/analytical process. If our data than are

---

[1] Unless one is independently wealthy, justification would be the first step in seeking financial support for such an investigation.

[2] One of the most influential books ever written on the logic of science is *The Logic of Scientific Discovery*, by Karl Popper, originally published in German in 1935 [128] and later translated into English in 1959 [6].

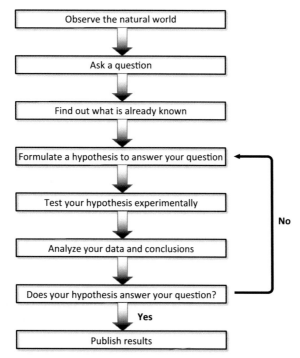

Figure 4.1 The scientific method. A simplified scheme showing consecutive steps comprising the scientific method. Hypotheses must be modified and retested if experimental results
are inconsistent with their predictions.

consistent with theory, and assuming our work is of sufficient novelty and importance, we
publish our findings in a peer-reviewed scientific journal.

The scientific method, as presented above, is an idealized version, one that is taught to
most science students and that represents how an informed public might define the term.
However, deeper examination of both the history and application of the scientific method
will reveal that our idealized version is insufficient to provide an apodictic explanation of
what it is in practice.[3] Paul Feyerabend, in fact, famously suggested in 1975 that *there is
no scientific method* and "anything goes" [130]. Many since then also have expressed their
own doubts [130–132]. However, very recently, Elliot Sober has argued, although *details*
of the method may differ among disciplines, there is in fact "a method of reasoning that
applies to all scientific subject matters" [133].

The scientific method does vary among different scientific disciplines. This is illustrated
in Fig. 4.2, a Venn diagram à la Sober [133], that there *does* exist a core scientific method
that is common to the conduct of scientific investigations not only in the natural sciences
but in the social sciences as well.

The scientific method may support the entire scientific enterprise, but it is important
to note that its use is not always an obligatory part of discovery. In physics, for example,

---

[3] For an interesting monograph on the subject, see [129].

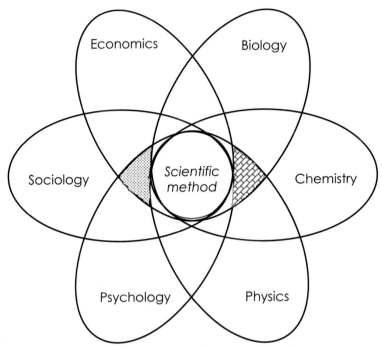

Figure 4.2 Venn diagram of scientific method in practice. The diagram illustrates the fact that many different fields apply the same core concepts of the scientific method in their practice. It also reveals subsets of shared methodological practices among nonscientific (dotted shading; e.g., economics, sociology, and psychology) and scientific (brick pattern; e.g., biology, chemistry, and physics) fields. (From Fig. 1.1 of Gauch [5], modified by the author.)

some of the most fundamental insights into the behavior of the natural world did not use the experimental method *per se*. Instead, those insights were obtained from *Gedankenexperiments* (German for "thought experiments"[4]), experiments conducted entirely in one's mind. Einstein's own words reveal part of what the thought experiment process involves.[5] He spoke of the

... search of those highly universal laws... from which a picture of the world can be obtained by pure deduction. There is no logical path leading to these ... laws. They only can be reached by intuition, based upon something like an intellectual love ('Einfuhlüng') of the objects of experience.

Einstein's intuition helped him perform a thought experiment in 1907 that provided a foundation for building the theory of general relativity [136].

---

[4] Interestingly, it is to Hans Christian Ørsted, the Danish scientist and philosopher, that the first use of this term is attributed. For more information on thought experiments, see Stuart [134] and Fehige [135].
[5] Taken from Popper [6].

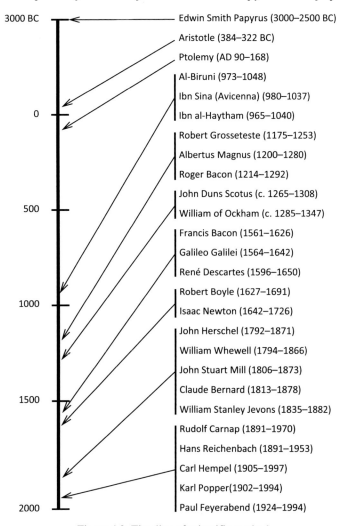

Figure 4.3 Timeline of scientific method

I was sitting on a chair in my patent office in Bern. Suddenly a thought struck me: If a man falls freely, he would not feel his weight. I was taken aback. This simple thought experiment made a deep impression on me. This led me to the theory of gravity.[6]

The theory of general relativity, published in 1915 [137], became one of the most universally accepted theories in the history of science. The first prediction of the theory, that light would be deflected by the sun's gravitational field, was proven correct in 1919 by Dyson *et al.* [138]. More than a century after Einstein's publication of the theory, its last

---

[6] Einstein said that the theory of gravity "could not be resolved within the framework of the special theory of relativity". [136]. It took Einstein eight years to incorporate the theory of gravity into the special theory of relativity, thus producing the general theory of relativity.

significant prediction, the existence of gravitational waves, was reported by Abbot *et al.* in 2016 [139]. It had taken this long to convert a *Gedankenexperiment* into what is considered a scientific fact using the scientific method.[7]

A timeline of the development of the scientific method is shown in Fig. 4.3 and brief biographical notes and contributions of each individual therefrom are discussed. Length constraints preclude detailed analyses of each contributor and some have not been included in the timeline if their contributions were considered to be of lesser significance.[8]

4.1 Egyptian Medicine and Greek *Scientia*

[T]he history of ... modern science should be devoted to its *theoretical* aspect at least as much as to the *experimental* one. .... The former is not only closely linked with the latter, but even dominates and determines its structure. The great revolutions of ... science – though based, of course, on the discovery of new facts (or on the failure to ascertain them) – are, fundamentally, *theoretical* revolutions of which the result was not a more perfect correlation of 'facts of experience' but, in my view, a new concept of the reality underlying these 'facts.'[9]

How did the modern scientific method develop through history? From where did it come and why? Just as the theory of evolution offers an explanation of how we, as humans, came from the primordial soup of a nascent earth, an examination of the history of the scientific method reveals the starts and stops, successes and failures, and societal forces that molded how we do science today. This is a history that is temporally non-linear and often characterized by irrationality, bias, dogmatism, conflict, and profound errors in thought but also by genius, perseverance, and a love of nature and its revealed truths. An appreciation of the development of the scientific method provides us not only with facts, but more importantly, and as argued in the quotation above, insights into its theoretical underpinnings that may be critical in guiding our own practice of science. We will learn not only the "whats" of the scientific method (explicit facts) but also the "whys" (concepts and philosophy).

4.1.1 The Edwin Smith Papyrus (3000–2500 BC)

Aspects of the scientific method are thought to have been practiced in Egypt as early as 3000–2500 BC [146]. Additions to, and refinements of, the method have been made by philosophers, natural philosophers, and scientists since that that time, first in ancient Greece

---

[7] All measurements have intrinsic limits in precision and accuracy. However, the closer an experimental result matches a predicted result, the more confidence we have in the result and its implications. For this reason, physicists have continued to test the general theory of relativity to determine if more precise measurements might reveal inconsistencies. One of the theory's principles, "the complete physical equivalence of a gravitational field and a corresponding acceleration of the reference system," [140] which had been confirmed on earth to be correct within an experimental error of 10^{-13} [141], was confirmed with 10-fold higher precision in space by the MICROSCOPE research satellite [142]. Scientists are hoping to have even better confirmation using a new satellite capable of errors of 10^{-18} [141]. So far, Einstein's predictions of the theory continue to be corroborated.

[8] For more detailed information on the scientific method, I refer the reader to the many books written on the subject, e.g., see [5, 143, 144].

[9] Taken from Koyre [145], p. 22.

and then in the Islamic world and Europe. I discuss here some of their most important contributions to the method.[10]

The first written record in which important features of the scientific method were contained may be from a papyrus on which case reports of 48 people treated for serious injuries or tumors were written[11] (see Fig. 4.3 for a timeline). Each case report describes the patient's malady (observation:diagnosis), how the patient was examined (hypothesis formulation and testing:physical and psychological examinations), a conclusion regarding treatment based on the exam evidence (prediction of outcome:prognosis), and a guide to treatment. The scroll is important because it is the earliest example of the practical application of key aspects of the scientific method, including observation, hypothesis formulation and empirical testing, deduction, and prediction [148, 149]. These aspects are discussed in the context of a rational and logical approach to understanding and acting upon information obtained by direct observation.[12] Remarkably, this approach is used to this day in clinical medicine. For example, "the diagnostic, prognostic and therapeutic rationale documented in the oldest known treasure on spinal injuries [i.e., the Edwin Smith Papyrus] can still be considered as the state-of-the-art reasoning for modern clinical practice" [146].

4.1.2 Aristotle (384–322 BC)

Two millennia would pass before Aristotle laid the foundation for the science and scientific method later developed in the Middle East and Europe. It is likely that Aristotle was familiar with the practice of medicine in Egypt, as he lived and studied there for approximately two decades. However, it was not until his writings that the seed of the scientific method germinated and began to develop. This is why many consider Aristotle the father of the scientific method. However, it may be more accurate to say that Aristotle was the bridge between pure logic, which was used by the Greek philosophers at that time (and especially by Aristotle's teacher, Plato) to understand the world and modern scientific practice. Among Aristotle's contributions to the development of the scientific method was his requirements for empirical observation, deep and systematic study of knowledge extant, the use of logic in the analysis of observations, and the necessity for continued consistency of new observations with inferences already made. Hypothesis testing *per se*, i.e., conceiving of and executing processes to test inferences, was not part of Aristotle's method. Aristotle thus wrote:

The facts on this subject have not yet been sufficiently ascertained; if ever they are, it will be necessary to trust our senses more than our reasonings, and the latter only when the results are in agreement with the new phenomena [150].

---

[10] An exhaustive discussion of the development of the scientific method is impractical in a single volume; thus, I have focused the discussion on those scientists and ideas I feel have been particularly important.

[11] The original document has not been found, but a copy apparently was produced in the seventeenth century. This papyrus scroll, the "Edwin Smith Papyrus," was named after the American art dealer who procured it in Luxor, Egypt, in 1862. It was written in hieratic, a script used by scribes that developed in parallel with, but was distinct from, hieroglyphics. An English translation was published in 1930 by Breasted [147].

[12] It should be noted that the scroll also contains mystical elements, as mysticism was a significant component of society at that time and was invoked by physicians to explain the unexplainable [149].

In essence, Aristotle is saying, "facts are more to be trusted than reasoning" [151]. It should not be inferred that Aristotle rejected reasoning, quite the contrary. He argued for a logic of interpretation of observation that was, in fact, the first formal attempt to systematize and codify logical processes [152]. This was his syllogistic approach to deduction. Suter may have captured best the novelty and significance of Aristotle's approach when he wrote "Aristotle's contribution to the world ... was his awareness of inference. This was a startlingly original appreciation, nor can one admire too strongly the mind which first became fully self-conscious here. To see clearly, for example, in the abstract, that if x is y, and z is x, then z is y, is a stroke of genius" [143].[13]

4.1.3 Ptolemy (c. AD 100–170)

How were Aristotle's "tools" actually used to investigate the natural world? Claudius Ptolemy explained how in the *Almagest*,[14] his *magnum opus* on studying the movements of the stars and planets. Ptolemy's writings about knowledge, observation, hypotheses, and theory validation would later become the bedrock of the scientific method [153]. He argued forcefully that knowledge only could come from observation of the natural world, not through any *a priori* concepts developed by philosophers or theologians. Ptolemy's detailed discussions of methods of observation, data analysis, predictions of planetary positions, and subsequent validation or invalidation of these predictions were a model for what we today would call hypothesis testing. Ptolemy assumed from the outset, as Aristotle had suggested, that the cosmos was geocentric. Nevertheless, Ptolemy's geocentric model of the stars and planets has been shown to remarkably accurate and to produce the same predicted positions of stars and planets as the subsequent models of Tycho Brahe (combined geocentric/heliocentric)[15] and Nicolaus Copernicus (heliocentric) [154]. A testament to the practical utility of the Ptolemaic system is its use today in planetaria, which reproduce it mechanically to display the movement of the sun and stars relative to the Earth [155].

4.2 Method in the Islamic World

The scientific method, as practiced by the Greeks, later was developed in the Islamic world between the tenth and fourteenth centuries into what we now understand the scientific method to be [156]. At that time, the Islamic world was a center of world learning. Its scholars translated many Greek writings into Arabic that otherwise would have been lost with the fall of Rome. These translations were themselves subsequently translated into

---

[13] Suter continued, "The word 'syllogism,' in fact, given by Aristotle to his three-step brand of inference, is the Greek equivalent of the Latin '*computo*'; but whereas 'to compute' has come to mean 'to think together' numbers or letters standing for them, 'syllogism' is used exclusively for the 'thinking together' of concepts such as 'man' or 'mortal,' or letters or other symbols representing them." Aristotle thus moved logic from the concrete (numbers and letters) to the conceptual and, in doing so, provided us tools to investigate a much broader range of questions both physical and metaphysical.

[14] The work was originally titled $M\alpha\Theta\eta\mu\alpha\tau\iota\kappa\eta$ $\Sigma\upsilon\nu\tau\alpha\xi\iota\varsigma$ (Mathēmatikē Syntaxis). It also has been called, in Latin, *Syntaxis Mathematica*, *Almagestum*, and *Magna Syntaxis*. The name Almagest is derived from the Arabic word al-majisiti, meaning "greatest." We discuss the tremendous contributions of Arab scholars to the scientific method in Section 4.2.

[15] Kuhn has written, "The Tychonic system is, in fact, precisely equivalent mathematically to Copernicus's system... [T]he Tychonic system is transformed to the Copernican system simply by holding the sun fixed instead of the earth. The relative motions of the planets are the same in both systems ..." [154].

Latin, which made them accessible to Europeans. The Latin translations became the foundation for the further development and use of the scientific method in Europe [144]. In fact, it has been suggested that the Islamic era produced a profound change in science in general – moving its practice from the ancient to the modern. Imad-ad-Dean [157] expressed this change as follows:

... this was an epistemological transformation, going from a pure rationalism ... into a complex epistemology in which reason, observation and experiment, and authority play an interactive role, each one checking on the other.

Observation and experiment were exemplified in the work and writings of Jābir ibn Hayyān (721–815),[16] also known as Geber the Alchemist,[17] one of the fathers of modern chemistry, who said:

The first essential in chemistry is that thou shouldest perform practical work and conduct experiments, for he who performs not practical work nor makes experiments will never attain to the least degree of mastery. ... do thou experiment so that thou may acquire knowledge.[18]

4.2.1 Ibn al-Haytham (965–1040)

It was not until Ibn al-Haytham (Latin: *Alhazan*), two centuries later, that the scientific method was, in essence, codified [159]. al-Haytham was a polymath who made important scientific contributions in many areas. It has been said that "Ibn al-Haytham should be regarded as the world's greatest physicist in the time span between Aristotle and Newton" [156]. He was especially well known for his systematic studies of optics, which were published in 1028 in the book *Kitab al-Manazir (Book of Optics)*. al-Haytham argued "that scholars should follow certain steps when dealing with scientific questions ... including observation, stating a problem, creating a theory, experimentation, analysis of results, interpretation of data, coming to a conclusion and publication of the findings" [160], steps that comprise the modern scientific method (Fig. 4.1).

4.2.2 Avicenna (980–1037) and al-Biruni (973–1048)

Two contemporaries of Ibn al-Haytham, also polymaths, were Ibn Sīnā (Avicenna) and Abū Rayhān al-Bīrūnī (al-Biruni). Each contributed substantially to the development of the scientific method, but they differed with respect to certain philosophical aspects of the method (see below). Avicenna may be considered a philosopher-physician. He was interested in establishing first principles of science and emphasized induction and experimentation as the means to achieve this goal. In his book *The Canon of Medicine* (1025), a medical textbook that would become a standard text for medicine for the next six centuries [156], Ibn Sīnā discussed the use of a "scientific method" to diagnose and treat

---

[16] It should be noted that some writings attributed to Hayyān may have been written by others.

[17] An interesting historical factoid is that the name Geber is the root of the word *gibberish*, used in reference to "obscure language" [156], as Geber's own writing was quite difficult to apprehend.

[18] Quotation in Holmyard [158], p. 60.

disease. The method involved careful determination of symptoms and signs (observation), a tentative diagnosis (hypothesis), evaluation of the diagnosis in light of knowledge of disease extant and clinical tests (experimentation), treatment options (conclusions), determination of treatment efficacy (hypothesis testing), and reevaluation of diagnosis and therapy (hypothesis testing/formulation) [161]. The logical methods developed by Ibn Sīnā to identify causes of disease included determination of agreement (finding a single feature, among all features, found to cause disease), difference (a single feature, among all features, missing in those not exhibiting disease), and concomitant variations (features co-varying relative to disease causation) [162]. These methods later would be promulgated by John Stuart Mill in his famous book *A System of Logic*, published eight centuries later.

Al-Biruni was an expert experimentalist with a keen appreciation of the need for instruments of high precision to reliably answer scientific questions. He recognized the fact that systematic errors can be introduced by instruments, as well as by observers themselves, and recommended *replication*, a key component of modern scientific method, as one means to deal with these errors (to enable determination of "common sense" averages). He also provided a logic for the exclusion of particular data in his data analysis, while still presenting excluded data in his publications (we shall address this thorny issue in Section 5.3). The precision of his astronomical instruments and experimental work is obvious from his determination of the radius of the Earth, 6,339.6 km, which is the same, *within experimental error*, as that determined by modern scientists (6,356.8–6,378.1 km).

From a philosophy of science perspective, Al-Biruni argued, with respect to theory creation, that experiments must come first, and only then should theories be formulated. This contrasts with Ibn Sīnā's argument that "general and universal questions come first and led to experimental work" [162]. It is understandable that such arguments should have occurred, as the historical period in which each great natural philosopher lived was one in which science was moving from an endeavor involving primarily pure logic (Plato) to one with a foundation in empiricism. In fact, viewed through the prism of modern science practice, both men were right. Consideration of theories extant is a key feature guiding the choice of experimental methods and novel experimental results can lead to new theories.

4.3 Science Moves to Europe

4.3.1 Robert Grosseteste (1175–1253)

Expansion of the Muslim world into the Mediterranean and Europe, especially in Sicily and Spain, facilitated a broad cultural interchange. The works of the great Islamic scholars were translated from Arabic into Latin, as were many Arabic translations of the ancient Greek philosophers (most importantly of Aristotle), thus providing European scholars with access to this knowledge. One of the first European scholars to incorporate this knowledge into his own scientific writings was the Englishman Robert Grosseteste, Chancellor of Oxford and Bishop of Lincoln. Grosseteste himself translated a number of the works of Aristotle from Greek to Latin and published important commentaries on them [163, 164]. The dissemination of Grosseteste's arguments and writings to his colleagues and students, and their own subsequent independent work, made Oxford a center for development of the scientific method in the thirteenth century.[19]

In Grosseteste's time, the Aristotelian approach to understanding the natural world, i.e., through the combination of empirics and logic, was the dominant perspective among natural philosophers. The methods through which this perspective was reduced to [scientific] practice were similar to those comprising the modern scientific method, but according to Crombie [164], these methods "remained *ad hoc* and rule-of-thumb. They [natural philosophers] had no conception of how to generalize problems and how to establish general proofs and explanations." Natural philosophers were only beginning to explore the deeper philosophical underpinnings of explanation, theory creation, theory choice, epistemology, and causation, all integral parts or results of the scientific method.

Grosseteste advanced these efforts in a number of ways. Grosseteste's commentary on Aristotle's *Posterior Analytics* addressed whether one could derive a single theory from an observation(s) (the problem of theory selection and un-conceived alternatives) and how to determine whether a theory was true (epistemology). Presaging Karl Popper's work on theory validation and falsification, Grosseteste suggested that "[D]emonstrations in natural science ... were 'probable rather than strictly scientific.'"[20]

The movement from observation of a phenomenon to development of a theory is an inductive process, and this is one aspect of the scientific method that Grosseteste emphasized. Once one has moved from the specific to the general and developed a theory, predictions (deductions) based on that theory could be made. These predictions would then be tested experimentally to validate or falsify[21] the theory. A key aspect of Grosstestes's method of theory testing was *controlling* experiments. "When one controls his observations by eliminating any other possible cause of the effect, he may arrive at an experimental universal of provisional truth."[22] Application of the scientific method to experimentation involving the creation of an experimental system cannot be done in the absence of controls. This is one of most important and fundamental aspects of the scientific method. Grosseteste also emphasized the importance of applying mathematics and its principles to experimentation and data analysis. Grosseteste argued, on epistemological grounds, that if "accurate knowledge of nature" was to be obtained [163], that knowledge must derive from the application of mathematics to the question at hand, a concept developed more fully by Galileo four centuries later (see below).

[19] It should be mentioned that not everyone agrees on the importance of Grosseteste's contributions to the development of the scientific method. For example, Clagett argues that Grosseteste's contributions were not substantially different or better than that of the ancient Greeks themselves [165]. By contrast, Koyre is of the opinion there is a "perfect and amazing continuity in the development of logical thinking: from Aristotle and his Greek (and Arabic) commentators to Robert Grosseteste, Duns Scotus and Ockham, the great Italian and Spanish logicians, and up to John Stuart Mill there is an unbroken chain of which the Bishop of Lincoln represents, indeed, one of the most important links: he revived this tradition and gave root to it in the West." [145]. Scotus, Ockham, and Mill, among others, are discussed later in this section.

[20] Quotation from Crombie [164], p. 108. It should be noted that Grosseteste only considered mathematics to be "science." Nevertheless, in modern science, much experimental work must also be considered "probable."

[21] Seven centuries later, *falsification* became a cornerstone of Karl Popper's logical framework of scientific discovery.

[22] Grosseteste, as paraphrased by Dales [163].

4.3.2 Albertus Magnus (c. 1193–1280)

A contemporary of Grosseteste was Albertus Magnus (Friar Albert the German or Albert of Cologne [166]), known even in his own time as Albert the Great. Albertus was a savant who wrote extensively on a broad range of subjects, both scientific (physics, astronomy, astrology, alchemy, physiology, psychology, medicine, natural history, logic, mathematics, etc.) and theology. [167].[23] Albertus wrote extensively on scientific method, using as a foundation the works, among others, of Aristotle, Euclid, and Al-Kindi.[24] Albertus's explications sometimes challenged Aristotelian methods, and in the process, led to a deeper understanding of the philosophical underpinnings of the scientific method *per se*. Albertus discussed extensively the fact that logic was an obligate part of the method.

In the process of moving from the known to knowledge of the unknown, the scientific investigator cannot simply begin to produce scientific demonstrations. ... This is because the process of demonstrating presupposes a discovery and description of something about the subject and in terms of which the subject will be understood. Scientific investigation, therefore, is a complex two-staged procedure of description (*narrativus*) and causal explanation (*causarum assignativus*). ... knowledge is built up from sense experience through the ordered application of a series of dialectical and demonstrative methodologies. ... Demonstration is necessary for science and the art of dialectics is exercised for the sake of scientific judgment made manifest in demonstrations.

Further, in his explication of Aristotle's *Topics*, Albertus specifically addresses the use of logic (dialectics) and inference in experimental planning, data interpretation, hypothesis creation and revision, and verification.

Demonstration involves inference to a necessary conclusion that is not known as necessary independently of being demonstrated. The only way in which an investigator could be in a position to demonstrate like this is by already knowing that it is possible to show the necessity of the conclusion. A conclusion cannot be known as demonstrable unless it is known to be probable – that is, if a reason can be given dialectically to accept it and there is no reason to doubt it. In an ongoing investigation of a subject, some conclusions that seem probable may come to have doubt cast on them. When this happens there must be a dialectical means by which this doubt is either laid to rest or a substitute probable conclusion suggested. Such procedures continue in the investigation until the conclusion is known to be probably demonstrable. Thus, dialectical methods stabilize the conclusions of demonstrations in a way that show them to be capable of being established as necessary.

When the scientific method is executed, scientists frequently (always?) seek to form a hypothesis from one or a few experiments and predictable behaviors of systems, and even laws of nature. This inductive process was part of Albertus's approach to the practice of science. Albertus discusses this in the following quote, which also provides one explanation of induction in science.

In investigations of nature ... it is necessary to get down to details so that the primary agent in each individual case may be ascertained ... because in investigations of nature we must discover the universal principles through singulars, since in such investigations the particulars are better known than

---

[23] The nineteenth-century edition of his Latin translations of, and commentaries on, the works of Aristotle amounts to more than 8,000 pages [167].

[24] Complete name: Abu-Yūsuf Ya'qūb ibn Isḥāq ibn aṣ-Ṣabbāḥ ibn 'Omran ibn Isma' il al-Kindī.

the universals. It is through the singulars that we come to believe that it is convenient and necessary for universals and their principles to exist, since it is only those universals which are exemplified in particulars that we accept, while those which are not so exemplified in particulars, we reject. …Universality, predictability of many, attaches to natures as abstracted by the intellect from the individuating conditions of matter.

Whether one can logically support the validity of induction has been a central feature of science and science practice since Hume famously addressed "the problem of induction" in 1739 [168]. Albertus certainly was aware of this problem and discussed the fact that exceptions could occur in nature [169].

   Albertus had a "desire for concrete, specific, detailed, and accurate knowledge concerning everything in nature,"[25] a desire that is part and parcel of the modern scientific method and that he made clear in his commentaries on Aristotle. He was not content, as most were, to consider natural phenomena the work of a divine being – work that was miraculous and unexplainable. Instead, he considered natural phenomena to be the mechanisms through which God accomplished his goals on Earth, which freed him from religious constraints to unfettered study of nature. Escaping the chains of dogma is critical to the practice and success of the scientific method and Albertus was one of the earliest and forceful proponents of this idea. Richard Feynman defined science as "the belief in the ignorance of experts" (see page 14). Albertus argued in a similar vein when he challenged the notion that the knowledge of the ancients (Aristotle *et al.*) was *not* inviolate. Regarding a tree which was "said to save doves from serpents," Albertus cautioned "But this has not been sufficiently proved by certain experience, … but is *found in the writings of the ancients* [Author's italics]." In questioning the validity of another assertion, Albertus argued "But this [assertion] is proved by no experience,"[26] and therefore is questionable. He also made clear that reputed facts should, in essence, be validated before being believed, when he said "As some affirm, but I have not tested this myself."[27] To Albertus, "accurate knowledge" could be established only through direct experience.

4.3.3 Roger Bacon (c. 1214–1292)

Roger Bacon[28] was a contemporary of Grosseteste[29] and of Albertus Magnus. Bacon was a prominent theologian who earlier in his life had been an academic in Paris and Oxford, two centers of learning in the thirteenth century and where experimental science and mathematics were being integrated into a scientific method [172]. Bacon is one of many who have been called "the father of the scientific method." However, according to some, it may be more appropriate to consider his role to be that of a midwife, who *assisted* in the delivery of a modern scientific method fathered by others. Like Albertus, Bacon argued "Without

---

[25] Taken from Thorndike [170], p. 535.
[26] That is, no empirical evidence was provided to support the assertion.
[27] Both quoted passages were taken from Thorndike [170], pp. 539–530.
[28] Not to be confused with Francis Bacon (1561–1626), who will be discussed later.
[29] Who he apparently detested for both academic and personal reasons [171, 172].

experiment nothing can be properly known."[30] Like Grosseteste, he studied Aristotle's *Posterior Analytics* and emphasized induction, experiment, and mathematics in his approach to science. However, Bacon was not an experimenter *per se*, and he incorporated magic and the occult in his scientific view, which *was* common in the thirteenth century but has no place in modern scientific method. This led Thorndike to opine "... [Bacon] gives no directions concerning either the proper environment for experimenting or the proper conduct of experiments. Of laboratory equipment, of scientific instruments, of exact measurements, he has no more notion apparently than his contemporaries."[31]

Others have been more critical in their assessments. Hackett [172] has suggested the basis for the "father" moniker was William Whewell's own opinions about Bacon, as expressed in Whewell's monumentally important three-volume work of 1859 entitled "History of the Inductive Sciences from the Earliest to the Present Time." Crombie also disputes a special role for Bacon in development of the scientific method [164], as he has stated "Bacon's discussion of the use of induction, experiment, and mathematics in science, and his detailed work on optics and the calendar, directly and explicitly followed the lines of inquiry [already] laid down by Grosseteste." The Center for Islamic Studies [173] has been blunter, arguing "It is absolutely wrong to assume that experimental method was formulated in Europe. Roger Bacon, who, in the West is known as the originator of experimental method, had himself received his training from the pupils of Spanish Moors, and had learnt everything from Muslim sources. The influence of Ibn al-Haitham on Roger Bacon is clearly visible in his works. Europe was very slow to recognize the Islamic origin of her much advertised scientific (experimental) method." In his book *The Making of Humanity*, Briffault states "It was under their successors at the Oxford School that Roger Bacon learned Arabic and Arabic science. Neither Roger Bacon nor his later namesake has any title to be credited with having introduced the experimental method. Roger Bacon was no more than one of the apostles of Muslim science and method to Christian Europe; and he never wearied of declaring that the knowledge of Arabic and Arabic science was for his contemporaries the only way to true knowledge."

What then are we to conclude about Bacon's contribution to the scientific method? Modern scientific method and practice seek to find truth through experimentation. Bacon's view of scientific method appears to have been the opposite, namely "It is not, like modern experimentation, the source but 'the goal of all speculation.' It is not so much an inductive method of discovering scientific truth, as it is an applied science, the putting of the 'speculative' natural sciences to the test of practical utility."[32] "Speculative natural sciences," according to Bacon, included optics, alchemy, astronomy, and astrology. These were classified distinctly from experimental science, which occupied a class on its own. Experimental science was the means through which the purported understanding of nature obtained through the speculative sciences might be demonstrated and thus verified as true. Bacon, in his strong arguments for the centrality of experiment in natural philosophy, thus

---

[30] *"Sine experientia nihil sufficienter sciri potest."* Note, for Bacon, "experiment" meant experience or observation.

[31] Taken from Thorndike [170], p. 650. Bacon's contemporaries included, in addition to Grosseteste and Magnus, Petrus Hispanus, William of Auvergne, Constantinus Africanus, Adelard of Bath, Pedro Alfonso, and Bernard Silvester.

[32] Taken from Thorndike [170], pp. 650–651.

saw experiment as a proof of reason rather than reason being proved through experiment. Regardless of this philosophical distinction in purpose, the practical result of Bacon's work was the broad promulgation of a scientific view in which mathematics and experiment, as opposed to pure logic, reason, simple observation, and divine elucidation, were central. The success of Bacon's proselytizing may have been summed up best by Koyré,[33] who said "...nobody praised experimental science as highly as Roger Bacon, who attributed to it not only the 'prerogative' of confirming – and disconfirming – the conclusions of deductive reasoning (*verificatio* and *falsificatio*) but also, and much more importantly, of being the source of new and significant truths that cannot be arrived at by any other means." Considering our birthing metaphor once more, it can be argued that Bacon did not father the scientific method at all but simply yelled "push" louder than others during its birth.

4.3.4 John Duns Scotus (c. 1265–1308)

In the late thirteenth and early fourteenth centuries, Aristotelian philosophy had become a mainstay of medieval thought and education. For example, in Paris "new masters of arts were required to swear that they would teach 'the system of Aristotle and his commentator Averroes'[34] ... *except in those cases that are contrary to the faith*" [Author's italics].[35] The European clergy, in essence, had begun the process of demarcating natural philosophy from theology. John Duns Scotus[36] was a Franciscan priest educated in Paris and Oxford who, "while not seeking a total separation between philosophy and theology, diminished the area of their overlap by questioning the ability of philosophy to address articles of faith with demonstrative certainty."[37] He differentiated knowledge obtained through divine revelation (theology) from that obtained by reason (philosophy) and developed detailed methods for approaching existential questions. His approach was characterized by depth of thought, fine distinctions, detail, and rigorous logic, all necessary components of the scientific method. However, his highly nuanced writing has been characterized by others as impenetrable and full of "arcane and trivial distinctions" [174]. Nevertheless, Duns Scotus promulgated an inductive method, the Method of Agreement, which was an important addition to the logical structure necessary for scientific discovery. The method sought to determine the element(s) within different sets of circumstances that necessarily had to be present to observe the phenomenon. One concluded, if the phenomenon was observed only when a particular circumstantial element was present, that this element *could* cause the phenomenon. In Duns Scotus's words, there was an "aptitudinal union" between circumstance and effect

---

[33] Taken from Koyré [145], p. 9.

[34] Averroes, the latinized name of Abū l-Walīd Muḥammad Ibn Aḥmad Ibn Rushd, was an Anadalusian Arab polymath. It was in large part through translation of his commentaries on Aristotle from Arabic to Latin in the second half of the twelfth century that Aristotelian philosophy became accessible in Europe.

[35] Taken from [167], p. 241.

[36] Duns Scotus's ideas came under attack in the sixteenth century and, according to the Oxford English Dictionary, one adhering to them, a *Duns* or *Duns man*, was considered a "dull obstinate person impervious to the new learning," or a "blockhead incapable of learning or scholarship." Prior to that time, Duns or Dunce already "was synonymous with [a] 'cavilling sophist' or 'hair-splitter.'" This characterization later was conflated with the conical hat, which Duns Scotus is said to have thought would increase learning, thus giving rise to the term Dunce Cap, which first occurred in print in 1841 in Charles Dickens's novel *The Old Curiosity Shop*. The Vatican had a different view and awarded Duns Scotus the scholastic honorific *Doctor Subtilis* (Subtle Doctor). Seven centuries later (1993), Duns Scotus was beatified by Pope John Paul II.

[37] Taken from [167], p. 242.

Table 4.1 *Method of agreement.*[a] *If one observes a phenomenon in instances with nonidentical sets of circumstances, those sets that contain a single element in common and produce the phenomenon allow the induction that this common element could be the cause the phenomenon. In the table above, that element is A.*

Instance	Circumstances	Phenomenon
1	A C D E F	*p*
2	A B D E F	*p*
3	A B C E F	*p*
4	A B C D F	*p*
5	A B C D E	*p*

[a] Table adapted from [176], p. 8.

(see Table 4.1). Duns Scotus did not argue that this element *did cause* the phenomenon because he believed that "uniformities in nature exist only by the forbearance of God"[38] and that only could produce the same phenomenon without the element present.

4.3.5 William of Ockham (c. 1285–1348)

William of Ockham, who was known as "The Singular and Invincible Doctor" (*Doctor Singularis et Invincibilis*[39]), also was a Franciscan priest and had been a student of Duns Scotus. Ockham argued even more strongly than his teacher that theology and philosophy were incommensurable. This world view was critical to the development of the scientific method because it relieved the prior necessity of natural philosophy to accord with religious canon. In doing so, Ockham was free to explore the philosophical and logical underpinnings of the study of the natural world. These underpinnings provided epistemological bases for methods of study of existential questions, a major pursuit of medieval theologians and philosophers. It was the incorporation of these methods into studies of nature itself that contributed to their realism and relevance.

In studying the question of cause and effect, Ockham argued that one could never know for sure whether an observed set of circumstances was *the* cause of a phenomenon, because one could never be sure that an unknown entity, x, was present and contributed to the phenomenon. He, therefore, considered such observed causes to be "immediate." Thus, when an immediate cause "is present the effect follows, and when it is not present, all other conditions being equal and dispositions being the same, the effect does not follow."[40] This method has been termed the "Method of Difference" and is illustrated in Table 4.2. Ockham

---

[38] See [175], p. 30.

[39] According to Thorburn [177], pp. 345–346, the quotation of Ockham most relevant to the etymology of Ockham's Razor (see below) was "*Sufficient singularia, et ita tales res universales omnino frustra ponuntur*" (Individuals suffice, and so it is entirely superfluous to assume these universals). This may be why Ockham was known as the "singular doctor."

[40] Taken from [178], p. 85.

Table 4.2 *Method of difference. If one observes a phenomenon in instances with nonidentical sets of circumstances, those sets producing the phenomenon should contain a particular element. This element should be missing from those sets not producing the phenomenon. In this case, that element is A.*

Instance	Circumstances	Phenomenon
1	A B C D E F	*p*
2	B C D E F	*not p*

required that every circumstance causally relevant to the occurrence of *p* be included in this scheme so as to minimize the possibility that an entity *x* might be missed [176].

Ockham may be best known for the eponymous metaphor of "Ockham's Razor,"[41] which has become entrenched as the popular expression of the principle of parsimony, "[T]he principle that in explaining anything no more assumptions should be made than are necessary" [180]. The metaphorical razor "cuts out unnecessary complexities from explanations,"[42] "slices and leaves aside a host of competing conclusions or arguments, leaving the simplest and most likely conclusion in place,"[43] or "shaves away unnecessary assumptions."[44] A modern, practical example of Ockham's Razor is the familiar pronouncement "If it looks like a duck, swims like a duck, and quacks like a duck, then it probably is a duck."

Interestingly, in digging a bit deeper into the origin of the metaphor, we find that the principle of parsimony that is its subject had been recognized since the time Aristotle [133, 177], who wrote "We may assume the superiority *ceteris paribus* of the demonstration which derives from fewer postulates or hypotheses." The term Ockham's Razor (Latin: *novacula occami*) did not actually exist until 1649, three centuries after Ockham, when the philosopher and theologian Libert Froidmont (Latin: *Libertus Fromondus*) used the term in his book *Philosophia Christiana de Anima* [182].

4.4 The Scientific Revolution (1500–1750)

4.4.1 Francis Bacon (1561–1626)

The person most often cited as the "father of the scientific method" was the Englishman Francis Bacon. He was neither a scientist nor an experimentalist but rather a brilliant man

---

[41] According to Adams [179] (Note 1, vol. 1, pp. 156–157), common renditions of Ockham's Razor include: (1) *Frustra fit per plura quod potest fieri per pauciora* (it is futile to do with more things what can be done with fewer), from Ockham, Summa Totius Logicae, i. 12; (2) *Quando propositio verificatur pro rebus, si duae res sufficiunt ad eius veritatem, superfluum est ponere tertiam* (when a proposition comes out true for things, if two things suffice for its truth, it is superfluous to assume a third); (3) *Pluralitas non est ponenda sine necessitate* (plurality should not be assumed without necessity); and (4) *Nulla pluralitas est ponenda nisi per rationem vel experientiam vel auctoritatem illius, qui non potest falli nec errare, potest convinci* (no plurality should be assumed unless it can be proved (a) by reason, or (b) by experience, or (c) by some infallible authority).

[42] Taken from [181], p. 1761.

[43] www.quora.com/Why-is-Occams-Razor-called-a-razor

[44] http://wordsmith.org/words/ockhams_razor.html

who was trained as lawyer, served in parliament, was knighted in 1602, appointed Lord Chancellor and ennobled as Baron Verulam in 1618, and made Viscount St. Albans in 1621. Bacon published a treatise on the scientific method (*Novum Organum*[45] *Scientiarum*, "new instrument of science" in English) in 1620. Surprisingly (or maybe not surprisingly considering it is an English publication), the Oxford English Dictionary[46] subscribes to the notion of Englishman Bacon as father because it defines the scientific method as "A method of observation or procedure based on scientific ideas or methods; spec. an empirical method that has underlain the development of natural science since the *17th cent.* [Author's italics]" [183]. As we shall see below, Francis Bacon did indeed discuss *a* scientific method in seventeenth century Europe. I use the article *a* in the previous sentence to emphasis the fact, as illustrated in Fig. 4.2, that there is no single, all-encompassing scientific method[47] and thus there can be no single "father"[48] thereof. Nevertheless, Bacon's prominence no doubt facilitated the wide dissemination of his writings, providing his contemporaries, as well as later scientists and historians, the opportunity to begin, and, unfortunately for many, end their study of the scientific method and its history with Bacon.

Let us now examine the "Baconian method" and its influence on science. Bacon began, as did many before him, by revisiting – and criticizing – the science of Aristotle, to wit:

Men fall in love with particular sciences and reflections ... Now if such then devote themselves to philosophy and general reflections, they distort and corrupt the latter in line with former fantasies; as we see most clearly in Aristotle who made his natural philosophy a mere slave to his logic, and so rendered it virtually useless and disputatious [Author's underlining].[49]

Bacon objected to the *methods* of Aristotle, which were based primarily on principles of logic and not empirics. In the following paragraph, Bacon contrasts the Aristotelian method with his own. The former involves initial observation (senses and particulars) of the natural world followed by the derivation of general principles that explain the observations. These general principles then can be used to derive less general, more specific principles (middle axioms) explaining the natural world. Bacon agrees that the search for the truth of the natural world begins with observations. However, he argues that these observations cannot be used immediately to establish general principles. Instead, they must generate axioms specific for the particular observations. Through a process of continued observation and axiomatization, one then proceeds (ascends) from specific axioms to more general, all-encompassing axioms. One cannot start with general axioms and then deduce from them more specific axioms (middle axioms), as Aristotle proposed, because the initial inductive process producing the general axioms is flawed by its primary dependence on logic alone. Any deductions derived from it thus also must be flawed.

---

[45] The word "organon" (Greek Οργανον, meaning "instrument, tool, organ") is the standard collection of Aristotle's six works on logic.

[46] Many, including this author, consider the OED to be the ultimate authority on the English language. This example shows that even this "authority" may not be authoritative.

[47] See page 61.

[48] The historical record requires use of the term *father*, rather than *parent* or another gender-neutral term.

[49] From Rees and Wakely [184], aphorism 54, p. 89. Bacon's use of the term "philosophy" should be understood in its historical context, i.e., as a reference to the process by which the natural world may be considered and understood, which historically was by philosophers using logic as their tool of enquiry.

There are and can be only two ways of searching into and discovering truth. The one flies from the senses and particulars to the most general axioms, and from these principles, the truth of which it takes for settled and immoveable, proceeds to judgment and to the discovery of middle axioms. ... The other derives axioms from the senses and particulars, rising by a gradual and unbroken ascent, so that it arrives at the most general axions last of all. This is the true way ... (IV, 50).[50]

Bacon's "true way" was to "... replace the Aristotelian organon ... with an entirely new logical instrument, a new method for the progress and profit of human science."[51] Bacon was convinced that natural philosophers, past and in his time, brought their own intrinsic biases to their quest for the truth of the natural world.[52] In doing so, this truth would be unattainable. Malherbe [185] expressed eloquently how Bacon's method would accomplish both the discovery of truths about the natural world and avoid the biases of the truth discoverers. The approach illustrates what may be the most important principle I try to enunciate in this book, namely that good science is built upon a bedrock of both practical and philosophical knowledge.

... the first act of the method, and its most fundamental act, is precisely to make us understand that if it is true that knowledge begins with sensible experience, it must obtain access to the invisible processes and structures of nature[53]. .... Although reliance on the senses acts as a bulwark against metaphysical systems and rhetorical uses of language, the empirical information thus gained is itself unreliable. We know that the immediate character of perception leads the mind too easily to consider the sensible content ... as the real nature of things. Therefore, senses should be helped with a methodological

---

[50] Taken from [185], p. 77.

[51] Taken from [185], p. 78. A translation of Bacon's actual words from the Preface of *Novum Organon* (see [184], p. 53) is "Now my plan is as easy to describe as it is difficult to effect. For it is to establish degrees of certainty, take care of the sense by a kind of reduction, but to reject for the most part the work of the mind that follows upon sense; in fact I mean to open up and lay down a new and certain pathway from the perceptions of the senses themselves to the mind. Now this was doubtless seen by those who have attached so much importance to dialectic – whence it is clear that they were looking for props for the intellect, distrusting the minds' inborn and spontaneous movements. But this remedy comes too late to a cause already lost once the mind has been invaded by the habits, hearsay and depraved doctrines of daily life, and beset by the emptiest of *Idols*. Thus (as I have said) the art of dialectic bolts the stable door too late and cannot recapture the horse, and does more to entrench errors than to reveal the truth. There remains but one way to health and sanity: to do the whole work of the mind all over again, and from the very outset to stop the mind being left to itself but to keep it under control, and make the matter run like clockwork. For if men really tackled work for machines with their bare hands, and without the help and force of instruments, in the same way as they have not hesitated to undertake work for the intellect with little besides the naked force of the mind, there would have been very few things which they could have got going or mastered, even if they combined to use their best efforts. ... it is perfectly obvious in every great work undertaken by human hand that individual powers cannot be strengthened nor the powers of all be combined without instruments and machines."

[52] Bacon discussed four types of intrinsic bias, which he termed "idols" (from the Greek *eidolon* (εἰδωλον), meaning image, apparition, phantom, ghost): (1) "tribe," which is "natural weaknesses and tendencies common to human nature" (From [9], aphorism 52), such as wishful thinking or seeing what is not there because of beliefs that are not due to external forces, e.g., socialization; (2) "cave," arising from "the peculiar nature of the individual ... as well as from education, custom, and accident" (From [9], aphorism 53), which do arise from external sources; (3) "market," which refers to the misuse and nebulousness of words and names, leading to confusion and misunderstanding ("names of things which do not exist ... or names which do exist but are muddled, ill-defined, and rashly and roughly abstracted from the facts" (From [9], aphorism 60)); and (4) "idols of the theatre," which refer to dogmatic beliefs imposed by religion, theology, civil governments, and philosophies. Bacon explained this idol quite entertainingly when he said, in aphorism 62, "*Idols of the Theatre*, or Theories are numerous ... For if men's minds had not now been seized for so many centuries with religion and theology, and if too civil governments (monarchies in particular) had not been hostile to suchlike novelties, even the contemplative ones, so that men dealing in them risk harm to their fortunes and not only go unrewarded but are open to contempt and spite, I have no doubt that many more philosophical and theoretical sects like the wide variety which once flourished amongst the Greeks would have sprung up. For just as many *Theories* of the heavens can be fabricated from the *Phenomena* of the ether, so much more can different dogmas be erected and established on the *Phenomena* of Philosophy. And such *stage-plays* share this with the plays of the dramatists, that tales got up for the stage are more harmonious and attractive, and to one's taste, than true stories from the historical record."

[53] Are "invisible processes and structures" real or are they just intellectual constructs? We address this question, the question of realism, in Section 5.7.

assistance: first, to correct their testimony and rectify their information which is biased by human receptivity; secondly, with the help of experiments, of substitutions or graduations, to try to submit to the senses themselves what escapes them because of their too narrow limits; at last, to extend and improve the collecting of facts, the content of which is very hazardous, by an experimentation that provokes nature where it has not yet informed the mind. Therefore, although it is the case that true science depends on the senses, we should not forget that their information must be corrected and enlarged. Method penetrates sensible experience itself, and stipulates the conditions according to which the senses can judge of the reality of things.

Bacon argued forcefully for the importance of *unbiased* observation and experimentation in the search for scientific truth. The latter is important not just as a practical method of confirming the observations of the senses, but more importantly, because it can reveal heretofore unobserved or unconsidered aspects of nature. Bacon expresses this idea as follows.[54]

...the nature of things betrays itself more readily under the vexations of art than in its natural freedom.

What do we do with the results of our "vexations?" According to Bacon, we use these results for confirmatory and inductive purposes. We move stepwise toward the proof or disproof of prior hypotheses and, especially importantly, we obtain further insights into nature and conceive of new hypotheses about nature's nature. Insights gained through the inductive process may be directly relevant to our understanding of aspects of nature that are not readily observable by the senses [186]. It is important to understand that Bacon distinguished *observation* and *experimentation* from *understanding*. The former were achieved through the senses, the latter through induction. Induction was the process through which nature could be axiomatized, a process that moves from instantial axioms to those of an ever more general nature. In this manner, one increasingly understands nature as one indivisible, all-encompassing, existential whole.

In executing his inductive process, Bacon employed the Method of Agreement (Section 4.3.4) and the Method of Difference (Section 4.3.5), practiced ≈300 years earlier by Duns Scotus and Ockham, respectively. To these methods, Bacon added a third approach, that of "Tables of Degrees or Comparative Tables,"[55] in which a table would be constructed listing circumstances under which phenomena occurred to different degrees as opposed to being totally present or absent. Any circumstance that did not vary proportionately with a phenomenon could be rejected as an explanation of that phenomenon. This approach may be viewed as "fine tuning" of the inductive process.

4.4.2 Galileo Galilei (1564–1642)

Our discussion of the development of the scientific method thus far has encompassed an historical period of ≈4,600 years, beginning in Egypt c. 3000 BC and extending into the

---

[54] Quotation taken from [185], p. 84. Here, "art" is construed as experimentation.

[55] Bacon explained the rationale for this table by saying, "...we must submit to *the tribunal of the intellect* instances in which the nature under investigation exists to a greater or lesser degree. ...it inexorably follows that no nature can be taken for the true form unless it always diminishes when the nature itself diminishes, and likewise always increases when the nature itself increases" (from [184], aphorism 13, p. 237).

early seventeenth century. We have seen our understanding of the natural world expressed in the language of pure logic by the ancient Greeks. We have observed in the Islamic world the addition of experimentation to the logical Greek world view and how these ideas later were translated from Arabic into Latin, in which form they were introduced into Europe. We noted how Grosseteste, and later Roger Bacon, emphasized the importance of experimental controls and mathematical analysis and how Francis Bacon vigorously argued the importance of induction, a process to which he made important contributions. We now come to Galileo Galilei, whose approach to science does indeed make him a father, and maybe even *the* father, of the scientific method. Einstein went further when he said "Propositions arrived at by purely logical means are completely empty as regards reality. Because Galileo realized this, and particularly because he drummed it into the scientific world, he is the father of modern physics – indeed, of modern science altogether."[56] Hawking agreed, writing "Galileo, perhaps more than any other single person, was responsible for the birth of modern science."[57]

Galileo was born in 1564. He studied mathematics in Pisa and joined the faculty of the University of Padua in 1592, where he remained until 1610, at which time Cosimo II de' Medici, a former student, invited him to Florence to become Chief Mathematician and Philosopher. It was here that Galileo produced his two most famous works, Dialogue Concerning the Two Chief World Systems (1632; Italian: *Dialogo sopra i due massimi sistemi del mondo*) and Discourses and Mathematical Demonstrations Relating to Two New Sciences (1638; Italian: *Discorsi e Dimostrazioni Matematiche, intorno a due nuove scienze*). In the former work, among other things, Galileo provides arguments supporting the accuracy of the Copernican model of the universe, in contradistinction to the model of Ptolemy. In the latter work, he discusses what later would become materials science and kinematics, including his famous experiment demonstrating that the force of gravity causes all objects to fall at the same rate, regardless of their mass.[58]

Before Galileo, many had argued for the importance of integrating mathematics with observation. Some of these arguments were philosophical in nature and some were applied. However, it was Galileo's integration of mathematics (including geometry) with experiment, for example, in his validation of the Copernican model of the universe, studies of gravity, and determination of the parabolic shape of the flight of projectiles, that illustrated most clearly how this integration could and should be done *in practice* [192]. Galileo did not want simply to observe and understand phenomena descriptively, but rather, as Crombie opines:[59]

---

[56] From [187], p. 271.

[57] From [188], p. 179.

[58] It is debatable whether Galileo did or did not actually drop two spheres of different masses from the Leaning Tower of Pisa and determine their transit times to the ground. Galileo's pupil, Vincenzo Viviani, writes that the experiment was performed for the benefit of Galileo's students and colleagues [189]. Many others argue that the experiment was merely a thought experiment on Galileo's part [190]. Interestingly, Galileo does not actually use the Italian word for experiment, "*esperimento*" or "*cimento*" in either of his two great works ([191], p. 85).

[59] Taken from [192], p. 136.

...I say that if it is true that one effect can have only one basic cause, and if between the cause and the effect there is a fixed and constant connection, then whenever a fixed and constant variation is seen in the effect, there must be a fixed and constant variation in the cause.[60]

Galileo was aware of observational errors and biases among natural philosophers. Biases of the sort described by the famous wordsmith Yogi Berra – "I wouldn't have seen it if I didn't believe it for myself"[61] – were and are common in scientific practice, but in addition, instrumental imprecision also may lead to erroneous conclusions. Instrumental precision is critical if one is to make use of methods of comparison, otherwise small but significant variations may be missed. This also was true for Galileo's astronomical studies, for which he modified a telescope to include a micrometer, which enabled him to accurately determine the dimensions of the moons of Jupiter and tables of their movements.[62] Four hundred years later, advances in scientific understanding remain dependent on ever better instrumentation, which reveals heretofore unimagined new questions. Answering these questions may then require fabrication of new instruments, which reveal new questions, which require new instruments ... and the never-ending cycle of discovery continues.

It is important to understand that Galileo's integration of mathematics and experiment did not manifest as rote method, but rather as a logic of science practice. Galileo sought to explain phenomena through mathematical abstractions, i.e., through propositions. These propositions were absolute. They required no exemplars and served as valid explanans[63] of the observed phenomena. Crombie suggests that "The momentous change that Galileo...introduced into scientific ontology was to identify the substance of the real world with the mathematical entities contained in the theories used to describe the appearances."[64] Galileo described this as

a method which uses the mathematical (geometrical) language in order to formulate its question to nature, and to interpret its answers; which, substituting the rational universe of precision for the empirically given world of the more-or-less, adopts measurement as its fundamental, and most important, experimental concept.[65]

Galileo emphasizes the role of mathematics in understanding the universe in an even more vigorous manner in 1623 in his book *The Assayer* (Italian: *Il Saggiatore*).

Philosophy [i.e., natural philosophy] is written in that great book which is continually open before our eyes (I speak of the universe), but no one can understand it who does not first learn the language, and come to know the characters, in which it is written. It is written in mathematical language, and the characters are triangles, circles, and other geometrical figures, without which means it is humanly impossible to understand a word; without these, one is wandering vainly in a dark labyrinth.

---

[60] Taken from [192], p. 137.

[61] This phrase, or a similar one, has been attributed to Yogi Berra, a Baseball Hall of Fame catcher for the New York Yankees (and to Marshall McLuhan).

[62] Galileo also constructed a rudimentary thermometer (thermoscope) to quantify temperature differences between two solids or liquids [193].

[63] *Explanans* are facts or observations that provide the basis for an explanation. The *explanandum* is the phenomenon, hypothesis, question, etc. that explanans explain.

[64] Taken from [194], p. 310.

[65] Taken from [145], p. 19.

Abstractions also served to simplify theory testing, as they required elimination of explanans thought to be irrelevant. Successful predictions of experimental results based on these abstractions supported the validity of the included terms and the irrelevance of the excluded terms. Galileo expressed a bent for simplicity by saying *"... nature ... does not act by means of many things when it can do so by means of few."*[66] He also argued forcefully that the absolute truth of a hypothesis built using abstractions is *"established ... by our seeing that other conclusions, built on this hypothesis, do indeed correspond with and exactly conform to experience."*[67] Galileo thus not only required hypotheses to explain observations extant, but also to be useful predictors of as yet unobserved phenomena.

We discussed above how Einstein and Hawking opined that Galileo was the father of modern science. Galileo's scientific discoveries, and the insights and instruments necessary to do so, certainly supported these opinions. However, Galileo also did something much more fundamental and far-reaching – he changed the philosophical underpinnings of science, which facilitated the scientific revolution of the sixteenth and seventeenth centuries. Koyré may have expressed most eloquently what was required to effect such a monumental change when he wrote, "... what the founders of modern science, among them Galileo, had to do, was not to criticize and to combat certain faulty theories, and to correct or to replace them by better ones. They had to do something quite different. *They had to destroy one world and to replace it by another* [Author's italics].[68] They had to reshape the framework of our intellect itself, to restate and reform its concepts, to evolve a new approach to Being, a new concept of knowledge, a new concept of science [viz., a mathematical explanation of nature] – and even to replace a pretty natural approach, that of common sense, by another which is not natural at all."[69] Galileo had made "a clear break from the past, deliberately replacing the scientist-philosopher with the mathematical scientist."[70]

4.4.3 René Descartes (1596–1650)

A generation after Galileo, René Descartes extended the application of reason and mathematics in natural philosophy through his works *Rules for the Direction of the Mind*[71]

---

[66] See [195], p. 135.

[67] Taken from [196], p. 164.

[68] In the mid-twentieth century, Thomas Kuhn [197] suggested much the same thing with respect to the effects of what he defined as "scientific revolutions."

[69] Taken from [198], p. 152. Shapere, responding to Koyré's statement, writes that "... a fundamental shift of philosophical viewpoint, rather than the refutation by experiment and observation of an old theory and its replacement by a new one, constituted the scientific revolution. It was not experiment or observation, but reason, which played the key role in the transition from medieval to modern physics; and the crucial question was that of the role of mathematics in understanding nature."

[70] Taken from [199], p. 456.

[71] Lat.; *Regulae ad Directionem Ingenii*. This book apparently was abandoned in 1628, but subsequent Descartes studies have suggested that work continued sporadically afterwards, which led to the posthumous publication of a Dutch translation (1684) and a Latin translation (1701). Whether Descartes adhered later in his life to the rules propounded in the *Regulae* is a question we will not address here, but one that has been discussed by others (e.g., see Flage and Bonnen [200]).

(begun in 1628), *Discourse on Method* (1637),[72] *Meditations on first Philosophy* (1641),[73] and *Principles of Philosophy* (1644).[74] Descartes had received a classical education at the Collège Henri IV La Flèche, at that time a new Jesuit school that later became one of the preeminent educational centers in Europe. Cress[75] has commented that Descartes found particular distaste in the Aristotelianism taught there, much preferring mathematics, a subject that was taught particularly well at La Flèche, relative to other European universities.[76] Descartes left La Flèche in 1614 and, paraphrasing his words, said that he "had become disillusioned with books and decided to educate himself by seeing the world."[77] In the process of doing so, Descartes struggled to find his path, but apparently did so, at least in part, through a series of dreams he had in Bavaria in the winter of 1619. According to Descartes, these dreams produced a "marvelous discovery" that likely gave Descartes confidence to pursue a grandiose mission that he had described in a letter to his friend Isaac Beeckman eight months earlier:

I want to expound ... an entirely new science, by which all problems that can be posed, concerning any kind of quantity, continuous or discrete, can be generally solved. Yet each problem will be solved according to its own nature, as, for example, in arithmetic some questions are resolved by rational numbers, others only by surd [irrational] numbers, and others finally can be imagined but not solved. So also I hope to show for continuous quantities that some problems can be solved by straight lines and circles alone; ... and finally others that can be solved by curved lines generated by diverse motions not subordinated to one another, which curves are certainly only imaginary .... I cannot imagine anything that could not be solved by such lines at least, though I hope to show which questions can be solved in this or that way and not any other, so that almost nothing will remain to be found in geometry. It is, of course, an infinite task, not for one man only, incredibly ambitious; but I have seen some light through the dark chaos of this science, by the help of which I think all the thickest darkness can be dispelled.

The "light" to which Descartes referred was his development of *calcul géométrique* (Eng., geometrical calculus), now termed "analytical geometry," which was a synthesis of geometry (lines) with algebra (symbols) that used a coordinate system in which a point could be localized within a plane relative to two fixed orthogonal axes (later know as a Cartesian coordinate system). Such a system allows visualization and modeling of complex relationships defined through algebraic equations and is one of the most fundamental tools in all of science. Descartes's conception of such a system was a profoundly important advance in mathematics. However, for our purposes, we must distinguish this scientific *tool* from the broader contributions Descartes made to modern philosophy, a field of which, Bertrand

---

[72] The complete title of this work, written in French, was *"DISCOURS DE LA MÉTHODE Pour bien conduire sa raison, & chercher la vérité dans les sciences Plus LA DIOPTRIQUE. LES METEORES. ET LA GEOMETRIE. Qui sont des essais de cete METHODE"* [Eng.; Discourse on the Method for Conducting One's Reason Well and for Searching for Truth in the Sciences. The Dioptric. The Meteors. The Geometry. Which are tests of this method].

[73] The complete title of this work, written in Latin, was *Meditationes de Prima Philosophia, in qua Dei existentia et animæ immortalitas demonstratur* [Eng.; Meditations on first Philosophy in which the existence of God and the immortality of the soul are demonstrated].

[74] Lat.; *Principia Philosophiæ*.

[75] See Cress [201], p. vii.

[76] See Russell [202], p. 558.

[77] Taken from Browne [203], p. 256.

Russell has opined,[78] Descartes is the father, and particularly to the philosophy of science practice (*méthode*).

During his lifetime, Descartes did not publish a text detailing his *méthode*, a word that at that time was used in the context of "a manner of proceeding in the sciences."[79] No step-by-step instructions were provided. In fact, many have suggested that Descartes had no method at all [205, 206]. Descartes himself said:

My present aim . . . is not to teach the method which everyone must follow in order to direct his reason correctly, but only to reveal how I have tried to direct my own. One who presumes to give precepts must think himself more skilful than those to whom he gives them; and if he makes the slightest mistake, he may be blamed. But I am presenting this work only as a history or, if you prefer, a fable in which, among certain examples worthy of imitation, you will perhaps also find many others that it would be right not to follow . . . .[80]

Leibniz agreed, when he said:

There have been many beautiful discoveries since Descartes, but, as far as I know, not one of them has come from a true Cartesian. . . . This is evidence that either Descartes did not know the true method, or else that he did not leave it to them [i.e., Cartesians].[81]

A tremendous amount of scholarly work on Descartes's method has been done by modern philosophers and historians of science. Yet they too have not arrived at consensus about "the method." Hatfield decries what he calls the "myth of method," by which he means "the belief that Descartes subscribed to a single method, announced in the Discourse on Method but only fully articulated in the posthumously published Rules for the Direction of the Mind, to which he credited his achievements in both metaphysics and natural philosophy . . . ."[82] Garber expresses a similar idea, poetically, when he states: ". . . one is hard pressed to find much evidence of the method at all after 1637, either explicit discussions of the method or explicit applications of the method in any of Descartes' writings, published or unpublished. . . . *Descartes in 1637 is, in a sense, like the butterfly, emerging from his cocoon, spreading his new wings to dry in the sun, not yet fully aware that he is no longer a caterpillar* [Author's italics]" [209].

Although the *Regulae* was indeed published after Descartes's death, he had planned originally to include within it 36 rules, in the form of 3 books of 12 rules each. Twenty-one eventually appeared in the published work of 1701, along with Descartes's explanatory comments.[83] Descartes did distill 4 rules from the 21 and published them in 1637 in the *Discourse*, along with a justification for presenting only 4:

---

[78] See Russell [202], p. 557.

[79] See Davies [204], p. 202.

[80] Taken from *Oeuvres de Descartes*, vol VI, p. 4.

[81] From a letter from Leibniz to Molanus, c1679, as found in [207], pp. 240–241.

[82] See Hatfield [208], p. 249.

[83] A number of rules constitute what Descartes referred to as "many others that it would be right not to follow." For example, Rule XVIII, advices that "Multiplication and division . . . should seldom be employed [in initial equation creation], for they may lead to needless complication, and they Can be carried out more easily later" and Rule XX states that "Once we have found the equations, we must carry out the operations Which we have left aside, never using multiplication when division is in Order."

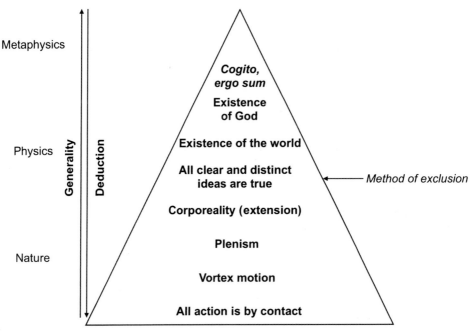

Figure 4.4 Descartes's Method. See text for an explanation. (Adapted from the figure *"Descartes's Pyramid"* in Losee [175], p. 69.)

Instead of the great number of precepts of which Logic is composed, I believed that I should find the four which I shall state quite sufficient, provided that I adhered to a firm and constant resolve never on any single occasion to fail in their observance.

1. The first of these was to accept nothing as true which I did not clearly recognize to be so: that is to say, carefully to avoid haste and prejudice in judgments, and to accept in them nothing more than what was presented to my mind so clearly and distinctly that I could have no occasion to doubt it.
2. The second was to divide up each of the difficulties which I examined into as many parts as possible, and as seemed requisite in order that it might be resolved in the best manner possible.
3. The third was to carry on my reflections in due order, commencing with objects that were the most simple and easy to understand, in order to rise little by little, or by degrees, to knowledge of the most complex, assuming an order, even if a fictitious one, among those which do not follow a natural sequence relatively to one another.
4. The last was in all cases to make enumerations so complete and reviews so general that I should be certain of having omitted nothing.

The philosophical foundation upon which these precepts were built is illustrated schematically in Fig. 4.4. The method is deductive (see downward arrow at left). One begins with a number of givens, beginning with Descartes's famous declaration *Cogito, ergo sum* (Eng., "I think, therefore I am"). Descartes moves from this metaphysical notion and the "fact" that God exists, to the supposition that God would not create man incapable of understanding his universe, and thus any idea that is both clearly and distinctly present to man's mind, through experience and rationality, must be true (Rule 1).

Rule 2 typically is described as breaking a complex problem into its component pieces. In Descartes's discussion of this rule, he argues that we cannot know what is unknown, other than through some relationship of the unknown to the known. He thus requires that *"the not yet known must be, in some way, marked out ... in relation to something that is already known ... But over and above this, if the question is to be perfectly understood, we require that it is made so completely determinate that we have no need to seek for anything beyond what can be deduced from the (already known) data."* These dictates suggest to answer a complex scientific question, one starts from first principles or laws that are inviolate (True with a capital T) and then, through a process of deduction, derives the necessary and sufficient parts of the problem.

Rule 3 then requires that these parts are assembled in logical order (one that may not have been apparent initially), one part building upon another, to yield the solution to the problem, a solution that must be grasped at once and not through disconnected consideration of each part on its own, even in sequence. Rule 4 may be considered a quality control measure in the modern sense, i.e., one should check and double check that every component of the problem that is necessary and sufficient for its solution has been recognized and enumerated in the proper stepwise order.

In some respects, Descartes's method was quite Platonic in nature, i.e., one understands the world through logic alone. Whewell states that Descartes's "labours ... were an endeavour to revive the method of obtaining knowledge by reasoning from our own ideas only, and to erect it in opposition to the method of observation and experiment."[84] Descartes thus has been considered a rationalist, in distinction from empiricists, who work from observation and experiment to general principles. In fact, Shouls opines *"Descartes consistently holds that* experimentation is extrinsic *to the method."*[85] Descartes viewed experiments merely as validations of the natural systems he rationalized. These rationalizations, or "hypotheses," as Newton would later refer to them when he categorically rejected their value, were *a priori* claims of knowledge that were *not* based on experience, but instead represented the prevailing understanding of nature and the world at the time and might have occult or religious underpinnings.

One of the best examples of such hypotheses was the "vortex hypothesis," which Descartes published in the *Principia Philosophiæ* in 1644 to explain the movements of the planets.[86] In Descartes's time, the nature of matter (*corporeality*; see Fig. 4.4) was very much at issue. Descartes used the term *extension* to describe inert matter, *"From the*

---

[84] Taken from Whewell [210], p. 256.
[85] Taken from Shouls [211], p. 38.
[86] Descartes presented his hypothesis thusly, *"... let us suppose that the matter of the sky where the planets are, turns without ceasing, as also a vortex [tourbillon] at the centre of which is the Sun, and that its parts which are near the Sun move more quickly than those which are further away up to a certain distance [the distance of Saturn] and that all the planets (in whose number we henceforth include the Earth) remain always suspended between the same parts of this matter of the sky; for by this alone and without employing other machines, we shall easily understand all the things that we notice in them. For as in the winding of rivers where the water folds back on itself, turning in circles, if some straws or other very light bodies float amidst the water, we can see that it carries them and moves them in circles with it; and even among these straws one can notice that there are often some which also turn about their own centre; and that those which are closer to the centre of the vortex which contains them make their revolution more quickly than those which are more distant; and lastly that, although these vortices of water design always to rotate in rings, they almost never describe entirely perfect circles, and extend themselves sometimes more in length and sometimes more in width, so that all the parts of the circumference which they describe are not equally distant from the centre; thus one can easily imagine that all the same things apply to the planets; and it needs only this to explain all their phenomena."*

sole fact that a body is extended in length, breadth, and depth; we rightly conclude that it is a substance: because it is entirely contradictory for that which is nothing to possess extension." The universe was corporeal and was imagined to exist as a plenum (*plenism*), a volume in which no empty spaces (vacuum) existed. By this construct, Descartes could explain the orbits of planets, which did not follow from one of his key clear, distinct, and true ideas, namely that *"each part of matter in itself never tends to move along curved lines, but along straight lines."* The answer was that the plenum was filled with areas of rotating matter, vortices (*vortex motion*) which, through direct collision (*contact*) with the planets, provided the force necessary to move them off straight lines.

The vortex hypothesis was widely accepted throughout the seventeenth century and even during Newton's life time. An important consequence of the hypothesis was the preclusion of any vacuum in the universe because every body within the plenum was in contact with another body. The plenum also precluded "action at a distance," as must occur if planets are not in contact with other particles and yet move orbitally. Newton would later challenge this hypothesis when he proposed gravitational forces.

It is instructive to compare Descartes's method with that of Francis Bacon. Bacon's approach is illustrated schematically in Fig. 4.5. In a way, it presaged Descartes's rules, but it moves in the opposite direction, from nature to metaphysics, by beginning with observations and then, through the process of induction, moving up in complexity and understanding to general principles ("forms"). Between the base and the apex of the pyramid are the establishment of relations among phenomena and causes. Bacon starts with apparent correlations and uses the Methods of Agreement, Difference, and Degree (see page 78) to eliminate spurious correlations. This process presaged Descartes's rules 2 and 3. In doing so, Bacon can then establish a second level of correlations, "more inclusive correlations," that reflects correlations of a more general nature. "Descartes sought to predict experimental observations by deduction from general principles developed *a priori* through reasoning."[87] Bacon, in contrast, establishes general principles by induction from experimental observations.

As discussed above, no consensus of a Cartesian scientific method exists. My own opinion, which may rightly be disputed by many, is that Descartes was not so much a contributor to method but rather an avatar of how method should be applied in science. If a scientific method could be axiomatized, then there would be no need for scientists because computers could execute the process from beginning to end. The contribution of Descartes was the demonstration that imagination alone (initial, *a priori* rationalization from first principles, even if they be wrong) could play a seminal role in the advancement of scientific knowledge. Although not his intention, Descartes showed that violation of his own rules (Rule 1) could lead to wholly unsupportable conclusions about the natural world. I speak specifically about Descartes's doxastic assumptions that the world was a plenum and action at a distance was impossible, assumptions that supported the vortex theory of planetary motion,

---

[87] Quotation from [175], p. 64.

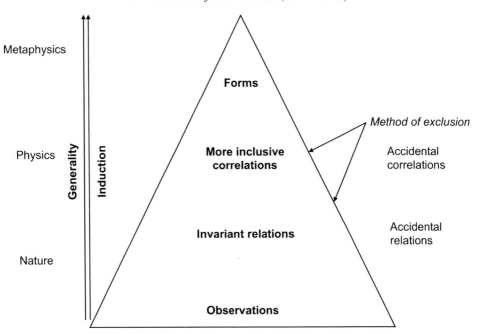

Figure 4.5 Bacon's Method. See text for an explanation. (Adapted from the figure *Bacon's "Ladder of Axioms"* in Losee [175], p. 58.)

a theory that would not be vanquished from science for almost century after Descartes proposed it. Should this violation be criticized on its merits? I think not, and for the same reasons that Ptolemy's notion of the movements of the heavenly bodies should not be criticized and discarded out of hand. Advancement in science does not require that *a priori* hypotheses be correct. It is the process of hypothesis creation and testing that moves science forward. If a hypothesis is correct, then it forms a basis for advancement of knowledge. If a hypothesis is found to be incorrect, it also forms a basis for the advancement of knowledge. It was only through careful testing and reassessment of the theories of Descartes and Ptolemy that our current understanding of the solar system and fundamental laws of nature, for example, gravity, was achieved.

4.4.4 Robert Boyle (1627–1691)

The continuing movement of natural philosophy in the seventeenth century from a speculative to a more rigorous, experimental, mathematico-mechanical enterprise is exemplified well by the work of Robert Boyle, who referred to himself as an "experimental philosopher"[88] and is considered one of the fathers of modern chemistry.[89] Boyle's use of an

---

[88] Boyle was quite lucky to have been the son of the richest man in England at the time, and after his father's death, Boyle was able to focus his attentions to science without the necessity of having to work or having a benefactor.

[89] In 1660, Boyle, along with 11 others, was a founding member of the Royal Society of London for Improving Natural Knowledge (Royal Society).

improved air pump developed by his assistant, Robert Hooke, allowed him to study the behavior of gases and create vacuums (disproving Descartes's notion that particle-free volumes were impossible). In doing so, Boyle discovered the inverse relationship between the pressure and volume of gases, $P \times V = k$, where k is a constant, later to be known as "Boyle's Law."[90] He also showed that sound could not be propagated in a vacuum and that air was required for combustion. Boyle's main contribution to chemistry was his foundational work on the "corpuscular" nature of matter, which led to our modern notions of elements[91] and the atomic[92] composition of matter.

In the *Sceptical Chymist*, Boyle defined "element" thusly:

I now mean by Elements, as those Chymists that speak plainest do by their Principles, certain Primitive and Simple, or perfectly unmingled bodies; which not being made of any other bodies, or of one another, are the Ingredients of all those call'd perfectly mixt Bodies are immediately compounded, and into which they are ultimately resolved.

Boyle was a Baconian with respect to experimental design and the inductive process and Cartesian in his application of geometrical principles to the study of natural phenomena. Boyle lauded the contributions of these others toward an understanding of nature when he said:

Des Cartes, Gassendus, and others, having taken in the application of geometrical theorems, for the explication of physical problems; ... have brought the experimental and mathematical way of enquiring into nature, into at least as high and growing an esteem, as ever it possessed when it was most in vogue among the naturalists, that preceded Aristotle.[93]

Boyle brought something else to natural philosophy, a preternatural eclecticism that enabled him to apply knowledge obtained from many sources to his construction of an experimental method. He utilized knowledge and ideas obtained from other scientists, as well as tradesman, artisans, and the legal system.[94] Students and practitioners of science would be well served by emulating such eclecticism. One never knows how information, either clearly relevant or *apparently* irrelevant, from one's experiences, sensations, feelings, or observations may seed novel ideas or insights.

---

[90] The law may also be referred to as Mariotte's Law (especially in France), after the French physicist, physiologist, and priest, Edme Mariotte, who in 1676, independently of Boyle, reported the inverse relationship between the pressure and volume of a gas, and specified in addition that the law requires a constant temperature. Boyle's own specification of the law is found in his 1662 volume "A Defence of the Doctrine Touching the Spring and Weight of the Air" [212], in which he wrote "... the same air being brought to a degree of density about twice as that it had before, obtains a spring twice as strong as formerly." The kinetic theory of gases was not propounded for another two centuries. Descartes and Mariotte worked under the assumption that air comprised tiny springs, which could be compressed or expanded depending on pressure.

[91] Davidson [213] describes this contribution as follows: "The physicists, Boyle called them "hermetick philosophers," upheld the Peripatetical or Aristotelian doctrine of the four elements – fire, air, earth, and water. The chemists, "vulgar spagyrists," were disciples of Paracelsus who believed in the tria prima – salt, sulphur, and mercury. Boyle showed that these theories were totally inadequate to explain chemistry and was the first to give a satisfactory definition of an element."

[92] The etymology of the term "atom" reveals that the Greek philosopher Leucippus and his student Democritus had discussed the concept of the *atomos*, an indivisible building block of matter, as early as the fifth century BCE. *Atomos* literally means "he who is unable to be separated or divided."

[93] Taken from Boyle [44], p. 59.

[94] Boyle apparently was quite willing to consider information taken from any source, regardless of whether the information was consistent with the then current received wisdom of experts. For example, with respect to physicians, Boyle argued that their knowledge "... *might be not inconsiderably increased,*" if they considered "*the observations and experiments, suggested partly by the practice of midwives, barbers, old women, empiricks, and the rest of that illiterate crew, that presume to meddle with the physick among our selves; and partly by the [Brazilian] Indians and other barbarous nations, without excepting the people of such parts of Europe it self, where the generality of men are so illiterate and poor, as to live without physicians.*"

Boyle, like many before him, was acutely aware of how mere sensory experience might mislead one with respect to conclusions regarding phenomena and their underlying causes. He said that to *"discover deep and unobvious truths ... intricate and laborious experiments"* must be performed. Theory, supposition, and speculation were insufficient if one wished to be confident in their understanding of natural phenomena. Boyle felt that chemistry was especially amenable to rigorous experimentation because of its inherent practical simplicity, i.e., the ability to study nature in controlled systems. These systems comprised components that were known (e.g., specific equipment, vessels, substances, volumes, temperatures, and pressures) and could be carefully assembled to create microcosms of nature that could be interrogated and understood in the absence of the many confounding variables found outside the laboratory in nature itself. This type of experimental system, as exemplified by Boyle's use of Hooke's improved air pump to study the spring of air (pressure), is considered by some to be the first example of controlled experimentation [214]. Whether Boyle, in addition to his suggested paternity of modern chemistry, also was the father of controlled experimentation is debatable, but what isn't is Boyle's clear demonstration of how such experimentation should be done. In fact, Boyle distinguished two types of experiments, "probatory" and "exploratory." The former was a type of control experiment that may be considered as system validation, i.e., one proves that one can use the system to answer the questions posed. For example, prior to his experiments on the spring and weight of air, Boyle first demonstrated that his air pump could indeed create vacuums in large vessels. The latter experimental type was designed to learn something new. It was the actual "test" experiment. Boyle also wrote extensively about a type of exploratory experiment that he termed *"fiat experimentum,"* which was an experiment that he wished to perform but could not due to a lack of appropriate technical means. In some ways, this presaged the thought experiments common in physics and so intimately associated with Albert Einstein (see page 62).

Boyle's most important contribution to the scientific method may have been the manner in which he carried out and reported his experiments. Reproducibility was a tenet of Boyle's method, for without it, Boyle could not have confidence in his experimental results.[95] In his publications, Boyle included detailed discussions of how he performed his experiments, including the instruments he used in doing so, something that often was not done at the time[96] but now is an intrinsic part of the practice of science. Boyle clearly recognized that not only could variability in results exist among different experimenters ostensibly doing the same experiment, but it also could exist in one's own experiments. Presaging the need for modern scientists to use pure materials, Boyle, in his 1661 book *Certain Physiological Essays*, included *"TWO ESSAYS, Concerning the Unsuccessfulness of Experiments, Containing divers Admonitions and Observations (chiefly Chymical) touching that SUBJECT,"*

---

[95] Boyle emphasized the importance of repetition of experiments, opining *"... the instances we have given you of the contingencies of experiments, may make you think yourself obliged to try those experiments very carefully and more than once, upon which you mean to build considerable superstructures either theoretical or practical, and to think it unsafe to rely too much upon single experiments."*

[96] The alchemists of his day, most of whom were focused at least in part on how to convert lead into gold, were quite secretive about their methods, for obvious reasons.

in which he discussed experimental variability and errors in data interpretation arising from the compositional heterogeneity of starting materials used in chemical experiments.[97]

In Boyle's considerations of error, he addressed known sources of error as well as unknown sources. Known sources included variability in time (the season in which an experiment was done and the duration of an experiment), glassware, and the dexterity of an investigator. Unknown sources were simply that – experimental failures or inconsistencies occurring in the absence of observable differences in experimental execution. Modern scientists encounter and, most importantly, must recognize these factors if they are to understand and interpret the results of their experiments correctly. Boyle also recognized problems due to scaling, i.e., the observation that the execution of experiments at large scales may not produce the same results as were observed at smaller scales.[98] Proper scaling is the basis for today's successful large-scale production of chemicals, solvents, pharmaceuticals, and a host of other products.

Problems are part and parcel of the conduct of science. Although Boyle warned his readers explicitly of the difficulties in doing science when he wrote *"... our way is neither short nor easy,"* and cautioned his readers to *"... a watchfulness in observing experiments, and wariness in relying on them,"* he also encouraged them to persevere by not letting problems produce a *"... despondency of mind, as may make you forbear the prosecution of ..."* experiments.

Boyle may have been one of the first to incorporate probability *per se* as an epistemological tool. Others before him had discussed extensively how one was to move from observations of phenomena to their causes, but the reasoning involved was somewhat qualitative in nature. Boyle, using as his model the method of decision making in the English trial courts, required a "concurrence of probabilities" to exist if a conclusion was to be believed based upon available experimental and observational evidence. Boyle expressed this epistemic reasoning as follows:

For though the testimony of a single witness shall not suffice to prove the accused party guilty of murder; yet the testimony of two witnesses, though but of equal credit ... shall ordinarily suffice to prove a man guilty; because it is thought reasonable to suppose, that, though each testimony single be but probable, yet a concurrence of such probabilities (which ought in truth to be attributed to the truth of what they jointly tend to prove) may well amount to a moral certainty, i.e. such a certainty, as may warrant the judge to proceed to the sentence of death against the indicated party.

A direct outcome of Boyle's insistence on experimental repetition and variation was the provision of many "witnesses," which would allow such a concurrence to be achieved. Boyle's invocation of *probabilities* as means to justify conclusions should not be considered as it would be now, namely as the use of statistics and probability theory to determine verisimilitude. Boyle's usage would better be likened to the phrase used in today's trial courts, "weight of the evidence." Nevertheless, *in practice*, both in the seventeenth century and now, and both in the trial courts and in science, statistics and probability theory *are* vital and obligatory components in experimental design and data analysis. In this sense,

---

[97] See Boyle [44], p. 318.
[98] See Boyle [44], p. 335.

Boyle's logic of epistemology also was part of the foundation of the modern scientific method.

In addition to "witnesses" in the metaphorical sense, Boyle also considered the readers of his work witnesses, i.e., they would judge the value and veracity of his experiments, both intellectually as well as experimentally. To do so required a detailed, well-organized presentation of methods, observations, interpretations, and any conclusions or new ideas derived therefrom. A codified format for the presentation of scientific results did not exist in Boyle's time. Boyle addressed this need in *A Proemial Essay*, contained within *Certain Physiological Essays* [215], in which he listed ten rules of good scientific writing, all of which remain relevant today.[99] These included:

1. The *"chief requisites . . . are candor and truth."*
2. The style should be *"philosophical [rather] than . . . rhetorical"* and be *"clear and significant."* These specifications were meant to convince writers to avoid rhetoric and other means of convincing readers of the correctness and value of ones work.
3. Convey information, both to the educated nonscientist (gentleman) as well as the scientist.
4. Present copious details allowing repetition and validation of his own experiments.
5. Discuss failures as well as successes, to encourage learning and correction of errors.
6. Do not eschew speculations, even sometimes ill-founded ones, so that readers may be encouraged to determine if the speculations are true or false.
7. *. . . carefully distinguish those things, that they know, from those, that they ignore or do but think, and then explicate clearly the things they conceive they understand, acknowledge ingenuously what it is they ignore, and profess so candidly their doubts, that the industry of intelligent persons might be set on work to make further inquiries.*
8. Compare and contrast one's own work with the work of others. *. . . divers particulars, which whilst they lay single and scattered among the writing of several authors were inconsiderable, when they come to be laid together in order to the same design, may oftentimes prove highly useful to physiology in their conjunction, wherein one of them may serve to prove one part or circumstance of an important truth and another to explicate another, and so all of them may conspire together to verify that saying, Et quae non prosunt singula, multa juvant.*[100]
9. Sacrifice conciseness for completeness.
10. *Be diffident in his presentation, so as to allow for his own errors and those of others.*

Boyle was acutely aware of the potential for error in experimentation, data interpretation, and hypothesis formulation. This was one reason for the prolixity often associated with Boyle. He wanted to present a complete experimental story, not just those experiments that worked or supported a particular hypothesis or idea, but a discussion of failures and their potential causes and implications as well. Such candor has become a casualty of modern scientific writing, which ironically, and considering Boyle's invocation of trial processes

---

[99] In addition, and importantly, Boyle considered writing a "part of the learning process and thus a crucial part of experimental activity itself." (Quotation from Sargent [33], p. 181.) The science student would be well advised to heed this dictum.

[100] Latin: "And what alone is not useful helps when accumulated."

into his experimental philosophy, often has the tone of a contentious legal argument rather than an open presentation and assessment of the successes and failures of one's efforts.

Boyle's work has been characterized by some as tentative or inconclusive because of his reluctance to "sell" it. I would argue strongly that the scientific enterprise, at least ideally, is about truth and not sales. It is up to the scientific community itself, through continuous repetition of experiments, and testing and retesting of hypotheses, to determine if particular results are conclusive or not. In this regard, Boyle likened hypotheses to "counterfeit pieces of money," which may appear to be real on their surface but may be found not to be upon deeper thought and inspection.

4.4.5 Sir Isaac Newton (1642–1727)

Sir Isaac Newton, who coincidentally was born almost exactly one year after Galileo's death, was the consummate "mathematical scientist" of his age. He was so gifted that he became the Lucasian Professor of Mathematics at Cambridge less than five years after receiving his bachelor's degree. Newton's mathematical brilliance is clear when one considers it was he who first invented the calculus (almost a decade before Leibniz did so independently). Newton is also known for his theory of universal gravitation and his work in optics.[101] Our goal here, however, is not to learn about all of Newton's discoveries, but to try to understand the methods that enabled him to make them and how these methods contributed to the development of the modern scientific method.

As we have seen before, and will see in later chapters, deep study of any issue, be it historical, philosophical, or scientific, often reveals complexities and subtleties not envisioned *pre facto*. This is certainly the case when we look deeply into Newton's scientific method. Newton never published a methodological treatise. In fact, it appears that Newton employed different methods in the experiments that he discussed in two of his classic works, *Philosophiæ Naturalis Principia Mathematica* (Mathematical Principles of Natural Philosophy), or simply the "*Principia*," and *Opticks*. The only discussion of methods is found in a six paragraph scholium (*General Scholium*) in the second and third editions of the *Principia*. Key elements of the methods were observation of specific phenomena, development of propositions (laws) underlying the phenomena through the process of inference, the use of induction to generalize the propositions from specific phenomena to all phenomena, and establishment of the validity of laws through demonstration of their experimental and predictive abilities. The best example of this approach may be Newton's studies of the movements of the sun, planets, and moons, which led him to propose a force, gravity, by which such movements could be explained and predicted. Newton did discuss, in general terms, the strategies he used in developing his theory of universal gravity.

... we have explain'd the phænomena of the heavens and of our sea, by the power of Gravity, but have not yet assign'd the cause of this power. ... But hitherto I have not been able to discover the cause

---

[101] Newton also is credited with inventing the color *indigo*, developing new machines at the Mint in London (first as Warden of the Mint and then as Master of the Mint), and determining the size of the biblical measure known as the *cubit*, which allowed him to model King Solomon's Temple.

of those properties of gravity from phænomena, and I frame no hypothesis.[102] For whatever is not deduc'd from the phænomena, is to be called an hypothesis; and hypotheses, whether metaphysical or physical, whether of occult qualities or mechanical, have no place in experimental philosophy. In this philosophy particular propositions are inferr'd from the phænomena, and afterwards render'd general by induction. Thus it was that the impenetrability, the mobility, and the impulsive force of bodies, and the laws of motion and of gravitation, were discovered. And to us it is enough, that gravity does really exist, and act according to laws which we have explained, and abundantly serves to account for all the motions of the celestial bodies, and of our sea.

Newton's method is quite similar to the modern scientific method, except for the fact that *his* sense of "hypotheses" sometimes differs. In fact, according to Cohen [216], at one time or another, Newton proffered nine different definitions. Domski has offered this reading of Newton and hypothesis.

Newton is urging us to read the methodology of the *Principia* as one that is incompatible with the use of 'hypotheses' in natural philosophy. On his account, either we investigate nature by relying on hypotheses, or we follow his 'experimental' program and investigate nature by appealing to the phenomena and properly marrying deduction and induction.[103]

Newton expounded further about "hypothetical philosophy" in a letter to Roger Cotes, the editor of the second edition of the *Principia*, in which he wrote:

Experimental Philosophy reduces Phænomena to general Rules & looks upon the Rules to be general when they hold generally in Phænomena... Hypothetical Philosophy consists in imaginary explications of things & imaginary arguments for or against such explications, or against the arguments of Experimental Philosophers founded upon Induction. The first sort of Philosophy is followed by me, the latter too much by Cartes [Descartes], Leibnitz & some others.[104]

It is important here to understand the historical context of Newton's work and writings. "Hypotheses" in the seventeenth century often were metaphysical in nature. They could include speculations, elements of which had no correspondence to anything in nature. A trenchant example of the type of hypothesis that Newton rejected categorically, and why Descartes [Cartes] was mentioned specifically in the quotation above, was Descartes's Vortex Hypothesis. No evidence existed for this hypothetical mechanical construct. Descartes found this system adequate because of its apparent explanatory value, which was all he required of his methodology. Newton did not. In excluding such hypotheses from the mathematical principles of natural science, Newton was following in Galileo's footsteps by, paraphrasing Koyré, "destroying one world and replacing it by another." Newton is propounding a "new concept of science [viz., a mathematical explanation of nature]" that appeared to many to be replacing common sense with mathematical abstractions disconnected from nature. Force, for example, was such an abstraction and because Newton would

---

[102] Koyré argues that Newton did not reject all hypotheses, but rather only those that could not be "proved or disproved by mathematically treated experiment." Newton, no doubt, was thinking of "global qualitative explanations" such as Descartes's Vortex Theory of the movement of planets and comets. (From [145], p. 16n3.)

[103] Taken from [217].

[104] Letter from Newton to Cotes, 28 March 1713, from [218], vol. 5, pp. 398–399.

not form a hypothesis (*"hypotheses non fingo"*) explaining it,[105] he was accused of dealing in the occult. Newton, in turn, contrasts his methodology with Descartes's, suggesting that his alone can provide a realistic representation of nature.

From my own perspective, both Descartes's and Newton's postulations about the dynamics of bodies in space are reasonable and experimentally testable *hypotheses* and therefore worthy of consideration and study. If so, how could Newton's methods be characterized as novel and as groundbreaking as they has been. The answer is not in the strategy writ large but rather in one key element, abstraction. Abstractions are incorporated into Newton's methods as mathematical constructs.[106] A very important point, and one that helps distinguish Newton's propositional method from the method of hypothesis, is his *pre facto* definition of certain features of empirical systems (e.g., mechanical systems) as *axioms*. Axioms are just true. They require no proof. It was the axiomatization of mechanics – in Newton's case his laws of motion – that forms the foundation for the reasoning in the *Principia* that gave rise to his theory of universal gravitation. Newton's axioms were general statements about phenomena that nevertheless were mathematically rigorous. It is in this sense that Newton's mathematical principles of natural philosophy are distinguished from the prevailing hypothetical philosophy of his day.

The use of mathematical constructs is freeing. It allows one to study a system using whatever parameters one feels would be useful, regardless of their initial correspondence with Nature. It also frees one from implicit constraints to discovery found in the scientific dogma in which one has been inculcated. As Cohen relates:

...the end product [of Newton's method] is a mathematical construct, a creation of the mind, in which Newton is perfectly free to consider whatever kinds of motions he pleases, subject to any type of force that he may imagine – because he is dealing with a mathematical construct and not with a physical situation. [...] Of course, the construct in question has been designed by Newton to be applied eventually to a specific end-use in natural philosophy, and so the construct has certain elements similar to the situation of the world of physics, the realm of natural philosophy revealed to us by our senses, by experiment, and by observation.[107]

Domski[108] emphasizes that the end result of Newton's method is to successfully connect his constructs/abstractions with the real world phenomena they are meant to represent. "...it is by relying on observed phenomena that Newton is able to bridge rationally intelligible

---

[105] This expression is among Newton's most famous and most quoted. However, most do not dig deeply enough into Newton, the man, to realize that one, if not the major reason, he writes *"hypotheses non fingo"* is because it would place him in a position open to criticism and ridicule, two things that this self-centered and secretive man could not tolerate. (See Chaudhury [219].)

[106] In a 1713, draft of a letter to Roger Cotes concerning emendations to the *General Scholium* for the second edition of the *Principia*, Newton summarized the steps followed in his method. "I like your designe of adding something more particularly concerning the manner of Philosophizing made use of in the Principia & wherein it differs from the method of others, vizt by deducing things mathematically from principles derived from Phaenomena by Induction. These Principles are the 3 laws of motion. And these Laws in being deduced from Phaenomena by Induction & backt with reason & the three general Rules of philosophizing are distinguished from Hypotheses & considered as Axioms. Upon these are frounded [sic] all the Propositions in the first & second Book. And these propositions are in the third Book applied to the motions of ye heavenly bodies."

[107] From [220], pp. 92–93.

[108] From [217].

principles to what is empirically actual, and in turn, to improve upon hypothetical philosophies such as Descartes's, which appear to promise only intelligible explanations. . . . the
presentations of the laws of motion, the law of gravity, and the qualities of bodies in the
Principia are meant to show us that the use of phenomena offer what hypotheses cannot:
evidence that the principles of natural philosophy are true and not merely imaginary."[109]

Our discussion of Newton thus far has revealed that Newton observed nature, formulated
questions, and sought ways to answer them. These all were, and continue to be, part of
the kernel of the scientific method. We also have seen that Newton, following Galileo,
integrated mathematics into natural philosophy.[110] Practitioners of each often were opposed
to their integration, both for historical (they've always been separate, and for good reason!)
and philosophical (nature is too complex to be understood using the rigid tools of geometry,
and mathematics itself is too abstract to represent any actual physical entities or behaviors
in nature) reasons. Fighting against such opposition, Newton revealed in the *Principia* and
in *Opticks* the methods enabling one to conceive of and understand gravity, the nature of
bodies and motion (e.g., planetary and terrestrial motion), and the spectral composition of
sunlight. Newton's methods provided the means to specify experiments and observations
through which empirical studies could provide answers to questions, ". . . in contrast to
conjecturing answers and then testing the implications of these conjectures."[111]

So, in a practical sense, what *were* Newton's methods and were they unique? It has been
argued that one can only answer these questions by inference from Newton's published
work. However, when one examines such inferences from Newton scholars, one finds no
consensus. Opinions have, and remain, varied (and often contentious). What most do agree
upon is the revolutionary manner in which Newton mathematized science. What separated
Newton from others was how he incorporated mathematics into theory creation and theory
testing – an incorporation providing a number of advantages relative to the extant scientific
method. Like Cohen, Domski emphasizes the elimination of conceptual hurdles in Newton's methods, and importantly, how Newton, unlike those before him, saw discrepancy as
a tool for discovery.

. . . adopting this mathematical view grants him greater freedom insofar as his treatment is not
restricted by considerations of the empirical circumstances that might give rise to the forces and
motions he is investigating. Just as the geometer can demonstrate propositions about triangles and
circles without taking into account the steps needed to construct such objects, so too Newton can
investigate 'those quantities of forces and their proportions that follow from any conditions that may
be supposed.' . . . by adopting a mathematical point of view, he can present rational mechanics as a
'science, expressed in exact proportions and demonstrations, of the motions that result from the any

---

[109] Shapiro [12], p. 188, expresses this idea thusly: "[Newton's method] is defined as a methodology with two essential
elements: the distinction of experimental from hypothetical philosophy, or the exclusion of hypotheses from natural
philosophy; and the requirement that propositions in experimental philosophy are 'deduced from the phenomena and are
made general by induction.' . . . By essentially equating 'experimental philosophy' with natural philosophy, Newton has
ruled out totally imaginary hypotheses from natural philosophy." Newton may be thought of as one of the first "scientific
realists," i.e., those who believe science is, in fact, dealing with reality and not metaphysical constructs thereof. See
Section 5.7 for a discussion of realism.

[110] It should be noted, at the time, mathematics and natural philosophy were considered two independent fields, the former of
which was felt by many to be inferior to the latter, as mathematics was considered the province of engineers and artisans.

[111] Paraphrasing and quoting Smith [221], p. 147.

forces whatever and of the forces that are required for any motions whatever.' With the reference to 'exact proportions and demonstrations,' Newton alludes to the second advantage of adopting a mathematical and specifically geometrical approach to the science of mechanics: geometry provides a framework in which motions can be characterized with exactness ... [and] allows us to reason with certainty about a wide-range of motions and forces.[112]

The *exactness* of Newton's methods, in addition to yielding, according to Newton, more secure knowledge, provided the means to discover and study phenomena that had not yet been recognized. This was possible because exactness allowed Newton to argue that differences between effects predicted by theory and those observed by experiment were meaningful. This contrasted with the interpretations of others who simply ignored differences as impediments to understanding or as irrelevancies. Huygens, for example, in trying to understand the movement of a circular pendulum, "had tried to find the corrected path required for strict isochronism with a physically real bob (the weight at the end of the pendulum rope), only to despair when the problem proved intractably complex. In the manner typical of pre-Newtonian science, the small residual discrepancies between idealized theory and the real world were dismissed as being of no practical importance."[113] Newton, however, saw "every systematic deviation from current theory automatically ... [as] a pressing unsolved problem."[114]

The *Principia* and the *Opticks* both were highly technical treatises and, as mentioned above, practical details of Newton's experimental approaches can only be gleaned from their careful examination. However, in addition to experimental detail, in Book 3 of the second (1713) and third (1726) editions of the *Principia*, Newton added four "Rules of Reasoning in Philosophy" (Latin; *Regulae philosophandi*). These rules codified how one should *think* about natural philosophy, as opposed to how one should *do* it (i.e., experimentally). Their value in the development of the scientific method is through the guidance they provide in the creation and consideration of theories. Motte's English translation [222] of the *Regulae philosophandi* is below. Each rule is followed by Newton's explanation of it (in italics).

Rule I

We are to admit no more causes of natural things than such as are both true and sufficient to explain their appearances.

To this purpose the philosophers say that Nature does nothing in vain, and more is in vain when less will serve; for Nature is pleased with simplicity, and affects not the pomp of superfluous causes.

Rule II

Therefore to the same natural effects we must, as far as possible, assign the same causes.

As to respiration in a man and in a beast; the descent of stones in Europe and in America; the light of our culinary fire and of the sun; the reflection of light in the earth, and in the planets.

---

[112] From [217], p. 15.
[113] From [221], p. 155.
[114] From [221], p. 159.

Rule III

The qualities of bodies, which admit neither intension nor remission of degrees [i.e., increases or decreases in qualities], and which are found to belong to all bodies within the reach of our experiments, are to be esteemed the universal qualities of all bodies whatsoever.

For since the qualities of bodies are only known to us by experiments, we are to hold for universal all such as universally agree with experiments . . . .

Rule IV

In experimental philosophy we are to look upon propositions collected by general induction from phænomena as accurately or very nearly true, notwithstanding any contrary hypotheses that may be imagined, till such time as other phænomena occur, by which they may either be made more accurate, or liable to exceptions.

This rule we must follow, that the argument of induction may not be evaded by hypotheses.

Rule I reiterates the principle of parsimony, in existence since Aristotle argued for the superiority of demonstrations derived from fewer postulates of hypotheses and Ockham's discussion of parsimony was accorded the name Ockham's Razor, the metaphorical implement that shaves away unnecessary complexities from explanations. What Newton has added to these principles, as foundations in natural philosophy, is the requirement that any postulated causes of effects must be both necessary ["true"] and sufficient.

Rule II is a universality rule that says, in essence, if cause A has been shown to effect B, then any future observed effect B must have been caused by A. This reasoning is clear in Newton's work on universal gravity and is a bedrock of modern physics. However, universality also is a vital principle of science in general. How else does one argue the value of *in vitro* experimentation if its results are irrelevant and inapplicable *in vivo*? If the bactericidal effects of penicillin only operated in Alexander Fleming's petri dish in his laboratory in London, there would be no reason to attempt to use the antibiotic in humans around the world. An even more intriguing example, one never envisioned by Newton or any scientists before the twentieth century, is the application in nature of predictions determined within the small, immobile chips of silica composites that form the CPUs of computers. So-called *in silico* experiments can produce accurate predictions, for example, about trajectories of missiles or protein folding, because of Newton's laws of motion, which are incorporated into algorithms controlling the behavior of imaginary particles and bodies within the virtual worlds of computers.

Rule III extends universality from the domain of cause and effect to that of qualities of matter. If we observe certain characteristics of matter in our observations/experiments, and we should observe new objects heretofore unseen, then those objects will have the same characteristics as those already known. Extension (space), hardness, impenetrability, mobility, and inertia were examples of the qualities to which Newton referred in Rule III. Examples also would include mass, which is immutable, and the quality of being affected by gravity, which operate on all masses. Magnetism of one body could not be induced to all bodies because simple observation showed that not all bodies were magnetic.

When the scientific method is used in modern scientific settings, it yields propositions that explain phenomena. Scientists seek to develop propositions that are general in nature. This goal is subsumed within Rules I–III. Rule IV is more epistemological in nature. It says that propositions derived through induction from phenomena, *viz.* through experimentation, should be considered true until and unless subsequent experimental studies produce propositions at variance with those originally promulgated. In this case, the original propositions may be modified or maintained as is, but with exceptions noted. This rule, and Newton's comments following it, seek to immunize experimentally derived propositions from attack by nonexperimentally derived arguments, such as hypotheses (as the term was understood in the seventeenth century). In practice, Rule IV provides a justification for using theories extant in the conduct of science, as well as for continuing efforts to extend their precision, breadth, and accuracy.

The quotation of Newton below shows that he was well aware of the intrinsic uncertainty of propositions derived through experimentation. This awareness must be made part of the scientific philosophy and intuition of all modern scientists if they are to advance a mission of science for truth about nature.

And although the arguing from Experiments and Observations by Induction be no Demonstration of general Conclusions; yet it is the best way or arguing which the Nature of Things admits of, and may be looked upon as so much stronger, by how much the Induction is more general. And if no Exception occur from Phænomena, the Conclusion may be pronounced generally. But if at any time afterwards any Exception shall occur from Experiments, it may then begin to be pronounced with such Exceptions as occur.[115]

Many have opined that the scientific revolution ended with Newton. There may be some truth in this statement, as the seventeenth century was a period in which fundamental changes in both the philosophical and experimental underpinnings of science were made. Science was mathematized, a "mechanical philosophy"[116] underlay views of nature and the universe, theories could be formulated using abstractions of reality, advanced techniques of algebra and geometry were incorporated into scientific practice, and methods of inference matured. All of "these were successful elements in a scientific method."[117] What certainly did not end was further development of the scientific method, especially in the Victorian era, during which substantial contributions were made by such scientists and philosophers as John Herschel, William Whewell, John Stuart Mill, and Claude Bernard. Butts has characterized science in this era as follows:[118]

---

[115] From Newton, *Opticks*, p. 404, as quoted in [223], p. 119.
[116] The "mechanical philosophy" comprised a number of distinct, interrelated elements, including the following: the world and its components behave like a machine so they can be described solely by the mathematical laws of mechanics; all causation is by contact action so that immaterial, spiritual agents are banished; matter is composed of invisible corpuscles; and hypotheses about the properties and motions of these invisible corpuscles may be formulated to explain visible effects. (Adapted from [12], p. 206.)
[117] Taken from [178], p. 96.
[118] Taken from Butts [224], p. 319.

...science is the ideal form of knowledge....Science as the ideal is identified with mathematical physics; astronomy is queen of the sciences and the highest realization of the quest for necessary truth. The methodology implicit in high science is, briefly stated, Baconian; it is some form of inductivism. As a Victorian methodology, it seems quite natural that it should accept that knowledge is power; as a British methodology, it is likewise fully understandable that it should think of itself as an improved image of classical Newtonianism.

4.5 Science in the Nineteenth Century

4.5.1 John Herschel (1792–1871)

It is during the *Victorian* era that we encounter Sir John Frederick William Herschel, who has been said to have been England's most famous scientist from ≈1830 to 1860. Herschel was a polymath with interests and expertise in a wide range of scientific areas, including astronomy, chemistry, physics, crystallography, mineralogy, geology, photography, photochemistry, and meteorology. His 1830 book, *Preliminary Discourse on the Study of Natural Philosophy (Discourse)*, was, according to Minto[119] *"...the first attempt by an eminent man of science to make the methods of science explicit."* William Whewell, who we shall encounter below, felt that "Herschel put methodology on the map."[120] As with most of the pronouncements we have encountered thus far, it is debatable whether this one is or is not true[121] Nevertheless, what is certainly true is that *Discourse* influenced a great number of the most famous natural philosophers of the time, including Whewell (see Section 4.5.2), Mill (see Section 4.5.3), Darwin,[122] Jevons (see Section 4.5.5), and Maxwell.

Discourse comprised three parts that covered a broad range of subjects in science, including its *"General Influence ...on the Mind,"* abstraction and empiricism, *"...its Application to the practical Purposes of Life, and its Influences on the Wellbeing and Progress of Society"* (all in Part I), its methods (Part II), and physical phenomena (Part III). We are interested here in Part II.[123] Here, Herschel discusses in more concrete terms his philosophy for "doing science." In Chapter IV, entitled *"Of the Observation of Facts and the Collection of Instances,"* Herschel espouses the importance of the initial accumulation of as much foreknowledge as possible regarding a scientific question, and especially knowledge obtained in different ways and from different perspectives:

Whenever, therefore, we would either analyse a phenomenon into simpler ones, or ascertain what is the course or law of nature under any proposed general contingency, the first step is to accumulate a sufficient quantity of well ascertained facts, or recorded instances, bearing on the point in question.

---

[119] Taken from Minto [225], p. 257.

[120] Taken from [226], p. 726.

[121] My own feeling is that the approbation of Herschel by Minto is not wholly deserved.

[122] Darwin quotes Herschel, with respect to Herschel's opinions on creation, on page 18 of *The Origin of Species* [227]. It has been noted that "Herschel...authoritatively established the naturalistic origin of species as a proper subject of investigation for Victorian Englishmen" such as Charles Darwin. [Taken from "Herschel, John (1792–1871)." Encyclopedia of Philosophy. Encyclopedia.com. 10 October 2018 www.encyclopedia.com.]

[123] Part II was entitled, "Of the principles on which physical science relies for its successful prosecution, and the rules by which a systematic examination of nature should be conducted, with illustrations of their influence as exemplified in the history of its progress."

Common sense dictates this, as affording us the means of examining the same subject in several points of view; and it would also dictate, that the more different these collected facts are in all other circumstances but that which forms the subject of enquiry, the bettor; because they are then in some sort brought into contrast with one another in their points of disagreement, and thus tend to render those in which they agree more prominent and striking.

Herschel argues for accuracy and precision of observation as well as for a keen aware-ness, during this process, of the danger of recording inferences instead of observations *per se*. This is a critical point in scientific practice, one discussed more fully in Section 5.3. Herschel also espouses the value of experimental repetition, both as a means of ensuring the veracity of conclusions therefrom, and in a statistical sense, to wit:

But how, it may be asked, are we to ascertain by observation, data more precise than observation itself? How are we to conclude the value of that which we do not see, with greater certainty than that of quantities which we actually see and measure? It is the number of observations which may be brought to bear on the determination of data that enables us to do this.

Herschel's approach to science is inductive. He wants to induce, from observation, laws governing natural phenomena. To do so, he must determine cause–effect relationships of these phenomena, and to do that he proposes ten *"... general rules for guiding and facili-tating our search, among a great mass of assembled facts, for their common cause, we must have regard to the characters of that relation which we intend by cause and effect."* These included the Methods of Agreement, Difference, Concomitant Variation, and Residues, which later comprised four of Mill's "five Canons of Induction."[124] These methods remain central to those aspects of the scientific method involving experimental design and data acquisition and interpretation. Hershel summarizes the approach when he writes "[The] ... successful process of scientific inquiry demands continually the alternate use of both the *inductive* and *deductive method.*"[125]

   The *Discourse* provided a philosophical foundation for the conduct and evaluation of experiments, as well as a wealth of practical examples illustrating the actual applica-tion of the philosophy to the determination of causes of phenomena and laws governing behavior in natural systems. These practical applications are subsumed within the modern methodological categories of observation, hypothesis, experiment, data analysis, and theory creation and prediction. Interestingly, although Hershel was a champion of the inductive method of discovery, he opined that one could also simply start in an *a priori* fashion *"... By forming at once a bold hypothesis, particularizing the law, and trying the truth of it by following out its consequences and comparing them with facts."*[126] This statement is almost metaphysical in its import and reminds one of Descartes top-down methodol-ogy. Newton certainly would have eschewed this type of *a priori* hypothesis creation, yet its "hypothetico-deductive" form, which has been said to have originated with Newton, underpins the modern scientific method.

---

[124] The Joint Method of Agreement and Difference was not enunciated.
[125] Quotation taken from Herschel [228], pp. 174–175. Italics are those of the author.
[126] How does this compare with "abduction?" See Section 5.4.1.

4.5.2 William Whewell (1794–1866)

It has been said of Herschel that his inductivism should be considered "not merely as a philosophical approach to understanding the sciences but as a methodological commitment in scientific practice."[127] This sentiment may apply even more strongly to William Whewell, another English polymath. Whewell entered Trinity College, Cambridge, in 1812 and was "identified as the best scholar of his year."[128] Five years later, he was made a Fellow of Trinity College. He subsequently was appointed Master and even served as Vice Chancellor of the University. Whewell's range of expertise was broad, including architecture, astronomy, economics, geology, mathematics, mechanics, oceanography, and physics. This range is reflected in the fact that he was a Professor both in Mineralogy and in Moral Philosophy, as well as an Anglican priest, historian of science, and poet. Whewell is credited with inventing the words "scientist," "physicist," "linguistics," "consilience," "catastrophism," "uniformitarianism," and "astigmatism," and suggesting to Michael Faraday the terms "anode," "cathode," and "ion!"

Whewell is probably best known for his focus on the way natural science was and should be conducted, i.e., the scientific method. It has been said that "For William Whewell . . . the methods of science merited scientific examination themselves."[129] Whewell did so in two monumental works, the three-volume *History of the Inductive Sciences, from the Earliest to the Present Times* (1837) and the two-volume *The Philosophy of the Inductive Sciences, founded upon their History* (1840). The earlier work was meant as an introduction to the latter, in which Whewell addresses the complicated and controversial question of how science progresses and produces knowledge. The answer, to Whewell, was methodological in the sense that progress depended on the manner in which observations were converted into truths. However, Whewell's methodology was not mechanical. It was philosophical. It had as its basis an inductive method, *Discoverers' Induction*,[130] which was designed to discover both phenomenal and causal laws and was based on, but distinct from, Baconian inductive processes [232]. Bacon's inductive processes were characterized by the accumulation of observations and the application of simple rules of logic – the various "Methods of" discussed in Section 4.4. No metaphysical component was involved. In contradistinction, to move from the general to the specific, or from observations to causes, Whewell required more than method and simple logic alone. He required a *logic of scientific discovery*,[131] and through this logic, he could establish inductive truths.

[I]nductive truth is never the mere *sum* of the facts. It is made into something more by the introduction of a new mental element; and the mind, in order to be able to supply this element, must have peculiar endowments and discipline.[132]

---

[127] Taken from Cobb [229], p. 21.
[128] Taken from Butts [230], p. 3.
[129] Taken from Cowles [226], p. 722.
[130] Whewell explains that "My object was to analyse, as far as I could, the method by which scientific discoveries have really been made; and I called this method *Induction*. . . . That it is not exactly the Induction of Aristotle, I know; nor is it that described by Bacon. . . . I am disposed to call it *Discoverers' Induction*. . . ." (From [231], pp. 416–417.)
[131] A century after Whewell, Karl Popper published "The Logic of Scientific Discovery," a seminal work in this area and a foundational text in the philosophy of science.
[132] Taken from Butts [230], pp. 170–171.

The "peculiar endowments and discipline" to which Whewell refers may be characterized as intuition and background knowledge. Intuition yields "Eureka moments"[133] in which seemingly *a priori* creation of new concepts, hypotheses, or theories occurs.[134] However, these intuitions depend on foreknowledge. We cannot intuit without it and, unlike Descartes, we cannot begin with *a priori* rationalizations [186]. Whewell argues, for example, that to "see that a convex surface of the earth is necessarily implied by the convergence of meridians towards the north" requires "the student [to] . . . have a clear conception of the relations of space, either naturally inherent in his mind, or established there by geometrical cultivation, – by studying the properties of circles and spheres." Louis Pasteur, one of Whewell's contemporaries, expressed this idea succinctly when he said, *"Dans les champs de l'observation le hasard ne favorise que les esprits préparés."*[135] "Discipline" (Whewell) and "preparation" (Pasteur) allow one to make connections among facts, observations, hypotheses, and theories extant – to intuit relationships not immediately obvious and to "see" what has not been seen before, i.e., unobservables – thereby enabling construction of new hypotheses and new theories through induction. Whewell states:

Facts are bound together by the aid of suitable Conceptions. This part of the formation of our knowledge I have called the *Colligation of Facts*; and we may apply this term to every case in which, by an act of the intellect, we establish a precise connexion among the phenomena which are presented to our senses. The knowledge of such connnexions, accumulated and systematized, is Science.[136]

Colligation, which Whewell considered an "act of thought," was the "new mental element" referred to in the quotation above regarding inductive truth. The role of conception was to provide "the true bond of Unity by which phenomena are held together."[137] Whewell created "Inductive Tables" of colligations as an aid to concept creation through induction. These tables looked like inverted genealogical trees in which facts were listed at the top level, above the colligations explaining them, and each level below comprised a more general set of facts and corresponding colligations. The most general inductive concept was found at the bottom. For example, in Whewell's inductive table of astronomy, top level facts such as the area of the moon illuminated at night and seasonal variations in the position of the sunrise and sunset eventually led down to the theory of universal gravitation. These tables provided a convenient visual summary of facts and subordinate colligations, but the elements included in the table, their connections, and the reiterative colligation at each level leading to a general truth were not automatic. According to Whewell, there

is always a new conception, a principle of connexion and unity, supplied the the mind, and superinduced upon the particulars. There is not merely a juxta-position of materials, by which the new

---

[133] The word "Eureka," signifying a moment of discovery, comes from the ancient Greek word $\epsilon\upsilon\rho\eta\kappa\alpha$, meaning "I found (it)." The use of the word in this context is attributed to Archimedes, who, after observing the level of water in a bath increase after he entered it, came to the sudden realization that the volume of water required to increase the level must be equal to the volume of the part of his body he submerged.

[134] It should be noted that Whewell rejected the notion that hypotheses could be formed *a priori*. He required a factual basis for hypothesis creation [232].

[135] English; in the fields of observation, chance favors only prepared minds.

[136] Taken from Whewell [210], p. 36.

[137] Taken from Whewell [210], p. 46.

proposition contains all that its component parts contained; but also a formative act exerted by the understanding, so that these materials are contained in a new shape.

Whewell considered the "mental element" of sagacity the engine for forming exact knowledge. Many others in the nineteenth century also emphasized sagacity – genius – as the key factor in scientific progress, as opposed to empirical observation *per se* [233]. Some even went so far as to say that "major discoveries do not 'depend upon method, but upon the genius of the investigator.'"[138] Can genius or sagacity be defined? Whewell says "no," and explicitly rejects the idea that knowledge formation by a "Discoverer" could be constrained by "rules, or expressed in definitions."[139] His historical discussion of great discoveries makes this quite clear:

> It would be difficult or impossible to describe in words the habits of thought which led Archimedes to refer the conditions of equilibrium on the lever to the Conception of pressure, while Aristotle could not see in them any thing more than the results of the strangeness of the properties of the circle; – or which impelled Pascal to explain by means of the Conception of the weight of air, the facts which his predecessors had connected by the notion of nature's horrour of a vacuum; – or which caused Vitello and Roger Bacon to refer the magnifying power of a convex lens to the bending of the rays of light towards the perpendicular by refraction, while others conceived the effect to result from the matter of medium, with no consideration of its form. These are what are commonly spoken of as felicitous and inexplicable strokes of inventive talent; and such, no doubt, they are. No rules can ensure to us similar success in new cases; or can enable men who do not possess similar endowments, to make like advances in knowledge.

Whewell proposed a logic of scientific discovery, not a rote method. He was well aware of the possibility that one's colligation might yield an incorrect concept/hypothesis. His methods thus should be considered a heuristic as opposed to an infallible formula for truth determination. In addition, Whewell recognized and accepted the fact that the discovery process often involved what he called "happy guesses," i.e., abduction.[140] As an example, he notes that Kepler's discovery of the elliptical motion of the planets only occurred after he conceived and rejected nineteen prior guesses (hypotheses).[141] Making bold hypotheses, guessing, is an important component of modern scientific method and progress, as is failure, something that Whewell suggests accelerates scientific discovery.[142]

It is important to note, especially with respect to a famous debate between Whewell and John Stuart Mill regarding inductive truth (see Section 4.5.3), that Whewell did not believe that induction alone could provide scientific truths. These could only be obtained by empirical tests of the induced truths. This two-step approach to scientific truth – first developing a hypothesis and then testing the hypothesis's prediction experimentally to determine if the hypothesis holds – came to be known as the hypothetico-deductive method. Whewell, in fact, used this method in practice, but he was not a deductivist *per se* [186]. His logic of

---

[138] John Tyndall, as quoted in Yeo [233], p. 272.
[139] See [234], p. 40.
[140] See Section 5.4.1.
[141] See Whewell [210], pp. 41–42.
[142] See Firestein [235] for an informative and entertaining discussion of the value of failure to science.

scientific discovery simply incorporated both inductive and deductive reasoning, or, as he states:

The hypotheses which we accept ought to explain phenomena which we have observed. But they ought to do more than this: our hypotheses ought to *foretel* phenomena which have not yet been observed. ...The truth and accuracy of these predictions were a proof that the hypothesis was valuable and, at least to a great extent, true.[143]

Science is an exceptionally broad undertaking and the number of experimental systems one can study are legion. It may be easy to demonstrate the validity of a hypothesis in a single system, but what scientists love to do is generalize. Whewell's logic of scientific discovery imparts greater confidence in hypotheses found to be generalizable.

...the evidence in favour of our induction is of a much higher and more forcible character when it enables us to explain and determine cases of a *kind different* from those which were contemplated in the formation of our hypothesis. The instances in which this has occurred, indeed, impress us with a conviction that the truth of our hypothesis is certain. No accident could give rise to such an extraordinary coincidence.[144]

In the last sentence of the quotation, Whewell argues that hypotheses are true if repeated observations and experiments are consistent with their predictions, which suggests that the entities involved are real, as opposed to being objects of our imaginations. The term "realism," in the scientific context, implies that what science reveals about the natural world is in fact what is actually there, i.e., the physical entities about which science deals exist. A century after Whewell, Putnam made a similar argument in support of realism – the "no miracles argument" – when he wrote *"The positive argument for realism is that it is the only philosophy that doesn't make the success of science a miracle."*[145]

Whewell's explicit invocation of intangible elements of scientific discovery, for example, the *"new mental element"* of Colligation of Facts, was a departure from the more mechanical inductive approach practiced earlier. In Section 4.5.3, we discuss how John Stuart Mill, Whewell's younger contemporary, developed an inductive method (although based largely on Whewell's work) that rejected Whewell's new mental element and, in doing so, produced one of the most interesting and controversial debates in the history of science.

4.5.3 John Stuart Mill (1806–1873)

John Stuart Mill has been described variously as a philosopher, political economist, civil servant, naturalist, utilitarian, liberal,[146] and empiricist. He also has been dubbed "the most influential English-speaking philosopher of the nineteenth century" [237]. Along with Herschel and Whewell, Mill was one of the most important contributors to the formulation of

---

[143] Taken from Whewell [210], pp. 62–63.
[144] Taken from Whewell [210], p. 65.
[145] See Putnam [236], p. 73.
[146] He was the first Member of Parliament to call for women's suffrage and argued for equality of the sexes in an 1869 essay entitled *The Subjection of Women.*

an inductive scientific method. Mill was not a scientist by training, but became interested in the search for truth as he studied and wrote about fundamental sociophilosophical issues. He read the published work of Whewell and Herschel assiduously and used it as a starting point for his own work on scientific method, *A System of Logic, Ratiocinative and Inductive; Being a Connected View of the Principles of Evidence and the Methods of Scientific Investigation*, first published in 1843.[147] In Book 3 of that text, Mill discusses induction, as well as a number of topics directly relevant to it and to the overall scientific method, including the laws of nature, causation, methods of experimental inquiry, deduction, empiricism, and bases for disbelief.

If we are to develop an accurate (true?) mechanistic understanding of the natural world, we must plan and execute enabling experiments. We must attempt to determine cause-and-effect relationships based on the data these experiments produce. This can be quite a difficult task, one that requires the application of rigorous procedures of logic. In the first millennium, Avicenna established three logical methods, the Methods of Agreement, Difference, and Concomitant Variations (see Section 4.2.2). These methods were expanded to ten by Herschel (see above). Mill then condensed these into "five Canons of Induction," which included the methods of Avicenna (by way of Herschel) along with the Joint Method of Agreement and Difference (also termed the Indirect Method of Difference) and the Method of Residues [238].[148] The Indirect Method incorporates both the Method of Agreement and the Method of Difference and reveals what causes are *necessary and sufficient* to produce a particular phenomenon. Mill regarded the Indirect Method *"as a great extension and improvement of the Method of Agreement . . . ."* The method was defined as follows:

If two or more instances in which the phenomenon occurs have only one circumstance in common, while two or more instances in which it does not occur have nothing in common save the absence of that circumstance; the circumstance in which alone the two sets of instances differ, is the effect, or the cause, or an indispensable part of the cause, of the phenomenon.[149]

The method of residues addresses the circumstance where one has a number of effects produced by their cognate causes and the causes for most of the effects are known. If one eliminates all the known causes, then the remaining effects must be due to the remaining "residue" causes. In the case where only one cause–effect pair remains, the approach is definitive.

Subduct from any phenomenon such part as is known by previous inductions to be the effect of certain antecedents, and the residue of the phenomenon is the effect of the remaining antecedents.[150]

---

[147] Mill acknowledges his debt to Whewell, Herschel, and others explicitly in the preface to the first edition of his own book, in which, he states "Whatever may be the value of what the author has succeeded in effecting on . . . [induction], it is a duty to acknowledge that for much of it he has been indebted to several important treatises, partly historical and partly philosophical, on the generalities and processes of physical science, which have been published within the last few years. To these treatises, and to their authors, he has endeavored to do justice in the body of the work. But as with one of these writers, Dr. Whewell, he has occasion frequently to express differences of opinion, it is more particularly incumbent on him in this place to declare, that without the aid derived from the facts and ideas contained in that gentleman's 'History of the Inductive Sciences,' the corresponding portion of this work would probably not have been written."

[148] See Ryan [239], pp. 43–48, for a more detailed discussion of the logic of these methods.

[149] From [238], Book III, Chapter viii, § 4.

[150] From [238], Book III, Chapter viii, § 6.

Mill's methods, which were designed to determine causes of phenomena, remain a core part of today's scientific method. Another part of the method's core, less methodological and more philosophical, although no less important, is the desire of scientists to move from the specific to the general, much as Newton did as he moved from the study of specific planetary orbits to the discovery of the universal law of gravitation. How do we know that the generalities we create are true? This question was central to Mill's consideration of scientific method and he sought to answer it by reducing "the conditions of inductive proof 'to strict rules and to a scientific test, such as the syllogism is for ratiocination.'"[151] It is interesting that Mill uses syllogism/ratiocination as an example of the type of process necessary for establishing inductive proof because, although in his discussion of inference[152] he distinguishes induction from syllogism/ratiocination quite strictly, it appears that the mechanism through which Mill ideally would establish inductive truth would be more syllogistic than inductive.

To understand Mill's reasoning requires us to examine, although briefly, Mill's knowledge base, his motivations, and how these motivations impacted his own ratiocinative process. Mill was not a scientist.[153] For this reason, the scientific foundation upon which his reasoning was built had to be provided by others, in particular Whewell and Herschel. Whewell had studied the history of science and how it *actually* had been done. He therefore had a keen sense of the practical, whereas Mill approached the subject from a more holistic philosophical perspective, one that was contaminated by tremendous *ab initio* bias.

Mill's bias was his overriding goal of "reforming moral and political philosophy,"[154] by eliminating "intuitionism" from the realms of society and its politics. The intuitionism to which Mill referred was characterized by the acceptance of knowledge based on one's intuition alone, as opposed to empirical knowledge. "I think it, so it is true" is an illustrative oversimplification of this idea. This *a priori* philosophy argues that some truths conceived of in the mind, without direct empirical support, may be necessary truths, i.e., absolute truths. Mill categorically rejected such intuitionism, writing in his autobiography:[155]

The German, or a priori view of human knowledge, and of the knowing faculties, is likely ... to predominate both here and on the Continent. But the "System of Logic" supplies what was much wanted, a text-book of the opposite doctrine – that which derives all knowledge from experience, and all moral and intellectual qualities principally from the direction given to the associations. I make as humble an estimate as anybody of what either an analysis of logical processes, or any possible canons of evidence, can do by themselves towards guiding or rectifying the operations of the understanding. Combined with other requisites, I certainly do think them of great use; but whatever may be the

---

151 Taken from Anshutz [240], p. 78.

152 See Mill [238], Book II, Chapter 1, § 3.

153 Interestingly, in a letter to his friend John Sterling written a decade before Mill published *A System of Logic*, Mill provided some indication of the nonscientific perspective from whence his work on scientific method derived. "The only thing which I can usefully do at present, & which I am doing more & more every day, is to work out principles: which are of use for all times, though to be applied cautiously & circumspectly to any: principles of morals, government, law, education, above all self-education. I am here much more in my element: the only thing that I believe I am really fit for, is the investigation of abstract truth, & the more abstract the better. If there is any science which I am capable of promoting, I think it is the science of science itself, the science of investigation – of method. I once heard ... that almost all differences of opinion when analysed, were differences of method." (See Mill [241], pp. 78–79.)

154 Taken from Snyder [242], p. 95.

155 Mill [243], p. 150.

practical value of a true philosophy of these matters, it is hardly possible to exaggerate the mischiefs of a false one. The notion that truths external to the mind may be known by intuition or consciousness, independently of observation and experience, is ... the great intellectual support of false doctrines and bad institutions. By the aid of this theory, every inveterate belief and every intense feeling, of which the origin is not remembered, is enabled to dispense with the obligation of justifying itself by reason, and is erected into its own all-sufficient voucher and justification. There never was such an instrument devised for consecrating all deep-seated prejudices.

Mill's inductive formalism is correct in that, by preclusion of intuitionism, guessing, speculation, or epiphany, it protects science from inferences that are not based on empirics or may have little or no empirical support. However, these preclusions also have the effect of stifling creativity, especially as it relates to quantum leaps in understanding and to the development of novel and important general truths. Science needs intuitionism of the sort propounded by Whewell if it is to move forward. Whewell showed quite clearly that the history of scientific advancement is rife with it. Whether truth was intrinsic to or supported such advancement remains an insoluble issue because of the difficulty in defining what truth is. Mill's rejection of the idea that intuitionism can produce inductive truth *per se* thus can be considered formally correct.

Mill and Whewell famously debated whether enumerative induction or the Hypothetico-Deductive method produced scientific truth.[156] The take-home message for the reader should be that science involves, and may require, multiple forms of inference, including induction and deduction. Mill's inductive process, followed by empirical testing, does produce scientific laws as does Whewell's method of superimposing concepts generated by the mind over facts. The advantage of Whewell's approach is that it considers causative entities that are not observable empirically.

4.5.4 Claude Bernard (1813–1878)

The French physician and experimentalist Claude Bernard has been said to have been the "father of experimental medicine and of scientific physiology" [244, 245]. He made a number of very important scientific discoveries, including: (1) the liver synthesizes sugar, (2) the pancreas is involved in digestion, and (3) there are specific nerves (vasomotor) that control the blood supply to the body. In addition, he developed the concept of cellular homeostasis,[157] originally termed "milieu intérieur" (inner world), when he wrote *"La fixité du milieu intérieur est la condition d'une vie libre et indépendante."*[158] The methods through which these discoveries were made have been applied in physiology [244], and in other branches of medicine and science, since then – in particular the logic of experimental design, most importantly as it relates to the creation of experimental controls. Bernard discusses these methods, as well as philosophical principles of experimentation, in his classic

---

[156] The debate is complicated, subtle, and deeply philosophical. It is beyond our purview to discuss it in more detail here. The reader thus is encouraged to access the vast literature on the subject.
[157] The term "homeostasis" was coined by Walter Cannon in 1932 [246].
[158] English; The constancy of the internal environment is the condition required for a free and independent life.

book, *An Introduction to the Study of Experimental Medicine* (1865; referred to here as *"An Introduction"*).[159]

Medical experimentation in the middle of the nineteenth century often was done by those who wanted to prove their ideas or hypotheses true, not to find the truth *per se*. Bernard, in contrast, viewed *"science as an enterprise moving inexorably to a final truth"* and the only way to establish this truth was to bring rigor to the entire biomedical enterprise.[160] This was done by dispassionate observation of phenomena, unbiased consideration of facts, and a strictly deterministic view of experimentation and the causal world, i.e., *"everything can, in principle, be explained,"*[161] as quoted below.

In medicine, we are often confronted with poorly observed and indefinite facts which form actual obstacles to science, in that men always bring them up, saying: it is a fact, it must be accepted. Rational science based, as we have said, on a necessary determinism, must never repudiate an accurate and well-observed fact; but on the same principle, it ought not to encumber itself with apparent facts collected without precision, and possessing no kind of meaning, which are used as a double-edged weapon to support or disprove the most diverse opinions. In short, science rejects the indeterminate; and in medicine, when we begin to base our opinions on medical tact, on inspiration, or on more or less vague intuition about things, we are outside of science and offer an example of that fanciful medicine which may involve the greatest dangers, by surrendering the health and life of the sick to the whims of an inspired ignoramus. True science teaches us to doubt and, in ignorance, to refrain ... Some physicians fear and avoid counterproof; as soon as they make observations in the direction of their ideas, they refuse to look for contradictory facts, for fear of seeing their hypothesis vanish. We have already said that this is a very poor spirit; if we mean to find truth, we can solidly settle our ideas only by trying to destroy our own conclusions by counterexperiments.[162]

... a simple and logical appearance is not enough to make us accept an experimental fact; we should still doubt and by a counter experiment should see whether the rational appearance is not misleading. This is an absolutely strict precept, especially in medical science which by its complexity conceals additional sources of error. ... even when a fact seems logical, i.e., rational, we are never justified in omitting a counterproof or counter experiment, so that I consider this precept a kind of order which we must blindly follow even in cases which seem the clearest and most rational.[163]

---

[159] It seems interesting that the acclaim accorded Bernard by modern physiologists was absent during his lifetime. Bernard's book, *Introduction à l'Étude de la Médecine Expérimentale*, did not receive much attention until the end of the nineteenth century (see Grmek [247], as discussed in Roll-Hansen [245], p. 72), likely because it was written in French and an English translation was not available until 1927. A German translation only appeared in 1961, almost a century after the book's initial publication! It is also possible that more attention was paid to the book after the publication of two additional elaborations of his method, one in 1867 (*Rapport sur le progrès et la marche de la physiologie générale en France*) and another in 1878 (*Leçons sur les phénomènes de la vie communes aux animaux et aux vegetaux*). Nevertheless, at the time, Louis Pasteur said of the book: "Nothing so complete, nothing so profound and so luminous has ever been written on the true principles of the difficult art of experimentation." Much later, Sir Peter Medawar (Nobel prize winner (1960) for his discovery of acquired immunological tolerance) wrote that Bernard provided *"The wisest judgements on scientific method ever made by a working scientist ... "* (From Medawar [248], p. 73.)

[160] Ernest Renan said of Bernard, "Truth was his religion" [249].

[161] Quotation from Stanford Encyclopedia of Philosophy [250].

[162] Taken from Bernard [251], pp. 55–56. Bernard presages Karl Popper, who argued the same thing, *viz.*, we must not seek proof of our hypotheses, but rather strive to disprove (falsify) them.

[163] Taken from Bernard [251], pp. 181–182.

The idea of a "counterexperiment"[164] is one of the most fundamental aspects of the scientific method. We can never be sure a finding, hypothesis, or theory is correct because we can never be sure that contradictory evidence may be revealed in the future. This was the central argument of the Scottish philosopher David Hume (1711–1776) with respect to the veracity of scientific inductions (see Section 5.7.8), and the argument remains valid to this day. The converse is not true. If we can devise and perform an experiment that produces results contrary to our ideas, then it is likely our ideas were wrong. Of course, it is important to qualify this statement by requiring that the counterexperiment be logically rigorous and technically sound, otherwise we could not be confident that any results emanating from the experiment were correct. Put crudely, a million confirmatory experiments do not prove us correct, whereas a single contradictory experiment can prove us wrong. Counterexperiments "falsify" hypotheses, a notion that was propounded by Sir Karl Popper a century later in his book *The Logic of Scientific Discovery*.

For any experiment, whether it be confirmatory or contradictory, Bernard emphasizes the necessity of including a "comparative experiment,"[165] what we today would call a "control." In the quotation that follows, Bernard provides an example from his work on the metabolism of sugar. In it, he also notes the necessity for repetition of experiments if their results are not to be doubted.

...I was once led to study the part played by sugar in nutrition and to investigate the mechanism by which this nutritive principle was destroyed in the organism. To solve this problem, I had to hunt for sugar in the blood and follow it into the intestinal vessels which absorbed it, until I could note the place where it disappeared. To carry out my experiment, I gave a dog sweetened milk soup; then I sacrificed the animal during digestion and found that the blood in the superhepatic vessels, which hold all the blood of the intestinal organs and the liver, contained sugar. It was quite natural and, as we say, logical to think that the sugar found in the superhepatic vessels was the same that I had given the animal in his soup. I am certain indeed that more than one experimenter would have stopped at that and would have considered it superfluous, if not ridiculous, to make a comparative experiment. However, I made a comparative experiment [Author's underlining], because I was convinced of its absolute necessity on principle: which means that I am convinced that we must always doubt in physiology, even in cases where doubt seems least allowable. ... So for comparison with the dog fed

---

[164] Bernard discusses these types of experiments on page 55 of *An Introduction*. "...*experimenters, who see their ideas confirmed by an experiment, should still doubt and require a counterproof. Indeed, proof that a given condition always precedes or accompanies a phenomenon does not warrant concluding with certainty that a given condition is the immediate cause of that phenomenon. It must still be established that, when this condition is removed, the phenomenon will no longer appear. If we limited ourselves to the proof of presence alone, we might fall into error at any moment and believe in relations of cause and effect where there was nothing but simple coincidence. As we shall later see, coincidences form one of the most dangerous stumbling blocks encountered by experimental scientists in complex sciences like biology. It is the* post hoc, ergo propter hoc *of the doctors, into which we may very easily let ourselves be led, especially if the result of an experiment or an observation supports a preconceived idea.*"

[165] According to Schaffner [252], p. 148, the method of comparative experimentation is identical to Mill's method of difference. However, Bernard clearly distinguished the two on the basis that Mill's methods and those of his predecessors were derived from the physical sciences, which were far less complex than the biological sciences. Thus, Bernard wrote on page 127 of *An Introduction*: "*Comparative experimentation, however, is not exactly what philosophers call the method of differences. When an experimenter is confronted with complex phenomena due to the combined properties of various bodies, he proceeds by differentiation, that is to say, he separates each of these bodies, one by one in succession, and sees by the difference what part of the total phenomenon belongs to each of them. But this method of exploration implies two things, first of all, that we know how many bodies are concerned in expressing the whole phenomenon, and then it admits that these bodies do not combine in any such way as to confuse their action in a final harmonious result. In physiology the method of differences is rarely applicable, because we can never flatter ourselves that we know all the bodies and all the conditions combining to express a collection of phenomena, and in numberless cases because various organs of the body may take each other's place in phenomena, that are partly common to them all, and may more or less obscure the results of ablation of a limited part.*"

on sugary soup, I took another dog to which I gave meat to eat, being careful moreover to exclude all sugary or starchy material from its diet then I sacrificed the animal during digestion and examined comparatively the blood in its superhepatic veins. Great was my astonishment at finding that the blood of the animal which had not eaten any also contained sugar.

We therefore see that comparative experiment led me to the discovery that sugar is constantly present in the blood of the superhepatic veins, no matter what the animal's diet may be. You may imagine that I then abandoned all hypotheses about destruction of sugar, to follow this new and unexpected fact. I first excluded all doubt of its existence by repeated experiments [Author's underlining], and I noted that sugar also existed in the blood of fasting animals.[166]

Bernard's implementation of the concept of comparative experimentation is further exemplified by his introduction of what we now refer to as "sham" experiments, which are an integral and obligatory part of rigorous biological studies. I defer to Bernard for the best explanation of what sham experiments are.

If . . . we wish to know the result of . . . ablation of a deep-seated organ which cannot be reached without injuring many neighboring organs, we necessarily risk confusion in the total result between the effects of lesions caused by our operative procedure and the particular effects . . . ablation of the organ whose physiological role we wish to decide. The only way to avoid this mistake is to perform the same operation on a similar animal, but without . . . [ablating] the organ on which we are experimenting. We thus have two animals in which all the experimental conditions are the same, save one, ablation of an organ whose action is thus disengaged and expressed in the difference observed between the two animals.[167]

Bernard's contributions to science and the scientific method are indisputable, but Bernard also has been given credit by some as the first to use blind experiments "to ensure the objectivity of scientific observations."[168] Descartes's admonition against believing anything you've been taught may apply here, as examination of the published record reveals no explicit discussion by Bernard of these types of experiments.[169] What study of the history of blind experiments does reveal is what may be considered the first blind experiments, literally, which were reported in 1784 [255]. These experiments were done to test the hypothesis that an imperceptible fluid, "animal magnetism," could be directed by a physician into a patient's body to cure that person of disease.[170] Patients were blindfolded

---

[166] Taken from Bernard [251], pp. 181–182.

[167] Taken from Bernard [251], p. 128.

[168] Quotation from [253].

[169] The attribution to Bernard of the idea of blind experiments often has been by Daston [254], but Daston never discusses blind experiments in her article. In addition, Daston herself suggests that one can only infer from Bernard's discussion in *A study* of "experimenters" and "observers" (p. 23), and "an uneducated man" (p. 38) that he is discussing what we now term "blind experiments" [Daston; personal communication]. Quoting from the former discussion: "In the experiment, we might also differentiate and separate the man who preconceives and devises an experiment from the man who carries it out or notes its results. In the former, it is the scientific investigator's mind that acts; in the latter, it is the senses that observe and note." Quoting from the latter discussion: "In this respect, indeed, an uneducated man, knowing nothing of theory, would be in a better attitude of mind; theory would not embarrass him and would not prevent him from seeing new facts unperceived by a man preoccupied with an exclusive theory."

[170] Franz Anton Mesmer, a German physician, was the one hypothesizing this effect and that his process, a form of hypnotism, was termed Mesmerism, which was the precursor to the eponymous verb "mesmerize." It is interesting that Benjamin Franklin was the chair of the committee charged with doing the experiments and that other members included Antoine-Laurent Lavoisier (1743–1794), considered the father of modern chemistry, and Joseph Guillotin (1738–1814), whose name later was eponymous with the instrument of execution known as the "guillotine," which, ironically, was used to execute Lavoisier during the French revolution ten years later. The guillotine with which Guillotin's name was associated actually was invented by Dr. Antoine Louis, secretary of the College of Surgeons, and was called the "louisette," after its inventor's name.

to prevent them from knowing whether they were being given animal magnetism or not. In some cases, the patients were mesmerized without their knowledge by situating the mesmerizing apparatus on the other side of a paper barrier. No effects of the treatment were observed when the patients were blindfolded, but when they knew they were being treated, they often felt better. These experiments thus not only were the first example of blinded experiments, but they were also the first example of a placebo effect. The first double-blind (neither patient nor physician was aware of whether the patient was in the control group or the test group), placebo controlled experiment was conducted in 1835 [256].

4.5.5 William Stanley Jevons (1835–1882)

We have seen thus far that the development of the scientific method proceeded through many steps and thousands of years.[171] What has not been encountered are methods for determining confidence in an experimental result, induction, hypothesis, theory, etc. Boyle invoked the legal principle of "preponderance of evidence" as a means to do so and others suggested that the more we observed a result predicted from hypothesis or theory the more confidence we would have in that hypothesis or theory. Grosseteste, commenting on uncertainly and confidence, said "[D]emonstrations in natural science ... were 'probable rather than strictly scientific'."[172] "Probability" in Grosseteste's time was a concept, not a mathematical property. No mechanisms for quantifying confidence existed. This changed dramatically in the eighteenth century, when William Stanley Jevons is suggested to have linked induction and probability.[173]

Jevons was an economist and logician who, through his use of mathematical and statistical methods, revolutionized the field that later (in the twentieth century) would be called "econometrics."[174] In addition, however, Jevons, who had received training in chemistry and mathematics (with Augustus De Morgan) at University College London, applied his logical and mathematical methods to the question of scientific induction. He discussed these methods in his book *Principles of Science*, published in 1874, and in doing so, established methods for the quantitation of confidence. These methods were based on probabilities and Jevons's approach to the question of inductive truth has been termed "probabilification."[175]

Jevons, as an economist, wished to understand dynamical economic systems. These systems, unlike scientific systems, were not amenable to experimentation because of the great

---

[171] Steps included those involving pure logic (Plato), logic and (Aristotle), codification of methodological steps (Ibn al-Haytham), mathematics as a tool for scientific discovery (Grosseteste, Galileo), induction (Magnus, Herschel, Whewell), genius (Whewell), establishing rules of causation (Ockham, Bacon, Mill), imagination (Descartes), reproducibility and experimental rigor (Boyle, Herschel), mathematical abstractions of nature (Newton), hypothetico-deductive method (Hershel, Whewell), unobservables (Newton, Whewell), experimental controls, including blind experiments, sham experiments (Bernard), and falsification (Bernard).

[172] Quotation from Crombie [164] p. 108.

[173] Whether this is, in fact, true has been the subject of debate [257]. Nevertheless, whether Jevons was the first, second, or *n*th to link induction and probability does not change the fact that he was a key player in the endeavor and his work had tremendous impact in economics and science.

[174] Defined by Samuelson *et al.* [258] as "the quantitative analysis of actual economic phenomena based on the concurrent development of theory and observation, related by appropriate methods of inference."

[175] See Laudan [259], p. 234.

many variables (especially the vagaries of human rationality) contributing to the dynamics. It is difficult, and often, impossible for one to perform controlled experiments in such social systems, yet economists still needed to establish models of causation to understand and predict economic behavior. One could conceive of models of economic behavior through induction, but one could never be sure these models were accurate, just as scientists could not be certain that their own models were correct. De Morgan expressed this fundamental problem thusly:

Complete induction is demonstration, and strictly syllogistic in its character . . . But when the number of species or instances contained under a name X is above enumeration, and it is therefore practically impossible to collect and examine all the cases, the final induction, that is, the statement of a universal from its particulars, becomes impossible, except as a *probable* statement . . . [176]

De Morgan's pupil, Jevons, characterized the problem in a similar fashion, but was more blunt in characterizing the probable statement as *"partial knowledge – knowledge mingled with ignorance, producing doubt."*[177] If we only are in possession of partial knowledge or probable statements, how then can we be confident in our understanding and in our predictions of the behaviors of dynamical systems, be they economic or scientific? What does confidence even mean? Jevons's answer, in both cases, was through belief expressed in probabilistic terms. Thus, instead of observing, hypothesizing, deducing the consequences of the hypothesis, and testing whether these consequences are actually produced experimentally, we observe, determine the *frequencies* of various observations, and then determine how the *observed frequencies* compare with the *predicted frequencies*. This method has been termed "inductive-probabilistic,"[178] and Jevons described it as follows.

. . . the theory of probability . . . in its simple deductive employment, . . . enables us to determine from given conditions the probable character of events happening under those conditions. But as deductive reasoning when inversely applied constitutes the process of induction, so the calculation of probabilities may be inversely applied [Author's underlining]; from the known character of certain events we may argue backwards to the probability of a certain law or condition governing those events. Having satisfactorily accomplished this hitherto; we may indeed calculate forwards to the probable character of future events happening under the same conditions; but this part of the process is a direct use of deductive reasoning.[179]

What is this "inverse method of probability?" It is a special case of Bayes's theorem,[180] which has traditionally been used for the calculation of "inverse probabilities," i.e., the

[176] From DeMorgan [260], p. 211.

[177] Taken from Jevons [261], p. 197.

[178] See Schabas [262], p. 71.

[179] Taken from Jevons [261], p. 240. Note that Herschel earlier had said, in essence, induction ". . . belongs to the inverse or *deductive* process, by which we pursue laws into their remote consequences." He also emphasized, as had many, ". . . that the successful process of *scientific enquiry demands continually the alternate use of both the inductive and deductive method* [Author's italics]."

[180] Bayes's theorem is $P(I|E) = \frac{P(E|I) \times P(I)}{P(E)}$, where $P(I|E)$ is "posterior probability," the probability of an induction I based on an observed event E; $P(E|I)$ is the probability of the event E given I; $P(I)$, "prior probability," is the degree of belief in I independent of E; and $P(E)$ is the degree of belief of observing E independent of I. It has been said that Bayes's theorem "is to the theory of probability what Pythagoras's theorem is to geometry." (Quotation from Jeffreys [263], p. 31.)

probabilities of causes given effects. Jevons estimates the probability of an induction based on the probabilities of the events upon which the induction was based.

Jevons's ideas were met with skepticism and resistance[181] by those who felt the logic and tools of induction extant were essentially infallible with respect to determining the truths of natural systems [264]. Why then would another approach, with its own inherent uncertainties, be of value? George Boole rejected "the principles of the theory of probabilities ... [as a] guide ... [to] the election of hypotheses," although with some diffidence.[182] John Stuart Mill was less charitable. In a letter dated December 5, 1871 [266], Mill expresses a sentiment that illustrates the views of a those for which the probabilification of induction was unnecessary and counterproductive, as well as a challenge to cherished beliefs. Such resistance was, and is common, in science practice. The reader should always be aware of such tendencies in themselves ... and resist them.

[Jevons] is a man of some ability, but he seems to me to have a mania for encumbering questions with useless complications, and with a notation implying the existence of greater precision in the data than the questions admit of. His speculations on Logic ... are infected in an extraordinary degree with this vice. It is one preeminently at variance with the wants of the time, which demand that scientific deductions should be made as simple and as easily intelligible as they can be made without ceasing to be scientific.

Mill, of course, had sought to provide, in his "Five Canons of Induction," a strict logical procedure for determining truth. Unfortunately, though of great use in simple systems, his approach fails, and fails miserably, in complex systems such as found in biology and medicine. It fails precisely for the reasons enunciated by DeMorgan and Jevons, namely that we can't know everything, especially all the variables in a system that may contribute to varying degrees to cause-and-effect relationships and upon which, in part through the inductive process, laws of nature may be based. Deductions, if they are to be made, require their premises be true and thus Mill's reference to scientific deduction misrepresented Jevons's aims. Jevons was pursuing inductive proof and he did so using the "inductive-probabilistic method," which assumes, for any two hypotheses H and H', $P(E/H\&H') < 1$. Mill instead sought to infer truth through deduction, for which $P(E/H\&H') = 1$.[183]

The argument between Mill and Jevons is a good example of the much larger question of Truth vs. truth or absolute vs. relative. Mill argues for the former while Jevons argues for the latter. Time, space, and matter are *not* absolute, and neither is their study using the scientific method. It is critical that the scientist appreciates this fact and incorporates it into their scientific practice. Jevons, in a sense, finally surrenders unconditionally to uncertainty, saying:

---

[181] An exception was the American philosopher and logician Charles Sanders Peirce.

[182] See Boole [265], chapter 5, p. 291.

[183] Symbols in probability and logic are found throughout this book. Table 4.3 provides a list of common symbols and their meanings. A complete lists of symbols may be found at www.stat.berkeley.edu/~stark/SticiGui/Text/gloss.htm. For an informative discussion of the application of Bayesian methods in inference (e.g., abduction), see Niiniluoto [267], p. S443.

Table 4.3 *Statistical and logical symbols.*

Symbol	Meaning
& or ∧	And
∨	Or
∀	For all
¬ or ∼	Negation of (not)
→ or ⟹	Implies (but not causes)
|	Given
∴	Therefore
P	A specific proposition (one and only one statement)
p	A general proposition (any statement within the context of P)
E	Evidence (data)
H	A hypothesis *H*
P(E)	Probability of evidence *E* prior to consideration of *H*
P(H)	Probability ("prior") of hypothesis *H* prior to consideration of *E*
P(E|H)	Probability of *E* given *H*
P(H|E)	Probability ("posterior") of *H* given *E*

…my strong conviction [is] that before a rigorous logical scrutiny the Reign of Law[184] will prove to be an unverified hypothesis, the Uniformity of Nature an ambiguous expression, the certainty of our scientific inferences to a great extent a delusion.[185]

Nevertheless, and as we shall encounter again in the succeeding pages, all is not lost or futile. We must give up on certainty in both its theoretical and practical aspects, but that does not mean we succumb to nihilism. Quite the contrary, we operate in an uncertain system, the natural world, accepting its probabilistic nature and at the same time being confident, backed by science's extraordinarily successful history of accomplishment, that our work and the knowledge it provides (such as they are) are of value.

4.6 Probabilification and Certainty in the Early Twentieth Century

Jevons's use of frequentist and Bayesian approaches to inductive truth remain a key part of the scientific method.[186] However, although the employment of the methods of probability may be straightforward and the conclusions derived from them may be considered meaningful, relevant, or even true in practice, deeper examination reveals complexities that are directly relevant to the scientific method and its epistemologic foundations. When we speak of the inductive-probabilistic method, we consider probability in the metrical sense.

---

[184] Meaning natural law.

[185] From Jevons [261], p. xi.

[186] Frequentists use simple ratios of specific events observed vs. all events observed to calculate probabilities. An infinite number of observations which is impossible, is required to define an absolute probability. Bayesians, in contrast, measure "degree of belief" as opposed to probability *per se*. Bayes's theorem requires a *posterior probability*, which cannot be determined absolutely but must be approximated through incorporation of both statistical and nonstatistical information.

For example, we can quantify the probability of error in a measurement or experiment using *p*-values, which are numeric. However, to do so, we must satisfy several criteria, including sufficient replication, instrumental precision, and low intrinsic system variance. What if these requirements cannot be satisfied? What can we do if we seek to evaluate nascent hypotheses and theories for which data and conceptual knowledge extant may be limited, a situation that is the norm in science as opposed to the exception? In these cases, we may only have recourse to our beliefs, which are expressed descriptively – "not confident," "little confidence," "on the fence," "very confident," "positive," etc. Do all scientists accord the same relative amount of belief to each of these terms? Clearly not, because descriptive terms are not metrical. It thus seems that we again require numbers if we are to appreciate the significance of our work, make choices about competing research ideas, or just understand each other.

4.6.1 Rudolf Carnap (1891–1970)

One of the most important modern philosophers of science to tackle such thorny problems of probability and inductive logic was Rudolph Carnap. Carnap was one of many prominent German philosophers and scientists, including Hans Reichenbach, Karl Popper, and Carl Hempel, who left Germany after Hitler came to power in 1933. Carnap was concerned with defining the *concept* of probability as a basis for a *theory* of probability. He sought to redefine prior concepts of probability – to replace them with "more exact" new concepts – and argued that this was important for "the development of science and mathematics." There existed many terms for "probability" at the turn of the twentieth century. Carnap discusses these, and the need for semantic clarity, in the quotation below.

... we find phrases as different as "degree of belief," "degree of reasonable expectation," "degree of possibility," "degree of proximity to certainty," "degree of partial truth," "relative frequency," and many others. This multiplicity of phrases shows that any assumption of a unique explicandum[187] common to all authors is untenable. And we might even be tempted to go to the opposite extreme and to conclude that the authors are dealing not with one but with a dozen or more different concepts. However, I believe that this multiplicity is misleading. It seems to me that the number of explicanda in all the various theories of probability is neither just one nor about a dozen, but in all essential respects ... chiefly two ...: (i) probability$_1$ = degree of confirmation; (ii) probability$_2$ = relative frequency in the long run.[188]

Probability$_1$ is a logical construct that has to do with the relationship between two sentences or propositions. It addresses the question of whether, for example, conclusions based on empirical evidence *logically* follow from that evidence – whether "the data support the conclusion(s)." Note, however, this type of probability has nothing to do with truths of the real world *per se*. It corresponds to our sense of credibility, i.e., it tells us whether the

---

[187] The Latin term *explicandum* refers to "that which is to be explicated," as opposed to *explicatum*, which is "the explanation itself."
[188] Taken from Carnap [268], p. 517.

way in which we have analyzed our data makes sense from a logical perspective. Incorrect application of logic may yield an inaccurate value for *probability*$_1$, but because this value depends only on semantics and not experimentation, a value will always be obtainable. *Probability*$_2$, in contrast, is a synthetic (empirical) construct that invokes absolute frequency $f = \lim_{i \to \infty} \frac{e_i}{i}$, where i is the total number of independent observations and e_i is the total number of observations in which a particular event is observed.

4.6.2 Hans Reichenbach (1891–1953)

Hans Reichenbach was born in Hamburg, Germany, in 1891. He studied physics, philosophy and mathematics with some of the most influential scientists, physicists, mathematicians, and philosophers in history, including Max Planck, Erwin Schrödinger, Albert Einstein, David Hilbert, Moritz Schlick, and Rudolf Carnap. He has been described as "the greatest empiricist of the twentieth century" [269]. Among his many contributions to the philosophy and epistemology of science was his work on induction and the foundations of probability. For Reichenbach, induction was not a generic inferential process characterized by the movement from the specific to the general. Reichenbach considered this "vague" and proposed what he referred to as a more precise formulation.[189]

The aim of induction is to find series of events whose frequency of occurrence converges toward a limit . . .

Putnam summarized this approach nicely when he wrote:

. . . the world consists of events. Those events form various sequences, and there are statistical relations among the events in those sequences. Those statistical relations indicate the presence of causal relations, and the detailed nature of the statistical relations fixes the nature of the causal relations.[190]

In some sense, induction was no longer a logical process involving statistical tools, but rather a probabilistic process on its own. Induction thus had been converted fully from the verbal to the mathematical. It is important to note that Reichenbach, like Carnap, recognized the logical (*probability*$_1$) and mathematical (*probability*$_2$) concepts of probability, but argued that instead of considering these to be disparate concepts they should be considered identical in terms of their structures. Reichenbach termed this the "identity conception" [270].

. . . we shall use . . . the term "identity conception" without always mentioning that there is, strictly speaking, a difference of logical levels involved. We use the word "identity" here in the sense of an identity of structure, and our thesis amounts to <u>maintaining the applicability of the frequency interpretation to all concepts of probability</u> [Reichenbach's underlining].

How did Reichenbach apply the frequency interpretation in induction? He did so through a process he called "concatenation of inductive inferences," to which he credited *"the overwhelming success of scientific method."*[191] An example of the use of this method is shown

---

[189] From Reichenbach [270], p. 350.
[190] From Putnam [271], pp. 68–69.
[191] Taken from Reichenbach [270], p. 364.

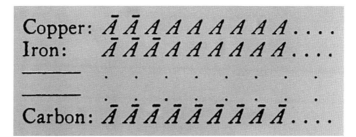

Figure 4.6 Probability lattice for the melting of metal. The column at left lists metals tested. Each series of "A"s represent temperatures, increasing from left to right. Symbols denote whether at each temperature the metal was a solid ($\bar{A}$) or liquid (A).

in Fig. 4.6.[192] The inductive question at hand is whether all substances will melt. At the time these experiments were done, experiments had revealed that copper, iron, and a number of other metals melt if heated to a sufficient temperature. However, within the temperature ranges studied, no melting of carbon had been observed. Based on these observations, we induce that carbon should melt at a high enough temperature. How confident can we be in our induction?

If we induce only based on the horizontal temperature series for carbon, we would predict it should not melt. However, we also could perform a higher level induction by considering as a series the data from all the series of substances tested. We do so by calculating the relative frequency of substances observed to melt, $f = (n_i)/n_j \approx 1$, where n_i is the number of substances that we observe melting and n_j is the total number of substances tested. We then would predict that carbon *would* melt, and indeed it does at a high enough temperature ($\geq 3{,}550°C$), except that, depending on pressure, it may sublime[193] instead [272]. The concatenation of inductive inferences, in this case by linking horizontal and vertical inductive series, allows us to develop inferences not immediately obvious through simpler inductive processes.

An important aspect of the probabilification of induction is, to a large degree, it solves the problem of induction raised by David Hume. Hume argued, correctly, that we could never be sure of the veracity of our inductions because we could never be sure that contradictory observations might be made in the future. The "sure" to which Hume refers means a probability of 1. For Reichenbach, this is nothing more than a special case in his general formulation of induction. We may never encounter *probability*$_2$ = 1 in our scientific work, yet we must still have the means to make decisions about what to believe, what projects to do, how to develop and choose hypotheses, and how to advise political and social bodies on issues that are intrinsically scientific (e.g., global warming, immunization,

---

[192] Reprinted from Reichenbach [270], p. 365.
[193] Sublimation is the process through which a solid, when heated, undergoes a state transition directly from solid to gas without going through a liquid phase.

Table 4.4 *The logical justification of induction.*

Either nature is uniform or it is not.
If nature is uniform, then scientific induction will be successful.
If nature is not uniform, then no method will be successful.
∴ If any method of induction will be successful, then scientific induction will be successful.

genetic engineering). Probabilistic approaches to induction allow us to actually do things in practice.[194]

What we usually call "foreseeing the future" is included in our formulation as a special case; the case of knowing with certainty for every event *A* the event *B* following it would correspond in our formulation to a case where the limit of the frequency is of the numerical value 1. Hume thought of this case only [Author's underlining]. Thus our inquiry differs from that of Hume in so far as it conceives the aim of induction in a generalized form. But we do not omit any possible applications if we determine the principle of induction as the means of obtaining the limit of a frequency. If we have limits of frequency, we have all we want, including the case considered by Hume; we have then the laws of nature in their most general form, including both statistical and so-called causal laws the latter being nothing but a special case of statistical laws, corresponding to the numerical value 1 of the limit of the frequency.[195]

It is vital to understand that the epistemic value of induction depends on nature exhibiting regularity. Hume presented this "uniformity principle" in *A Treatise of Human Nature*[196] in which he discussed justification for establishing causes and effects based on past experience and reason.

If reason determin'd us, it wou'd proceed upon that principle, that instances, of which we have had no experience, must resemble those, of which we have had experience, and that the course of nature continues always uniformly the same.

This principle underlies Reichenbach's logical justification of induction (Table 4.4).[197]

4.6.3 Carl Hempel (1905–1997)

Carl Gustav ("Peter") Hempel was another émigrée from Nazi Germany who made profound contributions in numerous areas, including methods of induction, scientific

---

[194] Reichenbach provides an informative and amusing example of the practical importance of investing confidence in inductive probabilities. "In any action there are various means to the realization of our aim; we have to make a choice, and we decide in accordance with the inductive principle. Although there is no means which will produce with certainty the desired effect, we do not leave the choice to chance but prefer the means indicated by the principle of induction. If we sit at the wheel of a car and want to turn the car to the right, why do we turn the wheel to the right? There is no certainty that the car will follow the wheel; there are indeed cars which do not always so behave. Such cases are fortunately exceptions. But if we should not regard the inductive prescription and consider the effect of a turn of the wheel as entirely unknown to us, we might turn it to the left as well." [Taken from Reichenbach [270], pp. 346–347.]

[195] Taken from Reichenbach [270], p. 350.

[196] See Hume [168], Section 1.3.6.4.

[197] Taken from Skyrms [273], pp. 46–47.

Table 4.5 *The deductive-nomological method. An explanation can be derived from a given set of circumstances (C) through deduction using a given set of laws (L). If the circumstance are true, the explanation also is true because it is produced solely through logical means.*

$C_1, C_2, \ldots, C_k$	Statements of antecedent conditions
$L_1, L_2, \ldots, L_k$	General laws
$\therefore$ E	Description of the empirical phenomenon to be explained

explanation, logic of confirmation[198] (e.g., of scientific theories), and empiricist philosophy. Hempel studied physics and mathematics at the Universities of Göttingen and Heidelberg. He was mentored by some of the greatest twentie-century philosophers of science, including Hans Reichenbach and Rudolf Carnap. After emigrating, Hempel was a faculty member at a number of eminent American universities, including the University of Chicago, Yale, and Princeton (where Thomas Kuhn was a colleague).

Hempel argues "The concept of logical probability, or degree of confirmation, is the central concept of inductive logic."[199] Hempel's concern was the establishment of logical rules of knowledge. These rules were not epistemological *per se*, i.e., they did not deal with *how* we know what we know, but rather *why* we think we know what we know. Hempel proposed two methods to address the latter issue [275], the "deductive-nomological" ("D-N"; also known as the "covering law model") and the "inductive-statistical" ("I-S").

The D-N method integrates two fundamental concepts, deduction and general laws of nature [275], to achieve an understanding of the causes of a phenomenon (Table 4.5). If the premises of the method are true, its conclusions are as well. Probabilistically, we are dealing with likelihoods of 0 or 1, only. Hempel illustrated the D-N method using the statement below [276]. It the statement is true, observing a substance that yields a yellow flame in a Bunsen burner means the substance is, in fact, sodium. This concept is schematized in Table 4.5 and shown as a syllogism in Table 4.6.

This crystal of rock salt, when put into a Bunsen flame, turns the flame yellow, for it is a sodium salt, and all sodium salts impart a yellow color to a Bunsen flame.

One problem with the D-N method is the fact that it is difficult, and sometimes impossible, to know if the initial conditions are all true. If not, the D-N method cannot provide us the

---

[198] One of Hempel's most amusing, but important, contributions in this area was "Hempel's paradox," commonly referred to as the "Raven paradox" [274]. Hempel starts with the hypothesis "all ravens are black." This proposition is logically equivalent to "all non-black things are non-ravens." Now, does this logic allow us to select instances that support the hypothesis? If we see a black raven, then we have support. However, logically, if we see a white shoe, this also lends support to the hypothesis. How can this be possible? This would mean if we see a bright yellow sun, we again would have support for the hypothesis. This is the paradox. Things we know are not relevant to the hypothesis nevertheless provide evidence supporting it. Hume's argument that we can never be sure of the correctness of our inductions operates here. If we examined a series of 1000 ravens, all of which were black, this certainly would support the hypothesis. But how do we know the *next* raven will be black? We don't. Yet if the 1001st raven had albinism (and thus was nonblack), it still would be a raven yet that fact would be inconsistent with the logical equivalent of our hypothesis, namely all nonblack things are non-ravens.

[199] See Hempel [275], p. 168.

Table 4.6 *The D-N method.*

All sodium salts produce yellow flames in Bunsen burners
This substance is a sodium salt
This salt will produce a yellow flame in Bunsen burners

Table 4.7 *The inductive-statistical method. Here a probabilistic "covering law," p(H/E) = r, coupled with evidence Ea comprising a set of antecedent facts a, predicts the hypothesis Ha with an inductive probability [r]. The double lines between the explanans and the explanandum indicate that our reasoning is not deductive in nature.*

$p(H/E) = r$	(r is close to 1)
Ea	
	$[r]$
Ha	

Table 4.8 *The I-S method.*

Almost all streptococcal infections are quickly cured with penicillin
John was given penicillin for his infection
John is almost certain to be cured

0 or 1 we desire. We instead would turn to the I-S method, which replaces certainties with uncertainties but does tell us the *likelihood* our induction is true (Table 4.7). If we have induced hypothesis (H) based on some prior evidence (E), and this evidence, in the vast majority of cases, supports our hypothesis ($p(H|E) \approx 1$), then when we encounter another instance of E_a, it is highly likely it is explained by H (H_a). Hempel provides the following example [276]. Here absolute certainty is replaced by "almost certain," which is appropriate, especially in medical practice, because the likelihood is close to 1. The syllogism is shown in Table 4.8.

John Jones was almost certain to recover quickly from his streptococcus infection, for he was given penicillin, and almost all cases of streptococcus infection clear up quickly upon administration of penicillin.

Hempel summarizes the differences between the D-N and the I-S methods as follows.

[The explanans of the I-S method] may confer upon the explanandum a more or less high degree of inductive support; in this sense, probabilistic explanation admits of degrees, whereas deductive-nomological explanation appears as an either-or affair: a given set of universal laws and particular statements either does or does not imply a given explanandum statement.[200]

The history of the scientific method up to Hempel's time can been characterized by its search for explanations of natural phenomena. We observe, we wonder, we hypothesize, and we do experiments to confirm our hypotheses. We analyze these experiments using the rules of logic and we accept or modify our hypotheses accordingly. Absolute certainty had been sought, and argued, using mystical, religious, metaphysical, and dialectical methods. Modern science, the science of the nineteenth and twentieth centuries, replaced absolutism with probabilism. Certainty no longer was simply binary, i.e., true or untrue, but existed in degrees established using statistical and probabilistic methods. In all cases, the goal was to prove or confirm our ideas – to establish truth. This positive ideal of science was turned upside down by Sir Karl Popper.

4.6.4 Karl Popper and the Vienna Circle

Karl Raimund Popper (1902–1994) was born in Vienna. He was trained at the University of Vienna in mathematics and theoretical physics and also was involved in social work (working briefly in a clinic for deprived children run by the psychiatrist Alfred Adler), politics (Socialism and Marxism), and music (in association with Arnold Schönberg). After a brief hiatus as a secondary school teacher of mathematics and physics, Popper returned to the university, where he received his Ph.D. in the Department of Psychology in 1928. The subject of his dissertation, "The question of method in cognitive psychology,"[201] presaged Popper's enduring interest in what Thorton [278] has characterized as "questions of method, objectivity and claims to scientific status." Popper fled to New Zealand before the German annexation of Austria in 1938 and remained there until moving to the London School of Economics in 1946, where he remained as Professor of Logic and Scientific Method until his retirement in 1969. Popper was knighted in 1965.

Popper was fortunate to have come of academic age during the heyday of the Vienna Circle (ca. 1923–1936), a discussion group comprising some of the most important and prominent philosophers, sociologist, mathematicians, and scientists of the twentieth century[202] and one with which he was able to interact. The Vienna Circle was famous for its promulgation of a philosophy termed "logical positivism" [280]. A similar group in Berlin, the Society for Empirical Philosophy (Ger., *Gesellschaft für empirische Philosophie*, also known as the "Berlin Circle"),[203] largely shared this philosophy, but Reichenbach preferred to refer to

---

[200] Quotation from Hempel [277], p. 51.

[201] Ger., Die Methodenfrage der Denkpsychologie.

[202] Members included Moritz Schlick (Chair, Philosophy and Inductive Sciences, University of Vienna and organizer), Rudolph Carnap, Herbert Feigl, Philipp Frank, Kurt Gödel, Hans Hahn (who Frank called "the founder of the Vienna Circle" [279]), and Otto Neurath (who coined the name "Vienna Circle" [279]). The Vienna Circle also had many occasional participants, including Paul Feyerabend, Carl Hempel, Hans Reichenbach, Alfred Tarski, Ernest Nagel, and Willard Van Orman Quine.

[203] The Berlin Circle was formed in 1928 by Hans Reichenbach and included his student, Carl Hempel, mathematician David Hilbert, and scientist Richard von Mises.

it as "logical empiricism."[204] Members of the two groups interacted extensively to promote their shared interests in scientific methodology, the integration of the methods of science (especially logic and mathematics) into philosophy proper (analytic philosophy), and the role their new philosophy could play in society. In fact, in 1930, Reichenbach and Carnap became co-editors of the journal *Annalen der Philosphie*, changed its name to *Erkenntnis* (Ger., "knowledge"), and made it *"the main mouthpiece"*[205] of the Vienna Circle.

One of the key premises of this volume is the necessity of integrating science practice with the philosophy of science. Scientists satisfy their curiosity and wonder about the natural world by seeking knowledge as to how the world works. They seek this knowledge by implementing the scientific method and believe by doing so that what they reveal will be true and real, i.e., display a one-to-one correspondence between nature *per se* and their understanding of it. Why do they believe this? Why *should* they believe this? What rules exist to justify their claims or guide them to the truth? In our examination thus far of the development of the scientific method we have seen that the bulk of scientists at any one time are guided by "conventional wisdom," which often is wrong. The best example of this might be a geocentric cosmos, the conventional wisdom for many centuries ... until Copernicus, *assuming nothing and considering only the empirical evidence before him*, postulated a heliocentric cosmos. A more modern example would be the shock of biologists when, in the late twentieth century, their beloved unidirectional "Dogma of Molecular Biology," i.e., DNA→RNA→protein (DNA yields RNA yields proteins), was shattered by the discovery of prions, infectious proteinaceous agents that self-replicate in the absence of nucleic acid (DNA or RNA) blueprints. No one wanted to believe it, but in the ensuing 30 years, prions have been found almost anywhere one looks and, importantly, they play important roles in *normal* organismal physiology. In each case, if scientists had, and had followed, a set of knowledge principles, science would have advanced, and society would have benefited, much more quickly. It is from the philosophy of science that such principles come and it was due in large part to the pioneering work of Carnap, Reichenbach, and Hempel on logical empiricism that the subdiscipline of philosophy of science emerged [283, 284].

I spoke above of the importance of integrating science and philosophy, and in fact, this integration was the goal of the Vienna Circle. However, whereas I argue that to "do science" well requires the scientist to be conversant in the philosophy of science, the Vienna Circle argued the converse, that to "do philosophy" well requires the philosopher to be conversant in the methods of science. In reality, the implementation of these goals achieves the same thing, which, in essence, is the integration of practice with theory – the *how* with the *why*. This was the main subject of discussion of a group of intellectuals (the embryonic Vienna Circle) who began meeting in 1907 on Thursday nights at "one of the

---

[204] Logical positivism has also been referred to as logical empiricism, logical neopositivism, or neopositivism. An excellent discussion of these different appellations may be found in Uebel [281]. However, note that "by the middle of the 1930s, 'logical empiricism' was the preferred term for leading representatives of both camps" (Taken from Richardson [282], footnote 1). For convenience, I will subsume all such terms under "logical empiricism," but this does not mean that each term is strictly synonymous.

[205] Quotation from Frank [279], p. 41.

old Viennese coffee houses"[206] to discuss "problems of science and philosophy."[207] The group, comprising Philipp Frank (physicist), Hans Hahn (economist), and Otto Neurath (mathematician), most often addressed the central problem how to "... avoid the traditional ambiguity and obscurity of philosophy?" The answer was to integrate the methods of the natural sciences into philosophy. Philosophy would become scientific. It would bring the empiricism of science together with formal logic to enable a clearer and more accurate understanding of the world and to provide "a brighter, less obscure and obscurantist future for philosophy."[208]

Herbert Feigl, one of the founders of logical positivism, and the American Albert Blumberg, sought to explain this philosophy in 1931 in a paper entitled, "Logical Positivism – A New Movement in European Philosophy" [280]. They opined that logical positivism was:

... distinguished from [earlier philosophies] ... by its results and by the fact that it embodies not the work of an individual, but the agreement of numerous logicians, philosophers, and scientists independently arrived at.[209] This is particularly encouraging in a field like philosophy in which anything approaching a general unanimity has seemed hopelessly unattainable.[210] The essence of this new development is its radically novel interpretation of the nature, scope, and purpose of philosophy – an interpretation gradually achieved through extensive inquiries into the foundations of logic, mathematics, and physics. ... it is precisely the union of empiricism with a sound theory of logic which differentiates logical positivism from the older positivism, empiricism, and pragmatism. ... Logical positivism has thus grown up in close contact with investigations into the foundations of the sciences.

Logical positivism presented a revolutionary revision of German philosophy, especially the philosophy of Kant [285] and his *analytic/synthetic* and *a priori/a posteriori* distinctions

---

[206] Quotations in this paragraph are from Frank [279], p. 1.

[207] It is important to understand the context of this focus. At the turn of the twentieth century, science was in crisis. According to Frank (See [279], p. 2), increasingly it was thought that *"... the scientific method itself had failed to give us 'truth about the universe;' hence nonscientific and even antiscientific tendencies gained momentum."* Frank was influenced greatly by the French historian and philosophers Abel Rey, who bemoaned the state of physics in 1907, saying, *"Traditional physics assumed until the middle of the nineteenth century that it had only to continue its own path to become the metaphysics of matter. It ascribed to its theories an ontologic value, and these theories were all mechanistic. Traditional mechanistic physics was supposed, above and beyond the results of experience, to be the real cognition of the material universe. This conception was not a hypothetical description of our experience; it was a dogma.*

The criticism of the traditional mechanistic physics that was formulated in the second half of the nineteenth century weakened this assertion of the ontologic reality of mechanistic physics. Upon this criticism a philosophy of physics was established that became almost traditional toward the end of the nineteenth century. Science became nothing but a symbolic pattern, a frame of reference. Moreover, since this frame of reference varied according to the school of thought, it was soon discovered that actually nothing was referred that had not previously been fashioned in such a way that it could be so referred. Science became a work of art to the lover of pure science, a product of artisanship to the utilitarian. This attitude could quite rightly be interpreted as denying the possibility that science can exist. A science that has become simply a useful technique ... no longer has the right to call itself science without distorting the meaning of the word. To say that science cannot be anything but this means to negate science in the proper sense of the word. The failure of the traditional mechanistic science ... entails the proposition: 'Science itself has failed.' ... We can have a collection of empirical recipes, we can even systematize them for the convenience of memorizing them, but we have no cognition of the phenomena to which this system or these recipes are applied."

[208] Quotation from Richardson [282], p. 4.

[209] Although this appears to be another example of conventional wisdom, it is not. Conventional wisdom in the modern sense generally presents a single view or opinion on a particular subject. There are rarely such things as "conventional wisdoms." The Vienna Circle, in contrast, produced what might be called an "instantial consensus" that, although amenable to publication in a fixed form, was actually temporally dynamic and constantly changing with the evolving thoughts and opinions not only of its authors, the "Ernst Mach Society," but of the entire community of philosophers, sociologists, mathematicians, physicists, logicians, *et al.* This dynamism is reflected in Philipp Frank's reminiscence, *"In 1929, we had the feeling that from the cooperation that was centered in Vienna a definite new type of philosophy had emerged. As every father likes to show photographs of his baby, we were looking for means of communication. We wanted to present our brainchild to the world at large, to find out its reaction and to receive new stimulation."* [Quotation from Frank [279], p. 38.]

[210] Indeed!

of propositions (logical sentences).[211] We shall not delve into these fundamental philosophical issues here but instead focus on the practical impact of logical empiricism on science practice. Logical positivism affected a number of the most important advances in physics and mathematics of the twentieth century, for example, developments in mathematics that led to Gödel's incompleteness theorems [283]. Most importantly, logical positivism addressed the fundamental epistemological question of defining knowledge. If scientists are to acquire knowledge about the natural world, it is incumbent upon them to understand what knowledge actually is.

Philosophy of science helps us to provide meaningful answers. It guides us in clear, logical thinking and provides the means to justify and validate our conclusions. In short, it provides a logico-theoretical framework for doing science properly. Wittgenstein provided a relevant definition of philosophy, one shared by the Vienna Circle [280], in his book *Tractatus Logico-Philosophicus* [20]. If we append the term "philosophy" with "of science" in the quotation below, we see how centrally important logical positivism and its offspring, philosophy of science, are to the conduct of science.

The object of philosophy is the logical clarification of thoughts. Philosophy is not a theory but an activity. A philosophical work consists essentially of elucidations. The result of philosophy is not a number of 'philosophical propositions', but to make propositions clear.[212]

It is interesting to note here an important distinction between the logically correct and the practically useful. Logical positivism considered metaphysical concepts such as Ockham's Razor (see Section 4.3.5) meaningless because "a proposition has meaning only when we know under what conditions it is true."[213] Ockham's Razor is metaphysical concept and thus empirically untestable. This emphasizes the movement's quest for an understanding and definition of knowledge *per se* versus the practicing scientist's need for heuristic means to enable discoveries and the formulation of hypotheses. In contrast, the metaphysical concept of induction *was* meaningful because, in theory, its predictions were empirically verifiable.

Verifiability is a central tenet of logical empiricism because it considers the meaning of propositions to be identical with the conditions of their verification.[214] Conditions, in this case, are empirical in nature, not logical or dialectical. Verifiability in the form of empirical proof is also a central tenet of science. It distinguishes science from pseudoscience, mysticism, or religion. We do experiments to test hypotheses and if the experiments support them, we consider them true (with a lowercase "t"). We may qualify truth probabilistically, but science historically has sought an affirmative answer to the question of truth. Although

---

[211] These important concepts in philosophy relate to knowledge, in particular to how to distinguish knowledge that is absolute and independent of experience, i.e., analytic or *a priori* (e.g., mathematics, semantic logic) from knowledge that requires prior experience, i.e., synthetic or *a posteriori* (e.g., science). Examples: "Green is a color" is *a priori*, "Grass is green" is *a posteriori*, "Triangles have three sides" is analytic, "The triangle is red" is synthetic. Blumberg and Feigl [280] summarized the views of the Vienna Circle thusly: "Against Kant the new movement maintains as a fundamental thesis that there are no synthetic *a priori* propositions. Basing its assertions upon recent developments in factual and formal sciences, it holds that factual (empirical) propositions though synthetic are *a posteriori*, and that logical and mathematical propositions though *a priori* are analytic."

[212] See Wittgenstein [20], 4.112.

[213] See Blumberg [280], p. 293.

[214] See Blumberg [280], p. 293.

Hume argues we can never be sure of our inductions, science cannot function *in practice* unless the scientific community accepts inductive truth, and the way to reach consensus is the continued demonstration of the predictive correctness of hypotheses. An excellent example of this is the repeated confirmation of the prediction of Einstein's general theory of relativity that gravitation bends light, first by Dyson *et al.* [138] almost a century ago, and most recently in 2017 by Sahu *et al.* [286].

Karl Popper vehemently opposed logical positivism[215] and its ideal of verifiability, arguing instead in his book, *Logik Der Forschung: Zur Erkenntnistheorie Der Modernen Naturwissenschaft* [128],[216] that the only logically sound epistemological option open to scientists is the ability to disprove, i.e., to *falsify*, their hypotheses. Verifiability is impossible and induction itself not only *"lacks the sanction of logic, but is actually in conflict with it."*[217] Einstein's theory was corroborated by Dyson *et al.* and Sahu *et al.*, but they did not *prove* it. In fact, if one or both had shown that Einstein's theories were not corroborated, then, according to Popper, the theories would be proven wrong, assuming the experiments testing the theories had been conducted properly.

Popper, in contrast to Carnap, Reichenbach, and most others, also categorically rejected probabilistic approaches to the establishment of scientific certainty. Rather than trying to prove something, which Hume had established was impossible [168], Popper argued scientists should try to falsify their hypotheses [6]. Inductive truth is achieved by proving every possible prediction of a theory is shown to be true, i.e., the frequency of occurrence of predicted events $f = 1$, a goal that clearly is never achievable because the set of all affirmative experiments is infinite. The best we can hope for is that $f \approx 1$, which means we are "pretty sure." The logical negation of $f = 1$, $\neg f$, the frequency of nonoccurrence of the prediction, is zero. Therefore, if even a *single* prediction is found to be false, then a hypothesis is disproven absolutely – it is falsified – because our approach is deductive not probabilistic. Popper had, indeed, turned the positive ideal of science upside down. Science, through the scientific method, now was about proving itself wrong! Thus, Popper opines:

It became clear to me that what made a theory, or a statement, scientific was its power to rule out, or exclude, the occurrence of some possible events – to prohibit, or forbid, the occurrence of these events. Thus the more a theory forbids, the more it tells us.[218]

Although Popper rejected the formal probabilism of the logical empiricists, he did argue that scientific theories could be "corroborated." However, corroboration was not proof of concept, nor was it related to the probability that the theory was true. It was the *failure to be falsified*. The greater the number, and the more demanding the tests of contradictory predictions of theories were, the greater corroboration the theories would enjoy, assuming these tests failed. Corroboration came from the failures of tests of failure. In this vein, scientists now often ask themselves or their students to think of the best experiment possible,

---

[215] He was nicknamed "the Official Opposition" by Otto Neurath.

[216] Eng., "Logic of Research: The Epistemology of Modern Science." Popper translated and republished the book in English in 1959 as *The Logic of Scientific Discovery* [6].

[217] Taken from Weinberg [287], p. 511.

[218] Taken from [288], p. 42.

the crucial experiment (Lat.: *experimentum crucis*;[219] literally, "experiment of the cross"), that would disprove their theories.

Conclusions arrived at through deduction in a semantic scientific world are absolute because they deal with the formal logic of propositions. They have nothing to do with nature and the real world *per se*. However, the "real world" is where we live and do science. This world is simple and complex, yielding and unyielding, predictable and surprising, which begs the question "how can we be sure that a single falsification of a theory does in fact falsify it?" For example, what if we could modify our theory, as the scientific method suggests, and then show that the modified theory does not fail a falsification experiment? Wouldn't this "save the theory?" The answer is "it depends." Although initially reluctant in his logic of discovery to allow theory modification, Popper later recognized that unrestricted modification could make all theories unfalsifiable, and therefore meaningless. In addition, modification(s) themselves could represent new scientific insights that could lead to better, more all-encompassing theories. Popper discusses this evolution in his autobiography[220] and then illustrates the latter point by discussing the presumptive falsification of Newton's theory of gravitation.

(1) My main idea ... was this. If somebody proposed a scientific theory he should answer, as Einstein did, the question: "Under what conditions would I admit that my theory is untenable?" In other words, what conceivable facts would I accept as refutations, or falsifications, of my theory?

(2) I had been shocked by the fact that the Marxists (whose central claim was that they were social scientists) and the psychoanalysts of all schools were able to interpret any conceivable event as a verification of their theories. This, together with my criterion of demarcation, led me to the view that only attempted refutations which did not succeed qua refutations should count as "verification".

(3) I still uphold (1). But when a little later I tentatively introduced the idea of falsifiability (or testability or refutability) of a theory as a criterion of demarcation, I very soon found that every theory can be "immunized" [read "modified;" Author's comment] ... against criticism. If we allow such immunization, then every theory becomes unfalsifiable. Thus we must exclude at least some immunizations. For example, the observed motion of Uranus might have been regarded as a falsification of Newton's theory. Instead, the auxiliary hypothesis of an outer planet was introduced ad hoc, thus immunizing the theory. This turned out to be fortunate; for the auxiliary hypothesis was a testable one, even if difficult to test, and it stood up to tests successfully.

Ironically, Popper "immunized" his own theory by introducing the concept of "degrees of testability," clearly a relativistic, as opposed to an absolutist, concept. Degrees of testability were related to "degrees of content," by which one could decide whether or not to adopt a modification to a theory. Popper explained this as follows.

... abstract theories, like Newton's or Einstein's theories of gravitation ... are falsifiable [but] prima facie falsification may be evaded ... by the introduction of testable auxiliary hypotheses, so that the empirical content of the system – consisting of the original theory plus the auxiliary hypothesis – is

---

[219] It is likely that this term originated with Bacon's use of the term *Instantiae Crucis*, or crucial instances, in 1620, which then was followed by Hooke's usage of the term *experimentum crucis* in 1665. This term was popularized in the scientific literature by Newton in his work on optics published in 1772 [289].

[220] Taken from [288], pp. 42–43.

greater than that of the original system. We may regard this as an increase of informative content – as a case of growth in our knowledge.[221]

We've come a long way and seen that an embryonic scientific method was practiced in Egypt as early as 5,000 years ago. We've seen the continual development of methods of scientific discovery and the logic underlying them, from alchemy, mysticism, and the close-mindedness of religious dictates, to the recognition of nature *per se* and the freedom to explore it. We've seen the conception of science as absolute truth and the eventual recognition that it actually is "a work-in-progress." We've seen logic employed to help scientists identify causal agents of phenomena, and eventually, to provide the means to ask if we really know what we think we know, … or if we *can* know at all. And … after all this, it appears no one can agree on what science and its methods are (some argue there is no method). Is the classical hypothesis-driven scientific process withering? Is it being replaced by robotic scientists and data-driven hypothesis formulation and testing? Let's see.

4.7 Senescence and Anarchy: Scientific Method into the Twenty-First Century

We would like to be *certain* of the foundations and discoveries of science and the quest for certainty has been part and parcel of the scientific method since antiquity. Certainty in the scientific community may have reached its zenith during the nineteenth century, as the logical empiricism[222] and mathematics of the scientific method flourished. However, these advances also presaged what many today consider to be the senescence of a normative scientific method. In anthropomorphic terms, the scientific method may be in its old age – and dying. Why do I say this?

Senescence may have begun with the integration of probability theory into science in the late nineteenth century and early twentieth century. Belief in, or dismissal of, observations, hypotheses, theories, and laws now no longer were categoric. The extremes of uncertainty and certainty, 0 and 1, were replaced by an open interval (0, 1).[223] Absolute certainty did not exist. Theories could not be proven, but they could be disproven [6]. Science practice was determined not by impartial, logical, and mathematical means, but by consensus of the scientific community [197], although it also was argued that this "consensus" actually was anarchy [290]. In fact, Hoyningen-Huene opined *"Historical and philosophical studies have made it highly plausible that scientific methods with the characteristics* [ascribed to them over the centuries; Author's insertion] … *do not exist."*[224] Lakatos has taken this sentiment a bit farther, suggesting that *"Scientific theories are not only equally unprovable, and equally improbable, but they are also equally undisprovable."*[225] Scientific method thus may already have died. Paul Feyerabend (1922–1996) went even farther, arguing that *there is no scientific method.*

---

[221] Taken from [288], p. 45.
[222] *Logical empiricism* in this context should be taken literally, *viz.*, as the requisite inclusion of a logic of scientific discovery and controlled experimentation into the scientific method. This use is distinguished from the school of philosophy of science known by the same name that was developed in Vienna in the 1920s and 1930s.
[223] Open number intervals, signified as (x, y), include all numbers between their limits of x and y, but not the limits themselves.
[224] See Hoyningen-Huene [21], p. 168.
[225] See Lakatos [291], p. 19

4.7.1 Paul Feyerabend (1924–1994)

. . . to some he was the court jester of philosophy of science, to others one of the most important philosophers of science of this or any other century.[226]

Paul Feyerabend was born in Vienna in 1924. After service in the German army during World War II, Feyerabend began studies at the University of Vienna. He initially chose to study history and sociology, but later switched to theoretical physics. His graduate work focused on classical electrodynamics,[227] but after some efforts in this area, Feyerabend switched to philosophy. As a graduate student, Feyerabend was able to interact with Karl Popper, Rudolf Carnap, Herbert Feigl, and Philipp Frank [294], interactions which greatly influenced him and certainly contributed to his focus on "basis sentences,"[228] considered by the Vienna Circle to be the foundations of scientific knowledge. Feyerabend himself had said "that in his philosophical work he had 'started from and returned to the discussion of protocol statements in the Vienna Circle.'"[229] In fact, Feyerabend was the student leader of another "Circle," the Kraft Circle (centered around Viktor Kraft, a former member of the Vienna Circle [295]), which considered "philosophical problems in a nonmetaphysical manner and with special reference to the findings of the sciences."[230] It is ironic, then, that Feyerabend studied with Karl Popper, the "Official Opposition" to the Vienna School and its logical positivism, at the London School of Economics during the academic year 1952–1953. Feyerabend was influenced deeply by Popper and said *"Falsificationism . . . seemed a real option, and I fell for it."*[231] Indeed, Feyerabend made it "the centerpiece" of his lectures when he started teaching. Feyerabend also subscribed to Popper's arguments against inductivism, *". . . the idea that theories can be derived from, or established on the basis of, facts."* The scientific method, through inductive principles, can produce laws. However, Feyerabend considered induction *"a sham."*[232]

In addition to Feyerabend's philosophical rejection of inductivism, he was concerned more generally that elements of many philosophies *"paralyze our judgment."*[233] It was this perceived paralysis that was at the heart of Feyerabend's philosophy and its consideration of scientific method. Scientific method, instead of facilitating discovery, actually could inhibit it, and in addition, might even be a threat to a free society [298] because of the myth of science's societal importance and primacy. These factors could result in the rejection of other philosophies, ideas, concepts, or methods that might be of greater importance and help to society, and to science. Feyerabend's exposition of his ideas came first in a paper entitled *Against Method: Outline of an Anarchistic Theory of Knowledge* [290], which began:

---

[226] Gonzalo Munévar, in [292], p. 6.

[227] Electrodynamics is "the branch of science that deals with the movement of electric charges in magnetic and electric fields and with the mutual interaction of electric currents." (From Oxford English Dictionary [293].)

[228] Feyerabend's thesis was entitled *"Zur Theorie der Basissätze"* (Ger., "The theory of basis sentences").

[229] Taken from Preston [295], p. 19.

[230] Quotation from Feyerabend [296], pp. 3–4.

[231] Quotation from [297], p. 89.

[232] He was convinced of this because of the observation that *". . . higher-level laws (such as Newton's law of gravitation) often conflict with lower level laws (such as Kepler's laws) and therefore cannot be derived from them, no matter how many assumptions are added to the premises."*

[233] Quotation from [297], p. 89.

The following essay has been written in the conviction that anarchism, while perhaps not the most attractive political philosophy, is certainly an excellent foundation for epistemology, and for the philosophy of science.[234]

Lee Hood, the visionary scientist we discussed in Section 3.3.3, actually put this philosophical idea to practice, and in doing so, became what some have considered the father of the genomics era. As one of Lee's postdoctoral fellows said in 1985, 15 years after the publication of *Against Method*, *"Lee thinks that anarchy brings out creativity."*[235]

Five millennia of history had shown that the practice of science, and the efforts of philosophers of science to develop prescriptive principles of scientific method to guide scientists, were rooted in rationality. Now comes Feyerabend who examined the same historical record and said, "no," science was *not* rational and its advances were *not* dependent on a universal, unitary method. Instead, science often involved "subterfuge, rhetoric, and propaganda." A scientific method did not and could not exist.[236] Scientists did science intuitively, not by following any method. Many have argued the modern scientific method is based on the experimental methods of Newton (Section 4.4.5). Feyerabend, in contrast, argues that Newton's own writings show that even *he* did not follow the scientific method.

Looking back into history we see that progress, or what is regarded as progress today, has almost always been achieved by counterinduction. Even Newton, who explicitly advises against the use of alternatives for hypotheses which are not yet contradicted by experience and who invites the scientist not merely to guess, but to deduce his laws from 'phenomena' ..., can do so only by using as 'phenomena' laws which are inconsistent with the observations at his disposal.

Feyerabend's ideas were controversial and iconoclastic, but provided a wonderful foundation for important arguments about how philosophy of science should address the reality of scientific practice. In fact, Feyerabend and his colleague at the London School of Economics, Imre Lakatos, were friends who had planned to publish a debate volume, *For and Against Method*, in which Lakatos would argue for a scientific method and Feyerabend would argue against it. Unfortunately, Lakatos died suddenly, precluding the effort. Feyerabend was forced to publish only his side of the debate in a book entitled *Against Method* [130]. According to Feyerabend, this book was a collage of his work over two decades, which he edited and in which he purposely replaced prior moderate statements with *"more outrageous ones."* Feyerabend opined that he *"... loved to shock people,"*[237] but in addition, he did so to purposely polarize the arguments that were to be published in

---

[234] In a review of Feyerabend's paper, Koertge said *"As usual I find it difficult to tell which of Feyerabend's comments to take at face value and which are attempts at ... 'heavenly maliciousness.' ... (Probably Feyerabend would be delighted at my distress because it may indicate that he has successfully mixed propagandising with case making!)"* [299]. We will find that this sentiment certainly was common, and unfortunately, many took his comments at face value instead of thinking dispassionately about their deeper implications.

[235] Quotation from Mitch Kronenberg in the Los Angeles Times in an article authored by Paul Ciottie entitled "Fighting Disease On The Molecular Front: Leroy Hood Built A Better Gene Machine And The World Beat A Path To His Lab." www.latimes.com/archives/la-xpm-1985-10-20-tm-14242-story.html.

[236] Feyerabend no doubt was influenced by Popper's famous statement *"I am a Professor of Scientific Method – but I have a problem: there is no scientific method."* (See Nickles [300], p. 177.)

[237] Quotation from [297], p. 142.

his book with Lakatos, stating that *"Imre wanted to have a clear conflict, not just another shade of gray."*[238]

Against Method created a firestorm of outrage and denunciation. Feyerabend was referred to as "the worst enemy of science" [301] and vilified by many. It is understandable, psychologically, why people would react in such a way to a prominent, internationally recognized philosopher who challenged long-held beliefs in such an important and cherished facet of existence as science. Unfortunately, in addition to antagonizing scientists and philosophers of science, Feyerabend's views were used by those who did not consider science to have any special intrinsic or societal value to support defunding government-sponsored research and support nonscientific ideologies (e.g., pseudoscience). The danger to society of Feyerabend's views were considered so serious as to be an existential problem [301]. John Horgan, who interviewed Feyerabend for *Scientific American* [302], might have summarized the feelings of many when he wrote:

It is all too easy to reduce Feyerabend to a grab bag of outrageous sound bites. He has likened science to voodoo and witchcraft …. He has defended the attempts of fundamentalist Christians to have their version of creation taught alongside the theory of evolution in public schools. Beneath these provocations lies a serious message: the human compulsion to find absolute truths, however noble, too often culminates in tyranny of the mind, or worse. Only an extreme skepticism toward science – and open-mindedness toward other modes of knowledge and ways of life, however alien – can help us avoid this danger.

I began our examination of science (Chapter 2) with the exhortation that the reader should not believe what *I* say, nor what anyone else says, but rather they should find out for themselves. Descartes expressed a similar view when he reminisced about his early education and said that he had accepted as true *"many false opinions,"* necessitating that he *"rid himself of all the opinions he had adopted … and commence anew the work of building … a firm and abiding superstructure in the sciences"* [303]. In thinking about Feyerabend, one may argue that what he had to say actually was nothing new… or outrageous. When he said *"The best education consists in immunizing people against systematic attempts at education,"* Feyerabend was, in essence but independently, repeating Descartes' sentiment. Later philosophers of science have re-examined Feyerabend dispassionately and recognized the deep, subtle, and fundamentally important messages in his writings [304, 305] and even published special journal volumes [306] and books (e.g., *The Worst Enemy of Science? Essays in Memory of Paul Feyerabend* [292]) devoted to a reexamination of Feyerabend. As we have seen throughout this volume, the truth, if I may be allowed to use the term, only can be unearthed by digging deeply, and even then, what we find is only our current sense of it, not Truth *per se*. The truth of Feyerabend will never be determined, but his vital and lasting contributions to science and its methods may ensure that scientists, and philosophers of science, always are aware of their assumptions, presuppositions, implicit biases, closed-mindedness, dogmatism, and even hubris. If science is indeed to remain scientific, then such everlasting vigilance is obligatory.

---

[238] *Ibid.*

4.8 Data-Driven Research

Revolutionary advances in science have been achieved through a number of mechanisms. Optics is a good example. Simple, but keen and thoughtful, *observation* of the world at a magnification of $1\times$, combined with logic, enabled Aristotle to lay the foundation for all of science and for Copernicus to propose a heliocentric solar system. Galileo's construction of the first telescopes allowed him to observe the world at $\approx 30\times$ and to confirm Copernicus's theory of a heliocentric heavens. It also enabled van Leeuwenhoek to study the microscopic world, at $\approx 250\times$, and in the process discover *"animalcules"* (from the Latin for "tiny animal"; *animalculum*) – including bacteria, protozoa, and spermatozoa – thus becoming the first microscopist and microbiologist. Electron and atomic force microscopes now allow scientists to see objects at magnifications up to $\approx 10^7 \times$, opening up an immense sub-cellular world in which even individual atoms can be visualized. These instrumental advances expanded the observable[239] universe, and the amount of information to which one had access, many orders of magnitude. The best example of a recent revolution in scientific method comes from computer science, a field that for all intents and purposes, did not even exist until the twentieth century. The development of mechanical computers (the "difference and analytical engines") and arbitrary, user-defined programs ("weav[ing] algebraical patterns") by Charles Babbage and Ada Lovelace, respectively, followed by the conception and development of electronic computers by John van Neumann and Alan Turing, ushered us into the current "information age." Materials science and nanofabrication methods then provided the means to miniaturize the switches (originally vacuum tubes[240]) that constitute the heart of a computer, enabling the construction of powerful desktop and super computers. It has been estimated that the exponential rate of growth of digital information is such that the amount of information generated every two days, 2.5 exabytes (i.e., 2.5×10^{18} bytes), is equivalent to the total amount of information produced during the last 20 centuries [307]. Figure 4.7 illustrates the growth of "knowledge" *per se*.[241] Although accurate estimation of knowledge doubling times is impossible, and knowledge in different areas grows at different rates, a useful rule-of-thumb estimate is 18 months. Not surprisingly, this rate of knowledge growth correlates with that of digital knowledge.

This data explosion has had substantial effects on science and the scientific method. In contrast to science in the past, it now is impossible to be aware of everything going on, even in one's own field or sub-field. There simply are too many journals, too many books, too many meeting, and too much information. This "abundance of literature implies an

---

[239] What is "observable" is quite a question in philosophy of science, one that will not be addressed here. Suffice it to say that whether what we see using instruments, and not with our own eyes, is a true representation of the reality of nature remains a question for many.

[240] Many reading this book may not know what a vacuum tube is (was). Information may be found here https://en.wikipedia.org/wiki/Vacuum_tube#Use_in_electronic_computers.

[241] These data have been attributed by many to the futurist Buckminster Fuller. I argue in this volume that it is incumbent upon each scientist to ensure that any information they convey to the world is accurate. One cannot trust that because someone is quoted in one source that the quotation is accurate. One must find the quote within the quoted individual's own published work. In this case, such a search has not revealed that Fuller actually produced these data, even within the presumptive source, his book *Critical Path*. This does not, however, alter the take home message.

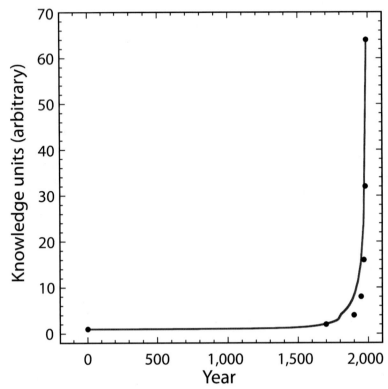

Figure 4.7 Growth of knowledge. The amount of knowledge accumulated on earth from prehistory to 1 AD is arbitrarily assigned the value 1. Each point signifies the total knowledge at that time in history. The line is a smooth fit to the data.

abundance of overlooked connections."[242] Scientists are forced to choose which information to acquire and study. This, in turn, requires methods to "capture, curate, and analyze data."[243] "Data science" is a new scientific field concerned with these methods,[244] and the practitioners of data science often are not the scientists producing the data, or even scientists. They are individuals with extensive training in statistics, mathematics, and informatics who apply their skills in these areas to answer questions posed to them by experimentalists. Historically, naturalists collected their own specimens, creating large repositories of analog information that they themselves curated and analyzed. They knew how to interrogate their data sets because they had the knowledge and familiarity with their subjects, and the intuition developed therefrom, to do so. They also could count on the consistency of their data

---

[242] Quotation from Sybrandt [308].

[243] Taken from [309], p. xiii.

[244] The National Science Foundation considers "data scientists" to be *"the information and computer scientists, database and software engineers and programmers, disciplinary experts, curators and expert annotators, librarians, archivists, and others, who are crucial to the successful management of a digital data collection"* [309]. Data scientists now have their own journals, including *Data Science Journal*, which is published by the Committee on Data of the International Council for Science.

because they presumably collected them in the same way. Data-driven science now requires collaboration among data providers, collectors, curators, and analyzers. It is thus susceptible to problems created by miscommunication, misunderstanding, and data variability. Each group of experts participating in the process may be superb practitioners of their own crafts but may not be conversant in those of their colleagues and thus may not appreciate how to properly tailor their work for integration into a knowledge creation process that will produce information, the veracity and usefulness of which, is indisputable [310]. Standards for curation and data analysis can be established by consensus among members of standards committees. However, it is more difficult to ensure that when the computational analyses of data are done that each datum is of equivalent quality to the next. In simple terms, one cannot compare apples to oranges and it is, in fact, impossible to completely eliminate variability among laboratories, experiments, experimentalists, cells, and animals. This fact does not invalidate data science, but it is something that must be considered when executing this new form of scientific method.

An outgrowth of data science is what some have called the "the fourth paradigm" of science [311, 312] – "data intensive" or "data-driven" science [313] – that follows the paradigms of *empirics*, *theory*, and *computation*. The most important feature of data intensive science is its inversion of the scientific method. One no longer begins with empirical observations that lead to testable hypotheses. Instead, it may be said that *scientists begin by doing no experiments and making no observations*. Hypothesis creation is initiated using computational (statistical, artificial intelligence [AI], machine learning) approaches to establish relationships among countless members of massive preexisting data sets. It is only after these relationships are established that one begins postulating and testing hypotheses. This approach has been likened to that associated with Kepler's laws of planetary motion.

It was Tycho Brahe's assistant Johannes Kepler who took Brahe's catalog of systematic astronomical observations and discovered the laws of planetary motion. This established the division between the mining and analysis of captured and carefully archived experimental data and the creation of theories.[245]

An illustrative example of data driven science is "-omics," an approach in which the initial step is to collect *all* data within a specific system and then to use computational methods to analyze (make sense of) the data. Hundreds of different kinds of omics exist.[246] A large percentage of -omics studies are in biology, for example, in genetics (structural genomics and functional genomics) and physiology (metabolomics, lipidomics, proteomics), but they also exist in psychology (behavioromics, humaninteractomics), economics (econonomics), health (healthomics), law (legalomics), and the humanities (literaturomics, religionomics).

---

[245] Quotation from Bell [312], p. xi. It is interesting that the work of Kepler, Galileo, and Copernicus on the heliocentric heavens also served as an argument for Feyerabend's thesis that some advances in science only occurred because the rules of scientific method were broken, to wit *"Copernicanism and other 'rational' views exist today only because reason was overruled...."* (Quotation from Feyerabend [290], p. 116.)

[246] For a complete listing of types of omics, see http://omics.org/index.php/Alphabetically_ordered_list_of_omes_and_omics.

The incisive reader no doubt may argue that data-driven science is not as removed from classical scientific method as one might think, and I would agree, because one had to have an idea, hypothesis, or theory in the first place, otherwise the data upon which discoveries depend would not have been collected. In functional genomics, for example, scientists are interested in determining relative levels of gene expression in particular cells under certain conditions. Gene expression is the transcription of DNA into messenger RNA, the type of nucleic acid that is used by the ribosome to synthesize proteins. Different genes are expressed at different times during the development and lives of all organisms, whether they be viruses, bacteria, plants, or animals. Differential gene expression allows organisms to produce specific proteins only when they are needed and in the right amounts (assuming the organism is healthy).

A biologist may wonder why fish living in waters below the freezing point don't freeze. The biologist may have no other information on which to base a hypothesis, but speculates that there must be some type of "anti-freeze" substance in the fish. This type of knowledge-free, *a priori* speculation was eschewed by Newton (see Section 4.4.5) but often is the basis for -omics-type investigations. In this case, the biologist may seek an answer to their ill-posed question by comparing the levels of expression of *all* genes within particular cells or tissues when fish are maintained at different temperatures. Data science methods then are employed to determine how the level of expression of each gene is correlated with temperature change, whether clusters of genes with similar responses to such environmental changes can be identified, and if so, the extent to which changes in one gene may be dependent upon or related to changes in other genes (gene ontology). In the process, the biologist may find that the expression of mRNAs encoding one class of proteins is markedly increased at low temperature. They may then purify and study the properties of the expressed proteins. In doing so, the scientist may find these proteins serve the same purpose as does anti-freeze in automobiles – they depress the freezing point. In fact, "anti-freeze proteins" do exist in fish [314] and many other species.

Are data *per se* now taking the place of empirical observations? What does "data" even mean? Is a datum real, like the observation of a rainbow, or is it simply a metaphysical representation of something else? If it does represent something, how does it do this and to what extent? Some may consider vast repositories of data to be disconnected from the natural world. These repositories are artificial digital worlds upon which mathematical and logical operation may be performed to deduce "things," but these things may remain metaphysical in nature and not represent reality. Leonelli [315] opines:

...two key features are often highlighted as pillars of [the data-driven science] ... approach: one is the intuition that induction from existing data is being vindicated as a crucial form of scientific inference, which can guide and inform experimental research; and the other is the central role of machines, and thus of automated reasoning, in extracting meaningful patterns from data.

Leonelli is saying, in essence, that instead of directly observing nature and then using inductive means to formulate hypotheses or theories, one begins by "observing" a digital representation of nature, a database. One then applies inferential methods, for example, induction, to this representation, as one would have applied to one's first-hand observations, thus enabling use of the H-D method. Chris Anderson, writing for *Wired* magazine [316], expressed this notion thusly:[247]

Kilobytes were stored on floppy disks. Megabytes were stored on hard disks. Terabytes were stored in disk arrays. ...[Exabytes] are stored in the cloud. ...[W]e went from the folder analogy to the file cabinet analogy to the library analogy to – well, at petabytes we ran out of organizational analogies. At the [exabyte] scale, information is not a matter of simple three- and four-dimensional taxonomy and order but of dimensionally agnostic statistics. It calls for an entirely different approach, one that requires us to lose the tether of data as something that can be visualized in its totality. It forces us to view data mathematically first and establish a context for it later.

...[T]he more we learn about biology, the further we find ourselves from a model that can explain it. There is now a better way. [Exabytes] allow us to say: "Correlation is enough." We can stop looking for models. We can analyze the data without hypotheses about what it might show. We can throw the numbers into the biggest computing clusters the world has ever seen and let statistical algorithms find patterns where science cannot.

Data driven science is, in fact, being employed increasingly and successfully across scientific disciplines, as well as in medicine [318–320], engineering [321], journalism, marketing, energy consumption, industrial process monitoring, economics, crime, traffic safety, and government policymaking. Computer-aided hypothesis generation (HG) may even become *de rigueur* in science [322, 323]. Humans would no longer participate either in the "hypothetico" part of the H-D method, but be replaced by a new scientific method, "machine science" [324], which would make "human supervision and conceptualization obsolete to the process of scientific discovery."[248] It is too early, though, to sound the death knell for humans doing science. Someone, *viz.* scientists, must choose a question to answer *a priori*. It is only then that *chosen* data are collected. Focused data selection is, in fact, critical to the successful use of current HG algorithms. Programmers must focus their information retrieval and analysis to limited information domains. According to Ilya Safro of Clemson University, this can make "cross-domain and 'not obvious' discovery impossible," especially when one wants to establish correlations between disparate domains of information, for example, between facial expressions (phenomenological) and enzyme expression (molecular biological).[249] Of course, the great promise of HG is precisely the revelation of the "not obvious" hypothesis!

Data-driven, *in silico* approaches do not eliminate the value of human-driven discovery because unless computer programmers can implement algorithms that are able to duplicate the thought processes of every human in existence (and they very well may in the future) and all those to come, "carbon-based entities" will still ask novel questions and conceive

---

[247] Not all agree with Anderson's statement. Pigluicci [317], for one, writes Anderson "...*doesn't understand much about either science or the scientific method. ... science advances only if it can provide explanations [which data-driven science cannot], failing which, it becomes an activity more akin to stamp collecting. Now, there is an area where petabytes of information can be used for their own sake. But please don't call it science.*"

[248] Taken from Leonelli [325], p. 112.

[249] Personal communication.

of unique hypotheses. Importantly, hypothesis creation is only the beginning. Who or what tests the new hypothesis? In the vast majority of cases, computers cannot perform the experiments necessary to do so, although robotic systems have been fabricated that follow the H-D method to investigate systems of limited complexity [37, 326].[250] Humans still will be necessary. In addition, why should scientists consider any computer-generated hypothesis to be reasonable, useful, testable, or informative? Does the output of a computer actually *explain* anything, which arguably *is* the goal of science, or does it just provide more questions to answer? Unbiased validation of HG systems is very difficult, as discussed recently by Sybrandt *et al.* [308].

HG systems are hard to validate because they attempt to uncover novel information, unknown to even those constructing or testing the system. For instance, how are we to distinguish a bizarre generated hypothesis that turns out to produce important results from one that turns out to be incorrect? Furthermore, how can we do so at scale or across fields? While there are verifiable models for novelty in specific contexts, each is trained to detect patterns similar to those present in a training set, which is conducive to traditional cross-validation.

Scientists are needed to answer questions such as those posed above, especially those that are more conceptual in nature and not amenable to the application of statistical and probabilistic methods. Interestingly, but maybe not surprisingly, among many methods data scientists use to validate HG data is the one central to classical scientific method, demonstrating that a hypothesis can *predict*. This typically is not done *pre facto* as scientists do the original HG using an extant data set from which they have removed a certain portion of the data, and then validate generated hypotheses using the previously excised data. It is likely, in the future, that computers will self-validate certain types of hypotheses *in silico* [308], for example, those that are strictly informational or involve highly focused experimentation in simple systems [37]. Machine science then simply will become another type of scientific method.

4.9 The Scientific Method: Yes, No, Maybe?

In some ways, it seems that we have come full circle. We began ≈5000 years ago with no scientific method. Around 2500 years later, the Greeks developed the logical tools necessary to reason about the world and the heavens and Aristotle argued the empirical nature of knowledge, i.e., science. Continued development of the logic of scientific discovery and of the empirical methods and instruments of science culminated in the scientific revolution of the sixteenth and seventeenth centuries and in the monumental advancement of science into the mid-twentieth century, when the scientific method, built with millennia of thought and effort and having succeeded so astoundingly well, was called into question by Popper,

---

[250] The system described by King *et al.* [326] follows the H-D method. It uses abduction to generate new hypotheses and deduction to test them. A laboratory robot performs experiments, the data from which are analyzed using AI methods to falsify hypotheses inconsistent with those data. It then repeats the cycle until it reaches an arbitrarily defined level of confidence in a specific hypothesis.

Lakatos, Kuhn, and especially by Feyerabend, who said "there was no scientific method" and "anything goes." We've gone from "no" to "yes" and back again – or maybe not.

"Anything goes" is where we began, but this certainly is not where we are. It also is true that there is no *single* scientific method. How science is conducted depends on the sub-field of science in which one works. Within each are standards of practice agreed upon by that community, and these standards may change. Nihilistic perspectives of the scientific method are thus just as untenable as are unificationist views. Scientific method, in fact, is pluralistic in nature. Gary Gutting may have summarized the state of our understanding of scientific method best when he said:

When we compare current discussions of scientific methodology with the long history of such discussions, perhaps the most striking feature of the current scene is its distance from the actual practice of science. ... [In the last century,] reflection on methodology has itself become a highly technical subdiscipline requiring so much specialization of its own that practicing scientists are no longer able to make significant contributions to it. Correspondingly, the intense demands of scientific specialization in the twentieth century made it impossible for scientists to look much beyond their disciplinary boundaries. As a result, the study of scientific method became autonomous in a way that it had never been before. ... There is no doubt that philosophical accounts of scientific methodology aimed at telling scientists how to proceed with their work are today otiose. ... But it is important to realize that ... scientists' practical knowledge of their methodology is implicit, not explicit, more a knowing *how* than a knowing *that*. Philosophical accounts of methodology have at least the advantage of explicitness, something that may be of value even to working scientists .... [Author's underlining].[251]

I have suggested that philosophical accounts of methodology not only are of value to working scientists but that understanding them is *obligatory* if science is to be conducted optimally. A scientist has to know "how," but also understand "that," the principles supporting and guiding the scientific enterprise are philosophical in nature. Thus, in the next section, we move from the theoretical to the practical and from the implicit to the explicit as we explore how scientists actually employ the scientific method in their quest to understand the natural world. But, before doing so, the practicing scientist would be well served by heeding Henk De Regt's caution [328].

... if a philosopher's favourite methodology is never used by real scientists (past or present) then the philosopher is probably wrong and not the scientific community.

---

[251] Taken from Gutting [327], pp. 430–431.

5

Science in Practice

5.1 Choosing a Project

...if you are too risky in your research you'll get nothing done. Or you can play it safe and reap rewards for doing essentially the same thing over and over again [but] ... you have to force yourself not to do that ... [but rather do] stuff that pushes the edges for you. [You have to ask] [w]hat can I personally tackle? Also you have to know the times you live in. Is there enough information for me to make progress here? When do you yourself say you're not going to be able to solve this? So you have to introspect and that's the good part. But you have to guess too and you could guess wrong. There are no guarantees.[1]

We now know that one universal scientific method does not exist nor do philosophers of science agree on how science *is* done and how it *should* be done. Given this fact, how does one actually learn to "do science?" We answer this question by reducing the question to practice. Our reductionist approach divides the process of science practice into five sequential steps, choosing a project, experimentation, observation and interpretation, inference, and dissemination. We discuss in turn how each is executed in the field of biomedical sciences, within which scientists and clinicians seek to understand the biology of diseases and to develop cures for them. Although some aspects of this process may differ in detail in other scientific disciplines, the general approach is constant. I will not discuss technique *per se*. One can learn how to use a pipette, determine a pH, set up primary cultures of neurons, take spectra, or run an SDS gel under the tutelage of those who have already been trained or from books and laboratory courses.[2] Our focus here, metaphorically, is on how to safely and efficiently navigate our scientific ship to its destination, not how to reef a sail.[3]

Science may be practiced differently depending on one's field and one's idiosyncrasies. What interests one person may not interest another. From an academic perspective, this is predictable, reasonable, and necessary. Diverse interests are a necessity, because without them, scientific discovery would be impossible. For scientific advances to be made, people must see the world from diverse perspectives, posit new hypotheses, try new methods,

---

[1] Quotation of Larry Abbott, William Bloor Professor of Theoretical Neuroscience and Professor of Physiology and Cellular Biophysics (in Biological Sciences), Department of Neuroscience, Columbia University, as quoted in Firestein [3], p. 136.

[2] Examples include books on gene cloning (*Molecular Cloning: A Laboratory Manual* (originally referred to as "The Maniatis Manual")), animal experimentation (*The Laboratory Mouse*), use of the CRISPR technique for gene editing (*CRISPR in Animals and Animal Models*), and courses in specific disciplines and techniques, for example, in systems biology (*Experimental methods in systems biology*, www.coursera.org/learn/experimental-methods).

[3] Definitions for nautical terms can be found here: www.christinedemerchant.com/nautical-terms-sail.html.

and explore directions that interest and excite them. This is the ideal. Unfortunately, this ideal is difficult to pursue in practice. Although it may be unimpeachable philosophically, scientists live in scientific, local, national, and international communities. Practical considerations based on the collective community gestalt are important determinants of scientific direction.

How does the science's gestalt affect its practice? Unless one is independently wealthy, the practice of science requires resources. These resources include the physical plant in which the science is done (institution, university, company, and garage), the scientific instruments and supplies necessary for experimentation, the staff necessary to carry out the work, and an environment permissive to scientific practice. A permissive environment is one in which no legal, social, political, or religious obstructions exist that would prevent or hinder experimentation. Until recently, such an environment has existed for scientists, one in which few impediments to a scientist's exercise of their creativity and curiosity have existed.

Let us assume, for the sake of this discussion, that one is functioning in a permissive environment. How are the resources necessary to do science acquired? The answer is one must convince others to provide resources to them. "Others" include private philanthropic institutions, government agencies, private citizens, businesses, universities, and private research institutes. It is the necessity of obtaining scientific resources that has the greatest impact on scientific project selection.

Those willing to fund research want something in return. In the United States, the National Institutes of Health (NIH), for example, "seeks fundamental knowledge about the nature and behavior of living systems and the application of that knowledge to enhance health, lengthen life, and reduce illness and disability." The National Science Foundation (NSF) seeks advances in the basic sciences and engineering. Although diverse in their objectives, both the NIH and the NSF must obtain their yearly funding through line items in the budget of the United States. Directors of each institution, in collaboration with scientific advisory boards, the larger medical and scientific communities, and others, determine foci for funding. However, in the final analysis, funding is a political process, not a scientific one *per se*. This is where the societal gestalt can affect the direction of science. It is what society wants, as reflected by the actions of its elected officials, that determines whether the programs of the NIH and NSF will receive support. Legislators in the House of Representatives, and the President, may seek the advice of scientists in creating a budget, but they also consider the opinions of many others in this process, including many that have no scientific training but represent powerful political factions. The politics of stem cell research and human cloning are examples [329–332].[4]

The impact of politics on science and science funding is no less important in other countries. The recent formation of UK Research and Innovation (UKRI), a central body uniting nine existing research-funding agencies, including the UK's seven research councils, has raised concerns among many about the loss of autonomy of the councils and how funds

---

[4] This is a statement of fact. I express no opinion on whether such studies should be funded by the NIH.

will be allocated with respect to different scientific and medical disciplines [333]. Similar concerns exist with respect to the European Union's massive €100 billion science funding program "Horizon Europe" [334]. One of the most egregious incursions of politics into science policy and funding occurred recently in Hungary, where its parliament passed a bill to create a government-run committee to determine science policy and which research institutes would receive funding – a role previously played by the Hungarian Academy of Sciences [335, 336].

In the United States, the Trump administration did much worse, systematically implementing a "fox guarding the hen house" strategy that contravened the missions of four agencies whose mandate is protecting and promoting the health of the nation, the NIH, the Food and Drug Administration (FDA), the Centers for Disease Control (CDC), and the Environmental Protection Agency (EPA). Political interference with the NIH disrupted or blocked research into such important areas as AIDS, diabetes, developmental disorders, and regenerative medicine using fetal stem cells [337], which has tremendous potential for curing disease and restoring damaged organs. Political appointees also may reject scientific knowledge and research outright, the two things most important for the function of any of the agencies mentioned above. For example, at the EPA, where former Oklahoma attorney general and state senate member Scott Pruitt, a lawyer and venture capitalist with no training or knowledge of ecology, and a person who rejects the existence of climate change, was appointed Administrator. Pruitt's deputy, also a climate change denier, had been a lobbyist for one of the largest coal companies in America. The result has been gutting of environmental protections, which endangers not only us, but succeeding generations as well [338].

By contrast, private philanthropic institutions and individuals have the power to fund most scientific research they desire without interference by governments. Nonscientific concerns may also be relevant here, but the mandates of these institutions is clearly specified in their charters. The American Heart Association seeks treatments and cures for heart disease. The Alzheimer's Association seeks treatments and cures for Alzheimer's disease. Private citizens often fund biomedical research in areas with which they have had personal experience, for example, because they or family or friends have suffered from particular diseases. Of course, some donors simply have academic interest in particular subjects.

Other sources of funding come from companies that wish to commercialize intellectual property created by scientists. Drug companies fund basic research and clinical trials. Equipment companies fund the development of specialized scientific and clinical instruments, instruments that often are developed to enable exploration of previously unaddressable questions or execution of novel clinical procedures. Scientists themselves may patent inventions, form their own companies, and thus self-fund their work. The wishes of resource providers thus have profound practical implications for research choice.

In the ideal, one must propose to do important, significant, novel, and exciting[5] research if they are to find funding for their work. This requirement is best met by asking the "big

---

[5] Of course, what the precise meanings of these adjectives are is nebulous. For example, "exciting" to one person may be "boring" to another. I learned this lesson from Seymour Benzer, one of the fathers of molecular genetics and molecular biology. After a long argument about a particular phrase in a manuscript we were writing, Seymour said *"One man's orgasm is another man's 'so what.'"* I couldn't agree more.

questions." What is the secret to long life? How can we cure cancer? How do we prevent heart attacks? Of course, the problem with asking the big questions is that they are unanswerable in the form in which they are asked. How *does* one discover the secret to long life or cures for diseases? What experiment can one propose to do so? The questions, as stated, are too difficult and complicated to answer. However, it *is* possible to answer simpler questions. This is called reductionism, which is the process of moving from the broad to the narrow, the complex to the simple, and most importantly, the unknowable unknown to the knowable unknown (see Section 3.5.1).

We make the assumption in this discussion that the broad subject area of interest has been decided. However, this is not a prerequisite. What is obligatory is the possession of general knowledge. We must know *something* to ask any questions at all. We observe that the sun rises and falls. Why? Newtonian mechanics does not explain the behaviors of subatomic particles or of large objects traveling near the speed of light. Why? We see that women sometimes give birth to boys and sometimes to girls. Why? These questions can only be posed when we possess prior knowledge, i.e., we know what the unknowns are. The broader our initial knowledge base, the greater latitude we have in selecting a question that is both interesting and fundable. However, one can also become a successful scientist if one is interested in only a single question, depending on what that question is.

Successful project[6] choice requires that the scientist be able to provide to themselves, and to the scientific community, compelling answers to a number of key questions. I use the term "successful" here in a practical sense to mean executing the project is possible (theoretically) and support for the research is obtainable (even though it may be difficult to do so). Projects that don't meet these criteria for success may be extremely important or interesting, but not practical, for a variety of reasons. For example, they may seek to identify unknown unknowns (low probability of success and difficult to evaluate impact), develop a time machine (not possible according to Stephen Hawking's "Chronology Protection Conjecture" [339]), or are in areas for which no funding is available (a sociopolitical problem). The following questions should be among the first one asks of oneself, but others also are important and may vary with field of study (Table 5.1). These questions are equally applicable to projects done in *in vivo* (in living organisms[7]), *in vitro* (literally "in glass," e.g., test tubes), *in hydro* (in an aqueous phase), *in vacuo* (in the vacuum, e.g., using mass spectrometry [MS]), *in caeli* (in air, e.g., optics), *in papyro* (using only paper, e.g., epidemiological studies or meta-analyses), or *in silico* (pseudo-Latin for the word *silicium*, "silicon," meaning in computers,[8] e.g., simulations).

Why do I want to do this project? The primary consideration here should be what excites you and about what you are passionate. You may spend years, and in some cases, a lifetime trying to answer the questions and test the hypotheses comprising your project. If you're not committed to this process, then you should choose another topic. Of course, if you're not ready for a lifetime of hard work (and joy), then maybe science isn't for you!

---

[6] Throughout this discussion the word "project" entails both the scientific question at hand and the means for achieving an answer to it. The "question" itself is what initiates and drives the project.

[7] I include prions in this category.

[8] Silicon is an allusion to the base material for making the semiconductors used in computers, hence the phrase means "in computers."

Table 5.1 *Project selection questions.*

Why do I want to do this project?
What gaps in knowledge would the project fill?
Why is the project important?
What is novel about the project?
What are the implications of answering the question(s) posed in the project – on the narrow field in which the question is asked, the larger subject area, and society?
Is the technology available to execute the project or must I develop it?
Can the project be completed in a reasonable amount of time?
Have I reduced the question(s) to the level of minutiae, so that the answer becomes trivial or unimportant?
Who cares?

What gaps in knowledge would the project fill, or in other words, about what are we ignorant? This is a question all funding agencies will ask, and it's pretty simple to answer …*if* you have recognized and cultivated your own ignorance (Section 3.5.3). You know your field (known knowns) and you know its known unknowns. Your project should be within the latter category.

Why is the project important? Filling a gap in knowledge is important, but the question is how important? Important is a relative term, i.e., relative to unimportant, and thus its meaning may vary in different milieus. A project may be important to an individual, but not to anyone else. If so, pursuing the work will be fraught with difficulty, especially with respect to obtaining support. Conversely, if the work is important to everyone, then it's likely that many are already ahead of you and, in addition, the competition for funding will be fierce. It is not unusual for scientists to change research directions based on the availability of funding.

It has been said the scientists are problem solvers. This means that for some, simply doing science, i.e., solving scientific problems, becomes more important than the problem *per se*. Scientists may decide upon a project, i.e., make it important, because it has particular relevance to society even if the project was not the scientist's passion *ab initio*. Filling a knowledge gap certainly can be important from an academic viewpoint, but this aspect of importance is only one of many and it is determined by the community of scientists in a particular scientific discipline, not with the help of an unbiased metric. Importance, as suggested above, can come from the perceived value of the project to society, writ large. It also can be determined by fiat, for example, if Congress decides the NIH or NSF should receive money to support projects that Congress wants, regardless of their intrinsic value.

What is novel about the project? A concept related to, but distinct, from importance is novelty. The two concepts overlap, but important projects may not be particularly novel, and vice versa. The most attractive projects are those that are both important *and* novel. If a project is very important, it need not be novel, but the converse is not true. A highly novel project that is unimportant is not likely to be funded.

What are the implications of answering the question(s) posed in the project – on the narrow field in which the question is asked, the larger subject area, and society? These are questions of extension, i.e., given the project will indeed produce novel and important information, how will this information be of benefit in the future? How can the information be used to further scientific discovery, improve or create new technology, or develop new diagnostics and therapeutics for disease? What other implications does the project have?

Have I reduced a question to the level of minutiae? A factor directly relevant to "implications" is the character of the question itself. Incremental advances in science are relatively easy because they build on what's already known and what is known to work. One can ask what effect a small alteration in an experimental variable might have, but this type of question is not going to open up new vistas of study or understanding. It will not revolutionize anything nor will it result in a publication in a high-impact journal (a very real practical concern for scientists and clinicians). This is not to say that incremental increases in knowledge are unimportant *per se*. Science continually expands its knowledge base, and this is a process necessary to support quantum leaps in scientific understanding. There is nothing intrinsically wrong in delving deeply into the complexities and nuances – the minutiae – of a particular system, be it biological, physical, cosmological, or theoretical, without particular concern for the scientific or societal impact of one's studies, *if* this is what one loves doing. Project selection is idiosyncratic, and different scientists may accord different weights to the factors involved in such decisions. However, in general, career advancement is better served by moving from the minute to the immense with respect to one's project choices and the questions therein.

Is the technology available to execute the project or must I develop it? We discussed earlier the "Cycle of Molecular Biotechnology" (Section 3.3.3 and Fig. 3.1) and how new questions may require the development of new technologies. If one loves methods development and instrumentation, they can search for scientific questions that are unanswerable with current technology and then propose to develop the tools that are needed. One also may decide to choose a project that already was known to be feasible. It also is possible to do a little of both. In my own scientific career, I was most excited about the questions themselves, not about choosing a problem for which I already had expertise, and thus I sought out the best technologies for answering the questions. This often required a vision of how techniques that were used in unrelated fields might be applied in my own. The search for, and the successful implementation of, these "new" techniques was exciting personally and rewarding scientifically because of the new insights that the techniques made possible [340].

Can the project be completed in a reasonable amount of time? If one hypothesized that the average speed of cars traveling on the M1 from London to Leeds exceeded the posted speed limit, one certainly could test the hypothesis within a few hours or days by standing on an overpass with a speed gun. However, if one chose to determine what consciousness was, the research likely would require many, many lifetimes (if the question were answerable at all). When one selects a project, the amount of time projected for the completion of the project is of tremendous practical importance because the scientist will commit themselves to answering the core question, even if it may take many years. It also is important because funding is never given carte blanche. A scientist is expected to create a timetable of progress and justify the amount of support requested at each point in it.

Who cares? This is my personal all-time favorite question, which I bluntly ask all of my students, and in a more diplomatic manner, of colleagues as well. It is a wonderful devil's advocate question, and it is one that all should ask as they ponder their research direction(s) for it encompasses many of the questions specified above. It is an open-ended question, one that has so many scientific, theoretical, practical, logical, societal, psychological, philosophical, and political components that attempts to answer it are tremendously challenging, but also stimulating and revelatory. It is revelatory in the sense that it offers one the opportunity to identify and evaluate their own motivations and reasons for selecting an area of study and then to determine whether, in fact, their motivations are reasonable and appropriate. It also prepares one for justifying their project in grant proposals, publications, and scientific meetings.

I note that the success of a project, regardless of its apparent importance, novelty, and feasibility, cannot be predicted with certainty. The converse also is true, projects that seem pedestrian, unfeasible, or irrelevant may turn out to be of monumental importance. Uncertainty should not be a reason *not* to do an experiment, nor should one's firm belief, based on knowledge extant and pure logic, that an experiment is pedestrian or unfeasible. If a student or colleague tells me "this experiment won't work," I invariably respond "this is why we do the experiment," namely to find out if this is so. If I'm told "this experiment is stupid," I respond in an identical manner.[9]

5.2 "Don't Just Do an Experiment, Sit There!"

We start here by accepting the proposition that we know something, the evidence extant supports the veracity of our knowledge, and the question(s) we seek to answer are reasonable (as discussed above). Now what do we do? We must transform the question from the realms of hypothesis and theory to practice, i.e., into a form amenable to scientific method. Big questions must be broken down into smaller questions, and even ostensibly simple questions often must be simplified further. These processes are reductionist in nature. We create "bite-sized" pieces (questions) that can be answered with single swallows (experiments). Reductionism is critical to science, but it is no less important in other fields, for example, in economics. Alfred Marshall, a giant in the field and one of the founders of neoclassical economics, discusses below how "the complex problem of value must be broken up."[10]

The element of time is a chief cause of those difficulties in economic investigations which make it necessary for man with his limited powers to go step by step; breaking up a complex question ... [segregating] those disturbing causes, whose wanderings happen to be inconvenient, for the time in a pound called *ceteris paribus*. The study of some group of tendencies is isolated by the assumption other things being equal: the existence of other tendencies is not denied, but their disturbing effect is neglected for a time. The more the issue is thus narrowed, the more exactly can it be handled: but also the less closely does it correspond to real life. Each exact and firm handling of a narrow issue,

---

[9] Merton [341] has opined *"Experience has shown that the most esoteric researchers have found important applications."*
[10] From Marshall [342], Chapter 5, §2.

however, helps towards treating broader issues in which that narrow issue is contained, more exactly than would otherwise have been possible. With each step more things can be let out of the pound; exact discussions can be made less abstract, realistic discussions can be made less inexact than was possible at an earlier stage.

Marshall breaks up a complex problem by putting confounding variables into a "pound" he describes as *ceteris paribus*.[11] This is the first step in experimental design. One must isolate a single variable from the constellation of variables so its role in system behavior can be ascertained. Marshall also addresses a critical caveat in reductionist approaches – when we reduce, are the results of our studies still relevant, i.e., to our system and to "real life?" Our hope as scientists is that our results do indeed elucidate elements of a complete system, a system that theoretically can be reconstituted by the recombination of all its elements.

In Section 5.1, we learned about the process of selecting a scientific question to pursue. This is the first thing any scientist does. However, regardless of the merit of the question, the method through which a question is reduced to scientific practice determines whether an answer can be obtained at all, and if so, the depth of new insights resulting from the answer.

Descartes has expressed this as follows:

So blind is the curiosity by which mortals are possessed, that they often conduct their minds along unexplored routes, having no reason to hope for success, but merely being willing to risk the experiment of finding whether the truth they seek lies there. As well might a man burning with an unintelligent desire to find treasure continuously roam the streets, seeking to find something that a passer by might have chanced to drop. This is the way in which most chemists, many geometricians, and philosophers not a few prosecute their studies. I do not deny that sometimes in these wanderings they are lucky enough to find something true. But I do not allow that this argues greater industry on their part, but only better luck. But however that may be, it were far better never to think of investigating truth at all, than to do without a method.[12]

Descartes's quotation emphasizes the fact that experimental conception and execution should be envisioned *pre facto* in one's mind, i.e., one should not just *do* an experiment, one should first *think* about it[13] and develop a logical plan of action. Through this method, the scientist might find truth. An excellent example of this method is the *gedankenexperiment*. One begins with a conception of nature and a question and then thinks through the entire experimental design, execution, and interpretation. This was the process Einstein followed in developing the general theory of relativity. Other famous examples include "Schrödinger's Cat" (see Schrödinger [344], §5) and "Maxwell's Demon" (see Maxwell [345], pp. 338–339). Table 5.2 presents a list of elements of good experimental design (many of which are shared with *gedankenexperiments*) that should be considered

---

[11] Lat., which can be translated as "other things being equal or held constant."

[12] Quotation from Descartes [117], p. 168, § "IV. There is need of a method for finding out the truth."

[13] Sylvia Boorstein, in her book *"Don't just do something, sit there. A mindfulness retreat with Sylvia Boorstein"* [343], was making the point that we need to *think* before we do. This point is no less important to the practice of science than it is to mindfulness and meditation in Buddhism.

Table 5.2 *Questions relevant to good experimental design.*

Consideration	Purpose
What is the experiment?	To reduce a question to a form that can be addressed using the scientific method.
What is the experimental design?	To determine what experimental system to employ and how the system will be manipulated.
How will the experiment be controlled?[a]	To take into account and eliminate confounding variables, ensuring the experimental system works and will provide a clear answer to the question.
How many replicates will be necessary?	To determine how many times an experiment must be done to ensure statistical significance.
What results may be expected?	To consider potential outcomes of the experiments and whether these outcomes can be interpreted clearly, and to be prepared for unexpected caveats or problems that may require modifications in the experimental system.
Do the technical skill and technology required to do the experiment exist?	To determine if the question can be answered with current technology or whether new technology must first be developed.
Should a "quick and dirty" experiment be done?	To first determine if the experiment has *any* chance of working by doing a relatively crude, simple, but informative experiment.
Are there orthogonal methods by which the experiment might be performed?	To better support a theory and the results of its tests.[b]

[a] It has been said that Abu Jābir, known as Geber (721–815), an Islamic scientist often referred to as the father of chemistry, was the first to introduce controlled experiments, dragging alchemy away from the world of superstition into one of empirical measurement.
[b] *"It is uncontroversial that some data support a theory more strongly than others. For example, heterogeneous evidence [i.e., obtained using different experimental methods] provides more support than the same amount of very similar evidence: variety is an epistemic virtue."* (Quotation from Lipton [348], p. 168.)

when planning a project, and why.[14] Allan Franklin, in a paper [346] and a book [347], both entitled *"What makes a good experiment?,"* has simplified this list to emphasize key factors making an experiment "good."

To sum up, then, a 'good' experiment is one which bears a conceptually important relation to existing theories ... or calls for a new theory. It must also measure the quantity of interest to sufficient accuracy and precision.

---

[14] These components also figure prominently in reviews of grant applications. Reviewers want to understand your goals, what experiments you propose to achieve them, what results your experiments likely will produce, how you will interpret those results, and whether any caveats (theoretical or practical (technical)) exist, and if so, how you would deal with them). Although many (most?) find grant writing time-consuming and difficult, many (most?) also will admit that the process helps crystallize their thoughts and conceive of new ideas and experiments they otherwise would not have.

What is the experiment? A wonderful example of reducing a huge question to one that is experimentally feasible it that of consciousness (or "awareness"), which Francis Crick and his collaborator, Christof Koch, long sought to understand. This problem was thought by some to be "too difficult to tackle,"[15] no doubt because of the enormity and complexity of the task. Where does one even begin? The answer is to employ reductionism. Crick and Koch did exactly this by establishing a framework for their investigations [350–352]. They began by invoking two assumptions: (1) that *"[i]t seems probable ... that at any one moment some active neuronal processes correlate with consciousness, while others do not"* and (2) that a common mechanism of consciousness exists for all its different aspects, for example, pain, seeing, qualia,[16] thinking, emotion, or self-consciousness [350]. Based on this logic, Crick and Koch chose to focus on the visual system[17] as a model of consciousness. This was the first step in their reductive process. A foundational assumption was consciousness was immediate, i.e., characterized by brain activities occurring in a time regime of less than one second. Crick and Koch then employed knowledge extant in fields such as psychology, physiology, neuroanatomy, and neurophysiology to reduce the problem further. They argued *"[a]t any moment consciousness corresponds to a particular type of activity in a transient set of neurons that are a subset of a much larger set of potential candidates."* This allowed them to ask a number of focused questions, the answers to which would come from well-designed ("good") experiments.

1. Where are these neurons in the brain?
2. Are they of any particular neuronal type?
3. How are they connected?
4. Is there anything special about the way they produce nerve impulses?

What is the experimental design? Two components of experimental design are selection of an experimental system and determination of methods of procedure. In manuscripts, these two components generally are combined and discussed in a "Materials and Methods" section. Crick and Koch proposed three different systems through which their questions might be answered: (1) "binocular rivalry" in humans, where neuronal activity is studied when one eye is given one constant visual input and the other a constant but rivalrous one; (2) using cats, where neuronal activity was compared during two different states, when the animal views visual signals first while it is awake and then view them again when it is in slow-wave sleep; and (3) studying the effect of anesthetics on awareness and recall in humans and on neuronal responses in monkeys under similar conditions.

One *can* use humans as research animals, and ideally one would want to do so, but ethical and practical problems often preclude this. Instead, rats are a frequently used system

---

[15] Quotation from Knight [349].

[16] This term describes how we experience subjective, conscious experiences, e.g., how we apprehend color, taste, pain, or emotions.

[17] Pun *not* intended. The visual system was chosen because it was considered *"the most favorable for an initial experimental approach ... [having] several well-known advantages as an experimental system for investigating the neuronal basis of consciousness. Unlike language, it is fairly similar in man and the higher primates. There is already much experimental work on it, both by psychophysicists and by neuroscientists. Moreover we believe that it will be easier to study than the more complex aspects of consciousness associated with self-awareness."*

to study a variety of questions, as are mice, fruit flies (*D. melanogaster*), worms (*C. ele-gans*), yeast, bacteria, and viruses. Each provides the means to study biological questions in the whole organism, which allows one to glean what may be described as "biologically relevant" information. However, if one wished to determine what retinal protein(s) were responsible for detecting light, and more importantly, how they did so at the atomic level, one then would need to study the process in isolation, i.e., in simpler systems in which confounding factors (neighboring cells, lens, vitreous humor, etc.) were eliminated.

Thus far, we have chosen a question to answer and a system within which to answer it. Now comes the methodological design of our experiment. What are we going to do in the laboratory, and how? These were questions Louis Pasteur faced in the mid-nineteenth century when he wanted to answer the question "do germs form spontaneously in the air?" or whether they were preexistent. He did so in what has been called the "swan-necked flask experiment" (Fig. 5.1). Before performing this experiment, Pasteur had to consider many

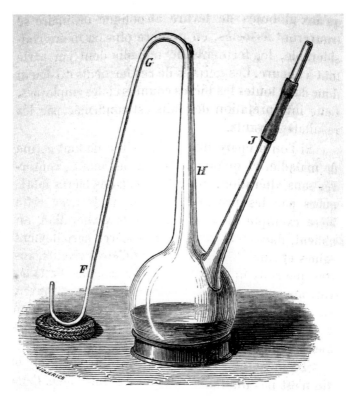

Figure 5.1 Drawing of Pasteur's swan-neck (*"col de cygnet"*) flask. The flask was filled with meat broth and then sterilized by boiling. Air could enter from the opening the left of F. Dust gradually accumulated over time at the bottom of the "U," but it, and any associated bacteria, could not pass "G" because of their size or adherence to vapors on the walls of the small tube "G-F." Broth could be introduced through the tube "J," which could remain sealed or opened to atmosphere to let air into the vessel. If J remained closed, no bacterial growth was observed. If not, rapid growth occurred.

technical factors, including what type of "test tube" to use, the material from which it was made, its shape and volume, the composition of the growth medium within it, how the medium could be freed of preexisting bacteria, the temperature at which the experiment would be conducted, the room in which the experiment would be done, whether the room would be shielded from sunlight and darkness, whether the tube would be irradiated with artificial light, and if so, what type of light, how long the experiment would last, and the dates the experiment would start and end (seasons have been shown to affect experiments). Each of these factors is no less important now than it was in 1859 when Pasteur performed the experiment. If experiments are to yield reproducible and interpretable results, such *attention to detail* is an absolute necessity.

A more recent illustrative example of attention to detail in experimental design is found in the field of cell biology, where cells grown in the laboratory are the objects of study. Interestingly, each of the 13 technical factors of Pasteur are also relevant here. Additional aspects of planning would include knowing: (1) whether the cells should be obtained directly from an animal (primary cell culture) or from frozen cell stocks, and if the latter, how the cells should be thawed and initially cultured; (2) whether the cells can be grown by themselves or if they need a "feeder layer" of other cells upon which to grow; (3) if the cells can divide, whether they will do so indefinitely, for example, as would tumor cells, or if, after a certain number of cell divisions, the cells begin to behave abnormally and then die (because experiments done under the latter conditions would be uninterpretable); (4) how often must the culture medium, which bathes the cells and contains nutrients and other factors the cells require to live, be replenished, and would this be done by, for example, removing and replacing *all* of the medium or just some fraction thereof?; and (5) what the atmosphere of the incubator that keeps the cells at an optimal growth temperature (usually $37°C$) must be, i.e., must it contain some percentage of CO_2 to maintain the growth medium at the proper pH (a measure of acidity) for growth. The 18 factors explicitly mentioned here are not the only factors to be considered during the experimental design process. Many others are implicit parts of the experimental system about which a scientist may or may not be aware. One implicit part is the manufacturer of the dishes in which the cells are grown. It is surprising how often plates composed of ostensibly identical materials produce different results!

Simulating real-world phenomena computationally requires another form of reduction. Here one reduces the reality of nature to the virtual world of silicon. The knowledge and thinking ability of the human brain, the most powerful thought processor in nature, is recapitulated in one of the dumbest and most ignorant technological creations in history, the computer, which has an alphabet comprising only two characters, 1 and 0^{18} and *ab initio* knows absolutely nothing and can do absolutely nothing – until the computer scientist begins entering information and programming the algorithms that instruct the *cpu* how to process that information.

The computer scientist must be even more attentive to details of experimental design than the bench scientist because they do not have the luxury of being able to use fully

---

[18] For the moment, we will ignore quantum computers, qubits, and superposition.

functional subsections of the real world to create their experimental systems. Even for extraordinarily simple systems (by biological standards), the number of variables and the parameterizations necessary to create a system are huge.[19] For example, to study the conformational dynamics of proteins in solution, one must model each of the atoms comprising each amino acid, the bond lengths and angles between atoms, how they change as the protein moves, the water molecules of the system, the system's temperature, and the size and geometry of the virtual world to be created. An order of magnitude estimate of the number of assumptions and parameters involved is 100, and there are few absolutes, if any, available to the scientist to guide them in this effort. This has two important consequences: (1) one cannot know for certain that their simulations represent what happens in the natural world and (2) each person's results likely will vary from another's, which makes reaching consensus difficult (and because of (1), that consensus could be wrong).

Simulation is an example of a situation in which the saying *"Don't just do an experiment, sit there!"* is particularly apt. Simulations may be the ultimate *gedankenexperiments* because one never physically carries out any experiments. This does not mean the considerations critical to physically executing experiments are ignored. It means that *all* imaginable variables in the system must be considered in the design of algorithms and their parameterizations. It does not mean that all will actually be included in a simulation. The coder must determine which variables are likely to be important in defining a system. This is when variables classified as "unimportant" are placed in Marshall's metaphorical "pound," to be ignored under the rubric of *ceteris paribus*. It also is when potential biases are introduced into the system, for how can one be certain *pre facto* that a particular variable will *not* play a major role in system behavior? And what happens if those creating the simulations disagree on this point?

Simulations need not be deterministic. If one runs a number of them, each will differ from the other if the algorithms randomize some parameters. However, it is possible to make inferences from the results of simulations if common dynamics are observed or if the simulations are run long enough, allowing them to reach the computational equivalent of chemical equilibrium, convergence (Fig. 5.2). A simple real-world example would be that of adding milk to coffee. The milk and coffee can be identified individually quite easily, but as time progresses, they mix until the milk is uniformly distributed and a homogeneous color is observed. Convergence has occurred. Formally, inferences such as these may be justified *in the context of the simulation,* i.e., they are warranted logically, but that does not mean the results reflect the actual processes found in living organisms. They may, and often do, but the scientist should be aware of this contextual caveat.

One should always seek to design the "ideal experiment," i.e., one that answers a narrow, specific question by minimizing the variability intrinsic to a system and its operator (the scientist). This is done through repetition of experiments and technical skill obtained through training and practice. Most importantly, *one must tailor the experiment to the question, not the converse.* I once was confronted by this issue during the planning of experiments to

---

[19] For one example of the complexity of a system for simulating the dynamics of protein structure, see Urbanc *et al.* [353].

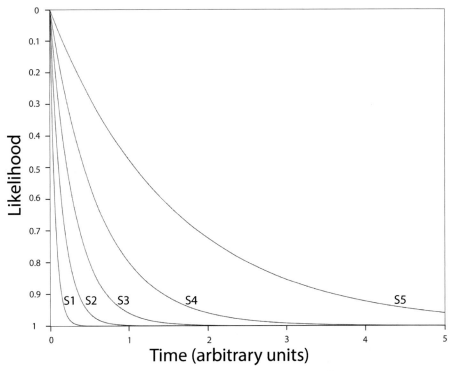

Figure 5.2 Convergence of simulations. Idealized data from five independent simulations (S_1–S_5) are shown. In practice, likelihood is replaced with a metric that correlates highly with system (equilibrium) state. All simulations begin with a low likelihood (≈ 0) that the simulation accurately represents the system state at that time, but as computational time (cpu cycles) increases, the likelihood also increases toward an asymptote of 1 – the simulations have converged. If the simulations were stopped at time 2.5, only S1, S2, and S3 would have converged, and thus yielded meaningful data.

determine the concentration dependence of cell death caused by a toxin. This is done by treating cells with different amounts of the toxin and observing the effects. One must design these experiments so that sufficient numbers of cell preparations are available to study a range of concentrations and to repeat the experiment a sufficient number of times to ensure reproducibility (generally a minimum of three times). In this regard, I was very surprised, and disappointed, when a postdoctoral fellow said to me, "I have enough petri dishes of cells that I think I can test three toxin concentrations." Here, a technical issue, the ready availability of cells, could be used to determine how an experiment was to be done. This was backwards experimental design. The fellow instead should have determined the number of dishes needed to test an entire series of concentrations and to allow testing of each concentration at least three times. Most professors, after hearing such a thing and trying their best to respond in a scientifically and socially acceptable manner, instead of yelling (not always possible under such circumstances), would respond simply by saying "grow more cells!" The reader may think this type of scenario occurs infrequently. Unfortunately, this is not the case.

How will the experiment controlled? When the experiment above was performed, the fellow observed no toxicity. I asked, "what did the positive control show?" The answer was, "Oh. There was none," to which I responded, "Then how do you know that these cells were even capable of being killed?" If in an experiment one does not include a positive control, a condition under which an effect is caused by a known toxic agent, then one does not know the system works, or even can work. Similarly, if one does not include a negative control, for example, a benign agent known to be nontoxic, then one does not know whether any effect observed using a putative toxic agent occurred due to the toxic agent itself or because of something else. Every experiment of this type *must* have both a positive and a negative control – the simplest and most fundamental of controls.

The known toxic agent just discussed is one type of reference standard. Reference standards are used to ensure that an experimental system, including any associated instruments, are performing as they should, i.e., producing data that are accurate and precise. The standard may come in many forms, depending on the system. Many different types of toxins may be used as standards (positive controls) for toxicity experiments. Geneticists and molecular biologists that seek to determine how various treatments of cells change the level of expression of a particular gene must have a standard with which to compare. By normalizing the expression of the gene of interest to this standard, one can determine how many "-fold" the treatment increased or decreased gene expression. This is frequently is done by measuring expression of a so-called "housekeeping gene," one that always is expressed at the same levels in a cell and is not affected by the treatment being studied. It this is not done, the investigator cannot determine whether changes in expression of a specific gene were due to effects of the treatment only on the gene of interest or to more general, nonspecific effects of the treatment on overall gene expression.

Each experimental system must be standardized because each will yield some type of metric, for example, gene expression level, amount of protein, light, energy, or radiation, that is expressed in standard units. Experiments that must precisely measure very short time intervals use time reference standards maintained at a variety of locations worldwide (see Section 3.2 for a more detailed discussion). Weight, expressed in kg (kilogram), has been standardized for the last 140 years using a cylinder of platinum–iridium (Fig. 5.3). However, beginning in 2019, a new standard kilogram based on Planck's constant $h/6.62607015 \times 10^{-34} m^2 s^{-1}$ was defined. This is the most precise weight standard. In practice, for experimental systems that don't require this level of precision, weight standards of sufficient accuracy for almost all purposes are readily available commercially. Avogadro's number, the number of molecules in a mole, an important standard in biology, physics, and chemistry, also has been redefined (now $6.02214076 \times 10^{23}$) as have the five other standard units (second, meter, ampere, Kelvin, and candela) [354].

The value of the scientific method depends, in part, on the verity of our knowledge. If some facts are false, the method can produce equally false knowledge. The way the scientist can mitigate this risk is to examine the assumptions inherent in the knowledge they use. These assumptions may be both explicit and implicit. For example, if one wants to

Figure 5.3 The kilogram. A cylinder of platinum–iridium kept in Sèvres, France, was the standard kilogram for all weight measurements from 1879 until 2019. (Credit: J. Lee/NIST, CC BY-SA 3.0 IGO.)

predict the path of an object thrown into the air, one must assume that the gravitational constant $G = 6.674 \times 10^{-11} m^3 kg^{-1} s^{-2}$ is correct. This assumption is explicit and thus can be evaluated during the process of experimental design, i.e., one can satisfy themself that the assumption is true. With few exceptions, for example, when one is trying to define a physical constant at higher precision, scientists accept the accuracy of G and of other constants. Problems occur when implicit assumptions are not recognized but have substantial effects on system behavior. If one does not realize this, experimental results may be incorrect or inaccurate without the investigator knowing it. This is a common occurrence that not only affects the scientist themself but may also compromise the work of other scientists who depend on the veracity of published work.

It is important to understand that an experimental result obtained in a controlled system may be justified by the experimental evidence, but wrong. This situation is, in fact, the norm in science. Although scientists, journal editors, reviewers, and the scientific community may be convinced of the validity and value of a manuscript that reports results from experiments done with proper controls and consideration of caveats – "the conclusions are supported by the evidence" – truth remains only within the province of that

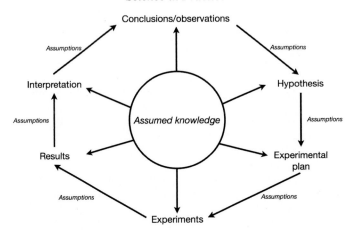

Figure 5.4 The danger of assuming. Each stage of the scientific method may be affected by assumptions (yours or others).

experimental system and its predicate logic. John Buridan (c. 1300–1360), a Parisian philosopher, provides an extreme example:[20]

...if someone, having seen and investigated all the attendant circumstances that one can investigate with diligence, judges in accord with the demands of such circumstances, then that judgment will be ...sufficient for acting well morally – even if that judgment were false on account of invincible ignorance concerning some circumstance. For instance, it would be possible for a judge to act well and meritoriously by hanging an innocent man because through testimony and other documents it sufficiently appeared to him in accord with his duty that that good man was a bad murderer.

The man was a murderer *within the legal system,* but not outside of it. The conclusion *was true within the experimental system,* but not outside of it. Experimental truth may be validated by additional experimentation in a variety of orthogonal systems, but it can never be considered Truth. *Assumed* knowledge and explicit and implicit assumptions affect every facet of the scientific method (Fig. 5.4).

How can the scientist avoid this type of pitfall? The answer is by identifying implicit assumptions and turning them into explicit assumptions, assumptions that then can be examined and potentially controlled. This is a difficult endeavor, but one that can be stimulating and enlightening, not to mention of great value to one's experimental work. The first step is simply to sit and think. One must play the role of sceptique extraordinaire and seek to poke holes in – actually try to destroy – one's own work, à la Popper. This requires, in essence, the trashing of everything in your memory and the reevaluation of your experimental plan, as if you never had thought about or created it. You could pretend you were an ignorant Martian who came to Earth with an interest in understanding your project. The Martian would have no preconceptions so they could evaluate your work in an unbiased fashion. Another, but less fantastic approach, would be to consider your plan as if you were

---

[20] Quotations from Buridan's *Questions on the Metaphysics,* II.1.f.9ra (p. 146 trans.).

a scientist from a different field. As a biologist, I find physicists and mathematicians to be excellent skeptics because they tend to question every assumption and point out assumptions you might never have considered. (They are remarkably good at asking "why" about seemingly everything!) *You* could do the same (or present your proposal to one or more of these scientists). The goal is to convert "unknown unknowns" into "known unknowns" and then plan an experimental approach that can control for them. This may not always be possible, in which case the experimental results will be subject to one or more caveats. If so, it is incumbent upon the investigator to discuss these caveats in their published work.

How many replicates will be necessary? Replication, in essence, is a control for variability. It provides the means to establish a statistical measure of the precision of your data and the ability to compare it with data produced in other experiments, at other times, or by different laboratories. Variability can arise from many things, including poor experimental technique, intrinsic system variability (as observed in human clinical trials), temperature, time, instrument performance, sample preparation and manipulation, vibration, and experimental vessel geometry (e.g., shapes of test tubes and glassware). It is reflected in the statistical variance of your results, as measured typically by the standard error of a mean. The smaller the error, the more likely it is that the scientist can establish (statistically) significant differences among experimental groups. If one is trying to determine a metric of some kind, for example, the boiling point of a liquid, replication provides the means to determine the precision of the measurement. The boiling point may turn out to be inaccurate, but it could nevertheless be precise, as determined by the variability of the average temperature observed.[21]

A number of means are available for determining how many replicates should be planned for a particular experiment. To allow you to conclude that differences in the data among different groups are statistically significant, one should use an appropriate statistical test and ensure that the data sufficiently "power" it, i.e., *n* is large enough. To do so, you stipulate the expected system variance and magnitude of group differences as well as the desired confidence level (typically 95%). These data are then entered into a relatively simple formula that provides you with the number of replicates you will need.[22] Many free online sites allow you to do this calculation.[23] One also may decide on replication number "on the fly," i.e., during a sequential set of independent experiments, by only doing as many as necessary to reach a desired level of statistical significance.

Replication and statistics go hand in hand, but what if the question isn't statistical *per se*? For example, what if one is considering testing a new hypothesis? How much evidence should one collect to make a decision? We know that sequential repetition of an

---

[21] We define "accuracy" as how close a measure is to its true value and "precision" as how finely the average is determined. Precision is reflected in the number of significant digits in a measurement. The greater the number, the smaller a difference between groups must be to distinguish one from another. Temperature provides a good contrast between accuracy and precision. We may have determined that the boiling point of our liquid was 100.0035°C, which is quite a precise measurement. However, the true boiling point may be 105°C. Of course, the converse also could be true. The average of our measurements may yield a boiling point of 100±10°C. Our average was very accurate, but we can't be sure whether the temperature is 100°C, 103°C, or 91°C.

[22] For a more detailed discussion of power analysis and the use, and misuse, of statistics, see [355–359].

[23] For example, see www.sample-size.net/sample-size-means/ or https://stattrek.com/survey-sampling/sample-size-calculator.aspx.

experiment to infinity is impossible, let alone desirable, but how do we know when enough is enough?[24] In an intriguing article discussing the relationship between hypothesis and evidence, Remco Heesen [362] suggests *"in most circumstances it is rational to do either zero, one, or a small handful of experiments!"*[25]

What results may be expected? When one develops an experimental plan, one does so hoping it will provide an answer to a question. Will it? Are there caveats in the conduct of the experiment that could compromise or complicate its execution or the interpretation of its results? Are alternative/undesired outcomes possible? Could one or more of these outcomes be uninterpretable or uninformative? If so, can the experimental plan be modified to make such outcomes interpretable and informative? If a previously unrecognized variable is identified, can this variable be controlled with new, different, or additional controls? These and related questions should be asked prior to execution of the experiment. The scientist must be ruthlessly self-critical if the experiment is to produce meaningful results. Ibn al-Haytham (Alhazen; see Section 4.2.1), apart from being one of the first to use a scientific method, also was a proponent of critical thinking and skepticism. He opined:

> The duty of the man who investigates the writings of scientists, if learning the truth is his goal, is to make himself an enemy of all that he reads, and … attack it from every side. He should also suspect himself as he performs his critical examination of it, so that he may avoid falling into either prejudice or leniency.[26]

Discussing an experimental plan with a colleague can be especially useful because that person likely will evaluate the experiment from a perspective different from your own. They may see things that you didn't and these could be problematic, but they also may help improve the experiment.

A practical benefit of imagining how an experiment will turn out is that such thoughts are required in many kinds of grant applications. Grant review boards want to know what you think your experiments may show, whether you have considered potential problems, and if so, how you would deal with these problems. If you have done so in experiments already performed (so called "preliminary data") and in the experiments you propose to do, your chances of being funded will increase.

Can the experiment be done? It is much easier to conceive of an experiment than to perform one. Sometimes a scientist may not know a necessary technique or have the right instrument. If so, they can learn the technique themselves or collaborate with an expert who will perform that phase of experimentation. Depending on the cost of a needed instrument, the scientist may choose to buy it or to collaborate with a laboratory that already uses such an instrument. For many types of experiments that require expensive or bulky equipment, for example, nuclear magnetic resonance spectroscopy (NMR) or magnetic resonance imaging (MRI), the scientist often will collaborate as opposed to purchasing an instrument for their own laboratory. Even if a laboratory did have the resources to purchase

---

[24] This is akin to the famous *Entscheidungsproblem* (Ger; decision problem) originally proposed in 1928 by Hilbert and Ackerman [360] and found by Alan Turing [361] to be impossible to answer.

[25] Note that Heesen's suggestion applied to Bernoulli trials. How applicable his suggestion is to other evidence-gathering processes is unclear.

[26] Quotation from Sabra [363], Book I, Chapter 1, [1].

the instrument, no one in the laboratory might have the required theoretical and technical understanding of the technique to allow them to successfully design and perform an experiment. For NMR or MRI, for example, it would be exceedingly rare and ill-advised for anyone who had not received graduate and postgraduate training in these techniques to try them on their own. No longer can one polymath do experiments in spectroscopy, biology, physics, optics, or cosmology, as scientists did before and during the scientific revolution. There simply is too much to know and learn. Specialization thus is required, which by force leads to collaboration within and across scientific disciplines.

How does one find the right collaborator? The answer is, by their scientific classification. Just as people may be classified by their political affiliation or country, scientists are classified by specialty. There are "worm people" (those who work with *C. elegans*), "fly people" (those who work with *D. melanogaster*), "zebra fish people" (those who work with *D. rerio*), electrophysiology people (those who stimulate and record nerve impulses from neurons), NMR people (a class that may be subdivided into solution phase, solid phase, and other types of NMR), "climate people" (climatologists), etc. As a neuroscientist, I'm an "Alzheimer's person" and my closest colleagues may be "Parkinson's people" (those working on Parkinson's disease), or "ALS people" (those working on amyotrophic lateral sclerosis, "Lou Gehrig's disease"). If I want to screen many different potential Alzheimer's disease drugs rapidly and efficiently in an *in vivo* system, a task that is not ethical or practical in humans, I may turn to *Drosophila*, an exceedingly well characterized system that has remarkable parallels with humans with respect to its genetics and behavior. To do so, I just reach out to "fly people,"[27] with whom I work to design synergistic experiments.

Scientists often must modify existing methods or develop new ones if they are to execute their experiments. Modifications may be as simple as changing the pH of a buffer from 7.4 to 8.0 or extending an incubation time from 1 to 2 hours. Circular dichroism spectroscopy, a commonly used technique to determine the secondary structure[28] of a protein, is an example. There is little need to modify the method to successfully carry out an experiment. However, new questions often require new methods and systems. It is not uncommon for graduate students in the biological and physical sciences to spend years developing "their system" and only weeks or months acquiring enough data to write their theses. Sometimes system development is so difficult that one reaches the point of diminishing returns and either tries to use a different system or simply gives up on the question. This also is not uncommon. However, at other times, a scientist may refuse to give up. In this case, as discussed with respect to Leroy Hood's creation of the "gene machines" (Section 3.3.3), the investigator changes their effort from answering a specific question to creating methods and instruments to answer entirely new *classes* of questions.[29]

---

[27] Most scientific work is done on a collaborative basis, by necessity.

[28] The peptide chains of proteins can be characterized with respect to shape (secondary structure), which may include helical, sheet-like, U-shaped, random, or others.

[29] Examples include the creation of the polymerase chain reaction (PCR) method by Kary Mullis (which led to his receipt of the Japan Prize and the sharing of the Nobel Prize in Chemistry in 1993) and the development of the gene editing method CRISPR/Cas9 ("clustered regularly interspaced short palindromic repeats/CRISPR-associated protein 9") by Jennifer Doudna (who received the Japan Prize in 2017) and independently by Feng Zhang, both of whom I predict will, in the future, be awarded the Nobel Prize in Physiology or Medicine for their method.

Should a "quick and dirty" experiment be done? Our focus thus far has been on design-ing and performing rigorous, well-controlled experiments in useful systems. We spend considerable time and effort doing so, sometimes years when we have to develop a new sys-tem. Quite often, however, in the interest of time efficiency and conservation of resources (supplies, labor, instruments, etc.), we just want to see if something *might* work. In these cases, we perform what are called quick and dirty experiments. These experiments are highly focused, simple in design, and often incompletely controlled. The principle is to "do something to something to see if we get a result." What we *do* is the technique and *to something* is the object of our interest (our sample). A *result* is an experimental finding, as opposed to no findings at all. Our experiment will fail if problems exist with either the technique itself or the sample. In these cases, we will get no result. As an example, let us consider implementing a chemical technique to irreversibly attach (cross-link) one protein chain to another. To be successful, the chemistry *per se* must work and the proteins to be studied must be capable of being cross-linked. The quick and dirty experiment would be to employ the technique to the test protein and determine if it is irreversibly cross-linked. It the result is "yes," we then may perform more controlled experiments to optimize the cross-linking reaction. If the test fails, we may decide to expend more effort to get the chemistry working for proteins or to see if the chemistry, although failing for our test protein, may work for other proteins. The point is that we place ourselves in a position to make these decisions having expended minimal time and effort investigating the approach.

Are there orthogonal methods through which the experiment might be performed? The fact that conclusions derived in a particular system are consistent with the data thus pro-duced does not mean that the conclusions are true. It is possible that further experiments in that system may produce different conclusions or that a system itself may produce mis-leading results. To address the first issue, it is good practice to use orthogonal methods in your system, i.e., methods that are based on principles distinct from those originally used. This is true both of experimental and computational methods. For example, if one seeks to determine the secondary structure of a protein, one can use circular dichroism spectroscopy. However, other methods, for example, attenuated total internal reflection infrared spectroscopy (ATF-IR) and fluorescence, also reveal secondary structure and they do so employing different principles. A concordance of data derived from each method provides strong evidence that a protein, in fact, contains a certain percentage of each type of secondary structure element.

Orthogonality of methods in computational studies, for example, in simulations, is par-ticularly important. Large numbers of assumptions are included in the algorithms used in simulations and the number of algorithms themselves is limited only by the imaginations of those creating them. This nonuniversality is complicated further by the fact that simulations are most useful when they reveal behaviors of systems that may not be determinable experi-mentally. What then is a standard against which the results and predictions of the algorithms can be compared? Here again, concordance of results from different simulations systems can increase one's confidence in the results. This may mean only that the results match quite precisely among systems. Whether the results are accurate, i.e., represent events in the real world, is an entirely different question, one that is the subject of debate not only among computational scientists but especially among philosophers of science.

Just as orthogonality in experimentation can increase one's confidence in the validity and correctness of their results, it also contributes robustness to explanations and theories. Robustness, as a confidence booster, was part and parcel of William Whewell's "consilience of inductions," as discussed in his seminal 1837 *magnum opus* [364] (see also Section 4.5.2). In Whewell's quotation below, "consilience" and "different classes" can be considered synonymous with "robustness of evidence" and "orthogonality," respectively.

The Consilience of Inductions takes place when an Induction, obtained from one class of facts, coincides with an Induction obtained from another different class. Thus Consilience is a test of the truth of the Theory in which it occurs." ...[It is axiomatic] that, other things being equal, the evidence for a generalization is strong in proportion as the number of favorable cases, and the variety of circumstances in which they have been found, is great.

Roughly two centuries later, Peter Lipton echoed Whewell when he discussed heterogeneous evidence [348] and its ability to provide *"a special epistemic oomph! to [an] hypothesis"* [365].

It is uncontroversial that some data support a theory more strongly than others. For example, heterogeneous evidence provides more support than the same amount of very similar evidence: variety is an epistemic virtue. Again, evidence that discriminates between competing theories is more valuable than evidence that is compatible with all of them.

Orthogonality is important in a more fundamental way, viz., in ensuring the validity of experimental results *per se*. Controlling whether an experiment has the potential to yield interpretable result requires, first and foremost, the inclusion of appropriate positive and negative controls, as well as other controls unique to the system. Each of these controls can be considered orthogonal to the primary experiment because they can provide, in essence, heterogeneous evidence for or against the hypothesis being tested. Orthogonality is particularly important in experimental design and evaluation of inferences developed from inter-species experimentation. In medicine, therapeutic agents typically are approved for human use by the FDA only after extensive testing in *in vitro* systems, animals, and finally, in humans themselves. A robust experiment or hypothesis is one that is reproducible *among* species, i.e., is successful yielding supportive heterogeneous evidence. It is not uncommon to observe the same phenomena in yeast, nematodes (roundworms), fruit flies, zebrafish, rats, mice, and nonhuman primates.[30]

Confidence is probabilistic not absolute. Nevertheless, the more experiments that support a tentative conclusion, the more likely one is to believe it. Confidence estimates (not confidence intervals) can be quantified using Bayes' theorem, which takes into account the prior probability of an expected result – in essence, one can use probabilities determined in *prior* experiments to predict the probability that new experiments will produce the same answer. The larger the *prior probability*, the larger the *predicted* probability will be.

---

[30] The reader may be surprised to learn how similar humans and flies are [[366], [367]]!

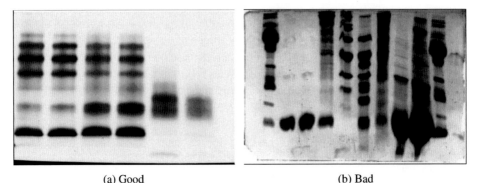

(a) Good   (b) Bad

Figure 5.5 A comparison of SDS gels done by people with "good" or "bad" hands. (a) This gel displays lanes that are rectilinear, not tilted, and of consistent width. The bands are horizontal, symmetrical, and clearly separated one from the other. (b) In comparison, this gel has lanes that may be trapezoidal in shape, tilted, and vary in width. Band themselves are tilted and asymmetrical, occasionally overlapping, and bulbous because of overloading. (Gel (a) courtesy of Dr. Robin Roychaudhuri. Gel (b) taken from www.ruf.rice.edu/bioslabs/studies/sds-page/gelgoofs/multi.html.)

Whether a probability estimate determined in this manner is believable, accurate, true, etc. cannot be determined with certainty because of assumptions inherent in the determination of prior probabilities. Nevertheless, one *can* obtain a sense of things through this approach.

One of the most wonderful aspects of science is the opportunity it provides to think, to imagine, to dream, to wonder, or as Newton opined, to be *"like a boy playing on the sea shore ...."* However, eventually, at least if one wants to *do something*, one must move from *gedankenexperiments* to *wirklichexperiments*.[31] One must use their hands, and in science, one wants to have "good hands," meaning that one's experimental technique is excellent so their experiments not only work, but their results are as precise and reproducible as their system will allow. The moniker good hands encompasses much more than just excellent small muscle control. It also requires attention to detail, exactitude, precision, patience, meticulousness, cleanliness (of workspace, materials, instruments, etc.), esthetics, and some degree of creativity (if alterations of methods or equipment are required to perform or improve an experiment). It also requires a clear understanding of the theoretical and practical principles upon which one's experiments are based.

Figure 5.5 provides examples of experimental results obtained from investigators with good or bad hands. Providing a primer on experimental technique is beyond the scope of this volume, but the figure does provide a sense of the importance of good technique. The figure shows SDS gels[32] that don't have (panel (a)) or have (panel (b)) technical issues.

---

[31] "Real experiments."

[32] Sodium dodecyl sulfate polyacrylamide gel electrophoresis (SDS-PAGE) is a standard biochemical technique for fractionating proteins according to their apparent molecular weights. When the proteins are introduced into a porous gel and subjected to an electric field, they migrate through the gel at rates proportional to their sizes, the smallest migrating fastest and the largest slowest. The proteins electrophorese as "bands," which are visualized by staining with protein dyes. Their sizes are determined by comparison with standard curves determined by electrophoresing proteins of known molecular weight.

The problems apparent in gel (b) are due to careless or inept pouring of the polyacrylamide gel, loading too much protein, an inhomogeneous electric field, and contaminating the gel surface with exogenous protein (likely from handling the gel without gloves). These problems could have been avoided if the investigator had chosen to do so. Laziness and haste, characteristics of bad experimentalists, are but two of the reasons they weren't.

5.3 Observation→Interpretation

Assume good strategic planning and technical execution of an experiment yields data. Frequently, the data will be in numerical or graphical form. For example, measurement of the absorbance of light by a protein solution at a given wavelength produces a number. Expression of green fluorescent protein in cells in culture produces a green fluorescence after irradiation, which can be seen and quantified with a fluorescence microscope. X-ray diffraction data produce patterns in a digital imager. Now what? What does one do with these data? The "what" is data analysis and interpretation, but to do so, we first must "see" the data. Seems simple, but seeing is distinct from analyzing, interpreting, thinking, hypothesizing, inferring, or any other action that requires processing of visual information. It is the simple act of allowing oneself to *experience* a virtual image of information sent from our eyes to our brain – nothing more. Posnock [369] suggests that this simple task is complicated by a bias taught to scientists during their training, viz., to incorporate logic into all that we do, to look past our results *per se*.

...logical investigation explores the nature of all things. It seeks to see to the bottom of things and is not meant to concern itself whether what actually happens is this or that. – It takes its rise, not from an interest in the facts of nature, nor from a need to grasp causal connexions: but from an urge to understand the basis, or essence, of everything empirical. Not, however, as if to this end we had to hunt out new facts; it is, rather, of the essence of our investigation that we do not seek to learn anything new by it. We want to understand something that is already in plain view.[33]

To be able to properly analyze and interpret experimental results requires the scientist to do the opposite, viz. to not try to immediately understand the essence of, or explain, what is in "plain view," for our view is anything but plain. Rather, it is full of facts that we could learn if only we would not think but just see.

I once did a simple experiment to get a sense of the ability of college students to observe and verbalize what they saw. I brought a wax apple and a wax banana into class. I held up the wax apple and asked the students what they saw. They were unanimous in reporting they saw an apple. I then did the same with the was banana ... and got the same result. They say a banana. When I then asked the students if they would like to eat either object, they were surprised and a bit confused. The objects, to them, obviously were wax fruit, *but that was not what they had seen*. Instead, they already interpreted what they had seen ... and they were wrong! I next reported to the class what *I* saw, which, in the case of the wax apple, was an object that appeared to be quasi-spherical with a diameter of ~4 inches

---

[33] Quotation from Wittgenstein [368], p. 42, §89.

and had an inhomogeneous surface, dull red color, along with an small elliptic green planar protuberance that was almost parallel with the object's surface. I then told them that my perception was of an object made to look like an apple with a leaf on top. I could not determine by simple observation of what the object was composed. I did pick it up and stick a fingernail in it, which revealed to me that the object weighed very little and was relatively soft. Therefore, there existed the *possibility* that it was made of wax. I would have had to further analyze the composition of the object to determine if that was true.

We usually interpret visual information unconsciously. We do not have to think about it. We do it automatically. If one is living in the African veldt and we see a lion running toward us, there is little time to consider what the image means. We must immediately associate the image with "lion" and run because during our lives we have been taught, or have observed, that the lion is an animal that possesses certain characteristics, including the ability to kill and eat us. Perceiving a lion has obvious survival value. However, automatic perception of scientific data is a formula for poor science.

Figure 5.6 is a famous optical illusion that illustrates the distinction between seeing and perceiving. And how an inability to see may lead not only to misinterpretation of data but to a lack of recognition of unexpected findings. Depending on one's perception, the figure is either a rabbit or a duck. It cannot be both simultaneously (unless it were a quantum image), because if it were, ears would have to be a beak, and vice versa, which is impossible by definition. This is the illusion. Let us say though that one does a cursory examination of the figure, perceives a duck, and concludes that this animal is what is represented in the figure. In this case, the viewer has missed something – perceiving a rabbit. Data are analogous to this example. In addition to the obvious, they often contain important information that may be neglected by the scientist because they have perceived before they have seen. The consequences are misinterpretation of the data and blindness to potentially novel, unexpected knowledge.

What *should* a scientist see in the figure? Among other things, they should see: (a) a series of slanted black lines, some continuous, some discontinuous, that vary in length; (b) a space between lines in the upper left quadrant; (c) the boundaries formed by the many lines appear to be contained within the confines of a thicker line that pokes slightly into the right margin of the figure; (d) a prominent discontinuity of lines in the upper right quadrant; (e) an annular black element (white at its center) that the lines do not cross; (f) a curved element of lines above the annulus that appear so dense that no white space can be seen; and (g) a thin white boundary between the bottom of the annulus and the lines. A scientist must first be able to *see* these elements before they can associate them with the outlines of an animal, a beak, ears, a head, a neck, a mouth, a tongue, an eye, an eyebrow, and an eyelid.

John Locke (1632–1704), philosopher, physician, and "Father of Liberalism," discusses the distinction between sensation (seeing) and perception in 1689 his *magnum opus An Essay Concerning Human Understanding*.

This [judgment], in many cases, by a settled habit, in things whereof we have frequent experience, is performed so constantly, and so quick, that we take that for the perception of our sensation, which

Welche Thiere gleichen einander am meisten?

Kaninchen und Ente.

Figure 5.6 Figure from the German humor magazine *fliegende Blätter* (Flying Leaves). The text above the figure reads *"Welche Thiere gleichen einander am meisten?"* (Which animals are most like each other?) Written below is *"Kaninchen und Ente"* (Rabbit and Duck). (Image from fliegende Blätter, 97.1892, Nr. 2465, p. 147.)

is an idea formed by our judgment; so that one, viz., that of sensations, serves only to excite the other, and is scarce taken notice of itself: as a man who reads or hears with attention and understanding, takes little notice of the characters, or sounds, but of the ideas that are excited in him by them.[34]

The "judgment" to which Locke refers is the interpretation of what we see. Scientists must see the "characters" or "sounds" first. If they don't, the interpretation of their data may be incomplete, incorrect, or biased. Their behavior then would be characterized quite accurately by an aphorism of New York Yankees Baseball Hall of Fame catcher Yogi Berra – *"I wouldn't have seen it if I didn't believe it for myself."*

What is relevant to observation is equally, and possibly more, relevant to interpretation. In analyzing data, one must not only free themselves from automatic thoughts, but they must also ensure that their interpretation of the meaning of the data is accurate and unbiased. Bias in interpretation can be passive or active. It is difficult for an observer to recognize, *ab initio*, their own (passive) biases. Passive bias is manifested in the application of deeply held beliefs or ingrained experience to the interpretation of data. This is particularly dangerous because without any sense of one's biases, how can they be eliminated? Passive bias most often comes from ignorance or dogma (although strict adherence to dogma also may be considered a type of active ignorance). One may not be consciously aware of this bias because one "grew up" with it. A good example is the notion of the Earth being flat, which was assumed from early antiquity until the classical period in Greece (c. 500–300 BC). This was common knowledge, and with good reason – it made sense, it seemed obvious by simple observation, and little evidence existed to suggest otherwise. Few questioned this "fact." It was not until the sixth (Pythagoras) and fifth (Parmenides) centuries BC, and

---

[34] Quotation from Locke [370], Chapter 9, §9.

most significantly through the work of Aristotle (384–322 BC) and Eratosthenes (276–194 BC), that the idea of a spherical Earth slowly emerged. This model then supplanted flat Earth models and, according to the Roman author and philosopher Pliny (Caius Plinius Secundus),[35] became the new common knowledge by the first century AD. It did so because astronomers allowed themselves to consider the existence of other models. They challenged the common knowledge by stripping "flat earth bias" from their interpretation of astronomical observations and thus concluded the data extant were most consistent with a spherical Earth. This model eliminated the many anomalies of flat Earth models. Knowledge had overcame bias.

A related historical example of bias comes from the work of Ptolemy (c. 100–170), who determined the positions and movements of the sun and planets using a geocentric model of the universe.[36] Ptolemy knew of prior studies of, and thoughts about, the Earth and heavens. Aristotle had proposed the Earth was spherical and the heavenly bodies displayed uniform circular motion. The notion that the Earth was at the center of the heavens had existed throughout history. This was what Ptolemy *knew* and had been *taught* as a student. Not surprisingly, Ptolemy's first model incorporated simple circular orbits of the sun and planets around the Earth. It was reasonable to do so because the model incorporated knowledge extant, and it was able to predict the movements of the sun and planets, although not without some anomalies (most notably the periodic, variable, retrograde movements of the planets). To improve the accuracy of the model, Ptolemy implemented epicycles ("orbits within orbits") within each planet's orbital path (deferent), made the Earth eccentric (moved it away from the center of all the orbitals), and created an *equant*, a point offset from the orbital center by an amount equal to the offset of Earth (Fig. 5.7). A viewer standing on the equant would observe the center of a planet's epicycle moving with constant angular speed in a circle. These modifications not only improved the model but also made it more complex (Fig. 5.8).[37]

Instead of trying to increase the complexity of the model to make it fit the data, why didn't Ptolemy consider other models? Alternative models that could have been proposed included those involving elliptical orbits of the sun and planets about the Earth as well as combinations of circular and elliptical orbits. These types of orbits also could have been considered in the context of heliocentric models. One such model was the heliocentric model of Aristarchus of Samos, who proposed it in the third century BC, and of which Ptolemy certainly was aware.[38] It is possible that Ptolemy's categorical rejection of the heliocentric model was because his passive bias made the idea unfathomable. How

---

[35] Pliny noted this in his book *Naturalis Historia*, where he wrote *"THAT the Form of the World is round, in the figure of a perfect Globe, its Name in the first Place, and the Consent of all Men agreeing to call it in Latin Orbis (a Globe), as also many natural Reasons, evidently shew."* Quotation from Pliny [371], Book 2, Chapter II.

[36] This example also has relevance to our discussions in Section 6.2 of psychological and sociological aspects of doing science.

[37] It should be noted that given a large enough number of variables, any system can be faithfully represented in a model, but that does not mean the model is realistic.

[38] *"However, they do not realise that, although there is perhaps nothing in the celestial phenomena which would count against that hypothesis [viz. heliocentricity], at least from simpler considerations, nevertheless from what would occur here on earth and in the air, one can see that such a notion is quite ridiculous."* Quotation of Claudius Ptolemy, from the Almagest [372], p. 44, H24.

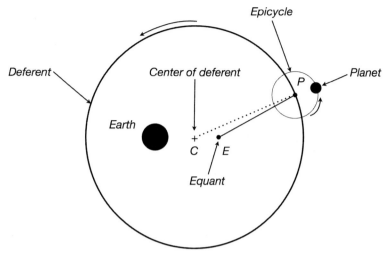

Figure 5.7 Ptolemy's model of planetary motion. (Used with permission of Professor Richard Fitzpatrick.)

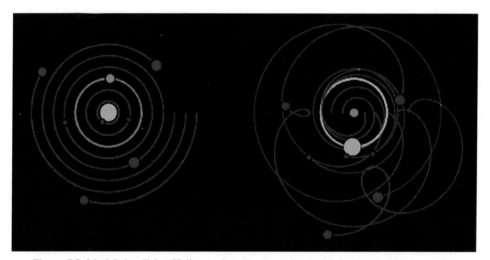

Figure 5.8 Model simplicity. Heliocentric and geocentric models both were able to predict the motions of the sun and planets, but the heliocentric model was much simpler. (Credit: Malin Christersson. www.malinc.se/math/trigonometry/geocentrismen.php.)

could the Earth *not* be the center of the universe? Ptolemy's geocentric universe received tremendous public and theological support and remained the standard for 1,300 years.

In practice, the model was quite good, which was one reason consideration of alternative models was not done vigorously. However, an active and overwhelming external bias against nongeocentric models was the greatest factor in retarding consideration of

Figure 5.9 Kepler's heliocentric model. Kepler improved upon the Copernican/Galilean heliocentric model by incorporating elliptical planetary orbits (although not including Uranus or Neptune, which were discovered later). (Credit: Storyboard That.)

heliocentric models. Bias imposed on natural philosophers from the public, and especially the Catholic Church, acted to stifle interest in such models and squelch discussion and publication of these models. The Church was particularly sensitive to interpretations of scientific results that produced conclusions contradicting the prevailing theological view that the Earth must be the center of the universe. One must remember that the suggestion of heliocentricity was heresy, an act that carried heavy legal penalties, including death. Galileo was, in fact, charged and convicted of heresy for his confirmation of Copernicus's results and further promulgation of the idea of heliocentric heavens (Fig. 5.9). His life was spared, but he spent the last nine years of it in his home in Florence under house arrest. Today, climatologists (at least those who support the notion of global warming), evolutionary biologists, and immunologists must deal with similar nonscientific, external biases from climate change deniers, creationists, and anti-vaxxers.

A modern example of bias comes in the form of the "central dogma of molecular biology (DNA→RNA→protein)," which codified a fundamental process that occurs in cells – how information encoded in the genome (DNA) is transcribed into a form (RNA) that can be translated into protein. This dogma was an intrinsic bias of the scientific community for almost 25 years, from the time Francis Crick originally defined it in 1958 [373], and redefined it in 1970 [374], until the discovery of "prions" by Stanley Prusiner. Crick's original dogma was termed "The Central Dogma" and was defined as follows.

…once 'information' has passed into protein it cannot get out again. In more detail, the transfer of information from nucleic acid to nucleic acid, or from nucleic acid to protein may be possible, but transfer from protein to protein, or from protein to nucleic acid is impossible. Information means here the <u>precise</u> determination of sequence, either of bases in the nucleic acid or of amino acid residues in the protein.

In 1982, Stanley Prusiner at the University of California, San Francisco, in his study of "slow virus diseases," diseases that have extremely long incubation periods (years to decades), suggested that these diseases were *not* caused by viruses, but rather by novel infectious agents termed "prions." Problem – no nucleic acids were found in purified preparations of prions, a finding that shook the foundations of molecular biology and genetics. How could an entity that violated the Central Dogma exist? Those for whom the Central Dogma was an implicit assumption – essentially everyone – balked. When these scientists examined the data produced by the Prusiner laboratory, they refused to interpret the data on its merits, which strongly supported the existence of prions, but instead, because of their biases, said Prusiner was wrong. There had to be a prion DNA encoding the prion protein, albeit a small one, that Prusiner must have missed. We now know, not only in mammalian cells, but in systems as primitive as bacteria, that prions do not contain nucleic acids (either DNA or RNA) and they have important functions in normal cells. They also can be toxic, as when they cause scrapie in sheep, bovine spongiform encephalopathy in cows (BSE; "mad cow disease"), or variant Creutzfeldt–Jakob disease in humans. Dr. Prusiner, who interpreted what he saw, *not* what he was supposed to see, was awarded the Nobel Prize in Physiology or Medicine in 1998 "for his discovery of prions – a new biological principle of infection."

The intrinsic bias that that Dr. Prusiner was able to overcome, viz., nucleic acids were the only vehicles for phenotype inheritance, had been taught to him at every level of his education, from K-12 through college and medical school and at the postgraduate level. The teaching of theory as fact is critical for the conduct and advancement of science, yet it also can contribute to ossification of the mind. A scientist must be acutely aware of dogmatic beliefs that could bias their interpretation of experimental observations. They must be able to conceive of new and even "blasphemous" interpretations of data if quantum leaps in knowledge, and also new dogmas, are to be created.

The intrinsic bias of the scientific community in its consideration of the prion hypothesis and its interpretation of experiments testing it evolved from its knowledge that the Central Dogma proscribed information "transfer from protein to protein." The problem was their knowledge was incorrect, they displayed "double ignorance." They argued strenuously that the prion hypothesis violated the Central Dogma and thus could not be true, but they did so because of their unconscious ignorance of the tenets of the Central Dogma. They thought that the Central Dogma prohibited the replication of an infectious agent in the absence of DNA or RNA, and thus prions could not exist. Had they considered the issue more carefully, and dug more deeply, they would have realized that prions do not replicate by transfer of *sequence* information from protein to protein, but rather by transfer of *structural* information from protein to protein. An existing prion acts as a template for converting a protein with nonprion structure into one with prion structure. This was indeed a new, exciting, and unexpected means of disease transmission, and it was entirely consistent with the Central Dogma.[39]

In contrast to intrinsic bias, extrinsic bias is introduced into the scientific enterprise in a premeditated or conscious manner. The investigator chooses to bias their own experiments and the interpretations therefrom. In a sense, this bias is the scientific equivalent of prosecuting a legal case in court or selling products to the public. The law paradigm is truth emerges from the legal dialectic produced by the argumentation of opposing views (prosecution/defense or plaintiff/defendant). It is the responsibility of each lawyer to argue in as cogent and compelling a manner as possible the merits of their own case. It is hoped, in doing so, that the court or jury will be able to integrate the information thus provided into a consistent, rational perspective on the matter at issue, and by doing so, render an impartial, fair opinion. There is no assumption of objectivity. Each side is concerned only with supporting their position and with working assiduously to weaken the position of the other side. Marketers do the same thing because they want you to buy their products (i.e., "rule in their favor"). Success for them is measured in sales, not in the level of truth intrinsic to their marketing strategies. Both behaviors are unacceptable in, and incompatible with, the scientific enterprise, which sees as its *raison d'etre* the provision of accurate descriptions of the universe.

A common example of scientific bias is selective data consideration – "cherry picking." For example, a scientist propounds a new hypothesis that predicts that inhibition of enzyme *Z* should prevent production of product *A*. In a set of carefully controlled experiments *in vitro*, the scientist notes that when an enzyme inhibitor was added to the test system, the level of product *A* dropped by 75%. The scientist then wrote in a manuscript that their experimental evidence showed that enzyme inhibition "blocked" production of *A*. The problem is product *A was* produced, albeit at 25% of its original level. Why was this so? Two reasonable explanations would be: (1) the enzyme inhibitor was not 100% efficient and (2) another enzyme existed that also could produce *A*. The scientist's interpretation was incorrect because they cherry-picked data in support of their hypothesis and left out conflicting information. If a manuscript were published reporting these results, the scientific community could be misled. Of course, the scientist could be right, but the data extant did not support their conclusion.

Cherry-picking may also be found in strictly verbal communication. Review articles often provide examples. A good review should provide the reader with a summary of the most important past and current work in a particular area and a narrative of the work that gives the reader a limited number of "take home messages." Each goal, however, is subject to bias because the author must make value judgements about which papers to include in the review and what, taken together, they tell us about the most significant findings, advances, and future directions of the field.

---

[39] Digging deeply would have revealed Crick had already thought about, and not ruled out, protein→protein information transfer [374], specifically mentioning replication of the scrapie agent.

Table 5.3 *Ballpark concentrations of Aβ42 in a variety of physiologically relevant contexts.*

Context	Ballpark concentration
Aβ42 "monomer in a fibril"	~221.5 mM
Aβ42 in a senile plaque	~50 mM
Neurotoxicity	~10–50 uM
Negative LTP modulation	~500 nM
Positive LTP modulation	~500 pM
Aβ42 generation by the brain	~330 fM s^{-1}

Table 5.4 *Ballpark concentrations of Aβ42 in a variety of physiologically relevant contexts (revised).*

Context	Ballpark concentration
Aβ42 "monomer-in-fibril"	≈221.5 mM
Neurotoxicity	≈10–100 uM
Negative LTP modulation	≈50–500 nM
Positive LTP modulation	≈50–500 pM
Aβ42 generation by the brain	≥330 fM s^{-1}

I recently encountered such bias during discussions with a colleague about preparation of a review article that sought to provide Alzheimer's disease researchers with reference values for important biological metrics. Just as blood tests have result ranges defining normal and abnormal clinical states, understanding what concentrations of the amyloid β-protein (Aβ), the protein that deposits in the brain in Alzheimer's disease, produce different types of neuronal injury and death is critical for both basic and clinical research into disease causation and cure. Put simply, how does a researcher or clinician know when a particular concentration of Aβ in a specific part of the brain, or in a culture of neurons *in vitro*, is normal or abnormal? What is the standard for comparison? Informing the field of such standards was the goal of the review, and after careful consideration, a table of values was created (Table 5.3). Five standard values were listed, along with a standard range ("neurotoxicity").

The potential biases in this case included data selection (which results should be believed?), and most importantly, the desire to provide the field with exact numbers. The problem in biology, in contrast with physics, is that very few, if any, highly precise standards (e.g., Planck's or Boltzmann's) exist. Instead, variability, and sometimes *huge* variability, is the norm. Recognizing that the *pre facto* desire to provide absolutes to "standardize the field" was not possible, the table was revised (Table 5.4) [375]. Conservative ranges of concentrations were listed for three effects of Aβ on neurons, "neurotoxicity," "negative LTP modulation," and "positive LTP modulation." Unbiased re-evaluation of data on the concentration of Aβ42 in a senile plaque (deposit) led to the conclusion that they were not rigorous or consistent enough to allow the suggestion of any concentration, so this metric was deleted from the table. By contrast, structural studies of Aβ fibrils and first principles of protein structure *did* allow a reasonable estimate of the Aβ concentration in brain deposits ("Aβ monomer in a fibril") to be made. Similarly, rigorous data on Aβ generation rate in humans were available, allowing inclusion of this metric in the table. This vignette provides a good example of researchers recognizing their own intrinsic biases and, in doing so, acting to eliminate them.

5.4 Inference

Solution of problems too complicated for common sense to solve is achieved by long strings of mixed inductive and deductive inferences that weave back and forth between the observed machine and the mental hierarchy of the machine found in the manuals.[40]

We've done the "hard work," having produced and analyzed our data. Now the fun begins because we can exercise our creativity and genius (however much we have) in understanding the meaning and implications of those data. This is the discovery stage in the scientific method. Scientists, by nature, have boundless curiosity and love the thrill of solving problems. Now is the time they may find the answer to a riddle they posed and sought to solve by experimentation, conceive of novel hypotheses or theories, or even realize what were thought to be correct answers to prior problems likely were wrong. We now can *explain* things that previously were unexplained. However, *to explain* means to understand how things work, a type of understanding that not only is practical in nature (two masses attract each other with a force $F = Gm_1m_2/r^2$) but also, and maybe more importantly, conceptual (observations, thoughts, and knowledge extant are synthesized into a concept – gravity) or abstract. How do we move from data to understanding? With the exception of flashes of genius or scientific epiphanies, we must have a rational process(es) to do so. That process is *inference*.

The Oxford English Dictionary[41] defines *inference* as:

1a. ...the drawing of a conclusion from known or assumed facts or statements; esp. in Logic, the forming of a conclusion from data or premises, either by inductive or deductive methods; reasoning from something known or assumed to something else which follows from it.

In addition to inductive and deductive methods, two additional, but related, types of inference, *abduction* and *inference to the best explanation* (IBE), are important parts of scientific practice. With the exception of IBE (see Section 5.4.2), each method of inference has existed at least as early as the time of Aristotle, when he discussed them in his book *Prior Analytics*. Each is distinct from the other,[42] although debate continues with respect to precisely what induction and abduction entail and the relationship between the two.

---

[40] From *Zen and the Art of Motorcycle Maintenance* by Robert M. Pirsig.
[41] See www.oed.com/view/Entry/95308?redirectedFrom=inference#eid.
[42] If you have trouble differentiating abduction, induction, and deduction, thinking about the Latin roots of these words might help. All three words are based on the word *ducere*, meaning "to lead." The prefix de- means "from," and deduction derives from generally accepted statements or facts. The prefix in- means "to" or "toward," and induction leads you to a generalization. The prefix ab- means "away," thus abduction helps you choose the best explanation from among many.

In this discussion, we will consider both "hypothesis" and "theory" to be *suppositions* that can be tested experimentally. In common scientific usage, hypotheses are focused suppositions about the mechanics of phenomena, whereas theories are broader in scope and often comprise laws formulated based on the results of extensive experimentation. For example, the *theory* of evolution provides an explanation for general aspects of speciation through natural selection, but it doesn't tell us explicitly and specifically why the beaks of hummingbirds vary so substantially in shape and length. We might *hypothesize* that because beaks are essential for feeding, different beaks might have something to do with the shapes of the flowers the birds feed on. This hypothesis could be tested simply by extensive observation of the behaviors of different types of birds, which would show, for example, that long, curved beaks are necessary for the animals to be able to feed on trumpet-shaped flowers in which the nectar is located deep inside.

Our working definitions of these types of inference are:

Abduction is hypothesis/theory creation by selection (educated guessing) of likely explanations. "[T]here is an implicit or explicit appeal to explanatory considerations"[43] without the need for accumulation of instances of the phenomena or dependence on likelihood estimates (statistics).

    Induction, in contrast, is "inference of a generalized conclusion from particular instances."[44] "There is *only* an appeal to observed frequencies or statistics."[45]

    Deduction is the logical process of deriving a conclusion from two or more antecedent premises. This process is strictly syntactic. If the premises are true, the conclusion must be true (as a matter of logical necessity). However, within this syntactic construct, lexical objects need not represent anything in the real world *per se*, although they generally do in science practice.[46]

Each method of inference can, and is, used singly in science practice, but usually the three are used in combination. Abduction and induction are particularly important in hypothesis/theory creation, selection, and testing. In contrast to deduction, neither depends on premises *assumed* to be true. Such premises are common in physics, where the high precision of experiments allows one to accept their truth confidently. However, the complexity and experimental imprecision characteristic of biological systems makes the use of deductive inference more difficult. Nevertheless, deduction is used routinely in the consideration of biological *sub-systems* (e.g., systems comprising a single type of cell [a sensory neuron] or a specific region of an organ [the pancreatic islets of Langerhans, which produce insulin]), for which premises *may* exist.

All three methods are used to establish cause-and-effect relationships, a primary aim of science. We *abduce* (make an educated guess) that one cause from many is responsible for effects observed empirically. We *induce* that a particular hypothesis or theory is a generalization of our observations. We *deduce* from observations of phenomena (premises) what might cause them. Peirce used coffee beans to exemplify these processes (Table 5.5).

---

[43] Quotation from [376].
[44] Quotation from *Merriam-Webster* dictionary www.merriam-webster.com/words-at-play/deduction-vs-induction-vs-abduction.
[45] Quotation from [376].
[46] Deductive principles are axiomatic so a separate section explicating deduction is not included.

Table 5.5 *Examples of inferential syllogisms.*[47]

Abduction	Induction	Deduction
All the beans from this bag are white	These beans are from this bag	All the beans from this bag are white
These beans are white	*These beans are white*	These beans are from this bag
∴ These beans are from this bag	∴ All the beans from this bag are white	∴ *These beans are white*

Implicit in our discussion thus far is the assumption that any inferences we make are derived rationally and supported by data extant. However, inferential processes are susceptible to errors of fact and the same types of biases discussed in Section 5.2. Outside of being a poor experimentalist, inferential errors are among the most common in science. Even if inferences are made thoughtfully and objectively, the same data may yield multiple, mutually exclusive inferences. This can lead to contentious debates within the scientific community, debates that may take substantial time to resolve (even up to millennia). A classic example is the contrasting inferences made from observations of the movement of the sun and planets. Observing the *same* sun and planets, Aristotle and Ptolemy inferred a geocentric heavens while Copernicus and Galileo, almost two millennia later, inferred a heavens that was heliocentric. A more recent example is that of high temperature superconductivity, where, using the same data, many different models have been proposed [378]. How could this be? The answer comes from differences in inferential procedure and theory formation as well as in methods of establishing theory truth, topics that we now discuss in more detail.

5.4.1 Abduction

Charles Sanders Peirce (1839–1914), a famous American philosopher, published extensively on scientific inference. At the time, deduction and induction were the primary accepted modes of inference. Peirce found these insufficient descriptors of the process of discovery. Deduction could only operate once premises were known, thus it could not support discovery. Similarly, induction, especially enumerative induction, was a process that also depended on the *a priori* existence of multiple examples. Neither inferential method addressed that creative moment when observation is transformed *directly* into explanation. Peirce termed this process "abduction" and considered it the third member of a triad of inductive methods that included deduction and induction.[48] An early example of Peirce's abductive logic dealt with the "Aha moment"[49] of scientific discovery when an hypothesis *H* is conceived based on a novel observation *E*.

---

[47] Examples taken from Peirce [377], 2.622.

[48] During his lifetime, he also referred to this type of inference as "hypothesis," "hypothetic inference," "presumption," and "retroduction." In addition to this terminological plasticity, Peirce's own conception of abduction was plastic, changing as he aged.

[49] According to the Merriam-Webster Dictionary, an "aha moment" is a moment of sudden realization, inspiration, insight, recognition, or comprehension .

The surprising fact, E, is observed;
> But if H were true, E would be a matter of course,
> Hence, there is reason to suspect that H is true.
> Peirce characterized abduction and its relationship to deduction and induction as follows.[50]

Abduction is the process of forming an explanatory hypothesis. It is the only logical operation which introduces any new idea; for induction does nothing but determine a value [**i.e, statistical**], and deduction merely evolves the necessary consequences of a pure hypothesis. Deduction proves that something **must be**; Induction shows that something **actually is** operative; Abduction merely suggests that something **may be**. Its only justification is that from its suggestion deduction can draw a prediction which can be tested by induction, and that, if we are ever to learn anything or to understand phenomena at all, it must be by abduction that this is to be brought about.

Peirce referred to abduction at various times as "...an appeal to instinct,"[51] "nothing but guessing,"[52] or a "suggestion [that] comes to us like a flash – ...an act of insight."[53] This "aha moment" produced a new hypothesis, but one that would be "on probation" until it was confirmed using inductive methods. Abduction and deduction were at "opposite poles of reason."[54]

Referring to Peirecean syllogisms in Table 5.5, we see that deductions are made using predicate logic. Formally, these deductions are purely syntactical, and are absolute, because they depend only on the truth of their major and minor premises. However, if the premises are true within a natural system, for example, in biology, chemistry, or physics, then the conclusions deduced therefrom have real world meaning. Inductions extrapolate from one or a few observations to the general case, for example, "all the beans from this bag are white," – a testable hypothesis. Abduction, by contrast, produces the logical fallacy "these beans are from this bag." Thus, abduction is logically fallacious because although the beans may, in fact, be from the bag, and this is a reasonable first hypothesis, they also may come from any number of other places.

One might ask, how does a method of inference that by definition comprises a logical fallacy be of any use? This type of fallacy cannot *exclude* any hypothesis, and thus one is faced with the prospect of testing an infinite number of them. The answer, as argued by Peirce, and later by many philosophers of science, is that means must employed to eliminate hypotheses that clearly are "fantastic"[55] or highly unlikely, i.e., one must parse "insights that come in a flash." Choosing hypotheses or theories for study is critical for scientific advancement. However, this does not mean that *correct* choices are necessary. We cannot know *pre facto* whether our hypotheses will turn out to have experimental support, which means that many (most?) hypotheses will be wrong. Well-chosen theories can yield valuable insights regardless of their veracity.

---

[50] Quotation from Peirce [377], 5.171
[51] Quotation from Peirce [377], 1.630.
[52] Quotation from Peirce [377], 7.219.
[53] Quotation from Peirce [377], 5.181.
[54] Quotation from Peirce [377], as quoted in Peirce [379], 7.218.
[55] Peirce's word.

In a sense, Peirce's arguments are muddled. After all, abduction and induction both involve facts extant and both yield hypotheses consistent with these facts.[56] The difference is that abduction considers a limited number of facts but an indeterminate number of hypotheses explaining them, from which the scientist selects (guesses) one that might be true. Induction begins with a limited number of facts extant and assesses probabilistically the value of a hypothesis, explaining them based on the consistency of prior and future experiments.[57] Additional experiments contribute nothing to our abduction *per se* because they had not been done when the abduction was conceived. By contrast, additional experiments may strengthen (or weaken) an induction depending on the number that are consistent (or inconsistent) with what has been observed before. Note the process here. Abduction focuses on the one whereas induction involves the many because it is theory selection, not theory creation *per se* (at least in Peirce's view).

Peirce's suggestion of a third form of inference generated tremendous debate over the next century. Was abduction a discovery process? Was it a logical process encompassing a defined series of steps? Was it both, or something else? Peirce argued strongly that abduction provided the means for scientific discovery, but he also argued that abduction was a logical process!

It must be remembered that abduction, although it is very little hampered by logical rules, nevertheless is a logical inference asserting its conclusion only problematically or conjecturally, it is true, but nevertheless having a perfectly definite logical form.[58]

Here, in the same sentence, Peirce appears to argue that abduction was *both* a method for discovery and a logical process. This complex nature of abduction was reflected in the work of later philosophers of science. For example, Anderson considered abduction to be *"both an insight and an inference"* [381] and Thagard has suggested that it often supports both *"discovery and justification of scientific theories."*[59] Frankfurt [383] suggested that abduction was paradoxical because its *"hypotheses are the products of a wonderful imaginative faculty ... [and] of a certain sort of logical inference,"*[60] which is essentially what Peirce argued. Plutynski, by contrast, wrote *"abduction is a strategic practice (involving merely plausible belief) that results in a proposed course of action – investigating a hypothesis further. ... Peirce is not ... advancing a new form of inference but, rather, a form of life, a practice for science."*[61] Whether one considers abduction to have a "logical form" or be a "practice of science," or not, Peirce did suggest that abduction should *explain* and be

---

[56] Peirce himself commented *"Concerning the relations of these three modes of inference to the categories and concerning certain other details, my opinions, I confess, have wavered."* Quotation from [377], as quoted in Peirce [379], 5.146.

[57] Peirce summarized the differences among abduction, deduction, and induction as follows (see Peirce [380]). *"I ask ... whether, if people, instead of using the word probability without any clear apprehension of their own meaning, had always spoken of relative frequency, they could have failed to see that what they wanted was not to follow along the synthetic [inductive] procedure with an analytic [deduction] one, in order to find the probability of the conclusion; but, on the contrary, to begin with the fact at which the synthetic inference aims, and follow back to the facts it [abduction] uses for premises in order to see the probability of their being such as will yield the truth."*

[58] Quotation from [377], as quoted in Peirce [379], 5.188.

[59] Quotation from Plutynski [382].

[60] According to Frankfurt, they can't be both, hence the paradox.

[61] Quotation from Plutynski [382], pp. 11–22.

testable, simple, plausible, and unifying (have broad applicability) [384]. He specifically rejected consideration of what he called "antecedent likelihoods" – what we now associate with Bayesian (statistical) approaches to theory selection.

Peirce's view of the dynamics of inference in practice is illustrated in Fig. 5.10. It comprises sequential application of all three inferential modes, abduction, deduction, and induction, beginning in the universe of all imaginable hypotheses (above the dashed line in the figure) explaining a phenomenon. The process then systematically reduces this huge number to one. The first step in the process is at the heart of abduction *per se*, namely conceiving of an hypothesis to explain a phenomenon. This gives the imagination *almost* free rein but rejects those hypotheses that clearly are useless or impossible *a priori*. Peirce then applies two additional filters, economic considerations and plausibility, to rank the imagined hypotheses. A close reading of Peirce's development of his own ideas reveals that only "insight" is a *sine qua non* for the abductive "process."

One may, and Peirce suggests one should, consider other factors, but any such consideration would only come later, after one's creativity was exercised in the absence of any constraints. In essence, the scientist lets the black box that is their brain "do its thing." One can argue that within the brain, data provided to it are processed in ways dependent on prior knowledge, attitudes, biases, personality, career considerations, impact, plausibility, value, significance, simplicity, elegance, testability, unifiability, etc. The scientist does not know, nor need to know, how these processes actually work. The only important thing is the output of the black box, namely hypotheses. Once a hypothesis is conceived, it then passes through stages of deduction, in which predictions are made based on the premises of the hypothesis, and induction, in which experiments are performed from which general principles can be induced. The end result is the acceptance of one hypothesis (the "best explanation") and the rejection of others. The triad of inferential methods is arranged linearly in Fig. 5.10, but, as discussed above, and as argued in the quotation of Minnameier below [385], these methods may, in practice, be unconsciously applied simultaneously.

What I am driving at is the fact that even in the simplest inductions we first abduce to some explanation that makes sense for us, and only then do we come to accept (or possibly reject) this explanation. ... abduction, deduction, and induction may, in the extreme, be merged into one single and immediate conscious judgement. Once again this makes clear that even in ordinary and habitual knowledge application the inferential triad must be operative, even though it is not consciously followed through. ... Without abduction we would apparently not be able to make sense of our experiences, without deduction we would not be able to derive suitable action schemes, and without induction we would never find fault with what we are actually doing. [Author's underlining.]

A feature of Peirce's view of inference that one might not at first expect was that of "economic considerations," which Peirce wrote depended on three factors: cost, the value of the thing proposed, in itself, and its effect upon other projects.[62] Peirce further opined:[63]

---

[62] See Peirce [386], section 7.220.
[63] See Peirce [377], sections 1.122–1.125 §23, The Economy of Research.

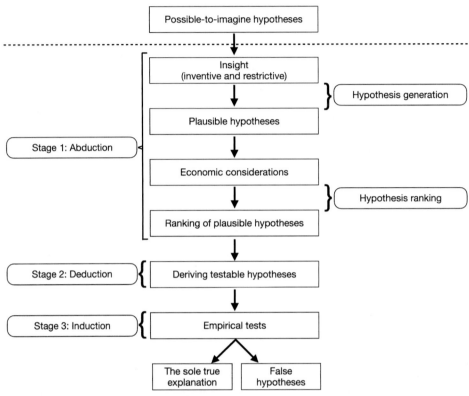

Figure 5.10 Stages of scientific inquiry. The diagram illustrates Peirce's sense of the dynamics of inference that lead from imagination/epiphany to considered fact. Above the dashed line is the universe of imaginable hypotheses. (Credit: Reprinted by permission from Springer Nature [396, Mohammadian], ©2019.)

The value of knowledge is, for the purposes of science, in one sense absolute. It is not to be measured ...in money; in one sense that is true. But knowledge that leads to other knowledge is more valuable in proportion to the trouble it saves in the way of expenditure to get that other knowledge. Having a certain fund of energy, time, money, etc., all of which are merchantable articles to spend upon research, the question is how much is to be allowed to each investigation; and for us the value of that investigation is the amount of money it will pay us to spend upon it. Relatively, therefore, knowledge, even of a purely scientific kind, has a money value.

Peirce argued that the more we know, the more valuable our knowledge would be, but he also suggested there was a point of diminishing returns at which the cost of doing the research was more than the economic benefit to be derived from it. Theory choice, in fact, was to be determined, in part, by comparison of projected remunerations among competing theories. Theories that "reinvented the wheel" were to be eschewed – as were parallel investigation of similar theories – only the most economically valuable was to be pursued. In scientific practice, these concerns often are of critical importance, especially when one is trying to convince a funding body to support one's research. Those who provide money expect valuable results. However, such concerns should only be considered after, not before, the process of hypothesis conception.

One contentious issue regarding abduction is its epistemological value, which is one means of comparing it to, and potentially distinguishing it from, induction and deduction. Does abduction tell us anything true about the world? Deductive inferences are absolute and unassailable, but only if their premises are absolutely true, which we will find may be impossible to determine. The truth value of inductive inferences can be quantified through likelihood estimates using standard or Bayesian statistical tools. We derive a number that represents a level of confidence. Abductions do not approach the levels of believability possible with deductions and inductions. Is abduction simply weak induction or is it not an inferential mode at all? Opinions vary, which is why the issue remains contentious.

Critiques of abduction may also center on its explanatory power and truth relative to other inferential modes. Quine has argued that abduction is useless because it cannot distinguish among many different hypotheses explaining a phenomenon[64] – but that was not Peirce's goal. Van Fraassen also raised the usefulness question [388], but with respect to the ability of an abduction to be applicable generally as opposed to only in certain contexts – but that was not Peirce's goal. Laudan's "pessimistic induction" [259], the notion that it is unwarranted to have confidence in the truth value of any inductively derived theory because of potential future falsification of the theory, is equally or more relevant to abduction. So, should abductive approaches to explaining our data and the larger world be rejected? I think not.

I argue for the importance and necessity of abductive inference based on the fact that it can be *useful* in normal scientific practice. If one seeks to justify, evaluate, or judge abduction, context is crucial. Philosophers of science rightly discuss abduction in contexts including epistemological, logical, explanatory, prescriptive, normative, and practical. These have been discussed above to help the reader develop an intuition about when and how to apply abductive inference. This application is practical, not theoretical. It is in the former context that abduction is most important because abduction is particularly useful in hypothesis creation and discovery. It can leverage nonverbal, nonphilosophical, quale-like things, such as epiphanies and hunches, to free the scientist from dogmatic or intrinsically biased thought processes that interfere with innovative thinking. Abduction provides the means to conceive of new things and simultaneously parse them so that obviously unlikely or "crazy"[65] ideas are not considered further. Whether abduction constitutes a logical process for doing so is largely irrelevant to the *use* of abduction.

---

[64] See Quine [387], p. 23.
[65] Of course, defining "unlikely or crazy ideas" is a subject for discussion itself, but for our purposes here we will stipulate that common sense will allow us to recognize these types of ideas.

5.4.2 Inference to the Best Explanation (IBE)

In 1965, Gilbert Harman proposed that enumerative induction[66] was a special case of a more general class of nondeductive inference he termed "the inference to the best explanation" (IBE) [389]. Although Harman stated that IBE corresponded *approximately* to abduction and other inferential methods,[67] IBE differed in important respects. The question of whether these were sufficient to distinguish it from abduction has been an active area of debate. Harman provided the following definition of IBE.

In [IBE] ... one infers, from the fact that a certain hypothesis would explain the evidence, to the truth of that hypothesis. In general, there will be several hypotheses which might explain the evidence, so one must be able to reject all such alternative hypotheses before one is warranted in making the inference. Thus one infers, from the premise that a given hypothesis would provide a 'better' explanation for the evidence than would any other hypothesis, to the conclusion that the given hypothesis is true.

The crux of Harman's definition, and a feature that distinguished it from other types of inference, was the premise that IBE was a method of selecting a single hypothesis from many competing hypotheses to establish *truth*. It was not simply for determining the frequency of particular observations (enumerative induction) nor was it a guessing or discovery process *per se* (abduction). It was, however, similar to other inductive methods in its incorporation of explanatory power and simplicity, among other things, as metrics for theory choice.

One of the most prominent advocates of IBE has been Peter Lipton, whose book, *Inference to the Best Explanation*, has a played seminal role in explicating IBE [348]. Lipton provided his own definition of IBE.[68]

... a new model of induction, one that binds explanation and inference in an intimate and exciting way. According to Inference to the Best Explanation, our inferential practices are governed by explanatory considerations. Given our data and our background beliefs, we infer what would, if true, provide the best of the competing explanations we can generate of those data (so long as the best is good enough for us to make any inference at all). Far from explanation only coming on the scene after the inferential work has been done, the core idea of Inference to the Best Explanation is that explanatory considerations are a guide to inference.

Of course, as you might have already wondered, don't these two quotations perforce raise the questions of "what is truth" and "what is explanation?" *Truth* (with an uppercase T) vs. *truth* (with a lowercase t) is a ubiquitous issue (see page 10, footnote 8) for which there can be no consensus, nor can there be one for explanation. However, for each issue, metaphysical, physical, statistical, and practical definitions have been proffered. We discuss these here in the context of IBE and its distinction from abduction (see Section 5.7 for more detail).

---

[66] A number of definitions exist for enumerative induction, all of which invoke a frequency of observation argument to support extrapolating from instances to general phenomena. Harman's definition was *"From the fact that all observed A's are B's we may infer that all A's are B's (or we may infer that at least the next A will probably be a B)."* Enumerative induction, at the time, also was referred to as "induction by simple enumeration," "simple predictive induction," or just "induction."

[67] These included "'the method of hypothesis', 'hypothetic inference', 'the method of elimination', 'eliminative induction', and 'theoretical inference'."

[68] See Lipton [348], p. 56.

Table 5.6 *Comparison of features of abduction and IBE. A checkmark signifies that an inferential method incorporates the feature. Shared features are grouped at the bottom. (See text for explanations of features.)*

Feature	Abduction	IBE
Single best "guess"	✓	
Ab initio	✓	
Short listing (Filter 1)		✓
Ranking (Filter 2)		✓
Lemmas		✓
Enumerative		✓
Likeliness		✓
Loveliness		✓
Economy	✓	
Explanatory	✓	✓
Verifiable	✓	✓
Simple	✓	✓
Unifying	✓	✓
Field aware	✓	✓

The field of philosophy of science (PoS) constantly evolves, and as it does, old controversies may be settled and new ones created. When Peirce argued abduction as a third fundamental class of inference, he did so at a time when a distinction between the contexts of discovery and justification, or of philosophers' need to define separate descriptive or normative functions of science, did not exist [390]. Peirce saw abduction as an explanatory and theoretical part of inference, as opposed to induction's descriptive, summarizing function. He also considered the idea of abduction something that should continue to be investigated, which explains why his sense of abduction later included its role in hypothesis testing [391]. This malleable view of abduction is one reason that its relationship to IBE has been controversial. A second reason is that simple definitions of IBE, as quoted above, cannot fully convey the complexities of the process and the contexts in which it operates. However, even with an historically accurate accounting of the meaning of abduction, opinions about the relationship of abduction to IBE have varied from synonymous with [389], to a special case of [392], to similar to [393], to distinct from [385, 388, 394–398]. Most philosophers of science now subscribe to this latter view, although they do admit similarities between abduction and IBE and thus may characterize their relationship as "similar to but different from."

Distinctions and similarities between abduction and IBE have been based on a number of considerations, including process, logic, temporality, and importantly, semantics [394]. How one defines truth, explanation, explaining, simplicity, loveliness, likeliness, plausibility, etc. shapes one's opinion in the matter. Table 5.6 is a partial list of qualities through which abduction and IBE may be characterized.

According to Lipton, although inference and explanation are considered contemporaneously in both IBE and abduction, in IBE this process is guided by an explicit logic of hypothesis selection – a ranking process – whereas, by contrast, abduction involves immediate insight, epiphany, or guessing – a single best guess. Abduction thus *discovers* hypotheses *ab initio*, whereas IBE *ranks* multiple, preexistent, competing hypotheses. According to Paavola [396], *"Lipton's own basis for IBE is the causal model of explanation and the model of contrastive explanations, that is, explanations that answer questions of the form 'Why this rather than that?'"* By contrast, abduction answers the simpler question, "Why this?" *directly* by conceiving *a* hypothesis that explains a phenomenon. The distinction thus may be considered "inference to the *best* explanation" vs. "inference to *the* explanation" [399].

How many explanations, i.e., *hypotheses that explain*, are considered in IBE? Lipton acknowledges that one cannot consider the entire universe of potential hypotheses because one cannot conceive of all of them and, even if one did, evaluate the merits of each. Lipton thus discusses a two-stage filtering process, the first step of which is creation of a "short list," from which the best hypothesis then is selected. Both filters involve explanatory considerations, but the first is distinguished from the second in its dependence on "background beliefs." These beliefs are explicit in IBE and implicit in abduction. Yet Lipton also admits that some hypotheses are rejected *"because they are never considered."*[69] For what reasons is not specified, but we assume that the reasons are implicit in our perspectives on science and nature and that they are employed reflexively rather than consciously. Once we have the short list, we rank (filter) them using many of the criteria listed in Table 5.6. These filters are not part of Peircean abduction because it is not a procedural process *per se*.

IBE is underpinned by the assumption that it produces true inferences about the natural world. However, this assumption is unwarranted because IBE (and abduction and induction) is a practical process for providing understanding. It is not deductive in nature and therefore can only guide us toward *truth*, but never to *Truth*. Lipton recognizes this fact when he says *"[o]ur inductive practice is fallible: we sometimes reasonably infer falsehood."*[70] Lipton argues, using simple logic, if we choose one explanation from many, then the ones not chosen cannot be true, so we are not selecting the best true explanation from among many. We are selecting the best *potential* explanation and thus are practicing "Inference to the Best Potential Explanation" (IBPE). Hempel, in discussing what he refers to as *potential deductive-nomological explanations*, provides two informative examples.[71]

We use the notion of a potential explanation, for example, when we ask whether a novel and as yet untested law or theory would provide an explanation for some empirical phenomenon; or when we say that the phlogiston theory, though now discarded, afforded an explanation for certain aspects of combustion.

---

[69] Quotation from Lipton [348], p. 149.

[70] See Lipton [348], p. 57.

[71] From Hempel [400], p. 338. "Deductive-nomological" is a model for scientific explanation. The name derives from the requirement that a given phenomenon is *explained by deducing* its description *from a law*, plus a description of the particular circumstances in which the phenomenon in question occurs.

Here there is no mention of truth – only the possibility of it. Instead, it is the practical value of the inference in explaining that is important. However, even this potential for truth is actually one step farther away from the truth than may at first be obvious because our assumption of the truth of the givens upon which we evaluate the potential for truth in our inference may, in fact, be unwarranted. In other words, what is intrinsic to IBE is the determination of the "truth of the truths" warranting the explanation. Harman expressed this more subtle point about IBE as follows [389].

...a necessary condition of knowledge is not only that our final belief be true, but also that the lemmas, or intermediate propositions between premises and conclusion, be true.

This is a restatement of one definition of knowledge that was considered as early as Socrates,[72] namely "knowledge was true opinion accompanied by reason."[73] For Harman, lemmas are the reasons or justifications upon which knowledge is based. On the surface, abduction does not involve determining the "truth of the truths," which distinguishes it from IBE. However, we must, eventually, satisfy Harman's "necessary conditions" for our abductions to be true. The difference between IBE and abduction is that this satisfaction must be achieved *ab initio* in IBE, whereas it is (or is not) achieved *after* we abduce and move forward to test our abduction (see Fig. 5.10).

By definition, IBE involves a selection process requiring a determination of "best," but what does best actually mean? What features characterize best? How does one weight the contribution of each feature to the overall determination of "best," i.e., how do we create a scale with a range worst→best? On what basis is this evaluation to be made? Metrics may be quantitative *and* qualitative, as exemplified, for example, by likeliness and loveliness,[74] respectively. But here we also run into problems because likeliness and loveliness also must be defined if we are to evaluate how "likely" or how "lovely" an hypothesis is. Lipton defines likeliest as the most probable, i.e., most warranted. Defining lovely is substantially more difficult. The expression "beauty is in the eye of the beholder"[75] illustrates this, and Lipton agrees.[76]

...explanatory loveliness is too subjective and interest-relative to give an account of inference that reflects the objective features of inductive warrant.

Lipton also asks a more fundamental question about the natural world, paraphrasing Voltaire.

What reasons do we have to believe that the loveliest explanation would also be the most likely to be true; why would we think that we live in the loveliest of all possible worlds?[77]

---

[72] See Plato, *Theaetetus*, 201.

[73] In more modern terms, knowledge has been defined as "justified true belief (JTB)" and expressed logically as "*S* knows *that p* if and only if *p* is true; *S* believes *that p*; therefore *S* is justified in believing *that p*."

[74] If one chooses to base their selection of potential hypotheses on their relative loveliness, one is performing Inference to the Loveliest Potential Explanation.

[75] This expression is first found in 1886 on page 142 of the book "Molly Bawn," written by Margaret Argles Hungerford. Lipton refers to this as "Hungerford's Objection."

[76] Quotation from Lipton [348], p. 70.

[77] Lipton refers to this as "Voltaire's objection."

The loveliest, if correct, would be that explanation that would be "the most explanatory or provide the most understanding."[78] Note that each criterion could result in selection of a different explanation from the same initial group of hypotheses. This possibility is one of the criticisms of IBE [401], as is the argument that the connection between loveliness and likeliness may be illusory because likeliness arises from principles distinct from those of loveliness. Further, the likeliest explanations usually are the loveliest because of their explanatory power. However, lovely explanations, although aesthetically appealing in a scientific or philosophical sense, still may be highly unlikely or impossible.

Returning to likelihood and the metrics by which it may be determined, it is clear that IBE is, in a fundamental manner, a probabilistic process. We may choose a hypothesis based on its likelihood, loveliness, simplicity, plausability, explanatory power, etc., but in each case there exists an implicit, and sometimes explicit, estimation of the probability (likelihood) of the selection factor. For example, we may say "hypothesis #1 is better at explaining than any other hypothesis," but "better" is a relative term that requires creation of a worst→best metric. Van Fraassen also believes that probabilism is a component in IBE.[79]

Despite its name, it is not the rule to infer the truth of the best available explanation. That is only a code for the real rule, which is to allocate our personal probabilities with due respect to explanation [and prediction]. Explanatory power is a mark of truth, not infallible, but a characteristic symptom.

A powerful statistical method for estimating likeliness based on *past* likelihoods uses Bayes's Theorem.[80]

$$P(H \mid E) = \frac{P(E \mid H) \cdot P(H)}{P(E)} \tag{5.1}$$

Here, we determine the *posterior probability* ($P(H \mid E)$) of an hypothesis H given new evidence, E, is true, by first calculating the product of the *likelihood* of observing E given the hypothesis H ($P(E \mid H)$) and the *prior probability* of the hypothesis H on its own ($P(H)$), and then dividing this product by the probability of E on its own ($P(E)$). Bayes' theorem uses estimates, i.e., *quantified beliefs*, to determine the future probability of an observation.[81]

Manipulation of Eq. 5.1 provides a number of take-home messages. The first is that if we eliminate $P(E)$ (a constant), then we find that the posterior probability depends only on prior probability and its newly acquired likelihood (Eq. 5.2). Here it is clear that the accuracy of probability estimates depends to a large extent on an estimate, $P(E \mid H)$, that is subject to bias and errors of fact. Work by Tversky and Kahneman (winner of the 2002

---

[78] See Lipton [348], p. 59.
[79] See Van Frassen [388], pp. 146–149.
[80] Please see Table 4.3 for an explanation of statistical and logical symbols.
[81] $P(H)$ and $P(E)$, termed "marginal probabilities," do not depend on the values of other variables.

Nobel Prize in Economic Sciences[82]) has, in fact, provided many examples of exceptionally inaccurate estimates, even by accomplished scientists [402, 403].

A second message, if we assume H entails E, i.e., $P(H) = 1$ and that the marginal probabilities of priors $P(H)$ and $P(E)$ are not 1 or 0, is the posterior probability must be greater than the prior probability because $P(E) < 1$ (Eq. 5.3). A consequence of this is if an unlikely prediction ($P(E) << 1$) of a theory is found to be true. In this case, the hypothesis is more strongly confirmed than would be a prediction that was highly likely ($P(E) \approx 1$). This provides statistical support for the notion that hypotheses that give rise to novel predictions are to be preferred over those that do not (see Section 5.5.9 for a discussion of this point).

$$P(H \mid E) \propto P(E \mid H) \cdot P(H) \tag{5.2}$$

$$P(H \mid E) = \frac{P(H)}{P(E)}. \tag{5.3}$$

Lipton argues that Bayesianism and Inference to the Likeliest Explanation are compatible, if likeliness is posterior probability as determined using Bayes' theorem [348]. Lipton also extends this compatibility to Inference to the Loveliest Explanation. Note that Lipton does not equate IBE with Bayesianism but rather focusses on the complementarity of the two procedures. Plutynski also does not consider IBE and Bayesianism to be two independent entities.

...IBE and Bayes are consistent, not competing accounts of nondeductive inference. IBE is a sort of heuristic, helping us to meet Bayesian standards. ...explanatory considerations are shorthand for what amount to appeals to priors.[83]

Okasha also subscribes to the notion of compatibility but goes even farther, arguing in fact that IBE can be modeled in Bayesian terms.[84]

The correct way of representing IBE ...views the goodness of explanation of a hypothesis vis-à-vis a piece of data as reflected in the prior probability of the hypothesis P(H), and the probability of the data given the hypothesis $P(E \mid H)$ [as in Eq. 5.2]. The better the explanation, the higher is one or both of these probabilities. Relative to this account, favouring a hypothesis on the grounds that it provides a better explanation of one's data than other hypotheses, and indeed making it a rule to do so, is perfectly consistent with Bayesian principles. ...No better reconciliation between Bayesianism and IBE could be hoped for.

It should be noted that strong arguments also have been made against any connection between IBE and Bayesianism, and even as to whether IBE should be considered an inferential method at all. Harman argued IBE was an autonomous entity with no connection to Bayesian methods [405]. Van Fraassen has written not only that there is no connection between IBE and Bayesianism, but that IBE itself is flawed, incoherent, and *"...fail[s] as [a] rational basis for opinion and expectation of the future"* [388].

---

[82] The Nobel Prize was motivated by Kahneman's integration of "...insights from psychological research into economic science, especially concerning human judgment and decision-making under uncertainty."

[83] Quotation from Plutynski [382], p. 17.

[84] Quotation from Okasha [404], pp. 703–704.

Thus far we have implicitly assumed that the best explanation exists within a set of competing hypotheses. However, this may not be the case and thus we may only be selecting what Van Fraassen [388] and Niiniluoto [406] term "the best of a bad lot."

As IBE is always restricted to a set of historically given or formulated hypotheses, it may lead to 'the best of a bad lot.' How could we know that the true hypothesis is among the so far proposed? And as this set is any case very large, its 'random member' must be quite improbable.

This argument is not directly relevant to abduction because it does not pretend to select the "best" explanation from an explicit pool of existing explanations. The pool of abduction has one member, the hypothesis at the moment, which is considered and then chosen or rejected. If chosen, the hypothesis then is tested. If rejected, a new hypothesis replaces the old one. It may, in fact, be true that an identical group of hypotheses may have been considered by IBE and abduction, but the process itself has differed.

Imagine a computer whose memory contains a set of hypotheses and whose programming subjects the members of this set to evaluation using a finite number of weighted variables, each of which characterizes some aspect of explanatory value. All the hypotheses are preexistent and the evaluation algorithm is fixed. After entry of existing data on a phenomenon, this computer determines the hypothesis that explains the data "best." Best is thus relative to the pool of hypotheses extant, which may or may not contain what, in nature, *is* the best hypothesis.

The differences in process between IBE and abduction thus include: (1) how each variable is weighted; (2) the evaluative algorithm into which these variables are placed; and (3) how many hypotheses are evaluated. Weighting and algorithms for IBE are givens that reflect accepted scientific practice (or at least an average form of it). In abduction, each is generated randomly for each hypothesis created and the number of hypotheses is arbitrary. These differences reflect a relatively constrained process in IBE vs. a relatively unconstrained process in abduction. These distinctions highlight the fact that *discovery* and *novelty* are key products of abduction, whereas *selection* of *preexistent hypotheses* are characteristic of IBE. Campos[85] has emphasized the importance of establishing this distinction.

... some important distinctions about the nature of scientific reasoning are lost when we blur the distinction between abduction and IBE, and so our own philosophical understanding of such reasoning is impoverished.

What is important for the practicing scientist to appreciate is not so much formulaic methods for hypothesis creation and selection, but rather the philosophical underpinnings of such methods. This understanding provides a framework upon which thoughtful, rigorous, significant, and valuable experimentation can be done. In fact, in practice, scientists employ a pastiche of strategic approaches to plan their work. The question the scientist must ask

---

[85] See Campos [384], p. 420.

pre facto is *why* any particular method should be employed at all, and it is this question that must be answered using philosophical means.

5.4.3 Induction

If a scientist observes that an effect *e* follows a cause *c*, they may *induce* that this cause-and-effect relationship is a *general* feature of the system and will be observed in future experiments. Philosophers of science love to use black ravens as an example. If one considers the color of a raven as a datum, one can do an empirical study by observing the numbers of black or nonblack ravens they see. In doing so, it is likely that one will never observe a white raven.[86] We don't know this for sure, but based on the limited data we have collected, we can induce all ravens are black.

Physicists, in particular, have been able to use the inductive process to establish universal theories – laws of nature – from relatively limited empirical information. Gravitation is a good example. Newton, using the limited information provided by temporal changes in the positions of the sun and planets, along with geometrical information developed by Kepler, calculated the force of attraction between two bodies of masses m_1 and m_2 with their centers separated by a distance r (Eq. (5.4)). The force of gravity, G (Eq. (5.5)), is a *universal* gravitational constant for the conversion of units measurable by experiment (masses m_1 and m_2, and distance r) into those of force (Eq. (5.4)). Newton's induction was that this relationship applied to *all* bodies with mass, not just those he observed in his initial experiments. He subsequently, and successfully, used this relationship to explain the paths of comets and the tides on Earth.

$$F = Gm_1m_2/r^2 \tag{5.4}$$
$$G = 6.674 \times 10^{-11} \, m^3kg^{-1}s^{-2}. \tag{5.5}$$

Biologists, in contrast to physicists, work in systems of immense variety, complexity, and indeterminacy. Biologists do use the laws of physics to understand biological processes because these laws *are* universal, i.e., they operate in, and are relevant to, all physical and biological systems. However, unlike physics, biological processes are, by nature, stochastic, which makes induction more difficult and determination of the truth of an induction particularly problematic. How can we predict in our next observation of a phenomenon that our conclusion $c \Longrightarrow e$[87] will be true? Logic tells us we cannot, because no matter how many times we find our induction confirmed by new experiments or observations, we can never know if the next observation will be inconsistent with our induction. What we would need to prove is:

$$P(e|c) = \lim_{i \to \infty} y/i = 1,$$

---

[86] Actually, white ravens *have* been seen, and photographed http://petslady.com/article/legendary-white-ravens-are-real-really, on Vancouver Island, British Columbia, Canada.
[87] The logical symbol $\Longrightarrow$ means "implies."

where y is the number of times the induction is confirmed in i experiments. This cannot be done, so the best we could do would be to establish P with a relatively small number of independent experiments. This fact was recognized by Ptolemy almost two millennia ago.[88]

...as for assertions of validity 'for eternity', or even for a length of time which is many times that over which the observations have been taken, we must consider such as alien to a love of science and truth.

The closest we can come to determining the truth of an induction is to produce a statistical measure of its probability. If P is "close" to 1, we are more likely to believe our induction is true. If P is closer to 0, we may not. Of course, this eliminates the possibility of the type of binary answer we would like, namely "yes" or "no." If there is no absolute, then how are we to interpret $P(e|c) < 1$? How close to 1 must the probability be for us to believe the induction is true? We discuss this, and related questions, in Section 5.7, but before doing so, let's take a look at how scientists conceive theories.

5.5 Theory Development

Theory development can be conceptualized as a circular three-step process comprising conception, choice, and testing (Fig. 5.11). Curiosity drives the first step. We are curious about nature. We have observed *something* in nature – a rainbow, an otter, the sun, a crystal, a mountain, water, an emotion, and myriad other *somethings* – that we want to understand and explain. We want to move from "what" we see to "how" and "why" we see it. There is no formula or normative process *per se* that we can follow to achieve this goal. Our brains process and interpret external stimuli such as sight (our observations), sound, touch, smell, and taste into forms we can understand. We then can act on this information, if we have the conscious desire to do so, by moving, thinking, or executing other actions. The neurophysiology of thought is complex, incompletely understood, and beyond the scope of this discussion, but for our purposes, it is sufficient to recognize that we can choose to think without knowing how we do it. Of course, this also means that we cannot teach anyone how to think – how to "conceive." We *can* infer or have flashes of intuition. We can even theorize counterfactually,[89] as Einstein did when he used *gedankenexperiments* to create impossible (imaginary) inertial systems. Myriad other theories may be derived in this manner. For example, if we think logically about the premise the sun exists, we might also think about this premise's negation, the sun does not exist. How might the sun not exist? We might theorize that an exothermic nuclear reaction ceased for some reason. What might this reaction have been? It might have involved nuclear fission or fusion, which then might lead us to wonder which elements were involved in these reactions, which might lead us to question how the elements were formed initially, how and why they coalesced in space

---

[88] Quotation from the *Almagest*, III, 1, H203, from the Greek edition by Heiberg [407] and the English translation of it by Toomer [372], p. 137.

[89] *Counterfactual thinking* "pertain[s] to, or express[es], what has not in fact happened, but might, could, or would, in different conditions." From the Oxford English Dictionary www.oed.com/view/Entry/42781?redirectedFrom=counterfactual #eid.

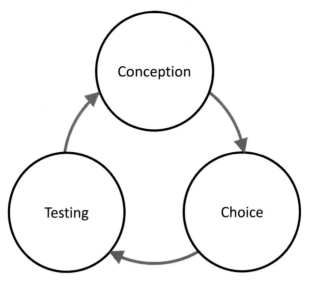

Figure 5.11 Theory development.

to enable creation of the sun, etc. *ad infinitum*. This chain of postulations created by counterfactual consideration of observations extant can be a powerful means of theory creation. Deduction also can yield a conception. Logical treatment of existing premises may produce conclusions heretofore unconsidered. However, if premises do not exist or are insufficient, then deduction can produce nothing. Thus, scientists conceive of theories in many different ways, ranging from the logical to the inferential and even to the metaphysical and fantastical, and they often do it in inexplicable ways.

Theory choice is a more difficult process, one without a unitary method or a consensus regarding approaches to it, which is why philosophers of science want to understand its rationale and logic. They do because an important goal of PoS is to provide a framework supporting science practice. Scientists want to answer the "how" and "why" questions of nature, but before doing so they must answer the question of why they would choose one strategy over another, one tactic over another, or one experiment or experimental system over another. Scientists should be able to justify their choices to themselves, other scientists, and, quite often, the public, because the success of their work and their prudent use of resources, both in time and money, depend on these choice(s).

Selecting the "best" theory is much harder,[90] and often impossible to do *ab initio*. Nevertheless, such choices must be made. Scientists do not have the time or resources to study everything. They *must* choose. How *does* one choose on which among many plausible theories one should focus? Again, no consensus exists, either among scientists or philosophers of science.

---

[90] In practice, according to Lipton [408], "...the scientist's problem is not to choose between equally attractive theoretical systems, but to find even one."

The final arbiter of successful theory development is testing. Testing is both an end and a beginning of theory development. It terminates the first cycle of theory creation and choice, providing the means to evaluate the success of initial theory selection, and it guides further iterations of the process that eventually may lead to what we consider the best theory choice.

On what basis can one theory be chosen from among many? Should it be the beauty, elegance, cleverness, or novelty of a theory? Is it logical – is there a logical process that can be followed to accomplish this goal? Is it metrical – can we assess the relative merits of different theories through mathematical or statistical means? Is it computational – can a computer analyze the myriad of information about the world and the phenomena in it, and then through some algorithm or application of artificial intelligence, come to a conclusion? We might decide that the theory that can accommodate the most data extant is the "right" one, or alternatively, that the chosen theory offers the greatest potential of yielding novel data or insights in the future. Practical value (usefulness) may be, and often is, the basis for theory choice. It is important to understand that bias and dogma affect choice. Also common is the sometimes tacit assumption that there is only one best theory explaining a phenomenon. Two or more theories may be equally good at explanation. Of course, how "goodness" is determined must be established if one is to accept this statement, which brings us back to our original question – what is the basis for theory choice? The answer, in a nutshell, is there is no *single* metric, means, or process for choosing a theory.

Scientists, and philosophers of science, consider a range of factors comprising relatively intangible (qualitative) and relatively tangible (quantitative) elements. I use the term "relatively" because it is difficult to determine absolute boundaries between these two elements – nor must we. For example, what quantity shall we use to characterize beauty and what quality shall we use to describe a likelihood estimate? We could say we find something *very* beautiful, beautiful, or *less* beautiful, but how are we to assign numbers to these terms? Similarly, if we obtain likelihood estimates of 0.31, 0.44, 0.63, 0.74, and 0.91, to which do we assign qualitative terms such as "ugly," "beautiful," "average," "striking," "stunning," or "unimpressive?"

5.5.1 The Underdetermination Problem

Before considering the justifications for theory selection, one caveat should be mentioned – we can never be sure we've selected the *only* theory that explains a system. Theories may be unitary in nature or encompass our entire universe, for example, the Theory of Everything (TOE).[91] However, in practice, one would like to choose a single theory. Are we warranted in doing so? Our warrant does come from historical information (data extant), experimental and observational data, inference, successful prediction, and corroboration. But is this enough? Underdeterminationists think not. They argue the we can never establish sufficient warrant for belief in a single hypothesis based on the evidence we have accumulated

---

[91] See Section 5.5.3.

because that evidence simply is insufficient. There may already be, or could be, equally plausible theories based on the same evidence.

Pierre Duhem proffered a *holistic* theory of undetermination in the period between 1892 and 1916 [409, 410].[92] The focus was physics and how to deal with apparent experimental failures to corroborate a theory. Duhem argued that any theory depends on many auxiliary theories, each of which, in turn, depends on foreknowledge *thought* to be true. As a result,

> ...a physics experiment is not simply the observation of a phenomenon. ...It is the precise observation of a group of phenomena, accompanied by the interpretation of these phenomena; this interpretation substitutes for the concrete given, actually gathered by observation, some abstract and symbolic representations which correspond to the given by virtue of the physical theories admitted by the observer.[93]

This is the "Duhem thesis." It connects experimental consequences of a theory with the theory *per se*. Underdetermination comes from the fact that

> ...the physicist can never subject an isolated hypothesis to experimental test, but only a whole group of hypotheses; when the experiment is in disagreement with his predictions, what he learns is that at least one of the hypotheses constituting this group is unacceptable and ought to be modified; but the experiment does not designate which one should be changed.

In the twentieth century, Willard Van Orman Quine [413, 414] developed the concept of *contrastive* underdetermination, which was distinct from, but had similarities to, the Duhem thesis. Quine's concept of underdetermination emerged from his theoretical consideration of empirical equivalence, the notion that two distinct theories can display the same observation conditionals.[94] He illustrates this idea by considering a theory in which the *purely theoretical* terms "electron" and "molecule" were substituted, one for the other.

> ...[t]he new theory formulation will be logically incompatible with the old: it will affirm things about so-called electrons that the other denies. Yet their only difference, the man in the street would say, is terminological; the one theory formulation uses the technical terms 'molecule' and 'electron' to name what the other formulation calls 'electron' and 'molecule'. The two formulations express, he would say, the same theory. ...the two theory formulations are empirically equivalent[95] – that is, they imply the same observation conditionals.

Contrastive underdetermination may be considered the converse of holistic underdetermination in that the latter addresses the question of how many experiments can corroborate *a* theory, whereas the former focuses on how *many* theories may corroborate the same set of experiments.

Scientists invent hypotheses that talk of things beyond the reach of observation. The hypotheses are related to observation only by a kind of one-way implication; namely, the events we observe

---

[92] Duhem's publications were in French. An English translation was not available until Duhem's book, *"The aim and structure of physical theory,"* was published in 1954 [411].

[93] Quotation from Stanford [412], p. 3.

[94] Observation conditionals are conditional sentences with boundary conditions as antecedent and an observation sentence as consequent or, put simply, "If p then q."

[95] Bandyopadhyay [415] consider theories *"empirically indistinguishable if and only if there is no possible evidence that would confirm one and not the other, or disconfirm one and not the other."*

are what a belief in the hypotheses would have led us to expect. These observable consequences of the hypotheses do not, conversely, imply the hypotheses. Surely there are alternative hypothetical substructures that would surface in the same observable ways.... Such is the doctrine that natural science is empirically under-determined; under-determined not just by past observation but by all observable events.

A fraternal twin of underdetermination is "unconceived alternatives." As the name implies, just as we can never be sure of the veracity of a theory based on instantial evidence, we also cannot be absolutely confident because there may exist alternative theories of which we are unaware. Stanford [416] put it nicely:

... we have, throughout the history of scientific inquiry and in virtually every scientific field, repeat-edly occupied an epistemic position in which we could conceive of only one or a few theories that were well confirmed by the available evidence, while subsequent inquiry would routinely (if not invariably) reveal further, radically distinct alternatives as well confirmed by the previously available evidence as those we were inclined to accept on the strength of that evidence.

If there can be no expectation of our ability to choose the "best" theory, how can we jus-tify inferring anything from our observations? In scientific practice, given the fact that it is exceedingly rare to be able to justify a theory on the basis of deduction, we are left with two inferential processes for theory choice: abduction and IBE.[96] Unfortunately, neither can provide the warrant we seek. By definition, both involve the *choice* of a theory and the type of choice we make is nondeductive and therefore unjustifiable (at least to under-determinists and Humeans). Here, however, we are talking about absolutes, of which there appear to be few or none in science. And this is alright. Relativism has worked wonderfully for millennia – as long as we work in systems amenable to it. We feel we have warrant because, in *this* system, but not necessarily in others, our theory can explain and predict system behavior. By restricting theory selection to *certain* systems for specific reasons, we can largely avoid the underdetermination issue. One could argue this is a tautological suggestion, and it is, but only *after* the best theory has been selected, a theory that serves our epistemic purposes. In this way, as Chang [47] has suggested, underdetermination may actually be a driving force toward explanation.

Scientists develop various systems of practice containing different theories, which are suited for the achievement of different epistemic aims. In this way the great achievements of science come from *cultivating* underdetermination, not by getting rid of it.

It should be noted that some philosophers of science do not accept the notion of underdeter-mination, at least not in its entirety or as commonly understood. Laudan and Leplin [417], for example, *"reject both the supposition of empirical equivalence and the inference from it to underdetermination."* Kitcher [418] not only recognizes the importance of considering underdetermination in science but also opines *"The Underdetermination thesis, in its usual guise, is a product of the underrepresentation of scientific practice."* Park, after exhaustive

---

[96] See Sections 5.4.1 and 5.4.2, respectively.

literature research, emphasizes that many distinct types of underdetermination exist [419], only some of which are problematic, a view shared by Stanford [420].

From a purely philosophical perspective, the jury remains out with respect to the concept and significance of underdetermination [412]. However, accepting, or at least acknowledging, the idea that all of our hypotheses or theories may be underdetermined helps the scientist avoid overconfidence and be more measured and contemplative in their own pronouncements. Scientists know that explanations for their observations may come from more than one theory, but to test a theory requires focusing on one theory at a time.[97] This is not to say that scientists can't simultaneously investigate multiple theories, given sufficient resources, for they may employ the strategy of "multiple working hypotheses" [421–423] to simultaneously investigate, or at least acknowledge the existence of, multiple potential explanations for phenomena. This is of particular value when either no reasonable hypothesis exists to explain one's observations, or many, equally plausible hypotheses exist. Now that we are aware of potential philosophical and practical caveats regarding theory choice, let us consider the many ways in which this process actually is executed.

5.5.2 Simplicity

A priori, there may seem to be little reason for simplicity being favored over complexity. For example, completion of a simple task is unlikely to yield the feel of achievement enjoyed from a complex task. Simple music, for example, music written in a single octave without sharps, flats, changes in the lengths of notes or tempi and played by a single instrument is unlikely to be appreciated as much as the complex sound of a symphony written by Bach or Beethoven and played by an orchestra of 90 or more members playing a variety of instruments. "Whodunit" novels must be complex by definition or they would not belong in the mystery genre. Tasks, music, and mystery are three milieus in which simplicity would be unsatisfying and boring. However, in the milieu of theory development, the opposite usually is true, and for simple reasons one of which is simple theories are easier to grasp and remember than complicated ones.

Is simplicity thus better than complexity? Aristotle, William of Ockham, and Isaac Newton certainly thought so. Aristotle opined *"We may assume the superiority ceteris paribus of the demonstration which derives from fewer postulates or hypotheses."* The writings of William of Ockham (the eponym of "Ockham's Razor"[98]) often included the Latin expression *"pluralitas non est ponenda sine necessitate,"* translated as "plurality should not be posited (assumed) without necessity," or other expressions with similar meanings.[99] Isaac Newton opined *"We are to admit no more causes of natural things than such as are both true and sufficient to explain their appearances."* Modern scientists and philosophers of science also subscribe to the notion, *ceteris paribus*, simpler theories are better [424]. Why?

---

[97] Wüthrich questions the idea that one can test even a single theory. À la the Duhem thesis, he says *"[w]hen a physical theory is put to test, it is an entire collection of theories and auxiliary hypotheses, rather than that single theory alone, which are tested."*

[98] See Sober [133] for an in-depth, more recent explication of Ockham's Razor.

[99] See Section 4.3.5 for a more complete discussion of Ockham and his "razor."

Simplicity makes theorizing easier and more likely to be accurate and useful. Simpler theories comprise only key explanatory elements. Elements that contribute insignificantly or negligibly to understanding a phenomenon are not considered (shaved off by Ockham's Razor). This is where practical value comes. Simple theories are easier to formulate and they tend to be more flexible than complex theories, allowing their application to a broader range of phenomenal instances. The more complex a theory, the more *precisely* it can model a phenomenon because it will account for its many elements. However, this does not mean that the model is an *accurate* representation of the phenomenon. The greater the number of elements considered, the less flexible the theory becomes because the likelihood of related phenomena comprising these exact elements is low. An example of this would be over-fitting data. In addition, an inherent assumption of complex models is that each element is important, otherwise why would it be included? This assumption often is later found to be wrong. Even if all elements are necessary in a theory, the more important issue for scientists is its practical usefulness. Does the number of assumptions constrain the use of the model to a single phenomenal instance? Answer – the likelihood of producing a relevant model is inversely related to the number of elements/assumptions.

The value of keeping assumptions to a minimum is cognitive, not ontological: It helps you to think. A theory is not 'better' if it is simpler – but it might well be more useful, and that counts for much more.[100]

If we posit that simpler is better, we then must be able to define simplicity itself and determine *how* it is better. We will find that defining simplicity is complicated! Derkse [426] has argued that limitation and reduction are general features of simplicity. This is an excellent starting point. However, as we dig deeper, we find that context, for example, use in theory creation vs. use in the laboratory, can change the details of any definition of simplicity. In science, for example, Baker distinguishes simplicity used for epistemological purposes (why should we *believe* one theory over another) from simplicity used for methodological purposes (why should we *use* one theory for guiding experimental work and not another) [427]. These different forms of simplicity may also include *semiotic* (intrinsic simplicity of each *explanans*), *semantic/syntactical* (the number of *explanans*, their simplicity, and the simplicity of their logical relations), or *ontological* (the simplicity of a theory's representation of nature.[101] Efforts to define simplicity are complicated further by the fact that each form may yield a different *degree* of simplicity [426]. So, is there consensus on a single role or form of simplicity? Pacer and Lombrozo [181] have argued "no," and the reason for doing so is the absence of a consensus definition of simplicity itself.

It is difficult to escape the circularity of definition↔role, but there may be no necessity to do so. Science is not monolithic operationally, and thus, there appears to be no compelling reason why we would expect this of simplicity. Simplicity, as discussed above, not only may have degrees, i.e., *quantitative* metrics, but it may also have different *qualities* in

---

[100] From Ball [425].
[101] As with Ockham's Razor; see Section 4.3.5 as well.

different milieus. Similarly, the reasons why we choose a theory may involve both quantitative (e.g., numbers of axioms, givens, process steps or the use of likelihood estimates) and qualitative (e.g., beauty, elegance, and novelty) considerations. However, when we plan to devote substantial time and resources to scientific investigation, whether we think something is beautiful or not doesn't help us justify why the simpler theory should guide our experiments rather than the more complex theory. A parallel to this would be how we justify eating/not eating something because someone told us it was good/bad. I'm sure we've all experienced one person's "good" being our "bad" and vice versa. A better understanding (if that were possible) of the bases for someone's taste would have helped us here. In this vein, Sober has suggested that justification for use of a simpler rather than a more complex theory can come if the simpler theory can answer three methodological/operational questions.[102]

1. Does it have a higher probability of being true?
2. Is it better supported by data extant?
3. Is its predictive ability superior?

Answering these three questions in the affirmative justifies use of the simpler theory and also suggests that the simpler theory is more plausible. Some suggest that, operationally, "simple" maybe be defined as one component of plausible. Plausible theories thus are simpler and simpler theories are more plausible. This definition, however, appears circular, as suggested by Sober when he opined "Just as the question 'why be rational?' may have no noncircular answer, the same may be true of the question 'why should simplicity be considered in evaluating the plausibility of hypotheses?'" [428]. This circularity may be true syntactically or epistemically, but it does not change the fact, in practice, simpler or more plausible theories have been easier to test and more likely to further efforts to understand, interact with, and manipulate nature.

Jeffery's illustrates why simple may be better than complex using as an example the equation for the position (*p*) of a falling object experiencing the force of gravity [429].

$$p = p_0 + vt + \frac{1}{2}gt^2. \tag{5.6}$$

This equation incorporates only *key* parameters in the system, namely where an object starts (p_0), how fast it is moving at that moment (*v*), the acceleration due to gravity (*g*), and time (*t*). In practice, for objects falling through the air for more than a few seconds, and near the Earth's surface, this simple quadratic becomes inaccurate. We could improve our fits of the equation to actual experimental data if we added additional terms, for example, air resistance, collision cross-section of the object, and continuous adjustment of the gravitational force (to account for the actual distance between the object and the center of mass of the Earth). However, whether we do this or not depends on our purpose. For ballistics purposes (e.g., with bullets or missiles), the additional terms would be necessary if precise projectile trajectories were to be calculated. However, if the goal is to teach basic physics,

[102] See Sober [133], p. 59.

adding these terms would not contribute significantly to one's conceptual understanding. The additional terms would be superfluous.

Superfluousness, also referred to as superfluity, is an important concept because it provides a rationale, other than that of quantity *per se*, for deciding whether a particular component of a theory should or should not be shaved off by Ockham's razor. The "anti-superfluity principle," as discussed by Barnes [430], mandates that positing superfluous theory components – components that are not required for the purpose of explaining the relevant data – should be avoided. This is *not* because of the "anti-quantity principle" espoused throughout scientific history, but rather, according to Barnes:[103]

... [because] parsimony does, in and of itself, make a theory more plausible, for it releases a theory from its commitment to components unsupported by the relevant data.

As one might expect, given the context of our discussion is PoS, diametrically opposed opinions also exist. Sober, for example, feels *"parsimony, in and of itself [i.e., as a strictly quantitative principle], cannot make one hypothesis more plausible than another."*[104] Recent work by Herrmann [432] lends support to Sober's argument. Herrmann examined, in the context of machine learning, the relationship between "probably approximately correct" (PAC)[105] learning [433] and "Occam Learning,"[106] the computational equivalent of Ockham's Razor. For each of these learning algorithms, the computer interrogated sets of data to create either an hypothesis with a high probability ("probably") of being a good hypothesis ("approximately correct"), or one that is the most simple (Ockham's Razor), respectively. PAC learning has been used to justify the epistemic value of Ockham's Razor [432, 434–436], but Herrmann provided a new analysis of this notion and concluded *"Occam's Razor is merely a pragmatic way to make it easier for us to work with our theories."* For the practicing scientist, regardless of on which side of this epistemic argument they are, Ockham's Razor remains useful in navigating the waters of theory selection.

Complexity also creates more difficulty in modeling system behavior mathematically and computationally. If data from one instantiation of a system behavior are used to create a mathematical model, that model should fit the data well. However, most systems display some level of complexity and variability, especially biological systems, thus subsequent experiments may produce data inconsistent with those used to construct the model and that are fit increasingly poorly by the model. By contrast, a simpler model may be less accurate, but it will remain useful for subsequent studies because it will be less sensitive to perturbations caused by inconsequential parameters. Douglas and Magnus [437] and Box [438] have said, respectively:

If a scientist overfits the data with a grotesque curve, then he will make poor predictions about future data. Contrapositively, if a scientist makes successful predictions, then we can infer that he has not overfit the data.

---

[103] See Barnes [430], p. 370.
[104] See Sober [431], p. 139.
[105] See https://en.wikipedia.org/wiki/Probably_approximately_correct_learning.
[106] See https://en.wikipedia.org/wiki/Occam_learning.

Since all models are wrong the scientist cannot obtain a "correct" one by excessive elaboration. On the contrary, following William of Occam he should seek an economical description of natural phenomena. Just as the ability to devise simple but evocative models is the signature of the great scientist so overelaboration and overparameterization is often the mark of mediocrity.

As discussed above, a fundamental problem in IBE and abduction is the fact that model selection depends on qualitative measures that are difficult to determine or standardize. This explains why the same data may yield different "best" models. However, if a quantitative metric could be developed that considered simplicity and "bestness," i.e., explanatory value, then one could compare models and select the one scoring best. A variety of statistical methods have been developed to achieve this goal, including the Aikake information criterion (AIC) [439, 440], the AIC_c (a variant of AIC), the "widely applicable information criterion" (WAIC) [441]), the "deviance information criterion" (DIC) [442, 443], and principal component analysis (PCA; see [444]). The AIC is far and away the most widely used method.[107] The AIC, and a later derivative, the minimum information theoretic criterion (AIC) estimate (MAICE), are "information-theoretic" methods that are based on the determination of the fit of a model to the data from which it was derived [439, 440]. Importantly, AIC and related methods of model selection differ substantially from null hypothesis testing using analysis of variance, the most common method of establishing significance in standard control group/test group experimentation. Burnham and Anderson emphasize this point, quite strongly, as follows.[108]

Information-theoretic criteria such as AIC, AIC_c, and QAIC are not a "test" in any sense, and there are no associated concepts such as test power or P-values or α-levels. Statistical hypothesis testing represents a very different, and generally inferior, paradigm for the analysis of data in complex settings. ... It is critical to bear in mind that there is a theoretical basis to information-theoretic approaches to model selection criteria, while the use of null hypothesis testing for model selection must be considered ad hoc (albeit a very refined set of ad hoc procedures in some cases).

The formula for AIC is given in Eq. 5.7, where k is the number of estimated parameters and $\hat{L}$ is the maximum value of the likelihood function (the fit of the model to the data):

$$AIC = 2k - 2ln\hat{L}. \tag{5.7}$$

The AIC is a *comparative* metric that allows one to choose the best among a number of models, not calculate a model's value *per se*. The "best" model has the lowest AIC, which means the more parameters (k; complexity) a model has, the more it is penalized, and the better the model fits the data ($2ln\hat{L}$; interpreted as explanatory value), the less it is penalized. AIC is useful in preventing over-fitting or under-fitting.

The AIC was a further development of the Kullback–Leibler divergence [446], *"a measure of the 'distance' or 'divergence' between statistical populations in terms of [a] measure of information."* The lower the divergence, the more similar are the two populations. This approach balances the simplicity of a model with the model's ability to fit

---

[107] So much so that some suggest it has been used to the exclusion of other methods [445].
[108] Quotation from Burnham [444], p. 84.

the data, thus producing an "Aikake weight" (w), a metric with a range $(0,1)$. This allows one to rank models so that the model with the smallest w could be considered the most likely [447]. I use the word "could" because, as with qualitative comparisons, one must determine whether the difference between two models is meaningful. This can be done by calculating an "evidence ratio" (w_i/w_j), the magnitude of which is a measure of how distinct model i is from model j.[109] One also may simply determine the difference between two weights, $\Delta w = w_i - w_j$. There is no specific value of w_i/w_j or $\Delta w = w_i - w_j$ that allows one to conclude that two models are alike or not alike, unless the absolute values of the weights or weight differences are very small or identical, respectively. Therefore, we must use relative terms such as "weak," "moderately strong," or "very strong" to describe the likelihood that a model is true.[110]

We conclude our discussion of simplicity with a quotation from Edsger Dijkstra, a Turing Award-winning[111] pioneer in computer science, which addresses the impact simplification had on his field [449]. His words are as relevant to scientific theory creation, choice, and testing as they are to computer science.

Simplifications have had a much greater long-range scientific impact than individual feats of ingenuity. The opportunity for simplification is very encouraging, because in all examples that come to mind the simple and elegant systems tend to be easier and faster to design and get right, more efficient in execution, and much more reliable than the more contrived contraptions that have to be debugged into some degree of acceptability ....

5.5.3 Unification and Causation

Unification is another measure, and part, of the definition (according to some) of explanantia. Kitcher has described unification as follow: *"The intuitive idea behind unification is the generation of as many conclusions as possible using as few patterns"* [450]. Here, *patterns* may be subsumed under *explanantia* or their interrelationships. In this sense, unification may be quantized in a manner similar to *AIC*.

$$U = aE - f(T). \tag{5.8}$$

Here, U is a unification metric, a is a constant, E is the number of *explanantia*, and $f(T)$ is a function of the number of theories unified by the explanation. U is penalized (increases) the more *explanantia* are involved and rewarded (decreases) the more unifications achieved. The process of *explaining* by revealing the causal dependencies of *explananda* on *explanantia* is often the first step toward understanding phenomena. Scientists seek a simplified world view and unification is a way to do so.

---

[109] See chapter 2.10 in Burnham [444].

[110] According to Royall [448], evidence ratios of 8 and 32 suggest moderately strong or strong evidence, respectively. In the case of Δw, some consider a $|\Delta w| > 2$ an indication that one model is better or worse than another. It should be noted that the highest ranking model may not be the best model. Further validation must be done.

[111] The A.M. Turing Award, sometimes referred to as the "Nobel Prize of Computing," was named in honor of Alan Mathison Turing (1912–1954), a British mathematician and computer scientist, who made seminal contributions in the fields of computer architecture, algorithms, formalization of computing, and artificial intelligence. He may be most well known to the general public for his work breaking the German and Japanese secret codes during World War II.

The quest for unification has a long history. Unification was discussed extensively in a number of contexts by Kant in the mid- to late eighteenth century when Kant was examining the *"unity of nature in time and space, and unity of the experience possible to us."*[112] He also commented on the subjective pleasure one could derive through unifications leading to novel theories.

... it is a fact that when we discover that two or more heterogeneous empirical laws of nature can be unified under one principle that comprises them both, the discovery does give rise to a quite noticeable pleasure, frequently even admiration, even an admiration that does not cease when we have become fairly familiar with its object.[113]

Continued admiration of the unifying effects of Einstein's General Theory of Relativity certainly attests to the accuracy of Kant's pronouncement.

In modern times, many have linked unification directly with scientific explanation [450, 452–455]. Hempel sought to establish logical rules for explanation through a deductive–nomological model. This normative model eschewed subjective, descriptive, intuitive, and related means of explanation. Instead, it emphasized a systematization of explanation, a process that used fixed rules to interrogate elements of the natural world and establish subsets of elements and their relations that were relevant to understanding. This type of parsing reduced the number of explanatory elements to a minimum – it unified. Hempel described the process thusly.

What scientific explanation, especially theoretical explanation, aims at is not this intuitive and highly subjective kind of understanding, but an objective kind of insight that is achieved by a systematic unification, by exhibiting the phenomena as manifestations of common underlying structures and processes that conform to specific, testable, basic principles.[114]

A particularly important proponent of the concept of explanation by unification was Michael Friedman, who published a seminal paper on the concept in 1974 [453]. Friedman was particularly interested in how scientific explanations help us understand the world. A key premise was useful explanations reduce the number of theories or ideas necessary to explain, i.e., understanding is engendered by the unification of multiplicities of causal elements or theories.

... this is the essence of scientific explanation – science increases our understanding of the world by reducing the total number of independent phenomena that we have to accept as ultimate or given. A world with fewer independent phenomena is, other things equal, more comprehensible than one with more.[115]

Friedman's ideas formed a foundation upon which further aspects of unification and explanation could be built. Unification was not *"simply a matter of reducing the [number of] 'fundamental incomprehensibilities'."*[116] Rather, it required what today we would call

---

[112] See Kant [451], IV, §210, "On Experience as a System for the Power of Judgment."
[113] See Kant [451], VI, §188, "On the Connection of the Feeling of Pleasure with the Concept of the Purposiveness of Nature."
[114] From Hempel [452], p. 83.
[115] Quotation from Friedman [453], p. 15.
[116] See Kitcher [450], p. 432.

"pattern recognition," the ability to decipher information from an ostensibly "incomprehensible" *gemish*[117] of observations, laws, and theories.

The heart of the unification approach is ... the idea of a systematization of the world in which as many consequences as possible are traced to the action of as small a number of basic mechanisms as possible.[118]

Salmon expressed a more theoretical, but similar sentiment, when he used the black box model to explain the unification aspect of the "ontic conception of explanation." Here, inputs could be of many types, including observations, laws, theories, or other *explanantia*. Unification was accomplished by reduction of the number of inputs necessary to explain the outputs of the Black Box process.

The ontic conception looks upon the world, to a large extent at least, as a black box whose workings we want to understand. Explanation involves laying bare the underlying mechanisms that connect the observable inputs to the observable outputs. We explain events by showing how they fit into the causal nexus. ... the ontic conception [takes] the unification of natural phenomena as a basic aspect of our comprehension of the world.[119]

The ultimate unification of natural phenomena may be the *"Theory of Everything [TOE], a complete and consistent set*[120] *of fundamental laws of nature that explain every aspect of reality"* [456], uniting all four natural forces, electromagnetic, weak, strong, and gravitational (Fig. 5.12). Existing laws would be special cases of the TOE. Reality in this context is physical in nature. It is the entire cosmos and all the matter within it. The TOE would explain how the elementary particles of matter, and the forces governing their behavior, produce the universe in which we live and the phenomena we observe.

    The ability to create the TOE exists because of its incorporation of the Grand Unified Theory (GUT). The GUT postulates, at extremely high temperature, unification of three fundamental forces, electromagnetic, weak, and strong, forms a single force in which each of the component forces has equal magnitude. These forces, and their associated elementary particles, are described by the Standard Model of particle physics (Fig. 5.13), which has been said to answer the questions "what is everything made of" and "how does it hold together?" The TOE and GUT certainly illustrate how valuable unification can be, but the reader should also understand that unification is but one tool of explanation and that circumstances often exist in which the most unifying theory, even if it were to be a complete and consistent TOE, does not yield the best explanation or even advance understanding [457].

---

[117] Yiddish; mishmash, mélange.
[118] From Kitcher [450], p. 497.
[119] Quotation from Salmon [454], p. 276.
[120] The phrase "complete and consistent set" may bring to mind Gödel's work on the completeness and consistency of axiomatic mathematical systems. He proved, for any consistent axiomatic system with sufficient expressive power to represent arithmetic, there were statements that were true by construction, but unprovable within that system, i.e., the system was incomplete. Even if a complete arithmetic system were to exist, Gödel showed that it could not be consistent, i.e., a theorem and its negation *could* be derived from it.

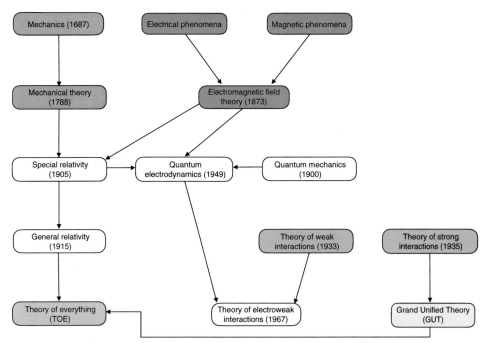

Figure 5.12 The Theory of Everything (TOE). The figure shows the historical development of the TOE and its component parts, which include all the particles and forces governing the physical manifestations of our reality. The penultimate theory, the Grand Unified Theory (GUT) of particle physics, predicts the existence of a single fundamental force formed from the unification of three of the four universal forces, electromagnetic (blue shading), weak (green shading), and strong (orange shading). If the three forces in the GUT (yellow shading) were unified with the force of gravity (pink shading), physicists will have created the TOE (cyan shading).

5.5.4 Abstraction

In "big science," a "more details are better" (MDB) concept is central. Here, one seeks to understand nature by computationally interrogating huge sets of data to find relationships among its elements. Can this goal be accomplished? Will the output of a computer *explain* or will it simply reveal statistical or probabilistic relationships of a complex system? This question, which can be expressed colloquially as "can we see the forest for the trees," has been debated intensely in the last few decades, especially among those dealing with massively complex systems. The debate is most apparent between those that consider mechanistic explanation (the trees) the "be-all end-all" (strict mechanicists) and those who think explanation should provide understanding at a higher level (the forest). As we will see, the two concepts are not necessarily mutually exclusive and, in fact, may be mutually dependent. Marder exemplifies this when she discusses the complexity of the brain and how to understand it [458].

Understanding how the brain works requires a delicate balance between the appreciation of the importance of a multitude of biological details and the ability to see beyond those details to general

Standard Model of Elementary Particles

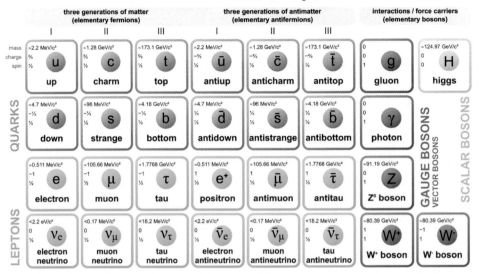

Figure 5.13 Standard model of particle physics. The model, in addition to its description of three fundamental forces (electromagnetic, weak, and strong) incorporates 30 fundamental particles subdivided into two major classes, *fermions*, particles composing matter, for example, protons, neutrons, electrons, and the elementary particles of which they are comprised, and *bosons*, particles carrying force, for example, gluons. Fermions comprise 12 different types of quarks and 12 different types of leptons. Six different types of bosons are postulated.

principles. As technological innovations vastly increase the amount of data we collect, the importance of intuition into how to analyze and treat these data may, paradoxically, become more important.

The debate is central to the question "what constitutes an explanation?" In the twentieth century (and earlier), the linear, asymmetric *explanantia→explanadum* account of phenomena underlay explanation. The details explained the phenomenon and these details were causative in nature. However, in the twenty-first century, as huge increases in computational power occurred, the study of complex systems was made feasible and the era of big science began. The most complex biological system in nature, the brain, now could be studied using reductionist mechanistic approaches (systems biology[121] – because its huge number of variables($\sim 10^{11}$ neurons, $\sim 10^{11}$ glial cells, and $\sim 10^{14}$ synapses) could be modeled mathematically and simulated computationally.[122]

But how would one understand the output of this process? A strictly deterministic (mechanistic) explanation of brain function and activity would remain incomprehensible for humans because of its complexity. We would indeed understand all the *parts* of the

---

[121] For current comprehensive introductions to systems biology and explanation in biology, the reader is referred to Green [459] 2017, *Philosophy of Systems Biology-Perspectives from Scientists and Philosophers)* and Braillard [460] 2015, *Explanation in Biology*, respectively.

[122] Estimates of cell numbers vary, but the relative amounts of each cell type are accurate.

brain but we would have no idea of what a brain was or did. This situation would be similar to the blind men describing the elephant. Each characterization was accurate and relevant, but no method existed to integrate them to explain what an elephant actually *was*. Explanation requiring comprehension and a strict mechanistic approach ($A \rightarrow B$) was incapable of achieving that goal – abstraction was required. The scientist thus would have to use an expansionist strategy to abstract macro-mechanisms from mechanistic details (*bottom-up*) and reductionist approaches to create micro-abstractions from descriptive ideas (*top-down*). Strevens calls this the "Goldilocks Problem" because it requires a solution to the problem of how much detail is necessary to make an explanation "just right." [461]

On the one hand, we say that an explanation is deep when it goes far down toward the physical level, the level of detail at which ultimate causal underpinnings are found. On the other hand, we also say that an explanation is deep when it has a certain striking generality when it attributes the phenomenon to be explained not to some very particular set of initial conditions, but to some high-level, abstract, often virtually mathematical state of affairs.

Systems biology seeks to provide a means to "explain" complex systems when pure causal-mechanistic accounts cannot. It does so by integrating *mechanistic* and *mathematical* forms of explanation. The mathematical aspect of explanation comes in the form of sets of differential equations that model time-dependent processes. They account for dynamical elements of systems, including functional-dynamical, feedback loops, distributed functionality, and robustness, none of which can be accomplished through classical mechanistic approaches. Target systems often are biological (cell and organismal development and physiology, evolution, ecology, cancer, and synthetic), but the philosophical idea of an integrative, interdisciplinary account of scientific explanation is applicable, to varying degrees, throughout science. This new integrative model of explanation has been termed "dynamic mechanistic explanation" (DME) [462] or "functional-dynamic explanation" [463]. The integrative aspect of the model comes from the anchoring of its mathematical equations (dynamics) to the pure causal–mechanical elements of a system. This dynamical approach has produced accurate, highly predictive models of system behavior that are said to be better suited for understanding certain phenomena in neuroscience, cognitive science, and psychology than are pure mechanistic explanations [464–466]. In his definition of DME (see quotation below), Bechtel bolded text to emphasize the temporal dynamics of a system.

Dynamic mechanistic explanation. A mechanism is a structure performing a function in virtue of its component parts, component operations, and their organization. The orchestrated functioning of the mechanism, **manifested in patterns of change over time in properties of its parts and operations**, is responsible for one or more phenomena.

Dynamics is what distinguishes DME from pure mechanistic explanation (*à la* Kaplan and Craver [467]). In many systems, mathematical treatment of the dynamics is essential if a phenomenon is to be explained. This essentiality, or relevance, is characterized the same way as mechanists do, namely a causal element in an explanation, including an equation, is relevant if eliminating or changing it severs the antecedent *explanans*→*explanandum* relationship. Thus, like mechanistic/causative *explanantia*, mathematical equations must be manipulable [468, 469] and counterfactually responsive to be considered relevant, i.e., if the equation had been formulated differently, the *explanandum* would be different.

A wonderful example of the use of mathematical formulations of system behavior is that of cell division in yeast. Yeast, like humans, are eukaryotes. Their cells contain organelles ("little organs"; membrane-delimited sacs) performing specific physiological functions. These include a nucleus, in which chromosomes are found; an endoplasmic reticulum and Golgi apparatus, which are involved protein synthesis and modification; mitochondria, power stations that provide energy for the chemical reactions in the cell; and lysosomes, the cellular garbage disposals that literally shred old proteins, DNA, RNA, etc. The physiologic similarity of yeast and human cells, and the relative simplicity of yeast, make it particularly useful for understanding human physiology.

Cell division in eukaryotes, in its simplest form, comprises a cycle of four stages, G_1, S, G_2, and M. To divide, a cell must manufacture all of the cellular contents its progeny need to function. This occurs in the G_1 and G_2 stages and causes an increase in cell size. The S phase is when a second copy of each chromosome is synthesized. Following G_2, the cell enters the M phase, during which one copy of each chromosome is pulled into each of two nascent cell, which then are sealed up to form daughter cells.

The regulation of the cell cycle is complex. It involves a variety of proteins that interact within a large network, the output of which is either division or stasis. The levels of proteins differentially oscillate over time – the system is dynamic. One of the mysteries of cell division has been why a cell will not move from G_1 to S unless it reaches a certain size. Systems biologists have studied this question by abstracting different elements from the system and writing ordinary differential equations to determine the time dependence of their activities [470]. Solving this system of equations revealed that cells must reach a critical size before dividing. This critical size was determined accurately by including 53 cellular elements in the system of equations (Fig. 5.14).[123] In addition, the dynamics of intracellular localization of the proteins between the nucleus and cytoplasm, and the rate of growth of protein production and cell size during G_1, were accounted for (see Table 1 of Barberis [470]).

The output of the system was *emergent*. A static consideration of the system's elements and their interactions, such as might be executed by classical mechanicists, could not have explained the qualitative observation that a cell must reach a certain critical size before entering S phase. Emergent properties also were a component of Hodgkin's and Huxley's Nobel Prize–winning exposition of the mechanism of nerve transmission. Levy[124] discusses why this exposition, through its ability to abstract from mechanistic detail, provided a better explanation of nerve conduction than did purely mechanistic approaches.

. . . mechanicists hold that a good explanation decomposes a phenomenon into underlying parts and their causal interactions. . . . I shall argue that this is not the right way to think about Hodgkin's and Huxley's work. For the model does not simply neglect the structure and functioning of ion

---

[123] The text in the figure is quite small, but legibility is unnecessary to illustrate complexity, which is apparent by the numbers of text boxes, colors, and arrows.
[124] From Levy [471], p. 470.

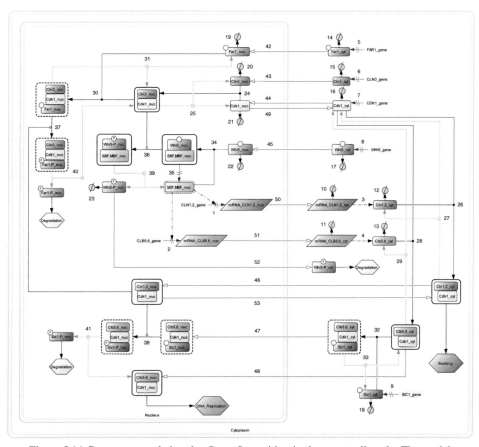

Figure 5.14 Processes regulating the $G_1 \rightarrow S$ transition in the yeast cell cycle. The model comprises transcription of genes coding for cyclins (reactions 1–2), mRNA translation for cyclins, Cdk1, and Ckis (3–9), degradation of mRNA (10–11) and proteins (12–23) (filled pink circles with a forward slash), reversible or irreversible formation of binary (24–29) and ternary (30–34) protein complexes, Cln3-independent formation of SBF/MBF (35), phosphorylation of protein complexes (36–38), and dissociation of phosphorylated protein complexes (39–41) followed by degradation of the phosphorylated protein. Transport of proteins and protein complexes with the cell (large yellow square box) occurs from cytoplasm (right side of figure) to nucleus (yellow rectangle at left) (42–48) and vice versa (49–53). Inhibitory interactions are signified by red arrows with double red backward slashes. (Credit: Figure and legend are from Figure 2 of Barberis [470], CC BY 3.0.)

channels. It deliberately abstracts from these molecular specifics. Hodgkin and Huxley employed a skeletal picture of underlying molecules and used it to explain whole-cell properties. Their achievement consisted in introducing a new form of explanation into neurobiology: an account that depicts cellular phenomena as the aggregate outcome of the activities of a large number of underlying constituents.

Strict mechanicists did not agree.[125] Kaplan and Craver [467], in particular, roiled the conventional wisdom of systems biologists seeking to explain the complex dynamical systems comprising the brain and nervous system with the following categorical statement.

…explanations lacking specific reference to causal mechanisms could not be considered explanations.

Notice that this statement focuses on "the causal structure of a mechanism," ignoring any system dynamics. Kaplan, Craver, and others were guided to this conclusion in part by common sense, which suggested that dynamical or mathematical models could not *explain* if the explanations did not incorporate the very causal mechanisms that were responsible for producing the phenomenon upon which the models were based. The normative expression of this argument was a *model-to-mechanism-mapping* (3M) requirement of explanation – "the mechanist's gauntlet."[126]

(3M) In successful explanatory models in cognitive and systems neuroscience (a) the variables in the model correspond to components, activities, properties, and organizational features of the target mechanism that produces, maintains, or underlies the phenomenon, and (b) the (perhaps mathematical) dependencies posited among these variables in the model correspond to the (perhaps quantifiable) causal relations among the components of the target mechanism. Although 3M was developed with neuroscience in mind, its principles are widely applicable.

The ability of a model to describe a phenomenon, confirm it, predict what would happen using counterfactuals, or be applicable to a variety of related systems was not sufficient to consider the model an explanation. Black box models [475] thus were rejected as explanations. These were considered phenomenal in nature, i.e., they were redescriptions of phenomena without explanatory value *per se*. Kaplan and Craver required that the causal mechanisms *inside* the black box be known. An oft-used example of a phenomenal explanation is knowing that a storm will come (black box output) from the observation of a falling mercury level in a barometer (black box input). This conclusion is substantiated when the storm does arrive, but it tells one nothing about *why* (inside the black box) this has occurred. The ideal gas law, expressed as a mathematical equation, shows an inverse relationship between pressure and volume at a given temperature.[127] Although it does have some deficiencies, it is remarkably accurate in describing phenomenal characteristics of an ideal gas. However, as with the barometer example, it does not tell us *why* these relationships hold. For this, the kinetic theory of gases is required.

Scientists, and philosophers of science, enjoy intellectual intercourse – exchanging ideas, arguing points of view, formulating new hypotheses or insights, etc. Some, like Feyerabend, took particular pleasure in inciting controversy. The one constant in this intercourse is ongoing consideration, acceptance, revision, or even rejection of ideas.

---

[125] See also Glenna [472], Machamer, Darden, and Craver [473], Woodward [474], and Bechtel [462].
[126] Quotation from Kaplan and Craver [467], p. 611.
[127] The law is $PV = nRT$, where P is pressure, V is volume, n is the number of gas molecules present in the given volume, R is the ideal gas constant (a constant of proportionality), and T is temperature.

Sometimes a consensus is reached. However, it is not uncommon that positional entrenchment and continued factionalization remain, at least temporarily. For mechanicists and dynamicists, this certainly has been true during the last decade, during which each group has softened its rhetoric and more explicitly recognized the value and importance of both mechanism and abstraction in explanation. Craver, one of the staunchest mechanicists, recently has moved from a nominal single, strong theory of mechanistic explanation ("no details, no explanation" (MDB))[128] to weaker, less complete theories that even abstract from relevant details. When should such abstractions be used? When should details be added or eliminated from an explanation to improve it? Craver and Kaplan,[129] like Marder, suggest that answering these questions:

...requires a sensitive balance between two competing and uncontroversial intuitions: first, that knowing more about how a mechanism works, on balance, improves one's understanding of how the mechanism works; and second, that unnecessary detail often obscures the abstract causal relationships by which the mechanism works. If one emphasizes mechanistic writings on one of these questions at the expense of their writings on the other, the balance is upset, and a lopsided view emerges.

What you want to know/explain places constraints on the balance between mechanistic and nonmechanistic components of explanations. In our example of the human brain, it is impossible, practically, to provide a complete mechanistic ("constitutive") explanation of this vast cellular network, i.e., an explanation comprising only the entities, activities, and features relevant to the *explanandum*. An *explanans* is constitutively relevant, in this regard, when its manipulation (*à la* Weinberg [468] and Woodward [469], see Section 5.5.6) can alter the *explanandum*, and conversely, when altering the *explanandum* results in alterations of the *explanantia*. By this definition, one could argue formally that a change in the velocity of a single neuronal impulse could change brain function (the *explanandum*), and vice versa.

If one assesses brain function by studying neuronal impulses, then one is dealing with a relatively low-level component of the *explanandum*. An even lower low-level component would be an individual ion channel within a neuron's membrane. These types of details become irrelevant as higher-level explanations are sought. For example, if we touch a finger to a very hot surface and our hand immediately jerks back, we explain the phenomenon as a "reflex arc," which comprises a sensory neuron in the finger that sends an impulse to an interneuron within the spinal cord that then signals an adjacent motor neuron to send an impulse from the spinal cord back to the relevant arm muscle to pull the hand back. This is a complete explanation of a high-level phenomenon. We don't necessarily need to delve into how these impulses ("action potentials") are created, how fast they traverse the reflex arc, what the effect of myelin in the axon is on the process, how the myelin is synthesized and laid down, etc. However, if you are a graduate student studying electrophysiology, this

---

[128] It should be noted that opponents of Craver sometimes have ignored qualifications to the strong theory, such as Craver's statement [476] that "A world viewed only at the fundamental level would be a world of gory details unfiltered by higher-level perspective" (Craver [477], p. 259).

[129] See Craver [477], p. 290.

explanation is vacuous. The student needs to have a deeper understanding that may include such things as neuronal membrane structure, placement of ion channels in the membrane, the arrangement of proteins that form the channel, and how the channel opens and closes depending on its binding of molecules that activate or inactivate the channel. The point here is that mechanistic explanations are not unitary. A plurality of explanations exist that enable one to understand a given phenomenon at the desired depth of understanding. However, mechanicists do not consider all forms of explanations valid or informative.

Mechanicists argue that no purely phenomenal explanation can be "constitutive." Black box explanations are good examples. Such explanations would be incomplete because they lack all of the entities, activities, and features relevant to the *explanandum*. They would merely redescribe a phenomenon. The thoughtful scientist recognizes the level of explanation necessary in a given situation and chooses an appropriate integration of low-level (mechanistic) and high-level (descriptive, abstract) components to comprise it. Machamer *et al.* used the sodium channel to exemplify the necessity of considering both low-level and high-level details in the formulation of intelligible explanations of phenomena [473].

The activity of the Na^+ channel cannot be properly understood in isolation from its role in the generation of action potentials, the release of neurotransmitters, and the transmission of signals from neuron to neuron. Higher-level entities and activities are ... essential to the intelligibility of those at lower levels, just as much as those at lower levels are essential for understanding those at higher levels. It is the integration of different levels into productive relations that renders the phenomenon intelligible and thereby explains it.

Our discussion thus far has considered different phenomena and the many ways to explain them. Implicit in these discussions has been the fact that each phenomenon was considered in isolation. We examined how explanations might be constructed to yield understanding of our cosmos, a reflex arc comprising three neurons, or cell division in one type of cell. But what if we wanted to select, from the many causal mechanisms involved in each system, those that were common to the different instantiations of each, i.e., to establish complete, unitary explanations covering[130] related phenomena?

Evolution is a good example of a covering explanation for the speciation and development of all organisms, not just humans. If we wish to explain the fundamental evolutionary process occurring in birds, elephants, humans, or bacteria, the existence of *explanantia* such as beaks, trunks, cerebral cortices, or flagella is irrelevant. These elements may be important for defining different organismal morphologies or physiologies, but each has nothing to do with evolution *per se*. Each is a *result* of the evolutionary process, not a cause of it. What is common among all organisms is the fact that sporadic mutations in genes that increase the chances of survival and reproduction are naturally selected.

Mechanistically complete sets of *explanantia* for the *explananda* birds, dolphins, or whales cannot be identical because, for example, many *explanantia* for birds would be irrelevant to dolphins, and vice versa. This means that the causal–mechanical requirements

---

[130] I use this term in analogy to the "covering law model" of explanation of Popper [128] and Hempel [275]. Our explanation should allow us to explain (cover) as many systems as possible.

of completeness and relevance of *explanantia* cannot be satisfied simultaneously if, in our efforts to explain the evolution of species, we simply unify all the component *explanantia*. Pure abstraction also will not do. To simply explain that evolution is the "evolution of specific traits" is uninformative. We at least would like to understand which *hows* and *whys* are common to the evolutionary process for each species. What is needed is a method of explanation that integrates static and dynamic mechanistic approaches, and descriptive approaches, to satisfy arbitrary criteria for complete, relevant explanations. I say "arbitrary" because, like *Truth* and *truth,* there are no absolute criteria (*Criteria*) for completeness and relevance, only relative ones (*criteria*).

Criteria are determined by the specific question(s) we wish to answer. We thus must first determine what features (behaviors, functions, milieu, etc.) of a system we wish to understand and what systems should be covered by this explanation. We integrate high-level abstractions with low-level causative mechanisms to "explain patterns of macroscopic behavior across systems that are [extremely diverse] at smaller scales." Batterman terms these "minimal model explanations" (MME) [478]. Surprisingly, what characterizes these types of explanation is the *rejection* of specific causal details in favor of acceptance of mechanistic abstractions, some of which may not be directly related to the *explanantia*. In some respects, MME and related simplifying approaches are the ultimate Ockham's razors. They slice away *all* causal elements not shared by *all explananda*, leaving behind a core of shared qualitative features. Batterman characterizes MME as follows:[131]

Perhaps the remarkable feature of minimal models is that they are thoroughgoing caricatures of real systems.

Do MME have any relationship to the real world or are they useless vehicles of explanation? Are MME the antitheses of explanations as they *usually are conceived?* How is it possible to reject the very *explanantia* required to understand the causal mechanisms of a phenomenon and replace them with irrelevant *explanantia*? Aren't MME just vacuous descriptions of phenomena and thus void of explanatory value? It certainly would seem so, except for the fact that MME are *not* designed to provide explanations as they *"usually are conceived."* Their remit is much broader than those of explanations of local, instantial, more circumscribed phenomena. They "latch onto the world" differently than do other types of explanation [478]. Representational accuracy is not required and, in fact, may preclude explanation (MME are the inverse of MDB). Typical explanations are applied to what Batterman [479] refers to as "type (i) why-questions," i.e., why does a given instance of a phenomenon occur? or "Type (ii) why-questions," i.e., why, *in general*, do we expect such phenomena to occur? With respect to evolution, MME are necessary if we want to explain features of the process that apply to *all* species.

The opinion expressed by Batterman in the quotation above had been proffered by Seger[132] two decades earlier when he addressed the question of the relevance and

---

[131] Quotation from Batterman [478], pp. 349–350.
[132] Quotation from Seger [480], p. 108.

usefulness of models for explaining real world phenomena. He, too, considered models to be caricatures, but caricatures that could help us understand the natural world, as do MME.

> ...models are intentional caricatures or cartoons, whose purpose is to help us strip away irrelevant complications so as to gain some insight about how a small number of key variables might interact. On this view a model is literally a toy world that we can manipulate and dissect in ways that we cannot directly manipulate or dissect the real world. Having understood the behavior of the toy world, we can ask whether it seems to mimic the real world in ways that suggest it embodies (in its highly abstracted and simplified way) interactions like those that occur in reality. To the degree that we persuade ourselves that it does, and that the modeled interactions are general, we have learned something potentially important about how the world might actually work.

A simple metaphor for the underlying principle of MME and other simplified explanations is that of bridges over rivers. If one is asked to *explain bridges*, one could offer mechanistic *explanantia* related to the materials used to build the bridges, design principles for the many different types of bridges (suspension, cantilever, truss, and cable-stayed), permissible soil composition, span, load capacity, etc. All are causal elements explaining bridges *per se*. However, what would be missing from these explanations is the simple abstraction, shared by all bridges, that what they have in common is they allow us to cross rivers.[133]

5.5.5 Relevance and Completeness

We now come to question of relevance. This is a difficult concept to define because the answer may depend on many things, including context, logic, intuition, prior knowledge, probability, prediction, and history. Even if a consensus about the importance of each of these elements of a potential explanation exists, different people may weight each differently. Some may assign particular importance to logic or probabilism, while others may minimize their importance and embrace intuition. Historical aspects of relevance may be considered in both Bayesian and intuitive senses, often without clear recognition of how each is being applied to the question. The opposite aspect of history, predictive value, may also be invoked.

We start here with three working definitions of explanatory relevance. Their features are subsumed under the concept *causal explanation*, which can be understood in simplest terms as E is a *direct* result of C ($C \implies E$).[134] Note that these definitions will involve temporality ("occurrence"), action ("makes a difference" and "act type"), logic ("an ordered pair $(p$, explaining $q)$"), type of relationship ("appropriate relationship"), as well as knowledge and counterfactuals. In addition, causal explanation may involve probabilistic or statistical reasoning. This leads to a somewhat more expansive definition of causation, "C causes E, if and only if, [E counterfactually depends on C] or [there is a causal mechanism by which C produces E] or [there is a positive probabilistic causal relevance relation between C and

---

[133] For the purpose of this discussion, we will not consider other methods of crossing, for example, boats, submarines, inner tubes, and swimming

[134] The perspicacious reader will now, of course, ask what "direct result" means, a question will we strive to answer in what follows.

E]."[135] The constant among the definitions is the notion that explanation and causation are inextricably linked. It has been said that explanation provides the information needed to establish causation.

Definition 1. "To provide an adequate explanation of any given fact, we need to provide information that is relevant to the occurrence of that fact – information that <u>makes a difference</u> to its occurrence. It is not sufficient simply to subsume an occurrence under a general law; it is necessary to show that it has some special characteristics that account for the features we seek to explain."[136]

Definition 2. "...[A]n explanation is an ordered pair consisting of a proposition and an act type. The relevance of arguments to explanation resides in the fact that what makes an ordered pair (*p*, explaining *q*) an explanation is that a sentence expressing *p* bears an appropriate relation to a particular argument."[137]

One may also define relevance as the opposite of irrelevance, which Woodward[138] defined as:

Definition 3. "...explanatory irrelevance is understood counterfactually: ...an explanans variable *S* is explanatorily irrelevant to the value of an explanandum variable *M* if *M* would have this value for any value of *S* produced by an intervention."

If we already have a valid explanation for a phenomenon, our explanation does indeed tell us the causes of the phenomenon. The question of how the explanation was established then looms, the answer to which is "through its causes." Which comes first, explanation or causation, i.e., does information flow linearly from explanation→causation or from causation→explanation? The answer is both, but at different moments. Information first flows one way at time t_i and then the opposite way at time t_j, as long as $t_i - t_j \neq 0$. This "commutative property" of explanation and causation is reflected in the use of "explains" and "causes" interchangeably by some [482].

In 1739, David Hume wrote extensively about causation in *"A Treatise of Human Nature."* His arguments are part of the foundation for modern discussions of this topic. Hume emphasized that cause and effects *qua* objects mean nothing. It is only through the *relations among objects,* i.e., their dynamics, that causation *per se* can be ascertained, and these dynamics are only relevant instantly, i.e., causes or effects are temporally "contiguous." Hume's own words express these ideas eloquently.

I find in the first place, that whatever objects are consider'd as causes or effects, are contiguous; and that nothing can operate in a time or place, which is ever so little remov'd from those of its existence. ...Now if any cause may be perfectly co-temporary with its effect, 'tis certain...that they must all of them be so; since any one of them, which retards its operation for a single moment, exerts not itself at that very individual time, in which it might have operated; and therefore is no proper cause.

Let us now consider temporality a bit more. If we believe $A \implies B$, then there must be some time interval between A and B, i.e., between the occurrence of A and the observation

---

[135] This definition comes from Weber [481], p. 305.
[136] Quotation from Kitcher and Salmon [450].
[137] Quotation from Kitcher and Salmon [450].
[138] Quotation from Woodward [469], page 14 of the chapter "A Counterfactual Theory of Causal Explanation Introduction."

Figure 5.15 A chain of causation. We start the chain by pushing over the first domino (flicking a switch), which then initiates the sequential tipping over of dominos (a chain of relevant events) ending with the light going on.

of *B*. For example, if we enter a dark room and turn on the light, there will be a finite lag between actuating a switch and sensing light. Based on *temporal proximity*[139] alone, we could explain the appearance of light by our action of flipping a switch, i.e., there is a one-to-one correspondence between our action (*A*) and its consequence (*B*). *A* and *B* thus are an "ordered pair" and their relevance to an explanation of the light appearing seems obvious. But is it? One could imagine that flipping the switch instantly sent a message to someone else to flip their own switch, which they also did instantly. In this case, it also seems obvious that someone else's action, but not our own, was the *proximate cause* of the light appearance phenomenon. This idea could be perpetuated *ad nauseam*, with many different intermediate steps involved in a chain of causation (see Fig. 5.15). Would all these steps be relevant? Our action made a difference, for without it, the room would have remained dark. Thus the relationship between switch (*explanans*) and light (*explanandum*) *appears* relevant.

---

[139] This term was used by Wesley Salmon to describe the fact that a cause must come before its effect.

Now imagine that our *explanandum* required an answer of greater depth. We might, for example, want to understand the mechanism(s) responsible for the ordered pair (*A*, *B*), i.e., what was signified by "→" and how did it happen – an explanation of an explanation. We could point to the fact that, by flipping the switch, electrical current was delivered to the lamp. Why? This would involve, among other things, electromagnetism and the contributions of electrons and electrical current to the phenomenon. Answers to all these *why* questions would comprise a temporal chain of causal events. Salmon [454] has referred to this as a "spatiotemporally continuous causal process." This process results in the transmission of "causal influence from one part of spacetime to another" and comprises two elements, causal processes and causal interactions.[140] The genesis of phenomena are causal interactions among different causal processes. Salmon wanted to differentiate causal from pseudocausal processes and genuine causal processes from noncausal interactions. Causal interactions among processes resulted in persistent modifications in those processes. Pseudoprocesses did not [484]. *Explanation* involves tracing the causal structure created by genuine interactions. This is exemplified by "an account that traces the subsequent motion of two billiard balls to their prior collision .... "[141]

Must we include *all* elements in our explanation of turning on a light? Aren't they all relevant? The answer depends on the context of the explanation, for example, for what field of science and for which scientist, or the level of detail we deem necessary in an explanation. If we were to explain to a child how to turn on a light, we need only include one element, the switch. All else would be irrelevant to the *explanandum*. If, by contrast, we wanted to explain the phenomenon to an alien, we might not be able to presuppose anything. Every element of the system, every *explanans*, would be a relevant and we thus would need to discuss materials science, electricity and magnetism, conductivity, chemistry, optics, etc. and how these elements were organized in a causal process – an endless, and impossible, process. Kaplan and Craver expressed this thought thusly.

[T]he idea of an ideally complete how-actually model, one that includes all of the relevant causes and components in a given mechanism, no matter how remote, negligible, or tiny, without abstraction or idealization, is a philosopher's fiction. Science would be strikingly inefficient and useless both for human understanding and for practical application if it dealt in such painstaking minutiae.

We have considered *linear* chains of causality. However, what we frequently encounter are causal *networks*. Figure 5.16 illustrates the difference between a simple linear chemical reaction, the combustion of hydrogen that propels the NASA space shuttle, and a complex, two-dimensional causal gene network [485]. In an analogy to logical components of explanation, the elements of the chemical equation may be described as initial conditions ($2H_2 + O_2$), event A (=, which represents ignition), and result B (H_2O plus energy). So we have the relationship $A \rightarrow B$ that represents "lighting a match" after mixing liquid hydrogen and liquid oxygen together. Our domino metaphor also represents a linear causal chain, albeit very long. When we move to the gene network diagram, we see many elements (728

---

[140] See Salmon [483], Chapter 8.
[141] Quotation from Woodward [469], Introduction and Preview, p. 4, citing Salmon [454].

(a) Combustion of hydrogen. A simple (*linear*) chemical reaction, combustion of hydrogen ($2H_2 + O_2 \rightarrow 2H_2O$ plus energy), powers the space shuttle. The image is of a test of the RS25 space shuttle engine. The large white plume is condensed water vapor. (Image credit: NASA.)

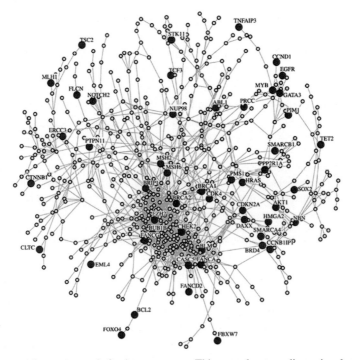

(b) Gene regulatory network for breast cancer. This complex two-dimensional network includes 51 consensus cancer genes highlighted in red, for example, BRCA1 and BRCA2. Overall, the network comprises 728 genes and 1671 interactions. (Credit: Reproduced from [485], CC BY 3.0.)

Figure 5.16 Linear and two-dimensional causal chains.

circles) and pairwise interactions (1,671 lines). Causal networks also can be created on descriptive bases. We may know that many elements or nodes exist in the network, but we have little information on the magnitudes of their causal dependencies (probabilities). Linking causality, and therefore relevance, to explanation and theory choice requires a decision process.

Ongaro has suggested that we determine relevance from two perspectives, ontological and cognitive [486], and that context is central to the selection of relevant explanatory elements.[142] Ontological relevance is the significance of a particular node in a network of dependencies to the network's product, i.e., relevance of an explanation. Significance here is a discrete entity. By contrast, cognitive relevance refers to justifications based on *"the non-discrete, continuous flux of life experience."* This continuous flux is reflected, in part, in prior beliefs and explanatory preferences (e.g., simplicity, unification, etc.) [488]. In essence, we have a distinction akin to *quantity* vs. *quality*.

Ontology and cognition work together to determine relevance. The presence of an element in a web of causation is not sufficient to deem that element explanatorily relevant. Science may establish networks, as shown in Fig. 5.16b, but relevance requires more than just presence; it requires a rationale. For Bayesians, this would be probabilistic. Relationships among elements that are especially likely would be considered explanatorily relevant. Of course, we now have the question of determining what "especially likely" means. Just as p-value < 0.05 has been used as an arbitrary limit of statistical significance, the demarcation line for making the binary decision of relevance or irrelevance based on the metric of "especially likely" will also be arbitrary. Thus, if our probabilities are not close to 0 or 1, a normative approach to selecting relevant elements from a web of causation cannot be established. Here is where cognition (common sense, experience, belief, inkling, "gut feeling") is involved, as it provides an orthogonal means of justifying element choice.

Although, theoretically, everything that now exists began with the Big Bang, we would not explain how to turn on a light by citing every element of the immense causal network comprising the sequelae of this cosmic event. We must make choices, and many times these choices involve metrics that are nondeterministic. Ongaro suggests that the mechanisms of action of such things have been determined during human evolution through the inherent need for survival. Survival demands that the organism be aware of its surroundings and respond quickly and appropriately (fight or flight) to danger, find food, or a mate (in most cases). Choosing a minimal number of particularly useful *explanantia* provides a survival advantage and can explain behavior. In this way, cognitive relevance becomes explanatory relevance.

Hempel and Oppenheim's foundational work on the deductive–nomological (D-N) account of explanation [275] was the received view of scientific explanation through the 1970s [484]. Immediately after its conception, the D-N theory became a central issue in PoS. Philosophers analyzed, discussed, and disputed the theory, as well as modifying it and creating alternatives. In doing so, many counterexamples (CE) were conceived that

---

[142] Salmon has expressed a similar view, but in which ontological and cognitive are expressed as empirical and logical, respectively [487].

showed contradictions, logical fallacies, or other problems with the D-N account. Many of these CEs are directly applicable to the question of relevance of *explanantia*. One of the most famous is "The man and the pill."[143]

A man explains his failure to become pregnant during the past year on the ground that he has regularly consumed his wife's birth control pills, and that any man who regularly takes oral contraceptives will avoid getting pregnant.

We cannot argue formally that a man *never* could get pregnant, for there is no reason a fertilized egg could not develop in a man if the right milieu were found. However, if we do accept the premises that a man regularly takes birth control pills and that men who do so do not get pregnant, then logic demands we deduce the man will not get pregnant. Remember, as we discussed before, pure logic may not have any connection to the natural world *per se* because it is a syntactic system. The relevance question here is answered by common sense and ontological and cognitive means.

An equally amusing, but insightful, example of irrelevance revealed by common sense is that of "The hexed salt," offered by Henry Kyburg in 1964 [490] and paraphrased by Salmon[144] as follows:

A sample of table salt has been placed in water and it has dissolved. Why? Because a person wearing a funny hat mumbled some nonsense syllables and waved a wand over it – i.e., cast a dissolving spell upon it. The explanation offered for the fact that it dissolved is that it was hexed, and all hexed samples of table salt dissolve when placed in water. In this example it is not being supposed that any actual magic occurs. All hexed table salt is water-soluble because all table salt is water-soluble. This example fulfills the requirements for D-N explanation, but it manifestly fails to be a bona fide explanation.

The irrelevance of the *explanantia* in these two examples is obvious. In science practice, this is not always the case. It sometime can be difficult, *a priori*, to eliminate irrelevant *explanantia* from putative explanations. The case of "The flagpole and its shadow" illustrates this (Fig. 5.17).

If an *explanandum* is "why is the shadow of that flagpole x meters long?" we can readily answer the question by citing the height of the flagpole, the position of the sun, and the laws of optics. However, if the *explanandum* was "why is the flagpole y meters tall?," we may run into a problem. Many philosophers of science consider this question to be of the form "why was the flagpole *constructed* with a length of y meters?" In this case, attempting an explanation whose *explanantia* are the elevation of the sun, the laws of optics, and the length of the shadow fails because the *explanantia* are irrelevant to the provenance of the flagpole. They did not cause the flagpole to be y meters tall, and causality is fundamental to explanation [454].

The explanation for the length of the shadow of the flagpole is asymmetric, i.e., we can explain in one direction but not the other ($\forall A \forall B \ (R_{AB} \ \& \ \neg R_{BA})$).[145] One could, however,

---

[143] From Salmon [489], pp. 22–23.
[144] From Salmon [484], p. 50.
[145] Read as "For all A and B, if A has relation R to B, then B does not have relation R to A."

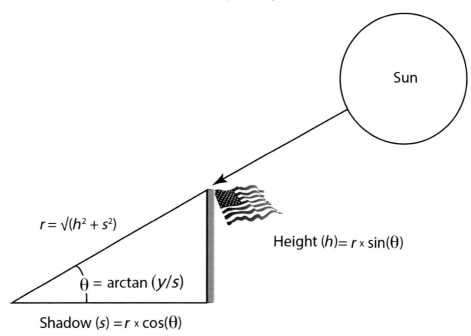

Figure 5.17 The flagpole and its shadow. Any two elements are relevant to determining the third. However, if the question is *"Why* is the height of the flagpole *h?"* then no elements are relevant because none have anything to do with the flagpole's provenance.

consider the question "why is the flagpole *y* meters tall?" from a deductive–nomological perspective, in which case the argument would be "the flagpole *must* have a height *y* because of the very same *explanantia*." From this logical perspective, the *explanantia* remain relevant and the explanations exhibit symmetry.[146]

When we come to completeness, we find a component of explanation for which it is exceedingly difficult (and likely impossible) to produce a formal definition. This is because completeness, like relevance, is context dependent (Fig. 5.18). For example, a strict mechanicist may require that for an explanation to be complete, they must know *all* elements of causation, from the time of the Big Bang until the moment they observed a phenomenon. This is clearly impossible. At the other extreme, as discussed in the context of a child in the "turning the light on" example (Section 5.5.5), a complete explanation may consist of a single *explanans*, for example, "I flipped the switch." Intuition suggests that the more complete an explanation is, the better. Thus, in analogy to MDB (more data are better), we have MCB – more completeness is better. But is it? Relevance informs completeness because it defines a subset of the set of all causal elements from which we can formulate a useful explanation that is uncomplicated and not obfuscated by irrelevant

---

[146] Another example of relevance asymmetry has been offered by Levin [491] : "Suppose someone asks 'Why are apples red?' This could either be a request to know why apples are red *rather than some other color*; or, it could be a request to know why things are red – *a request for a reductive explanation of color*."

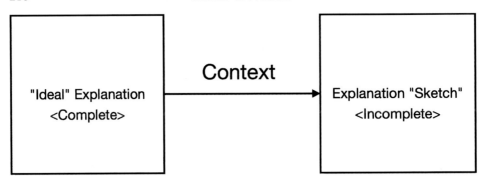

Figure 5.18 Explanations, contexts, and completeness. A simple block diagram illustrating
how context operates on the complete (ideal) set of causal elements to produce explanation
sketches that comprise only a subset of causes (hence incomplete).

minutiae. We want to avoid the inclusion of *theoretically* relevant, but practically unimportant, details in our explanations. This is an appeal to pragmatism, for which far-fetched or
unlikely, but theoretically possible, causal elements would be eschewed because we deal
with the natural world as we perceive it, using our natural intelligence, cognitive abilities,
and senses – not as a fantasy. We *use* our sense of importance in a similar practical manner.
We don't *philosophize* about what importance means *per se*, but we establish a real world
context within which we can decide between importance and unimportance. Of course, we
must establish a normative basis for such a decision and at the same time recognize that
our bases, *ipso facto*, are in no small part, and unfortunately, arbitrary. Nevertheless, they
are necessary and thus we do as best we can.

Woodward [492] provides a practical example that illustrates the contrast between complete (ideal) and incomplete (sketches) explanations. It is the phenomenon of knocking over
an inkwell. How do we explain this phenomenon? We could produce a complete explanation, i.e., one comprising all causal elements or we could "sketch out" an explanation. The
complete explanation would require recognition of many causal elements and their relationships, implicit assumptions, and actual events. For example, for the explanation to be
complete, it might include antecedent conditions such as "there is a desk" and "an inkwell
sits atop it." It must also stipulate that a causal event did occur, i.e., someone ran into the
desk. The elements of the two types of explanation are shown below.[147]

Ideal Explanation
  i. Whenever knees impact tables on which an inkwell sits and further conditions
     K are met (where K specifies that the impact is sufficiently forceful, etc.), the
     inkwell will tip over. (Reference to K is necessary since the impact of knees
     on table with inkwells does not always result in tipping.)

---

[147] Adapted from Woodward [492].

ii. My knee impacted a table on which an inkwell sits and further conditions K' are met.

iii. The inkwell tips over.

Explanation Sketch

i. Someone knocked into the desk, causing the inkwell to tip over.

Further condition K' might specify the geometry of the inkwell, its center of gravity, and the gravitational force. K' might specify that the table legs do not have wheels underneath and the knee delivered a force within a minimal amount of time (because if the total force was delivered in a slow, continuous manner, it would not result in the inkwell being knocked over). These elements of explanation are implicit in the explanation sketch, which, as the name implies, is a skeletal (incomplete) rendering of an ideal explanation. Are we now saying that both complete and incomplete explanations produce understanding? In essence, the answer is "yes," *if* we now subsume sketches under a new category of completeness defined by the ability of explanations to comprehensibly relate *explanantia* to *explananda* within specific contexts. What "relate" means also is context dependent, which leads us to an operational definition of explanation.

A complete explanation explains a phenomenon to the depth determined by the context in which the explanation is required.

Depth and completeness are correlated, i.e., the more complete an explanation is the deeper the explanation will be, and vice versa. Craver has suggested three metrics for completeness [477].

These metrics are particularly useful when one wishes to choose the "best" explanation from among many because, *ceteris paribus*, the enumerative natures of the metrics enable quantitative comparisons to be made.

i. The number of what-if-things-had-been-different questions answered (w-questions).
ii. The number of how-does-that-work questions answered (h-questions).
iii. The number of relevant elements (r-questions).

Test (iii) requires specification of relevant elements. This specification is plastic. It will vary due to context. For example, if one seeks to explain how an enzyme works, multiple levels of relevance may exist, some simultaneously. The atomic-resolution structure of hemoglobin, the protein that carries oxygen to all the cells in our bodies, is relevant if we wish to understand how oxygen binds to the protein and how the oxygen can be released to the tissues. However, this question becomes irrelevant if we wish to answer the high-level question "how is oxygen carried to our tissues." I note, relative to a single explanation of some *explanandum*, calculating completeness as the sum of $w + h + r$ is not meaningful in the absolute. It may be that a complete explanation may comprise only a few components whereas another may require numerous components. To say that the latter is more complete than the former is not possible because the contexts of the *explananda* are unique.

The important thing is not to seek a rigid definition of relevance (which does not exist), but to have an understanding of the concept and its application in science practice. In the final analysis, one must decide for themselves what is considered relevant or irrelevant by evaluating the relationship between *explananda* and their corresponding *explanantia*.

5.5.6 Explanatory Power

Explanatory power, as a concept, has been cited many times earlier as one metric for theory choice, but what *is* this concept? We may assume, *a priori*, that the better a theory or hypothesis *explains*, the more powerful it is and the more it contributes to understanding.

Explanatory power is the ability of a hypothesis or theory to effectively explain the subject matter it pertains to.[148]

If explanatory power is "how well an explanation explains," then perforce, the questions "what is an explanation?" and "what does it mean 'to explain'?" arise. After *these* questions are answered, one then may determine how well an explanation explains. Many and varied definitions exist. They contain such ideas as "makes clear or intelligible," "accounts for something," "a method of," "giving reason for or cause of," "elucidation," and "elaboration." Duhem has provided a more metaphorical definition.

To explain (explicate, *explicare*) is to strip reality of the appearances covering it like a veil, in order to see the bare reality itself.[149]

Hempel and Oppenheim addressed this issue more analytically in a landmark paper in 1948 [275] in which they conceptualize explanation as means to answer *why* questions (as opposed to *what* questions). What they could not do, however, was to provide *an* answer to the question of what scientific explanation was because the question was too complex.[150]

To explain the phenomena in the world of our experience, to answer the question "why?" rather than only the question "what?," is one of the foremost objectives of all rational inquiry; and especially, scientific research in its various branches strives to go beyond a mere description of its subject matter by providing an explanation of the phenomena it investigates. While there is rather general agreement about this chief objective of science, there exists considerable difference of opinion as to the function and the essential characteristics of scientific explanation.[151]

Explanation *per se* may be viewed from the perspectives of *what?* and *why?* For example, one can define scientific explanation[152] in a descriptive manner by saying an explanation answers a *why* question. That is, you define *what* explanation is. On the other hand, one can ask *why* and *how well* an explanation explains, in which case one must establish mechanisms and metrics for answering these questions, a process for which there do indeed exist

---

[148] https://en.wikipedia.org/wiki/Explanatory_power.
[149] See Duhem [411], p. 7.
[150] More recent work also has failed to provide an answer because, according to Halpern [493], *"getting a good definition of explanation is a notoriously difficult problem."* Ruben has gone farther, saying *"... it is an open question whether any of the explanations actually given in science are scientific explanations at all"* (See Ruben [494], p. 15).
[151] From Hempel and Oppenheim [275], p. 135.
[152] There exist many types of explanations. We restrict the discussion here to explanation in the context of science.

"considerable difference[s] of opinion." This is not surprising considering the diversity of elements that may define, or determine the value of, explanations.

Van Fraassen viewed explanation as *contrastive* in nature. Instead of asking "Why *explanandum* P_k?" one asks "Why P_k *instead of* P_i?" P_k is the topic of the question. P_i would be a member of the "contrast class" X, which also contains P_k and other *explananda* $P_1, P_2, P_3, \ldots P_n$. A question, Q, can be represented as an "ordered triple" $< P_k, X, R >$. Q then can be restated as "Why P_k *rather than* any other relevant (R) members of X?" X is not exhaustive, but is restricted only to those *explananda* that have an acceptable causal relevance relation, R, to P_k. R is determined in large part by the context of the question,[153] i.e., the background knowledge (K) extant and the idiosyncratic motivations of the questioner.

Salmon provides the following example of context.[154] The question at issue is "Why did Adam eat the apple?" This seems a simple enough question, but consider how the inflection or emphasis of the speaker, or other contextual clues, might yield three different contexts and thus three different answers.

1. Why did *Adam* eat the apple? where the contrast class = [Eve ate the apple, the serpent ate the apple, the goat ate the apple, etc.].
2. Why did Adam *eat* the apple? where the contrast class = [Adam ate the apple, Adam threw the apple away, Adam gave the apple back to Eve, Adam fed the apple to the goat, etc.].
3. Why did Adam eat the *apple?* where the contrast class = [Adam ate the apple, Adam ate the pear, Adam ate the pomegranate, etc.].

Philosophers of science typically consider "explaining" to be a process through which is understood the causes of an observation or proposition. To explain the observation that the planets have regular orbits around the sun, we invoke Newton's laws of motion and the principle of gravity. Hempel and Oppenheim generalized such examples in logical terms.

We divide an explanation into two major constituents, the *explanandum* and the *explanans*. By the explanandum, we understand the sentence describing the phenomenon to be explained (not that phenomenon itself); by the explanans, the class of those sentences which are adduced to account for the phenomenon.

In its original formulation, which was relevant to the logic of the deductive–nomological model of explanation, the *explanans* comprised both "antecedent conditions" (data) and "general laws." In addition, to be a valid explanation, a number of adequacy conditions were necessary. These included the logical condition *explanans* $\implies$ *explanandum* and *"The explanans must have empirical content; i.e., it must be capable, at least in principle, of test by experiment or observation."* Empirical conditions also were specified, in particular that the *explanans* must be true, at least when the explanation is first proffered. A more general account of adequacy, one that remains relevant today, especially to predictivists, also was provided.[155]

---

[153] For further discussion of relevance in explanation, see Section 5.5.5.
[154] Text reproduced from Kitcher [450], pp. 139–140.
[155] From Hempel & Oppenheim [275], p. 154.

It may be said... that an explanation is not fully adequate unless its explanans, if taken account of in time, could have served as a basis for predicting the phenomenon under consideration.... It is this potential predictive force which gives scientific explanation its importance: Only to the extent that we are able to explain empirical facts can we attain the major objective of scientific research, namely not merely to record the phenomena of our experience, but to learn from them, by basing upon them theoretical generalizations which enable us to anticipate new occurrences and to control, at least to some extent, the changes in our environment.

The notion that adequate explanations allow us to "control ... our environment" was further developed in the next half century by Weinberg [468] and Woodward [469]. Weinberg discussed the effects of new techniques and instrumentation on the development of molecular biology, which he argued moved that field from merely a descriptive endeavor to one in which the molecular mechanisms causing biological phenomena could be established. In doing so, molecular biologists could then *"... manipulate things [e.g., DNA] they will never see."* The ability to *"change critical elements of the biological blueprint"* allowed scientists to *"explain, at one essential level, the complexity of life ... and create versions of life that were never anticipated by natural evolution."* Causative elements in complex biological systems now not only could be identified but modified at will.

Woodward also was concerned with the distinction between descriptive knowledge and explanation. Manipulability of a system was a key element of this distinction, apart from prediction, unification, systematization, organization, or "tracing spatiotemporally continuous processes."[156]

... we are in a position to *explain* when we have information that is relevant to manipulating, controlling, or changing nature, in an 'in principle' sense ... We have at least the beginnings of an explanation when we have identified factors or conditions such that manipulations or changes in those factors or conditions will produce changes in the outcome being explained.

Woodward qualifies his view of explanation using the words "in principle" because sometimes it is impossible to manipulate a system, as, for example, the "Big Bang." The *explanatia* thus could be factors and relationships such that *"if (perhaps contrary to fact) manipulation of these factors were possible, this would be a way of manipulating or altering the phenomenon in question."*

Closely related to manipulation and control is counterfactual dependence, a property of explanations that allows them to answer "what if things had been different" questions. Understanding provided by an explanation is recognized through a pattern of counterfactual dependence between *explanans* and *explanandum*.[157] A valid explanation must enable us to understand what would happen to an *explanandum* if its *explanantia* had been different (i.e., counterfactual). Counterfactuality is one of the most important principles driving scientific discovery. It is what we do when we wonder, i.e., when we pretend that things are or could be different from what we now know and imagine how these differences could

---

[156] Quotations taken from Woodward, *Making Things Happen*, [469], pp. 8–9.
[157] For more detailed consideration of counterfactual dependence, the reader is referred to the seminal works of Lewis [495, 496].

alter the world in which we live. It can provide a framework for experimentally answering *what if* questions.

Causality is the defining element of explanation, but what about so-called non-causal explanations, for example, those that purportedly explain through the creation of mathematical formulations of the behavior of systems? How can one argue that a mathematical relationship among variables causes a phenomenon. In Section 5.5.4, we discussed and contrasted the involvement of mechanism or abstraction in explanation. We have similar issues here. Equations are mathematical abstractions representing reality (and sometimes purely imaginary worlds), but in and of themselves, they have no physical form. Mathematical explanations thus must be considered noncausal in a formal sense. As an example, we can describe the height of a bullet shot from a gun using the following equation,

$$y = x \, tan\theta - \frac{gx^2}{2v_0^2 \, cos^2\theta},$$

where y is vertical position, x is horizontal position, v_0 is initial velocity, g is the acceleration due to gravity, and θ is firing angle relative to the ground. The terms of the equation *represent* physical entities, but they are not themselves these entities – they are imaginary, thus how does the equation explain? Woodward [497] and Kuorikoski [498] have, in fact, argued that there may *not* be such a thing as explanation in mathematics. In physics, Nerlich [499] implicitly disagreed when he proposed a discrete class of noncausal explanations, geometrical explanations, to explain what happens to a cloud of unconnected particles moving through a non-Euclidean space, such as a curved space. They change shape, something that does not happen in Euclidean spaces. Newtonian mechanisms requires that a change in trajectory of objects requires the provision of force, yet no such forces are found in space. This precludes a mechanical causal mechanism and leaves us with the fact that the phenomenon only occurs in curved space, which did not cause it but does explain it.

However, there is a theory of explanation that can account for explanations by mathematics and other non-causal components, the counterfactual theory of explanation (CTE). The core requirements of this theory are not only that antecedent *explanantia* "explain" *explananda*, but that these explanations can answer "what if things had been different" questions (*w*-questions), i.e., counterfactuals. In the example above, if we imagine, counterfactually, that the movement of particles occurred in a Euclidean space, we would see no changes in particle trajectories, therefore, *ipso facto*, the explanation must lie in the existence of curved space.

An explanation cannot be logically self-contradictory. It must be both consistent and coherent. Although these terms may sometimes be used synonymously, I agree with Haack that the terms should be distinguished by their use; consistent as *favorable logical appraisal* and coherent as *favorable epistemological appraisal* [500]. Here, consistent is used in its Latin etymological sense, i.e., several objects (*con*) in agreement (*consistens*). The objects could be axioms, laws, equations, statements, concepts, hypotheses, theories, experimental results, formal mathematical systems, etc. A consistent explanation requires that all of its elements could be true simultaneously.[158] Thus, in logic, *"a set of propositions is ... [consistent] if and only if a contradiction, the conjunction of a formula and its negation, p and ¬p, [cannot] be derived from it; otherwise, it is logically inconsistent."*[159]

Coherence, by contrast, is less precise and more descriptive. Definitions of this term include words such as esthetically ordered, clarity, intelligibility, unified, making up a whole, or harmonious. Coherence thus may be contrasted with consistency by saying that instead of several objects being in agreement, several objects work well together or appear naturally suited to each other. The rigid agreement criterion is minimized and a more holistic approach is used that characterizes coherent explanations as more than simply the sum or their elements. Agreements among elements do not have to be absolute but can be assessed probabilistically and expressed as degrees of belief. Thagard, in his book *Coherence in Thought and Action* [501], presents a metaphor for JTB (knowledge) that is equally applicable to explanation.

Our knowledge is not like a house that sits on a foundation of bricks that have to be solid, but more like a raft that floats on the sea with all the pieces of the raft fitting together and supporting each other.

It thus can be said of explanation that it is justifiable, which it must be to be effective, not because it is indubitable or is derived from indubitable elements, but because it reflects the coherence of individual elements that jointly support each other. How can we determine the level of coherence? Thagard has developed algorithms for assessing "maximum coherence" [501, 502]. This computational approach is analogous to that used to determine consistency in that it seeks to divide the set of all elements into two groups, in this case, coherent and incoherent. However, instead of considering these groupings as one would in enumerative induction (counting), one evaluates combinations of elements from each group to determine which *combination* yields the maximal value of a coherence metric W. Every proposition influences every other one and influences are set up systematically to reflect explanatory relations. This means that some nominally coherent or incoherent elements may be reallocated, after computation, to the sets of incoherent or coherent elements, respectively (Fig. 5.19). Coherence thus is a *"satisfaction of multiple interacting constraints."* These constraints (principles) are used as parameters in the algorithms that evaluate explanatory coherence, as are preferences for simplicity and unification (Table 5.7).

Positive and negative elements may be included in evaluations [503], and these elements may be subsumed under Thagard's principles. In speaking about justification of beliefs (or in our case, using coherence as a metric for justification of explanation), Haack addresses the *explananda* and *explanans* involved.

---

[158] In logic, a sentence ϕ is consistent with a sentence ψ if and only if there is a truth assignment that satisfies both ϕ and ψ. A sentence ψ is consistent with a set of sentences Δ if and only if there is a truth assignment that satisfies both Δ and ψ.

[159] Quotation taken from Haack [500], pp. 168–169, with modifications.

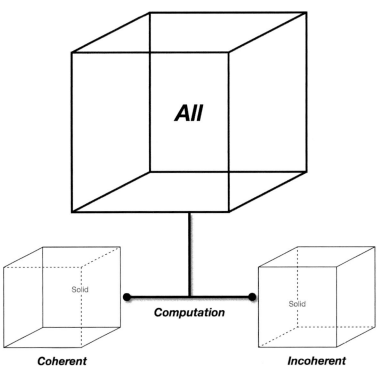

Figure 5.19 Coherence sets. The illusion of the Necker Cube is used to illustrate the grouping of the set of *all* elements of coherence (above) into subsets (below) comprising *coherent* or *incoherent* elements. For every pair of elements, a coherence value w is determined. The sum of the w values for all element pairs is calculated and the combination of elements in that set (the intersection of sets) yielding the highest coherence value W defines the overall level of coherence of the explanation. The word "solid" is located at the centers of opaque surfaces to help the reader perceive the two different geometries of the cube.

The idea is that our criteria of justification are neither simply atomistic nor unqualifiedly holistic: they focus on those elements of the whole constellation of explanans ... at [time] t which bear a causal relation, sustaining or inhibiting, to the particular ... explanandum in question.

Haack, instead of using the metaphor of a raft, likens *explananda* to crossword puzzles and *explanantia* to the letters therein [503]. One's confidence in the coherence of an explanation depends on one's confidence in the crossword's entries, their intersecting letters, and how much of the puzzle is completed. Coherence of elements of an explanation may be viewed in the same way.

We have addressed many of the conceptual components of explanatory power. When one employs explanatory power as a metric for theory strength, especially in the context of theory choice, these concepts come into play. However, how an individual or

Table 5.7 *Principles of the theory of explanatory coherence (TEC). The table describes seven principles incorporated into algorithms measuring coherence, as well as the relationships among propositions P, Q and $P_i \ldots P_j$ that define each principle.*

Principle	Description
Symmetry	If P and Q cohere, then Q and P cohere. If P and Q incohere, then Q and P incohere.
Explanation	Coherence arises from explanatory relations: (a) a hypothesis coheres with what it explains, which can either be evidence or another hypothesis; (b) hypotheses that together explain some other proposition cohere with each other; and (c) the more hypotheses it takes to explain something, the lower the degree of coherence.
Analogy	Similar hypotheses that explain similar pieces of evidence cohere.
Data priority	Propositions that describe the results of observations have a degree of acceptability on their own.
Contradiction	Contradictory propositions are incoherent with each other.
Competition	If p and q both explain a proposition, and if p and q are not explanatorily connected, then p and q are incoherent with each other (p and q are explanatorily connected if one explains the other or if together they explain something).
Acceptance	The acceptability of a proposition in a system of propositions depends on its coherence with them.

a scientific community *weights* these ultimately helps determine whether a theory is or is not favored. Weighting is intrinsically determinative and idiosyncratic. We assign a value, either implicitly or explicitly, to each component of explanation and then determine explanatory power. Each scientist, as well as the larger scientific community, may value each explanatory component differently, which leads to differences of opinion as to which explanation has the greatest explanatory power.

A number of enumerative and statistical approaches[160] have been suggested as means to more precisely determine explanatory power. Kyburg [490] has suggested enumerating the number of "logical possibilities that are ruled *out*" and according those theories that leave out the largest number to be most powerful. This is logical because an explanation that explains everything, such as the TOE, is incomprehensible in practice. To understand anything, such an explanation must be parsed extensively so that only elements relevant to the depth of explanation desired can be achieved. By contrast, if all logical possibilities were excluded, except one, that explanation would the most powerful. Peirce [377], and later Hempel and Oppenheim [275], have suggested using the level of surprise as a metric, i.e., one assesses how much *less* surprising an observation would have been if a particular explanation had been available. Peirce provided the follow syllogism to describe the deductive process involved. Implicit is the fact that one must determine the *magnitude* of the "reason to suspect" if one is to render an opinion about the explanatory power of hypothesis H.

---

[160] For example, see Okasha [404] or Lipton [348].

$l(H, E) = p(E \mid H)/p(E \mid \neg H)$	Good (1950)
$\hbar(H, E) = \big(p(E \mid H) - p(E \mid \neg H)\big)/\big(p(E \mid H) + p(E \mid \neg H)\big)$	Kemeny and Oppenheim (1952)
$\imath(H, E) = \big(p(H \mid E)/p(H)\big) - 1$	Finch (1960)
$c(H, E) = p(H \wedge E) - p(H)p(E)$	Carnap (1962)
$d(H, E) = p(H \mid E) - p(H)$	Carnap (1962)
$n(H, E) = p(E \mid H) - p(E \mid \neg H)$	Nozick (1981)
$m(H, E) = p(E \mid H) - p(E)$	Mortimer (1988)
$\delta(H, E) = p(H \mid E) - p(H \mid \neg E)$	Christensen (1999)
$g(H, E) = 1 - p(\neg H \mid E)/p(\neg H)$	Rips (2001)

Figure 5.20 Standard Bayesian measures of probability. Each measure, for example, $x(H, E)$, aims to clarify the precise formal relation that is attained between hypothesis H and evidence E, to the extent that the former is explanatory of the latter. Symbols are: $\mid$, given; $\neg$, negation (not); $\wedge$, and.

The surprising fact, C, is observed;
  But if [hypothesis] H were true, C would be a matter of course,
  Hence, there is reason to suspect that H is true.

Weighting, determining how many logical possibilities are excluded, or establishing the magnitude of a reason to suspect are all in the province of probabilities and statistics. Statistical philosophers thus have sought to eliminate, or at least lessen, the doxastic character of hypothesis assessment by employing Bayesian statistical methods to quantify an expectation of an *explanandum* given a hypothesis. Figure 5.20 lists a variety of logical constructs used by statistical philosophers to determine explanatory power E.

If an explanation is compelling, it is reasonable to infer the probability of the hypothesis underlying the explanation is high. This is not surprising because the determination of E requires *pre facto* likelihood estimates and the more confident one is in a hypothesis or explanation, the larger the estimates would be expected to be. The logical relationship explanatory power $\Longrightarrow$ probability often is implicit in Bayesian analysis of explanation. However, no formal method exists for defining the mechanism "implies" ($\Longrightarrow$). Different people conversant with the same data may have different estimations. Many factors contribute to this variability, including biases, dogma, perspective, age, and psychological state of the scientist. The reverse logical relationship, probability $\Longrightarrow$ explanatory power, also occurs, as one might expect. Here again, the mechanism "implies" remains enigmatic. Colombo *et al.* [504] have investigated experimentally how preexistent probability estimates may influence one's consideration of explanatory power. They did so by presenting a number of vignettes to study participants and then asking them to rank each of seven elements of explanation (Table 5.8).

Vignette: There are two urns on the table. Urn A contains 67% white and 33% black balls, Urn B contains only white balls. One of these urns is selected. You don't know which urn is selected, but you know that the chance that Urn A is selected is 25%, and that the chance that Urn B is selected is 75%. From the selected urn a white ball is taken at random. Please now consider the hypothesis that Urn A has been chosen.

Table 5.8 *Elements of explanation.*

Logical implication	The hypothesis logically implies that a white ball has been taken out.
Causality	The hypothesis specifies the cause that a white ball has been taken out.
Confirmation	The hypothesis is confirmed by a white ball being taken out.
Posterior probability	The hypothesis is probable given that a white ball has been taken out.
Explanatory power	The hypothesis explains that a white ball has been taken out.
Understanding	The hypothesis provides understanding of why a white ball has been taken out.
Truth	The hypothesis is true.

Some elements were interdependent, so factor analysis[161] was used to decompose them. This yielded three factors highly influenced by probabilistic information; cognitive salience (clusters explanatory power together with related cognitive values such causality and understanding), rational acceptability (captures cognitive values such as probability, confirmation, and truth that are relevant to the acceptance of an hypothesis), and logical entailment (logical relation between hypothesis and evidence $(p(E|H))$. The authors concluded:

[E]xplanatory value is …a complex psychological phenomenon related to several other cognitive processes such as deductive and inductive reasoning, causal reasoning, and reasoning about truth.

The complexity added by consideration of psychological factors makes it even more difficult to conceive of a simple explication of explanation or explanatory power, for many reasons, not the least of which is:

…the relationship between explanation and other cognitive capacities remains largely opaque.[162]

We may not have been able to define the term "explanation," but the value of explaining is immense. As Phillip Kitcher has stated:[163]

The growth of science is driven in part by the desire for explanation, and to explain is to fit the phenomena into a unified picture insofar as we can. What emerges in the limit of this process is nothing less than the causal structure of the world.

5.5.7 Explanatory Integration, Pluralism, and the Kairetic Account

Scientists and philosophers of science, when confronted with a new phenomenon or idea, seek *the* explanation of it. Temporally, the best explanation will be the first because no others exist. If the explanation is reasonable, informative, and promotes understanding, it may function as *the* explanation, much as Ptolemy's model of the heavens was for ≈1,500 years. Like Ptolemy, Hempel's D-N theory was paradigm-shifting and considered *the* theory of

---

[161] Factor analysis is similar to, but distinct from, PCA. It creates "factors" comprising related interdependent variables, thereby decreasing the dimensionality of complex data sets.
[162] See Lombrozo [488].
[163] Quotation from Kitcher [450], p. 500.

explanation, the one to which all competing theories were compared and evaluated. This remained the case for ≈ 3 decades, during which careful consideration of, and vigorous debate about, the D-N theory revealed several weaknesses and inconsistencies.[164] Better theories of explanation then were promulgated, resulting in a change from the prior unitary perspective of the D-N theory to a more pluralistic one [483, 505]. The most important of these competing views was causal explanation [454, 455, 481].

In general, unified theories display the optimal combination of generality (maximal theory diversity), simplicity (minimal causative elements), and cohesion (consistent relationships among elements) [506]. By contrast, the *sine qua non* of causal explanations is their ability to reveal a causal chain of events (*explanation*) that ends with the production of a phenomenon (*explanandum*). This chain is asymmetric. Explanation is produced by traversal of the chain in only one direction, *explanantia→explanandum*. Results can be explained by causes, but not the converse (see the flagpole example in Section 5.5.5).

In science and its philosophy, differences of opinion are vital. Intellectual intercourse among interlocutors leads to improved understanding, and sometimes, a melding of positions. An example of this melding is Strevens' synergistic combination of causation and unification [506] to create a "kairetic account" of explanation.[165] This melding eliminates the rigidity of accounts of explanation such as the D-N model by allowing one to balance unifying and causative factors in different ways. It does so by selective weighting of *explanantia* with respect to their contributions to a unified world view (unification) or to local deterministic or probabilistic events (causality). The key feature of this weighting process is the determination of relevance,[166] which Strevens explains as follows.[167]

A good causal explanation ... abstracts away from causal details that play a role in determining <u>how</u> the phenomenon occurs but that make no difference to <u>whether</u> the phenomenon occurs. Explanations balance two goals, then: to give enough causal information to imply that the phenomenon occurs, but to be as abstract as possible while respecting this constraint. They tell the least detailed causal story that is nevertheless a story that necessitates, or at least highly probabilifies, the thing to be explained.

If elements of causation are somewhat vague or abstract so as to maximize the generality of an explanation, the kairetic account can still produce a valid explanation by overweighting its unifying elements. The same type of compensatory mechanism operates when causative elements provide a precise pathway of causation at the expense of generality (unification). Precision then dominates over generality. How does one implement such a system of weighting? Strevens does so by defining relevance in terms of "difference making" *explanantia* that determine *whether or not* a phenomenon occurs, i.e., that make a difference, are relevant. If, on the other hand, an *explanans* contributes only to understanding

---

[164] Hempel and Oppenheim, in their original paper (see [275], p. 135) already had acknowledged the difficulty in establishing their theory when they wrote *"... there exists considerable difference of opinion as to the function and the essential characteristics of scientific explanation."*

[165] The word kairetic is derived from the Greek word *kairos*, meaning "a decisive point."

[166] See Section 5.5.5.

[167] Underlining by author.

how a phenomenon occurs, but not to whether it does or doesn't, it would be considered irrelevant for explanatory purposes. Relevance also may be established counterfactually (see Section 5.5.6) or probabilistically [507].[168]

The kairetic account, in practice, determines difference-makers (what is to be unified) in a similar, but nonidentical, manner to that used in the unification account. What the unification approach contributes to the kairetic approach is its superior ability to weed out causative elements that compromise generality without increasing predictive power. The kairetic account, By contrast, emphasizes minimization of causal detail at the expense of generality. However, this may be mitigated by the fact that generality in the kairetic account may include numerous *possible* systems, whereas unification is restricted to *actual* systems. The difference between possibility and actuality is exemplified by a goal unique to the kairetic account – to create an "explanatory abstraction" similar to a mathematical system.

The invocation of abstraction in the kairetic account of explanation reflected a progressive movement toward less rigid and more flexible conceptions of causation, conceptions that better reflected the definition and use of explanation by scientists themselves. An example is Railton's pragmatic account of causation [455], which some even characterized as *laissez-faire*. Like Hempel, Railton viewed explanation as a quest for laws of nature (nomothetic). However, he took issue with the D-N requirement that every explanation must either be based upon existing laws, or suggest their existence, because many satisfactory explanations do neither [455]. They do, however, provide accurate information that enables understanding, just by using a different way of explaining. How well an explanation explains, its "explanatoriness," is evaluated in terms of the ability of the explanation to provide accurate information relevant to an "ideal explanation,"[169] where this information does not come in the form of laws. A set of ideal explanations may be postulated, but it never can be defined because, according to Railton, it is *"non-denumerably infinite."* What the scientist can do, however, is to abstract subsets of *explanantia* from the ideal explanation that are most relevant to a specific *explanandum* (see Section 5.5.5 for more discussion of this point).

In a wonderful recent book, *Understanding Scientific Understanding* [508], de Regt provides a comprehensive treatment of scientific understanding, a primary component of which is explanation. He too recognizes the primacy of unification and causation as types of explanation and discusses how these, and other types of explanation, may be integrated into a unified theory of scientific understanding (UT), a theory that would provide

---

[168] The problem with probabilistic approaches is their failure to establish absolute measures of confidence, i.e., 1 or 0. So, in the range (0,1), where does confidence begin and end, and why? What would be our justification for any limits?

[169] Railton [455] describes the ideal explanation as "an inter-connected series of law-based accounts of all the nodes and links in the causal network culminating in the explanandum, complete with a fully detailed description of the causal mechanisms involved and theoretical derivations of all the covering laws involved. This full-blown causal account would extend, via various relations of reduction and supervenience, to all levels of analysis, i.e., the ideal ... would be closed under relations of causal dependence, reduction, and supervenience. It would be the whole story concerning why the explanandum occurred, relative to a correct theory of the lawful dependencies of the world. Such an ideal causal D-N text would be infinite if time were without beginning or infinitely divisible ...." [Definition of supervenience: "A set of properties A supervenes upon another set B just in case no two things can differ with respect to A-properties without also differing with respect to their B-properties, i.e., 'there cannot be an A-difference without a B-difference'." From https://plato.stanford.edu/entries/supervenience/#Intr.]

"superunderstanding" – an understanding of all global (unification) and local (causation) modes of explanation, in all contexts. However, he does not agree that combining unificatory and causative accounts of explanation is synergistic, i.e., yields superunderstanding. He rejects the notion that Strevens' kairetic model exhibits complementarity between accounts. He also argues against Salmon's thesis that unificatory and causative accounts are complementary as defined below.[170]

Complementarity Thesis. Causal–mechanical explanation and explanation by unification produce complementary types of understanding that are mutually exclusive and jointly exhaustive components of complete understanding (superunderstanding).

The PoS purportedly guides us in the experimental and theoretical conduct of science. Among other things, it tells us how we should look at the natural world, why it is that we do and believe certain things, how to determine whether what we find in our experiments is true or not, the best way to produce and test theories, and methods of inference, the key to discovery. However, from a philosophical perspective, the model of the PoS that has been created by its practitioners throughout history, and even now, suffers from the difficulty (impossibility?) of creating normative explications of the huge variety of elements that comprise science practice *per se*. Just as we can never determine *Truth* but are stuck with *truth*, the explications PoS can produce are transitory and relevant only in specific contexts. These contexts may be broad, but they are never all-encompassing. Models of explanation suffer from the same malady, which is complicated by the subjectivity and idiosyncraticity of actual scientists who don't adhere to one unified model of science practice. If these things are not considered, then philosophers of science and its devotees may experience great satisfaction and stimulation in their machinations, but these could be irrelevant to science *per se*. If subjectivity and idiosyncraticity are considered, then these same people may also be stimulated and gratified by deep understanding within specific contexts, but not by some unified TOE. de Regt proffered a related caution when he wrote:[171]

If we do not want to brush away the opinions of scientists (past or present) as irrelevant, we should acknowledge contextual variation in criteria for understanding and, accordingly, in the ideal explanatory text.

What do scientists seek in explanations? Their goal is to understand nature, but what does "understand" mean? There are at least as many types of understanding, and therefore, explanations, as scientists themselves. Some scientists want or need a mechanistic explanation of a phenomenon. What are the root causes of the phenomenon? What physical or mechanical properties of the nature are responsible for it? Some scientists want or need an explanation that deals with the general principles underlying a phenomenon. Some are interested in the sociological or psychological underpinnings of phenomena. As mentioned above, "one size does not fit all." The existence and varied importance of different explanations precludes superunderstanding, but scientists don't need this. They need explanations and understanding within their unique realms of science, be it physics, astronomy, biology,

---

[170] See deRegt [457], p. 130.
[171] Quotation from de Regt [457], p. 142.

computer science, etc. de Regt has provided a summary of types of explanation (unification, causation, kairetic, etc.) and their roles in understanding:[172]

... instead of two complementary types of explanation, one finds a plurality of tools to understand the world scientifically. ... Different explanatory strategies do not complement each other (in the literal sense of adding up to a greater whole [viz. superunderstanding]) but are merely alternative tools to achieve the same goal. Accordingly, it can be concluded that seemingly rival theories of explanation, such as causalism and unificationism, can indeed be reconciled, although not in the way Salmon envisaged [viz. complementarity].

5.5.8 Testability and Falsifiability

Theories, at their most basic, are educated guesses about mechanisms of nature. We conceive of theories to explain and we hope they are true. How do we prove it? Ideally, we apply the scientific method in many orthogonal systems to determine if predictions of our theory can be confirmed. Each confirmation provides more evidence we are right. Robert Millikan, after winning the Nobel Prize in Physics in 1923 "for his work on the elementary charge of electricity and on the photoelectric effect," discussed how he had established Einstein's work on the photoelectric effect, which yielded the "Einstein equation," was correct [509].

... accurate experimental measurement of the energies of emission of photoelectrons ... resulted, contrary to my own expectation, in the first direct experimental proof in 1914 of the exact validity, within narrow limits of experimental error, of the Einstein equation.[173]

Millikan then recounted experiments of many others, using orthogonal systems, to test Einstein's theory. In doing so, Millikan was led to conclude:

In view of all these methods and experiments the general validity of Einstein's equation is, I think, now universally conceded, and to that extent the reality of Einstein's light-quanta may be considered as experimentally established.

To be of immediate practical value, a chosen theory must be testable, otherwise it would be at risk of becoming a metaphysical entity with little or no connection to reality. In this latter case, strong justification for consideration of the theory in isolation, or for its selection from within a group of competing theories that were experimentally testable, would be difficult. It is possible that a theory may become testable in the future as new experimental methods are developed. However, until then, the theory may be considered interesting, provocative, etc., but there would be no means to study it experimentally or to use the theory for any practical purpose. This was the situation in the seventeenth century when philosopher-scientists like Descartes theorized that space was a plenum in which the movements of the planets could be ascribed to collisional effects from matter within a series of independent

---

[172] Quotation from de Regt [457], p. 145.

[173] The Einstein equation is $\frac{1}{2}mv^2 = hv - P$, where $\frac{1}{2}mv^2$ is the kinetic energy of a photoelectron, hv is the energy associated with the light quantum expressed as the product of Planck's constant (h) and the frequency of the light, and P is the work required to release an electron from a metal surface.

vortices, each associated with a specific planet (see Section 4.4.2). No methods existed for direct testing of this theory, which remained a mainstay of astronomy well into the eighteenth century, when it was supplanted by Newton's theory of gravitation operating within an empty space, which itself did not receive experimental support until a century after it was propounded. Note that in Descartes' time, theories about the movements of the planets were not directly testable, so choices only could be made from among other nontestable theories.

When a theory is tested experimentally, the possibility exists that the results will confirm or discredit it or prove uninformative. However, the veracity of such conclusions depends on how the experiment was designed. Was it designed to prove, or to disprove, the theory? These are critical questions intrinsic to any experiment, the answers to which determine our level of belief in a theory. Hume famously argued that although past experience (experiments) may allow us to make inductions about causality, there is no guarantee that future events will be consistent with our inductions. This idea is at the core of the oft-heard statement by investment firms that "past performance is no guarantee of future results." It can be argued that the history of science shows that many inductions have been, and continue to be corroborated, thus providing strong evidence for a belief that a theory is in fact true. However, Hume's caution reminds us that our inductions should not be considered deductive proof for any theory. The future may show us to have been wrong. According to this logic, experiments designed to *prove* a hypothesis thus must be considered failures even before they are performed because repeated confirmations remain susceptible to Hume's caution. This is not true of experiments designed to *disprove* a theory, because except for some complications (see Section 5.5.1 and below), a single contradictory experiment can falsify it.

The validity of this asymmetry between proof and disproof is supported by logic, an example of which is found in the recent book *The Meaning of Science* by Tim Lewens.[174] Here the issue was whether a new drug, *Veritor*, effective in patients in a small clinical trial, was efficacious in the general population. Lewens argues that "... deduction will never tell you that all people respond positively to *veritor*. On the other hand, if you find just one person who responds badly to *veritor*, you can conclude – with deductive certainty – that the statement 'All people respond positively to *Veritor*' is false." Einstein certainly agreed with this logic when he said *"No amount of experimentation can ever prove me right; a single experiment can prove me wrong."*[175] In the clinical arena, Lewens's statement certainly is true, but this may not be so in physics. In the following paragraph from the publication *The Confrontation between General Relativity and Experiment* [511], Clifford Will eloquently addresses issues of proof and refutation and the question of how many attempts at refutation must be performed before a theory, for example, general relativity, is accepted as true. Will's ideas likely would be welcomed by David Hume and those subscribing to the notion of pessimistic induction.

---

[174] See Lewens [510], p. 14.
[175] It is not clear if Einstein actually uttered these words, but the quotation is attributed to him.

General relativity has held up under extensive experimental scrutiny. The question then arises, why bother to continue to test it? One reason is that gravity is a fundamental interaction of nature, and as such requires the most solid empirical underpinning we can provide. Another is that all attempts to quantize gravity and to unify it with the other forces suggest that the standard general relativity of Einstein may not be the last word. Furthermore, the predictions of general relativity are fixed; the pure theory contains no adjustable constants so nothing can be changed. Thus every test of the theory is either a potentially deadly test or a possible probe for new physics. Although it is remarkable that this theory, born 100 years ago out of almost pure thought, has managed to survive every test, the possibility of finding a discrepancy will continue to drive experiments for years to come. These experiments will search for new physics beyond Einstein at many different scales: the large distance scales of the astrophysical, galactic, and cosmological realms; scales of very short distances or high energy; and scales related to strong or dynamical gravity.[176]

I state above that a single contradictory experiment *may* disprove a theory and Einstein states that such an experiment *can* disprove a theory. However, this is not always the case. Just as it is important not to overstate one's belief in any experimental "proof," it is equally important not to be overly secure with disbelief engendered by a single experiment. A contradictory result may be caused by unseen or unconsidered factors, for example, unrecognized technical or instrumental errors. An instructive example comes from an experiment that measured the velocity of neutrinos and concluded that they moved faster than the speed of light in a vacuum [513], a result not possible according to the principle that nothing can move at superluminal velocities. This theory had been tested in many different ways and no exceptions to it had been found, thus physicists had become increasingly confident about its validity. Therefore, when the finding was published that neutrinos produced in Geneva, Switzerland traveled 730 km arriving in L'Quila, Italy $\approx$61 ns before the predicted arrival time of light, a furor ensued [514]. Comments included "If it's true, then it's truly extraordinary"[177] and "... *extraordinary claims require extraordinary evidence.*"[178]

Adam *et al.* [513] had reported that "... the results of the analysis [using the OPERA neutrino detector] indicate an early neutrino arrival time with respect to the one computed by assuming the speed of light of ... 60.7 $\pm$ 6.9." The report, produced by the highly qualified and respected OPERA group, had been published after six months of effort to find any technical problems that could have produced it. None had been found. Nevertheless, the group was rightly cautious. Much as Newton had distanced himself from creating hypotheses about force when he said *"hypotheses non fingo,"* Dario Autiero, a member of the OPERA group who presented the finding at the European Organization for Nuclear Research in 2011,[179] showed in his closing slide the following bullet points:

---

[176] Quotation from Will [511], p. 88. A wonderful recent (April 7, 2022) experiment illustrates the prescience of Will's statement *"every test of the theory is ... a possible probe for new physics"* [512]. The experiment has shaken the foundation of particle physics, the Standard Model, which is the theory explaining three of the four fundamental forces (electromagnetic, weak, and strong interactions, but not gravitational) of the universe. The new experiment produced the most precise measurement of the mass of the *W* boson, a charge carrier of the weak nuclear force (see Fig. 5.13) – and it was *too heavy* relative to that predicted by theory and observed in prior experiments. This means that the standard model, a theory that has been used, corroborated, and accepted for more than half a century, might be wrong (at least parts of it)! The achievement required >400 scientists (all of whom are listed as authors of the paper), ~10 years, and ~450 trillion particle collisions.

[177] Quotation of John Ellis, a theoretical physicist at CERN, taken from Brumfiel [515].

[178] Quotation from Martin Rees, a cosmologist and astrophysicist, taken from Lewens [510], p. 18.

[179] Dario Autiero (2011) CERN, "New results from OPERA on neutrino properties."

- We cannot explain the observed effect in terms of known systematic uncertainties.
- We do not attempt any theoretical or phenomenological interpretation of the results.

Immediately after the finding was made public, other groups (e.g., Antonello *et al.* [516] and Bertolucci *et al.* [517]), as well as the OPERA group itself [518], began working to corroborate the finding. No significant differences between the predicted and actual arrival times of the neutrinos were found.[180] Investigation of the OPERA system revealed two possible sources of error: a previously unidentified delay in signal transmission from the GPS system used to synchronize the system (start time and arrival time) to the master clock and a problem with the oscillations in the clock itself (Fig. 5.21).

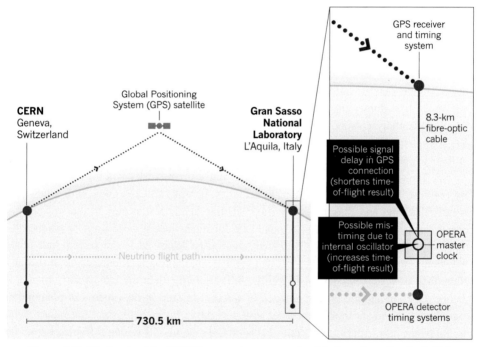

Figure 5.21 Timing trouble. OPERA researchers reexamined their system after the initial report of "faster than light" neutrinos. They identified two previously unrecognized technical problems that produced an anomalous arrival time in their original experiments. (Figure taken from [514], with permission.)

The "neutrino story" is a wonderful example of the importance of sufficient justification for belief/disbelief in theories and scientific results and of the need for caution in interpreting single experiments that purport to corroborate or refute. It illustrates a critical element

---

[180] The actual differences (ns) were 6.5 ± 7.4, -1.9 ± 3.7, and -0.8 ± 3.5 [518], 0.3 ± 4.9 [516], 0.8 ± 0.7 (Borexino), -0.1 ± 0.7 (ICARUS), and -0.3 ± 0.6 (LVD) [517]). (The different neutrino detectors used are in parentheses.) Bertolucci concluded "[The] measurements exclude neutrino velocities exceeding the speed of light by more than $1.0 \times 10^{-6} c$."

of the scientific method, its self-correcting ability, and it illustrates the necessity for empirically useful ("good") theories to be testable. Sergio Bertolucci, one of the scientists at
OPERA, discussed these points at an international conference on neutrinos in 2012.[181]

The story captured the public imagination, and has given people the opportunity to see the scientific method in action – an unexpected result was put up for scrutiny, thoroughly investigated and
resolved in part thanks to collaboration between normally competing experiments. ...That's how
science moves forward.

Sir Karl Popper, in his autobiography,[182] equates his notion of *falsifiability* with testability
and refutability. Popper's focus was on distinguishing science from nonscience or pseudoscience. One of the methods for doing so was determining if a theory/hypothesis[183] could
be falsified, i.e., shown to be inconsistent with empirical observations. If a prediction of the
theory is "the sun rises each morning," then one could readily appreciate that this hypothesis was testable experimentally – just wake up before dawn and observe. If, in fact, the sun
rose, then the theory would be corroborated.

Popper, like Hume, argued that a theory never could be shown to be *True*, but it could
be experimentally corroborated, and the more it was, the greater confidence in its *truth*
one could have. Theory selection for Popper depended on an initial evaluation of whether,
in theory, the theory was falsifiable. He did not consider unfalsifiable theories to be scientific. The sun theory, for example, *was* falsifiable theoretically and thus was scientific.
By contrast, the theory that the sun, in actuality, is replaced each night with a new one,
is not falsifiable and thus not scientific. Falsifiability is a first step in theory selection
because if a theory, theoretically, cannot be falsified, then it would not be science but
rather metaphysics or fantasy. This idea, vigorously discussed in the second half of the
twentieth century, does not now represent mainstream thought in science or its philosophy.
Nevertheless, if one equates falsifiability with *testability*, then the idea is fundamental to
theory selection and the proper practice of science. In fact, in the process of planning an
experiment, scientists should not ask "how do I prove my hypothesis?" but rather "what
would be the best experiment I could envision to prove myself wrong?" Popper expressed
this idea thusly:[184]

...our dreams and our hopes need not necessarily control our results, and that, in searching for the
truth, it may be our best plan to start by criticizing our most cherished beliefs [author's underlining].
This may seem to some a perverse plan. But it will not seem so to those who want to find the truth
and are not afraid of it.

---

[181] Sergio Bertolucci (Research Director, CERN) at the 25th International Conference on Neutrino Physics and Astrophysics,
2012, Kyoto, Japan.
[182] See Popper [288], p. 43.
[183] For our purposes here, I again consider the words "theory" and "hypothesis" to be synonymous, so anything relevant to one
is relevant to the other.
[184] From Popper [519], p. 6.

The "best experiment" or "best plan" would be what Deborah Mayo terms a "severe" test of the theory.[185] The more severe the test, the less likely the hypothesis would be confirmed. Thus, if the results of the test *did* corroborate the hypothesis, one could have confidence in its validity.

I have suggested that theories that are nonfalsifiable should be rejected. However, this statement has a temporal underpinning, namely the theory is rejected because no means to falsify it are available *now*. This raises the thorny question of whether we could envision the future development of a means to verify the theory? If so, why then should we reject this theory now? These are thorny questions because no one, at least to my knowledge, has the capacity to look into the future. How then can we know now that a theory truly is nonfalsifiable and that we should not choose it as an avenue of discovery? The answer is "we cannot." However, this does not mean that one should automatically throw up their hands and give up on the theory. The history of science is replete with theories that could not be tested at the time they were conceived. However, these theories were not rejected outright because scientists instead envisioned potential ways to test the theories, worked to develop them, and were able eventually to do empirical testing. Even if no potential means to test a theory can be envisioned, should we, perforce, throw a theory out? The answer to this question depends on both philosophy and practicality. We may take the philosophical position that no one can know, *pre facto*, whether or not an hypotheses is falsifiable. In this case, we archive the hypothesis and wait and see what happens in the future. Practicality provides a more definitive answer. How would we study a hypothesis that, *ab initio*, could not be tested? Even if we thought the means to do so could be developed in the future, what are we to do in the laboratory *at this moment*? The answer is something that does not involve the theory.

5.5.9 Accommodation or Prediction?

In . . . accommodation, a hypothesis is constructed to fit an observation that has already been made. In . . . prediction, the hypothesis, though it may already be partially based on an existing data set, is formulated before the empirical claim in question is deduced and verified by observation.[186] Well-supported hypotheses often have both accommodations and successful predictions to their credit. Most people, however, appear to be more impressed by predictions than by accommodations.[187]

People that are "more impressed" are predictivists. They believe *"evidence confirms theory more strongly when predicted than when accommodated"* [522, 523]. Predictivism has a long history that can be traced back to Descartes [25] and many justifications for this

---

[185] Mayo defines "severe test" as follows: *A passing result **e** is a severe test of hypothesis **H** just to the extent that **e** results from a procedure that would rarely yield such a passing result, were **H** false. Were **H** in error, then, with high . . . probability, the test would either have failed hypothesis **H** or would have produced an outcome more discordant from **H** than **e** is.*

[186] In logical terms, accommodation and prediction may be defined as follows: *Prediction:* Let H be a scientific theory, let B_0 represent some subset of the background information available at the time H was developed, and let Q be some phenomena or data not entailed by B_0 alone. Theory H *predicts* Q if H & B_0 entails Q, and H (when conjoined with B_0) was *not designed* to entail Q. H *correctly* predicts Q if H predicts Q and data, or phenomena, Q exist. *Accommodation:* Same as prediction except that H (when conjoined with B_0) was specifically designed to entail Q. (Definitions according to Collins [520]. The term accommodation was coined originally by Horwich in 1982 [521].)

[187] Quotation from Lipton [348].

idea have since been argued. Horwich [521] has suggested this notion relates to criticism of Freudian psychoanalytic theory in that Freudians *"can concoct explanations afterwards of someone's behavior, but ... this is too easy."* Collins [520] has argued that there is general agreement from both predictivists and accommodators that *"knowing that a theory predicted, instead of accommodated, a set of data can give us an additional reason for believing it is true by telling us something about the structural/relational features[188] of the theory."* Whether there is, in fact, general agreement, is something about which disagreement exists! Maher *et al.* have argued against a strong thesis of accommodation [524] whereas Howson *et al.* have argued for it [525, 526]. If we consider accommodation an inductive process and prediction a deductive process [527], then a theory that predicts should be favored because its predictions are unassailable (at lease logically). By contrast, induction yields propositions to be corroborated by future experimentation. Scientists do indeed employ this logic in choosing one theory from among many that seem equally good at explaining given phenomena. Of course, *at the time* a theory is chosen, no one can know whether it will be corroborated by future experiments. It is possible a theory based on accommodation will be superior to one based on prediction, or vice versa. It also is possible both theories will be equally useful. Achinstein agrees that all three possibilities may obtain, but he rejects the idea any of these outcomes depend on prediction or accommodation *per se* [528].

Temporality is a problem intrinsic for efforts to justify the desirability of prediction over accommodation. When one selects a theory to guide their research, any predicted evidence has not yet been observed. Instead, the value of predictivism must be determined *post hoc*. It thus appears illogical, but not necessarily wrong, to use this as a metric for theory selection. Now let us consider events occurring after our choice of theory. A predictive theory, plus one new experiment inconsistent with its prediction, might be replaced by a new theory that accommodates the new data. If the revised theory's prediction is corroborated, then more data have been produced that can be used for new accommodated theories. If a theory produced through accommodation is corroborated by one new experiment then, in hindsight, the original theory must have been predictive. In practice, no one would use the method of accommodation to create a theory and then *not* test its predictions. Accommodation *does* enable prediction, even if the original mechanism for producing it focused only on data extant. In turn, prediction may be based on accommodation because it required knowledge of data extant that was fitted to a model, logically evaluated, or used to foster creation of new concepts. John Maynard Keynes, a founder of modern macroeconomics, emphasized temporality and a union of, not a competition between, accommodation and prediction.[189]

The peculiar virtue of prediction or predesignation is altogether imaginary. The number of instances examined and the analogy between them are the essential points, and the question as to whether a particular hypothesis happens to be propounded before or after their examination is quite irrelevant.

---

[188] These are defined as features pertaining to the structure of the theory, its relation to other background theories, and its relation to the known data. From Collins [520], p. 213.

[189] Quotation from John Maynard Keynes [529], pp. 305–306.

...If a hypothesis is proposed à priori, this commonly means that there is some ground for it, arising out of our previous knowledge, apart from the purely inductive ground, and if such is the case the hypothesis is clearly stronger than one which reposes on inductive grounds only. But if it is merely a guess, the lucky fact of its preceding some or all of the cases which verify it adds nothing whatever to its value. It is the union of prior knowledge, with the inductive grounds which arise out of the immediate instances, that lends weight to any hypothesis, and not the occasion on which the hypothesis is first proposed.

In contrast to Achinstein and Keynes, Herbert Simon discusses a justification for attributing more value and confidence to predictivism than to accommodation that also involves temporality [530].

Let us suppose that the hypothesis h implies the two observational propositions e_1 and e_2. Scientist A makes the observation e_1, formulates the hypothesis h to explain his observation, predicts e_2 on the basis of h, and subsequently observes e_2. Scientist B makes the observations e_1 and e_2, and formulates the hypothesis h to explain them. Most persons ... assert that h is confirmed to a higher degree in the first case than in the second. They justify their assertion by arguing that, given the evidence, it is always possible to frame a hypothesis to explain it, and that the acid test of any hypothesis is its prediction of a phenomenon that was not known when the hypothesis was formulated.

Lipton agreed, writing *"Evidence that was available to the scientist when she constructed her theory may provide less support than it would have, had it been predicted instead"* [348]. These opinions raise the recursive question of whether the predicted evidence would not itself precede a subsequent theory or hypothesis. As "predicted evidence," Lipton argues that the evidence would be more persuasive, whereas once the evidence is established, and then used to support a hypothesis, it suddenly becomes less persuasive, even though it is precisely the same evidence. Worrall [531] has argued that it is not *when* evidence exists but *whether* it was used in theory creation that is most important epistemically.

...a piece of evidence e that follows deductively from T (plus relevant auxiliaries) is predicted by T just in case it was not accommodated within T by fixing some initially free parameter on its basis. Hence a piece of evidence that was known (perhaps long known) at the time some theory was formulated may perfectly well be predicted by that theory in what I claim is the epistemically important sense. What matters is whether or not the evidence was used in the construction of the theory (or rather the particular version of the theoretical framework/programme that entails it).

A preference for predictivist theories can be argued for another reason – they may be less likely to be biased. If a variety of observations preexist, and one believes that from these observations a true hypothesis can be formulated, then one can cherry-pick data from these observations so as to produce a theory most consistent with them. By contrast, when little is known about the ontology of a phenomenon, the predictivist has nothing to try to prove. They can choose subsets of data to fit *ad hoc* theories, but this is not done to support a preconceived notion of what a theory should be. A particularly unscientific and dangerous type of cherry-picking is "p-hacking" (p-value hacking). Scientists are under immense pressure to publish, climb up the professorial ladder, earn a good salary, provide the "right" advice to politicians and public policy makers, ensure that the FDA approves a new drug, etc. At the core is the pressure to publish, and one cannot publish unless their work has sufficient scientific rigor. So, if the results of a study are not statistically significant, some will selectively remove data until the p-value becomes significant ("hack" the p-value). It has been difficult to determine how widespread p-hacking is, as the percentage of scientists and statisticians responding to polls of this type is poor. Nevertheless, Wang *et al.* [532], with the support of the American Statistical Association, were able to poll 390 consulting biostatisticians. An astounding 84% of them reported that they had been asked by scientists to hack the p-value or falsify the data so that a desired outcome was achieved.

A different conclusion may be reached if one employs Bayesian statistics, for which temporality is central. Bayesian statistics considers probabilities determined by past events in its calculation of future probabilities. This means that the probability $P(H)$ of a person tossing a coin and seeing heads in that one toss is a function not only of the expected probability (0.5), but of the posterior probability of heads. If prior tosses have resulted in heads most of the time, then a future toss, according to Bayesian statistics, will have $P(H) > 0.5$. If we reconsider Simon's justification in a Bayesian sense, we can reformulate the question taking into account whether observation e_2 was or was not known at the time of theory creation, as follows:

$$P(T|e_1 \& \neg e_2) > P(T|e_1 \& e_2). \tag{5.9}$$

The probability P of theory T given knowledge of evidence e_1 but not of e_2 ($\neg e_2$) is greater than P when both e_1 and e_2 were known. According to Maher [524], there is no reason this inequality cannot be true and thus support the superiority of predictivism in theory choice. However, Maher, instead of actually providing evidence for the supremacy of prediction, has merely restated the case in probabilistic terms. Horwich also has used Bayesian methods, but comes to the opposite conclusion because he considers not just Bayesian logic *per se*, but the *a priori* plausibility we assign to *T*. We know that

$$P(e_1 \& \neg e_2) + P(e_1 \& e_2) = 1 \tag{5.10}$$

and

$$P(e_1 \& \neg e_2|T) + P(e_1 \& e_2|T) = 1; \tag{5.11}$$

thus, according to Bayes' theorem,

$$P(T|e_1 \& \neg e_2) = \frac{P(T) \times P(e_1 \& \neg e_2|T)}{P(e_1 \& \neg e_2)}. \tag{5.12}$$

If we believe that T is true, and thus plausible, than even before e_2, $\frac{P(e_1 \& \neg e_2|T)}{P(e_1 \& \neg e_2)} > 1$ thus

$$P(T|e_1 \& \neg e_2) > P(T) \tag{5.13}$$

and therefore, as in Eq. 5.10, i.e., prediction is to be desired over accommodation.

$$P(T|e_1 \& \neg e_2) > P(T|e_1 \& e_2). \tag{5.14}$$

However, if we cannot infer the plausibility of *T pre facto*, i.e.,

$$P(e_1 \& \neg e_2 | T) = P(e_1 \& \neg e_2), \tag{5.15}$$

then no advantage can be assigned to the predictive theory T:

$$P(T | e_1 \& \neg e_2) = P(T). \tag{5.16}$$

Achinstein, Keynes, Simon, and others include temporality and the number of events observed (either in the past or future) in their consideration of the accommodation/prediction question. Keynes, in particular, emphasizes the number of instances examined, but considers the timing of these events irrelevant. Nevertheless, it is intuitively obvious that the greater the number of observations of a phenomenon (events), the more likely any theory explaining it will be true (assuming each observation is independent of the other and one is not simply doing the same experiment *ad infinitum*). Figure 5.22 illustrates this idea. Here arbitrary mean probabilities ($\pm$SD) are plotted against log (number of events) for theories constructed using accommodation (blue) or predictivism (red). Three elements, in particular, should be noted: (1) initial values ($P(T|e_1)$); (2) shapes of the curves; and (3) asymptotes. The greatest uncertainty ($P(T|e_1)$) (lowest probability) exists when only one event has been observed, but it decreases steadily after that for both types of theories. The increases are nonlinear because, as the number of experiments increases, the amount of new relevant information decreases, reaching a point where, theoretically, we know all possible relevant data so $P(T) = 1$.

The intrinsic variability of a phenomenon determines the initial slope and the rate at which $P(T)$ approaches 1, two quantities that are inversely related. If we assume the first event is representative of a real phenomenon, then accommodation of those data should produce a theory with some level of truth, i.e., $P(T|e_1) > 0$, and if we assume that the intrinsic variability of the phenomenon isn't huge (otherwise it would be highly unlikely that we would observe another event), then we can assign, though arbitrarily, a SD. The magnitude of the SD decreases as more events are observed. Because the accommodated theory is informed by existing data, and again assuming that the data reflect a true phenomenon, its initial probability likely will be greater than that of the predictivist theory,[190] for which fewer initial constraints existed. As the number of events increases, the SD for each type of theory will decrease, but those for the predictivist theory should always be larger than those of the accommodated theory because of its fewer constraints. This is an advantage of the predictivist theory, because its intrinsic variability allows discovery of things outside the constraints of the accommodated theory, which is one justification for its favored use in theory selection. Of course, as *e* increases, both theories converge.

We have discussed a number of factors affecting our decisions about the use of accommodation or predictivism in theory choice, including inference, temporality, precision, simplicity, flexibility, corroboration, and statistics. However, human psychology also plays a role. The most obvious and ubiquitous example of this is how many predictivists consider their theories more beautiful than those derived through accommodation. Beauty was

---

[190] It is possible that the predictivist theory chosen may be excellent from the start, in which case its probability could be higher than that of the accommodated theory.

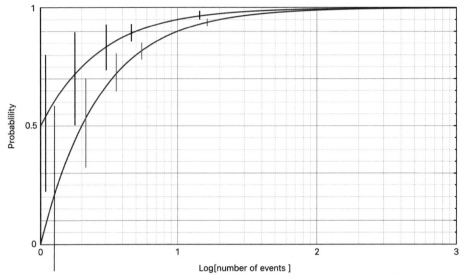

Figure 5.22 $P(T)_i$ vs. event number. The probability of theory T after i events, $P(T)_i$, is plotted versus the log of the number of events observed. $P(T)$ for the first event is arbitrary and is designed only to illustrate that $P(T)$ determined by accommodation likely is larger than that determined by prediction. $\lim_{i \to \infty} P(T)_i = 1$ for both methods of theory construction. Top curve: accommodation. Bottom curve: prediction. Bars are estimated standard errors.

discussed extensively in Section 5.5.10, so I will not address all of its aspects again here. I do argue that the combination of broad explanatory power with minimal constraints, which is characteristic of predictive theories, is considered more "beautiful" than an accommodated theory that typically is more highly constrained, less flexible, and thus less likely to reveal scientific novelties.

A more insidious example of poor human behavior is data manipulation ("fudging"). Scientists seeking to accommodate data may choose to omit certain of those data to develop a theory that better accommodates the data remaining or to add auxiliary hypotheses to achieve the same goal. These behaviors commonly are observed when scientists seek to prove their own theories correct rather than seek the truth. They may also occur if scientists must argue, for example, in grant applications, for the significance, novelty, or importance of a theory. Another type of fudging, HARKing (Hypothesizing After the Results are Known) [533], may occur during theory creation and testing. The following limerick gives an amusing account of HARKing.

HARKing

A reader quick, keen, and leery
Did wonder, ponder, and query
When results clean and tight
Fit predictions just right
If the data preceded the theory

(Attributed to Pat Laughlin)

Those who commit HARKing do not cherry-pick data, conceive of auxiliary hypotheses to save their own cherished ones, or commit unethical acts such as data falsification or manipulation. Their "crime" is temporal in nature, i.e., they alter their own preexistent hypotheses based on new data but present the hypothesis as if it were conceived prior to that. In this way, scientists can propose a "novel" hypothesis, make a prediction that they already know has been validated experimentally, and then later corroborate their own prediction. HARKing has been studied most in psychology and sociology, where it is common, but it exists throughout the scientific enterprise.[191]

Is *all* fudging bad? Maybe not. In discussing theory choice, Lipton argues that fudging may be an appropriate and useful strategy for dealing with new data inconsistent with a chosen theory.[192] Instead of simply rejecting the theory and looking for a better one, which might, at that moment, be impossible, Lipton proposes two types of fudging, "auxiliary" and "theory." The first approach involves modifying auxiliary elements of the theory without changing the theory proper. The second approach retains the auxiliary elements but changes the theory itself. In this manner, the scientist creates a theory that accommodates the data and provides a structure for experimental design and testing of that theory. Lipton justified this approach as follows.[193]

Often, the scientist's problem is not to choose between equally attractive theoretical systems, but to find even one. Where sensible accounts are scarce, there may be a great temptation to fudge what may be the only otherwise attractive account the scientist has been able to invent. The freedom to choose a completely different system is cold comfort if you can't think of one. And, in less dire straits, where there is more than one live option, all of them may need to be fudged to fit. This brings out an important point that the pejorative connotations of the word 'fudge' may obscure: a certain amount of fudging is not bad scientific practice. At a particular stage of research, it may be better to work with a slightly fudged theory than to waste one's time trying unsuccessfully to invent a pure alternative.

In this section, we have seen unconvincing, persuasive, and heretical arguments for and against the relative value of theories created through accommodation or prediction. These arguments, usually by philosophers of science, raise important and interesting questions that should be considered by practicing scientists before, during, and after their studies. But how does a scientist answer such questions, which philosophers of science themselves cannot? Is there *an* answer? What should be the take-home message for scientists? I think, as we have seen repeatedly before, that the answer, if one is possible at all, lies somewhere between the extremes of accommodaters and predictivists and incorporates both views. Stanger-Hall, in a *Science* article entitled *Accommodation or Prediction?* [534], may have provided the most reasonable, practical answer possible, one in which she recognized the

---

[191] Rubin [527] has distinguished three different types of HARKing: (1) CHARKing, constructing hypotheses after the results are known; (2) RHARKing, retrieving hypotheses word-for-word from published literature after the results are known; and (3) SHARKing, suppressing *a priori* hypotheses after the results are known. He concluded that it is never acceptable to CHARK, always acceptable to RHARK, and only acceptable to SHARK if the suppressed hypotheses are both poorly tested and unrelated to the research conclusions.

[192] See Lipton [408], p. 142.

[193] Quotations from Lipton [348], p. 174.

distinct but complementary contributions of accommodation and predictivism to theory creation, choice, and for practicing scientists, use.

Together, accommodation and prediction will lead to scientific progress where either one in isolation would be incapable of doing so.

5.5.10 Beauty, Elegance, Cleverness, and Novelty

[I]]t is widely agreed, both in thought and in practice, that science's exclusive focus on empirical evidence is its greatest strength.

Thus has opined Michael Strevens [535] and many others[194] in ruminating about why science is considered "special," i.e., particularly authoritative relative to other human endeavors. Science's authority comes from its experimental and logical methods, which yield true statements about the world and how it works.[195] Nonscientific endeavors, for example, art, don't require experiments (in the scientific sense) and logic. All we have to do to create a work of art, for example, a painting, is to transfer images from our sight and imagination onto a blank canvas of some type. To appreciate art, all one must do is look. To do science at the highest levels usually requires a doctorate or M.D. degree, many years (sometimes decades) of study, specialized training and instrumentation, and the ability to make logical and rational inferences. Training in art (and other humanities) is conspicuously absent in science, and for good reasons, the most important and obvious of which is it is not scientific. It is not a part of the scientific method. Beauty has no place in scientific publications. Beauty just is not what we scientists "do." Or is it?

... it is impossible to follow the march of one of the great theories of physics, to see it unroll majestically its regular deductions starting from initial hypotheses, to see its consequences represent a multitude of experimental laws down to the small detail, without being charmed by the beauty of such a construction, without feeling keenly that such a creation of the human mind is truly a work of art.

Thus wrote Pierre Duhem in *"The aim and structure of physical theory."*[196] Duhem is talking about feelings here, not about choosing strategies or planning experimental methods, hypothesis selection and validation, mathematics, or proof. Three Nobel Prize winners, Murray Gell-Mann, Paul Dirac, and Steven Weinberg, respectively, agree.

[W]e live in a world ... where beauty, simplicity and elegance are a chief criterion for the selection of the correct hypothesis.
    It is more important to have beauty in one's equations than to have them fit experiment.
    [W]e would not accept any theory as final unless it were beautiful.

Why would beauty, for example, be a criterion for anything? Isn't this irrational [535]? Milena Ivanova recently addressed this question brilliantly [536]. She suggests *"a beautiful theory is more likely to be true,"* and history has shown often this is the case. This is how

---

[194] See Chapter 2.
[195] For the moment, I will stipulate that the products of science are true. See Section 5.7 for a full discussion of truth.
[196] See Duhem [411], p. 24.

the "average" scientist feels and what they practice. Why they feel this way has been an active area of discourse in PoS, but we cannot address this very interesting psychological question here. We will focus on the practice of science *in fact* not in theory.

In practice, whether it is stated explicitly or is implicit, beauty is an important metric for theory development, selection, and evaluation. One reason is beautiful theories have turned out to be true more often than ugly ones. It also is true that theories felt to be beautiful at their conception, but not felt to be rational, relevant, or even possible, *can* turn out to be true. This was the case with the hypothesis that gravitational waves exist. The mathematics underlying the hypothesis was beautiful, but physicists thought an experiment to test the hypothesis could not be performed. They were right, initially, but after 40 years of effort and a cost of $750 million, gravitational waves *were* detected [535].

We have linked beauty to scientific practice in a conceptual sense, but is there any direct evidence that the opinions of Nobel laureates and the study of history are relevant to this question? Indeed there is. Zeki *et al.* [537] were aware of many instances in which the experience of mathematical beauty had been likened to the experience of viewing great art. They hypothesized that regions of the brain responsible for these experiences might be shared. To test this hypothesis, highly trained mathematicians were asked to look at various equations and decide if they were beautiful, ugly, or neither. Functional magnetic resonance imaging (fMRI), a technique that shows areas of the brain active during certain activities, then was performed while the mathematicians were shown those equations they had previously felt were beautiful. What Zeki *et al.* found was the same brain region activated experiencing great art or music, the medial orbito-frontal cortex, an area at the front of the brain involved in decision-making and pleasure, reward and hedonic states, was activated when beautiful equations were seen. There now was direct experimental evidence for a common neurophysiological basis for mathematical and artistic information processing. Maybe the Nobel laureates were right about the necessity and importance of beauty in theory selection, mathematical formulations, and theory acceptance.

It is undeniable that most people prefer beauty over ugliness, elegance over crudeness, and cleverness over foolishness. Magazines for women, and men, usually show photographs of models, and of celebrities, of preternatural beauty unattainable by the vast majority of people. "High society" magazines popularize elegance in all things, be they cars, homes, art, yachts, clothing, or vacations. Scientists see science in the same way. They consider theories that are beautiful, elegant, or clever to be preferable to those that are not.

Sometimes beauty is ascribed not only to a scientist's work product, but to the scientist themself. This was the case for John Nash, the mathematician famous for, among other things, his development of the Nash embedding theorem, the Nash functions, the Nash–Moser theorem, and the Nash equilibrium. The latter accomplishment was recognized by the awarding of the 1994 Nobel Prize in Economic Sciences to Nash, along with John Harsanyi and Reinhard Selten, "for . . . pioneering analysis of equilibria in the theory of non-cooperative games." Although Nash suffered from paranoid schizophrenia, Lloyd Shipley, a friend of Nash during his graduate school days at Princeton, characterized him as having a "keen, beautiful, logical mind." This beauty was the basis for the title of a biography of Nash by Sylvia Nasar entitled "A Beautiful Mind."

Science in Practice

(a) The golden ratio ($\frac{a+b}{a} = \frac{a}{b} = \phi$).   (b) "Ugliness is Dead" by Mihai Tenovici.

Figure 5.23 Beauty or ugliness? (Credit for Fig. 5.23a: eightonesix/Freepik.)

When you read the paragraph above you likely had an intuitive sense of what is meant by beauty, elegance, or cleverness. However, it also is likely that your intuition might differ from someone else's. This is illustrated in Fig. 5.23, which shows a red rose on the left and a painting by the Romanian artist Mihai Tenovici on the right. A red rose is considered beautiful from a strictly esthetic perspective. However, for a geometer, mathematician, engineer, or scientist, it may not be the rose that is beautiful, it's the fact that the geometry of the petals approximates the shape of the golden ratio (Fig. 5.23a, thin black curve), a number that has held people in thrall since the time of ancient Greece. In contrast to the rose, the painting in Fig. 5.23b, at least in the minds of many people, exemplifies the ugly, crude, affected, or psychopathologic. Surprisingly though, expressionist paintings such as this have been purchased at remarkably high prices. For example, William De Kooning's "Woman III" fetched $137.5 million. We must assume the buyer either thought the picture *was* beautiful, possessed an unusually bizarre artistic sense, or was an astute investor. Clearly something intangible is operating here, otherwise two people seeing the same object (read "theory") should not have different opinions.

Similar considerations operate with respect to elegance and cleverness. If one compares the penmanship found in the US Declaration of Independence (Fig. 5.24a) with that found in one of Isaac Newton's laboratory notebooks (Fig. 5.24b), one sees a difference. In the former case, calligraphy is seen, whereas in the latter case, what is seen might aptly be described as scribbles. I don't think the notebook is elegant, but because of its context – the image of the notebook brings me closer to Newton and his time – I see the writing as historic and ground breaking, but not beautiful.

Aspects of cleverness may include simplicity, utility, and cost, among others. One of the best examples of cleverness of which I am aware is a fantasy about the solution to the problem of writing in space. The fantasy begins by noting that ball point pens will not

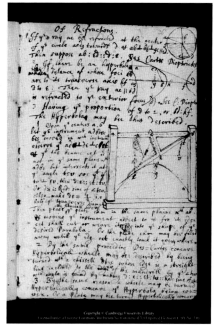

(a) US Declaration of Independence

(b) Newton's Lab Notebook

Figure 5.24 Elegance or crudeness? (Credit for Fig. 5.24b: Reproduced by kind permission of the Syndics of Cambridge University Library.)

function under the weightless conditions of space (which is true). NASA[197] had a problem. How were their astronauts to write? According to this fantasy, the Americans spent $2 million to develop a pen that *would* work in space. This pen, the actual name of which was the "AG-7 'Anti-Gravity' Space Pen" (Fig. 5.25a), comprised an extra strong cylindrical body, a rubberized grip for easy writing with gloves on, a refillable cartridge for oil-based ink, and a piston that pressed against a chamber to send compressed (35 ppi) nitrogen gas to the ink cartridge, thus ejecting the ink onto a writing surface through a tungsten-carbide ball at the tip of the pen. The ink existed as a gel within the ink cartridge but was liquified by the tip of the pen as it moved over a writing surface, allowing up to 492 feet of writing. The Russians, instead, devised a clever solution to the problem – they used pencils (Fig. 5.25b) – which were simple, useful, easy to obtain, and cheap to buy! This is a nice fantasy – patently false[198] – that does illustrates the point.

As a final exemplar of the use of intangibles in theory choice, we turn to the question of the geometry of the "heavens." Figure 5.26 shows geocentric (panel *a*) and heliocentric (panel *b*) models. The first thing to notice is how simple the paths of the planets

[197] National Aeronautics and Space Administration.

[198] In reality, both space programs began with pencils, but the wooden detritus from sharpening the pencils, and the fact that the pencils were flammable, made this undesirable. In addition, it was not NASA, but rather the Fisher Pen Company that developed the pen, at an actual cost of $1 million. At that time, in 1967, both NASA and the Russian space program were able to buy these pens at the discounted price of $2.39 each. You yourself can have one now, and for the "low, low price" of only $8.06 on Amazon.com.

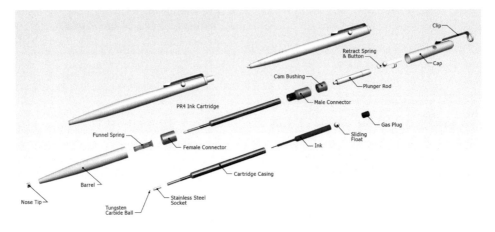

(a) Fisher AG-7 "Anti-Gravity" Space Pen.

(b) #2 Pencil

Figure 5.25 Cleverness or foolishness? (Credit: www.cultpens.com/news/product-news/q/date/
2013/07/01/the-fisher-space-pen-myths-and-magic.)

are in the latter model relative to the former. Only ellipses exist. It is easy to visualize
the sun, the planets, and their motions. This is not the case with the geocentric model.
Similarly, the mathematical formulation of the paths of the planets around the sun is easier. It involves fewer variables and less complicated mathematics, which means that the
chances of introducing error into the system are lower. Here, beauty, simplicity, and elegance all contribute to choosing the Keplerian theory for the movement of the planets over
the Ptolemaic theory.[199]

[199] Question: If the heliocentric universe with elliptical orbits is the best theory explaining the movement of the planets, why do
some planetaria engineer their star balls (a type of projector) using the Ptolemaic theory in the design? The answer is the
theory of Ptolemy accounts for the relative positions and movements of the planets quite well. Maybe Ptolemy's theory was
accepted for as long as it was because people found circles more aesthetically appealing than ellipses?!

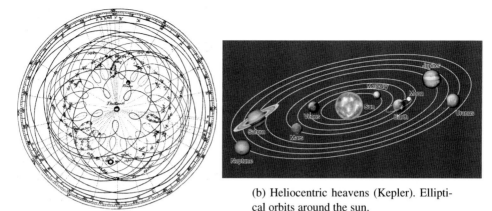

(b) Heliocentric heavens (Kepler). Elliptical orbits around the sun.

(a) Geocentric heavens (Ptolemy). Circular orbits, with epicycles, around the Earth.

Figure 5.26 Theories of the geometry of the heavens.

These four examples of theory choice show how the intangibles of beauty, elegance, and cleverness may guide theory choice. They also show that this guidance is not uniform. Different interpretations of each intangible may result in different theory choices. Why then do most scientists allow these types of considerations to influence their scientific practice? One reason may be evolution. According to this theory, the overriding principle of existence is existence itself, i.e., survival. Those with the best abilities to deal with their environments, writ large, have a better chance of survival, and the intangible of what may be called "aesthetic processing" is an important part of these abilities. Brown *et al.* [538] have studied the emotional aspects of aesthetic processing and suggested that the evaluation of "what is good for me" and "what is bad for me" is the determining factor controlling how we *feel* about a particular choice. A theory that is beautiful, elegant, or clever will produce pleasurable feelings, leading to its choice. By contrast, ugly, inelegant, or foolish theories are unlikely to be pleasurable and likely will be rejected.

Humans process sensory information and determine whether it is or isn't already known, i.e., whether it is new. New information must be evaluated so that an organism can react to it. A proper reaction can determine whether the organism survives, and if so, how well. Humans tend to like new things, whether they be foods, clothes, cars, songs, movies, books, or news. Nonhuman primates also are intrigued by new things, as are cats, mice, rats, and porpoises. Science, in particular, is driven by newness. Unless data are new, they generally are not published. Unless hypotheses or theories are new, they are ignored because they provide no additional insights, explanations, or predictions compared to those extant. The OED, in its first four definitions of the word *new*, uses the words "not previously" three times and "first" once.[200] Temporality thus is a key element in this definition.

---

[200] Oxford English Dictionary, 3rd edition, Oxford University Press 2014 www.oed.com/view/Entry/126504?rskey=FbyW6y&result=1#eid.

Novelty is different. Something must be new to be novel, but the converse is not true. For example, an invention may be *new*, but unless it is *novel*, it cannot be patented. "Novel," according to the OED,[201] is defined not by time but by nature. Something is novel if it is "Of a new kind or nature; not previously imagined or thought of; (now) esp. interestingly new or unusual." A new theory is not novel because it is new, for it may follow from existing theories by addition of auxiliary hypotheses or predict prosaic outcomes. It is novel because it makes predictions that could not be made by existing theories. It is novel if its explanations of phenomena had not been known or considered before. It is novel because it was previously unimagined or unusual.[202] Examples include the theories of cells, evolution, gravity, and relativity, as well as the discovery of the structure of DNA or the existence of prions. Scientists prefer to choose novel theories because of their exceptional usefulness for understanding natural phenomena. Of course, once these novel theories are accepted and become a part of the scientific entity extant, they no longer are novel.[203]

5.6 Theory Testing

We have conceived of several theories consistent with a phenomenon and chosen one or more that appear best at explaining it. Now we want to know whether a theory is "right" or "wrong." These terms are written within quotations to signify their definitions vary with context and there are *degrees* of rightness and wrongness. Note that right $\neq$ true. Right is a relative term, i.e., it answers the question of whether a theory is *consistent relative to* one or more propositions under consideration. If not, the proposition(s) is wrong. In contradistinction, Truth and False (with capital letters) are *absolutes*. True and False proposition are unitary states in which gradations do not exist. Something is or isn't and there is no "almost" or "nearly." The caveat here is that even if all propositions of the theory are true, i.e., consistent with all data extant, and all predictions of the theory are observed, no *formal* proof of a theory *can* be proffered because the possibility exists that the theory may be refuted in the future [168, 259] (see Sections 4.5.4, 4.6.2, and 5.5.8). However, formality and practicality are two different things. Formality provides a structure for thought, whereas practicality enables action, i.e., doing experiments, which helps us assess a theory and decide how much confidence we have in it.

Our confidence in a theory should never be absolute, nor should we become "married to hypothesis." This is a dangerous form of self-delusion that closes the mind and produces bias. We one actively ignores any existing or potential problems or caveats, for example, counter-hypotheses, contradictory evidence, inconsistencies, or predictions of the hypothesis that are not observed experimentally.

The astronomer Raymond Lyttleton, a Fellow of the Royal Society and winner of the Society's Royal Medal, created an amusing, but accurate, pictorial representation of the

---

[201] www.oed.com/view/Entry/128758?rskey=BZYlnZ&result=2#eid
[202] What is "unusual" is a semantic question that we put aside for the moment.
[203] Lest the reader be led astray, there are *degrees* of novelty [539], but consideration of these does not change the fact that novelty, regardless of its degree, is an important part of theory choice.

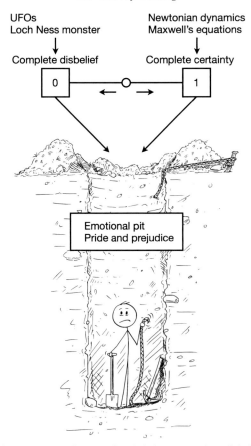

Figure 5.27 Belief range. A continuum of belief, from complete disbelief (0) to complete certainty (1). The magnitude of one's belief in a theory is indicated by the position of the bead. At the top are examples of things one may have beliefs about. If the bead were to reach the ends of the string, the scientist would fall into an extraordinarily deep pit from which they could never escape. This may occur if one has calcified opinions (old scientists), hubris (young scientists), and any number of other nonscientific reasons for belief. (Lyttleton [540].)

range of confidence (belief) one may have in a theory, and what to do with it (Fig. 5.27). Here, a movable bead is attached to a short string representing the range of belief, from 0 (absolute disbelief) to 1 (absolute certainty). Lyttleton's take-home message was simple:

Never let your bead ever quite reach the position 0 or 1.

How a theory is tested depends to a significant degree on the theory's provenance. Did it come from *gedankenexperiments*, examination of the historical record, or direct observation and experimentation? Some theories can be tested experimentally, but others may be difficult or impossible to test and thus corroboration of these theories must be obtained through other means. For example, in the size regimes of quantum mechanics and cosmology, testing theories about the behavior of matter and space-time[204] often is impossible. One must imagine universes in which the existence and behavior of unobservable components of subatomic and cosmological matter, and their forces, are determined by a combination of existing and novel laws of physics, and then think about what might happen if components of these universes were manipulated. This is a very difficult proposition because these systems are preternaturally complex. Complexity also complicates the testing of theories in the fields of ecology and evolution.

The following quotation from William Harvey,[205] famous for his elucidation of the circulatory system in humans, provides a wonderful framework for doing science, not with respect to practical methods *per se*, but with respect to avoiding marrying your hypotheses, being close-minded, stubborn, biased, etc. One would be well-served if they modeled themselves after Harvey's "true philosophers."

True philosophers, who are only eager for truth and knowledge, never regard themselves as already so thoroughly informed, but that they welcome further information from whomsoever and from wheresoever it may come; nor are they so narrow-minded as to imagine any of the arts or sciences transmitted to us by the ancients, in such a state of forwardness or completeness, that nothing is left for the ingenuity and industry of others. On the contrary, very many maintain that all we know is still infinitely less than all that still remains unknown; nor do philosophers pin their faith to others' precepts in such wise that they lose their liberty, and cease to give credence to the conclusions of their proper senses. Neither do they swear such fealty to their mistress Antiquity, that they openly, and in sight of all, deny and desert their friend Truth. But even as they see that the credulous and vain are disposed at the first blush to accept and believe everything that is proposed to them, so do they observe that the dull and unintellectual are indisposed to see what lies before their eyes, and even deny the light of the noonday sun. They teach us in our course of philosophy to sedulously avoid the fables of the poets and the fancies of the vulgar, as the false conclusions of the sceptics. And then the studious and good and true, never suffer their minds to be warped by the passions of hatred and envy, which unfit men duly to weigh the arguments that are advanced in behalf of truth, or to appreciate the proposition that is even fairly demonstrated. Neither do they think it unworthy of them to change their opinion if truth and undoubted demonstration require them to do so. They do not esteem it discreditable to desert error, though sanctioned by the highest antiquity, for they know full well that to err, to be deceived, is human; that many things are discovered by accident and that many may be learned indifferently from any quarter, by an old man from a youth, by a person of understanding from one of inferior capacity.

5.6.1 The Theory and Practice of Theory and Practice

It is also a good rule not to put overmuch confidence in the observational results that are put forward until they are confirmed by theory.
—*Arthur Eddington (1934)*

---

[204] Remember our discussion of gravitational waves and how difficult it was to create an experiment able to test the theory.

[205] From Harvey [83], pp. 7–8, "Dedication To His Very Dear Friend, Doctor Argent, The Excellent And Accomplished President Of The Royal College Of Physicians, And To Other Learned Physicians, His Most Esteemed Colleagues."

Theorists play critical roles in science, including the *de novo* conception of theories, some of which can be tested experimentally, and the creation of *in silico* systems that model aspects of nature and can be used to test theories computationally. This can be especially helpful in testing theory predictions that are difficult or impossible to test experimentally. For some theories, physical experiments (*in vitro* or *in vivo*) are possible. For all these experimental types, results may contribute to understanding the implications, generality, or universality of a theory. They may uncover profound new theories or hidden caveats and contradictions that would require theory revision or rejection.

A general feature of all tests is the determination of system behavior after manipulation (see Section 5.5.6). If we alter the initial conditions or system variables, how does the system output change? The alterations usually follow a counterfactual prescription, i.e., what would have happened, or will happen, if the initial conditions or postulated properties (what goes on inside the "black box") of the system were varied appropriately, reasonably, or rationally? These three qualifiers of system variation all require that any perturbations imposed on a system are realistic and relevant. One does not learn much of anything if they are not, or if they are beyond the realm of possibility. Here, possibility does not mean only physically possible. Changes in virtual systems must meet the same requirements.

Although a theory may have strong theoretical (mathematical or logical) support, it remains *just* a theory unless it is corroborated, ideally by using orthogonal experimental methods. A linkage between theory and physical reality helps validate our belief, and increases our confidence, in a theory. However, our levels of belief or confidence may be irrelevant *in practice*. The disbelief of nineteenth-century race horse breeders in the theory of evolution by natural selection did not change the fact that they unknowingly were accelerating the evolution of fast horses by creating a natural environment in which slow horses were not allowed to procreate – they were selected against.[206] Another example is the movement of the sun relative to the Earth. Formally, one does not know, and cannot prove, that the sun will rise every day in the future. This has not prevented humankind from using the periodicity of the sun's movement (24 hours) each day to tell time. Nor has it prevented farmers from planting seeds that will require sunlight once they germinate. These examples emphasize the fact that theories do not have to be proven correct to be useful. The seminal expression of this concept is attributed to the statistician George E. P. Box [541], who in 1978 wrote *"All models are wrong but some are useful."* (Box also is credited with introducing the word "robust" into the statistics vocabulary [542].) He expanded on this idea the following year [543].

Now in the real world it will be very remarkable if any system existing could be exactly represented by any simple model. However, cunningly chosen parsimonious models often do provide remarkably useful approximations. For example, the law $PV = RT$ relating pressure P, volume V and temperature T of an "ideal" gas via a constant R is not exactly true for any real gas, but it frequently provides a useful approximation and furthermore its structure is informative since it springs from a physical view of the behavior of gas molecules. ... For such a model there is no need to ask the question "Is

---

[206] Ironically, it was Darwin's knowledge of selective breeding that helped him formulate his theory.

the model true?". If "truth" is to be the "whole truth" the answer must be "No." The only question of interest is "Is the model illuminating and useful?".

We use the scientific method to establish truths within the contexts of model systems, yet these systems are microcosms of the world that cannot possibly represent the world's complexity and the dependence of each microcosm on another. One is wise to understand this point and to always be cognizant of the limits within which one can declare with any certainty that their findings are true.

5.6.2 Theory Corroboration and Refutation

> ... evidential support is not a straightforward matter of logical or probabilistic connections between theory and observation, but a complex relationship mediated by epistemic values, which can be divergent and contextual.[207]

If we are bench scientists, conception and selection of a theory is but the beginning of what might be a lifetime of experimental work studying the system the theory explains. This is true whether the theory has emerged from observation and inference in the experimentally accessible natural world or from worlds less accessible, such as quantum or cosmological. Experimental testing can move a theory from tentative to accepted (or rejected), strictly theoretical to factual, nominally informative to universally important and valuable (or not), or potentially useful *in fact*. When we develop a theory, we hope that it is correct. Logic and evidence might strongly support the theory. However, it is not until the theory is tested that we may (or may not) develop confidence in its veracity, and this confidence may progressively increase if new data corroborate the theory. John Herschel noted this well in his 1831 publication *"A preliminary discourse on the study of natural philosophy."*[208]

> The surest and best characteristic of a well-founded and extensive induction, however, is when verifications of it spring up, as it were, spontaneously, into notice, from quarters where they might be least expected, or even among instances of that very kind which were at first considered hostile to them. Evidence of this kind is irresistible, and compels assent with a weight which scarcely any other possesses.

Striving to falsify a theory may be the most important method of testing it.[209] Understanding and using this strategy in practice enables identification of suspect hypotheses and theories, which can save substantial time, effort, and money otherwise wasted on scientific dead ends. Falsification may be viewed as one method of counterfactual experimentation. Instead of manipulating a system to see how relatively small or benign changes in initial conditions affect its output, we design an experiment whose conditions are nominally consistent with the theory but require its weakest element to function properly. We want to create the severest test so that if the experiment is a success, we might react by saying "if

---

[207] See Chang [47], p. 29.
[208] See Herschel [228], p. 180.
[209] The concept of falsification is central to experimental design and to demarcating science from pseudoscience (see Section 4.5.4).

that experiment worked, all experiments should work." We aren't striving to determine how *well* the theory works, but rather whether it works at all. If it doesn't, then our theory likely is false. I use the term "likely" because it is possible the design or execution of the experiment may be faulty, leading people to believe the theory is false when it isn't.

A wonderful example of instrumentation failure leading to the wrong conclusion was the Gran Sasso project (see Section 5.6.2). The Gran Sasso investigators worked diligently to design, fabricate, and perform an experiment that was technically sound. They had every reason to believe the results of their experiment did, in fact, demonstrate that neutrinos travel faster than the speed of light. The scientific community, by contrast, strongly believed this result was impossible, but *they couldn't prove this experimentally.* If we look at this controversy counterfactually, imagining that only one or a few experiments suggested that the upper bound of velocity is the speed of light, there would be no reason *a priori* to accept one hypothesis over the other. The two sides of the controversy could argue, each with experimental support, they had corroborated or falsified a prediction of relativity theory. This situation often occurs in scientific research. To resolve it, the first thing competing factions usually do is examine their assumptions and methods. This can be done within each laboratory, and if no problems are found, the laboratories may carry out *each other's* experiment to answer the question "what happens if I do the experiment with my own hands." The Gran Sasso episode was resolved when a flaw in their timing system was discovered. Thus, falsification of a fundamental hypothesis in physics was itself falsified.

This episode illustrates the fact that auxiliary conditions, including those that are unrealized or purely technical, may affect the outcome of a falsification experiment. Hume's problem of induction thus operates here, as it does in inductive processes, because we can never be sure a falsifying experiment may itself be falsified in the future. Einstein said that one falsification of his theory could prove him wrong, but he assumed, as scientists often do, that no auxiliary factors existed. They may.

5.6.3 Practical and Philosophical Implications of Theories

The necessity of theory testing may seem intuitively obvious. However, this determination has important practical consequences outside the constrained realm of theory *per se.* Acceptance may result in the application of a theory in varied scientific and non-scientific domains, for example, in biology, physiology, chemistry, physics, engineering, medicine, agriculture, economics, sociology, education, economics, athletics, or law. This application, however, may or may not work. Examples include the success of vaccination and the failure of communism. Edward Jenner's 1796 immunization of a child with cowpox virus to produce immunity to smallpox was based on the theory that injecting specially prepared disease-causing bacteria or viruses into otherwise healthy individuals would protect them from the cognate diseases. In this case, immunity was observed, supporting the theory, which then was corroborated by myriad other types of immunization around the world. By contrast, application of Marxist theory, in the form of communism, produced severe societal damage and conflict. This was realized in the forms of oppression, failed economies, low standards of living, and the Cold War. From a philosophical–economical perspective, the USSR, China, and North Korea were continuing tests of Marxist theory. These "tests" have not provided support for the theory, but rather increasing evidence of its failure.

In the realm of science, theories may be beneficial, detrimental, or neither, depending on their veracity. Newton's theory of gravitation is an example of the value of a true theory. Acceptance of this theory fundamentally changed the way we thought about the cosmos and physical processes within it. Newton's theory explained myriad theoretical and physical entities, for example, the bending of light by the sun, ocean tides, ballistics, children's slides in the park, and thrill rides at amusement parks (where one can experience the force of gravity as one is dropped from a high tower). Another example is the use of the gravitational force of the moon as a slingshot to fling spacecraft to the outer reaches of our solar system, and beyond.[210]

Theories can be detrimental in a number of ways. The most obvious is the retardation of scientific development. If a scientific community bases its studies on a false theory, it will be misled and waste resources doing pointless work. This is especially bad for graduate students, postdoctoral fellows, and assistant professors who must be productive to be successful, i.e., publish papers, and have limited resources and time to do so. But how does one know they are being mislead, i.e., a theory is wrong? This poses a conundrum. How do we determine the absolute validity of a theory while at the same time knowing, theoretically and practically, this is impossible? Answer: We can't. However, we do have a proxy for theory veracity – time. It is unusual for a scientist to execute a project quickly. Time scales are measured in months, years, decades centuries, and millennia. However, the longer one's project continues to fail, the more likely it is that: (1) the theory guiding the work is wrong; (2) the question to be studied is irrelevant, impossible, or ill-posed; or (3) the methodology is flawed. If one rules out possibilities (2) and (3), then one is left with the conclusion that the theory has problems. At this point, the scientist has to decide whether to "fish or cut bait." Sometimes this decision is mandated by an end to funding or a job change. The rationale for discarding a theory often comes simply from impatience or experience, each of which involve *time*. Experience also comes into play if the researcher has had profound failures in the past, which helps to establish a time limit for success. Note that experimental failure is the norm in science and it provides us with important information.[211] It can reveal problems or caveats in our experiments, change the types of experiments being done, provide new questions to answer, and most importantly, reorient our scientific journey away from dead ends to useful paths.

Many examples of false theories that were embraced for extended periods of time by the scientific community exist. Historical examples include phrenology (the shape of the skull reveals anatomic areas of the brain that control one's personality, character traits, and other elements of brain function) and phlogiston (the fire-like element responsible for the burning of substances in air). Recent examples are "ulcers come from hyperacidity" (they

---

[210] This "gravitational slingshot" (also gravity assist, swing-by) method was applied in space missions as early as 1959, enabling Luna 3 to photograph the far side of the moon, and continuing with missions to Venus, Mercury, Jupiter, Saturn, and outside the solar system (Pioneer).

[211] For an informative and amusing account of failure in science, see the book *"Failure–Why Science Is so Successful"* by Stuart Firestein.

are linked to stomach infection), "niacin is an effective dietary supplement to decrease heart attack risk" (it isn't), and "Alzheimer's disease is caused by simple protein deposition in the brain" (it may be caused by the body's response to infection by microorganisms, not just genetically). It can be argued that the acceptance of these theories to the exclusion of others slowed scientific progress in astronomy, physiology, chemistry, and medicine. However, there also were much more immediate and practical effects. For example, ulcer patients were treated with antacids and dietary changes, which accomplished nothing with respect to eliminating their underlying disease. These treatments were standard in gastroenterology and were not questioned until progress in understanding the etiology of ulcers identified *H. pylori* as the causative agent. Oral antibiotic treatment now could rapidly cure patients, saving them from years to decades of suffering.

Would advances in the treatment of ulcers have occurred faster if the acidity theory had not been accepted by the field? Would advances in understanding brain physiology or chemistry occurred at the rate they did if Gall[212] or Stahl[213] had *not* propounded their theories? We cannot be sure of the answers because we can't change history. Arguments for both retardation and acceleration have been proffered by interpretation and extrapolation of the historical record. One certainly can argue if Ptolemy's geocentric theory had not been accepted by rote then it would not have taken 1,500 years for it to be rejected in favor of heliocentricity. On the other hand, Ptolemy *did* provide a starting point for consideration of the question of the movement of the heavens. It is possible, without this stepping stone, that advancement in astronomy might have faltered. The phlogiston theory may be considered in the same manner. If this theory had not been *de rigueur* for almost a century, would the phenomenon of oxidation have been discovered sooner? Again, some say "yes" and some say "no," depending on whether they have a retrospective or prospective view of history. It is easy to look at history and argue counterfactually that if *x* were not embraced then *y* would have been discovered sooner. But one must start somewhere with a notion of what is going on and such notions rarely are correct the first time around. Discovery takes time. Notions are vehicles for discovery, much in the same manner as Newton characterized his success, in a 1675 letter to Robert Hooke in which he wrote *"Pigmaei gigantum humeris impositi plusquam ipsi gigantes vident."*[214] Many have argued the phlogiston theory retarded the development of chemistry. However, a prospective consideration of the theory reveals it as a key advance in the movement from ancient alchemy to modern chemistry. Although wrong, the theory incorporated experimental observations and logic that would later support the oxidation theory of Lavoisier.

---

[212] Franz Joseph Gall, a German anatomist/physiologist, proposed the phrenology theory in 1796. See www.britannica.com/biography/Franz-Joseph-Gall.

[213] Georg Ernst Stahl is credited with the development of the theory that substances that burn in the air contain phlogiston, a fire-like element. See https://en.wikipedia.org/wiki/Georg_Ernst_Stahl.

[214] Commonly translated from the Latin as "If I have seen further it is by standing on the shoulders of giants." The idea, according to John of Salisbury [544], is attributed to Bernard of Chartres (d. 1124), who wrote "We are like dwarfs on the shoulders of giants, so that we can see more than they, and things at a greater distance, not by virtue of any sharpness of sight on our part, or any physical distinction, but because we are carried high and raised up by their giant size." However, Priscian (Priscianus "Cesariensis"), a famed Latin grammarian of the sixth century, is credited with the first use of the metaphor as we now understand it [738]. The original context was from Greek mythology and the story of the giant Orion, who, while temporarily blinded, was guided by a youth, Cedalion, standing on his shoulders (Latin: nanos gigantum humeris insidentes, translated as "dwarfs sitting on the shoulders of giants").

5.6.4 Testing Theories Impossible to Test

Some theories cannot be tested contemporaneously because of methodological or monetary considerations. However, tests considered impossible at one time may become possible in the future, if and when new methods and instruments are developed. For example, Einstein theorized in 1911 that light would be bent in a strong gravitational field and later suggested that proof of this would provide support to his general theory of relativity. At the time, this prediction could not be confirmed experimentally, though it had been suggested as early as 1801 by von Soldner [545, 546] based on consideration of Newtonian mechanics and gravitation.[215] However, eight years later, and only with the assistance of nature (a total eclipse of the sun), Eddington and colleagues found that light was indeed bent by the sun's strong gravitational field [138]. The experimental observations of Eddington are an example of a theory waiting for a method or opportunity for testing.

Another example is Darwin's theory of evolution,[216] which he termed "descent with modification," a process involving "natural selection." In the nineteenth century, genetics, let alone genes themselves, were unknown. Darwin, akin to Aristotle, collected many samples of organisms and then systematized them. He theorized about *what* evolution was and *why* it existed, but he could not determine *how* it worked, i.e., its mechanism. This had to wait for the birth of molecular genetics in the second half of the twentieth century. Scientists now could determine the DNA sequences comprising the entire genome of an organism and use computational tools to determine the ontology of organisms varying in size and complexity from viroids and viruses to whales and elephants, and humans. Untestable theories in biology and physics now *could* be tested experimentally. Most of Darwin's work now has been confirmed, although some problems did emerge [550]. What may be most heartening for those looking for corroboration of Darwin's theories is the recent finding by van Holstein and Foyer [551], 140 years after his death, that lineages with higher diversification rates evidence this capacity at both the species and subspecies level, as Darwin predicted. Van Holstein acknowledged Darwin's prescience,[217] saying, *"My research investigating the relationship between species and the variety of subspecies proves that sub-species play a critical role in long-term evolutionary dynamics and in future evolution of species. And they always have, which is what Darwin suspected when he was defining what a species actually was."*

---

[215] Note that von Soldner's prediction of the angle of bending was half that of Einstein's.

[216] Note that Darwin was not the first to propose the theory of evolution by natural selection nor did he conceive the theory independent of others. In 1831, Patrick Matthew proffered the idea of natural selection resulting in speciation. Darwin acknowledged this in a later edition of Origin of Species, and in 1838, discussed how he had been influenced by the work of Malthus. He wrote *"I happened to read for amusement Malthus on Population, and being well prepared to appreciate the struggle for existence which everywhere goes on from long-continued observation of the habits of animals and plants, it at once struck me that under these circumstances favourable variations would tend to be preserved, and unfavourable ones to be destroyed. The result of this would be the formation of new species. Here, then, I had at last got a theory by which to work."* Two decades later, Wallace proposed a theory almost identical to that later published in *The Origin of Species*. Wallace, in fact, had submitted a paper on the subject to the Linnean Society [547, 548], which, ironically, had been sent to Darwin for comment! The two scientists then did something that would be highly unlikely in the twenty-first century; they combined their work into a single paper that was indeed read in 1858 [549].

[217] Scientists should make a practice of citing relevant work by those that came before, as well as by one's contemporaries.

Figure 5.28 A British POLICE Public Call BOX. (Credit: Lee Avison/Alamy Stock Photo.)

5.6.5 The Value of Experimental versus Historical Theories

We have seen how acquisition of knowledge, development of new experimental methods and technology, and the cooperation of nature, enable theory testing. However, this is not the case for the study of historical events, for example, as in paleontology or cosmology. We are not able to step inside a TARDIS[218] (Fig. 5.28) and travel through time to observe and do experiments at the beginning of the universe, nor can we board a "TARDIS Local" and perform experimental studies on the different plants, animals, and environs existing at each stop in time. We must do "historical science." We study instances of phenomena described in the historical record or we observe the results of prior events. For example, paleontologists (paleobiologists) study organismal life in the past by examining fossils in the present. Each fossil provides a substantial amount of data, enabling the scientist to infer what life was like in the distant past. Although each datum is a static picture of prior life, dynamic information can be obtained by characterizing and comparing fossils through time. The Big Bang theory is another example of historical science, and of abduction, as cosmologists had to create theories by, in essence, guessing about how the universe came into existence. The guesses were "educated," as cosmologists did have data on the

---

[218] The TARDIS (Time and Relative Dimension in Space) allows travel through time and space and is used by Dr. Who in the long-running eponymous BBC science fiction television series.

organization of bodies in the cosmos and their trajectories, but these data were obtained by observation of a system for which elements of causation could not be manipulated or elucidated directly. Nevertheless, regardless of how the theory was conceived, many of its predictions, i.e., what one would expect to see if the theory were true, were amenable to testing and many have been observed. The Big Bang theory thus is considered the best at explaining the origins of the universe. This does not mean the theory is correct. It does mean that acceptance of the theory by the scientific community reflects the results of the processes of theory choice discussed above.

Should we accord greater significance and veracity to theories based on experimentation and observation than on those derived historically, for example, by paleontology? Many believe so. Gee, for example, has written a scathing denunciation of the entire field, characterizing it as "unscientific" [552].

... palaeontology is either scientific, or it is not, and you may ask whether the particular problems that palaeontology has with its subject – Deep Time – should be allowed to mitigate its inability to reproduce experiments in the approved scientific manner. They should not. To see palaeontology as in any way historical is a mistake in that it assumes that untestable stories have scientific value. But we already know that Deep Time does not support statements based on connected narrative, so to claim that palaeontology can be seen as an historical science is meaningless: if the dictates of Deep Time are followed, no science can ever be historical. Palaeontology read as history is additionally unscientific because, without testable hypotheses, its statements rely for their justification on authority, as if its practitioners had privileged access to absolute truth ....

Cleland categorically disagrees [553].

... *there are no grounds for claiming that the hypotheses of one [experimental science] are more securely established by evidence than are those of the other [historical science].*

Her argument centers on the distinct approaches of the two sciences in executing the scientific method. Experimental science seeks to posit a single hypothesis and then tests its predictions experimentally. Historical science, by contrast, considers multiple hypotheses that then are parsed by seeing how past events support one or another. The data considered for each process thus can be characterized as prospective or retrospective in nature, but of equal value, the only substantial difference being that one set comes from experiments executed by scientists in the present versus experiments executed by nature in the past. The former may be considered corroborative while the later would represent "smoking guns." To which does one accord the greater importance and on what basis? Over time, these question may become moot as more experiments or smoking guns are observed supporting or contradicting a theory.

Testing a theory results in three things: confirmation, refutation, or neither. If a theory is confirmed, it then is accepted by (most) scientists and is used as one basis for explaining the natural world writ large. If not – if it is refuted – then the opposite happens. The theory is discarded and, eventually, forgotten. Logic then tells us that incorporating this theory into one's research cannot be a good idea. The third result, which is quite common, is the inability to conclude anything from the results. Those who believe in the theory may continue to seek corroborating experiments. Those not believing will hunt for experiments refuting it. It also is possible, after many attempts to perform a conclusive test, that the scientific community may decide proving or disproving the theory is difficult or impossible, in which case theory testing may cease. We should not be so concerned about future scientific efforts after inconclusive tests because the scientist then simply makes a binary decision, continue testing or not. It is conclusive tests that are of greatest concern, for a simple reason. What if our conclusions are wrong, but we don't realize this? This is a fundamental and never ending concern in science. It is to this concern that Mark Twain's quotation on page xix, *"It ain't what you don't know that gets you into trouble, it's what you know for sure that just ain't so,"* is particularly relevant.

5.7 Are Our Theories *Really True?*

In the 1997 science fiction film, "Men in Black" (MIB), Agent K (Tommy Lee Jones) and agent J (Will Smith) open locker C18 in Grand Central Station. They see that an entire civilization, the C18 Locker Aliens, lives inside. The aliens have no idea they live inside a locker, let alone what a locker is. They live in a self-contained world, and unless philosophers of science, cosmologists, or theoretical physicists lived among them, they would never question their perceived reality. We also live in what, in essence, is a self-contained world – the universe. This is all most us know about our existence. Thus, it may come as a monumental shock that our *uni*verse may be only one of many comprising a *multi*verse [554]. If so, our observations of the natural world might not be exclusively those of *our* universe but may include aspects of parallel universes. How can we determine how much of our reality comes from which universe and if, in fact, what we deem real is only a relative (*real*), not an absolute (*Real*), term, akin to the relationship between *truth* and *Truth*?[219] Are the laws of physics always *True* or do they change to *true* if quantum events place us in an alternate universe . . . that may or may not be [R/r]eal? These questions may seem fantastic or a waste of thought, but many argue it *is* possible that a multi-verse exists [218, 555]. The point is we cannot be sure in the absolute if our theories are or are not "really true." In practice, however, most would argue that we can, do, and must accept our universe as we see it if we are to function within it. This pragmatic approach has and continues to provide us with immense knowledge of the natural world, knowledge we have used for the benefit (and, unfortunately, to the detriment) of humankind. In the sections that follow, we examine scientific aspects of our own Grand Central Station locker and ask whether what we experience is real, a fiction derived from our senses and imaginations, or something in-between.

5.7.1 Scientific Realism – An Undefinable Pseudoissue?

The fundamental question of assuming the reality of what we experience – realism – is a vast area of study. If one begins with the question of the reality of the multiverse and

---

[219] By the way, at the end of MIB, Agent J opens a door that should "never be opened" and sees that his universe also is in a locker!

Table 5.9 *Realisms. A partial alphabetical list of realisms. Our discussion focuses only on scientific realism. (Other realisms are listed to illustrate their diversity, but they are beyond the scope of our discussions.)*

Aesthetic	Liberal	New
Art[a]	Literary	Offensive (structural)
Christian	Modal	Organic
Classical	Moderate	Platonic
Critical	Modern	Political
Defensive (structural)	Moral	Representative
Epistemological	Mystical	Rise and fall
Human nature	Neo-classical	**Scientific**
Indirect	Neo	Transcendental

[a] Different realist movements in art include hyper-, fantastic, magical, photo, poetic, social, and socialist.

then progressively examines the reality of smaller and smaller volumes of it, for example, our universe, the solar system, the Earth, and the macroscopic, microscopic, atomic, and subatomic entities that comprise it, as well as the metaphysical worlds of concepts, ideas, semantics, abstractions, etc., a myriad of different realisms will be found (Table 5.9). Here, we will restrict our scope to those types of realism relevant to philosophers and practitioners of science, so when we use the word "realism," we will consider it synonymous with "scientific realism."

We begin by discussing the general features of *scientific realism*,[220] which in a simple sense, is the belief that things, states, and qualities exist independent of our own thoughts, perceptions, etc. Things might include the moon, sugar, the Internal Revenue Service, cells (biological, as well as cell phones), or atoms and subatomic particles. States may be of hibernation, vegetative existence, or war. Qualities could be color, taste, or pressure. Scientific realists usually exclude supernatural or spiritual concepts, for which no empirical means exist to demonstrate their existence or nonexistence. Realists thus also may be characterized as *naturalists*.[221] Dubray has characterizes scientific realists as those ...

... looking upon nature as the one original and fundamental source of all that exists, and ... attempting to explain everything in terms of nature. [556]

Naturalists believe that the forms and behaviors of the natural world are a function only of natural laws related to material objects and their properties, for example, mass, energy, velocity, shape, or chemical reactivity.

Let us now attempt to define realism in more specific terms. We might say "what is real exists," but then we must deal with the circular argument "what exists must be real." We

[220] We subsume *classical* and *traditional* realism under *scientific realism*.
[221] It has been suggested that naturalism can be broken down into two components, ontological and methodological, the former associated with more metaphysical notions of nature, for example, teleology, and the latter characterized by acceptance of naturalism as a practical paradigm for experimental science.

might also include the concept of *being* and say what *is* is real and exists. The differences among these concepts are subtle. Past efforts to establish a consensus definition of realism have failed miserably. In 1947, John Wild described this failure as follows [557]:

> ... 'realism' once referred to a systematic mode of thought with distinctive principles clearly opposed to every form of subjectivism and idealism. ... this is no longer the case. Contemporary philosophical usage has allowed this word to fade into a murky cloud of ambiguity in which nothing very clear or distinct can be discerned.

Seventy-four years later, Mario Alai [558] opined:

> There is probably no universally accepted comprehensive definition of 'realism' in philosophy, and perhaps the various doctrines called by this name do not share a common set of properties, but only Wittgensteinian family resemblances.
>   ... [realism remains] "a murky cloud of ambiguity."

Family resemblance, *familienähnlichkeit* in German, was a term Ludwig Wittgenstein used in discussing "the general notion of the meaning of a word." With respect to the meaning of realism, instead of trying to define a *single* set of necessary and sufficient properties defining it, Wittgenstein would see "a complicated network of similarities overlapping and criss-crossing ...." The term realism thus may be considered a rubric under which many types of realism exist with no one thing in common, but with the many types being related in different ways, just as members of a family may share, or not share, characteristics such as "build, features, colour of eyes, gait, temperament, etc."[222] The scientific realism "family" has, or has had,[223] many members (see Table 5.10).

The fact that so many types of scientific realism have been proffered may explain why a unitary definition does not exist and why debates about realism have been so vigorous and protracted – so much so that many suggest just living with the uncertainty of the term. Chang [564] provides an appealing opinion on this matter when he asks:

> Why should anyone care about the seemingly interminable philosophical debate concerning scientific realism? Shouldn't we simply let go of it, ... especially in relation to the endless arguments surrounding realist attempts to show that the impossible is somehow possible, that empirical science can really attain assured knowledge about what goes beyond experience. It is time to face the fact that we cannot know whether we have got the objective Truth about the World (even if such a formulation is meaningful).

Fine [565] went a bit farther when he discussed what he termed the *"schizophrenia over realism"* in quantum theory. In a sense, he considered the question of realism a pseudoquestion.

> ... quantum theory has an integrity of its own which conforms to neither of these opposing images; that is, that quantum theory neither supports [the] realism of atoms and molecules, etc., ... nor does it deny it. ... [R]ealism is a metaphysical doctrine that finds neither support nor refutation in scientific theories or investigations.

---

[222] See Wittgenstein [368] §65, §67, and §71.
[223] Some of whom are now estranged!

Table 5.10 *The realism family and its familienähnlichkeiten. The rubric of scientific realism is in bold. Types of scientific realism are listed below it. Types of realism following metaphysical all have a metaphysical component.*

Name	Description
Scientific realism[a]	A natural world, comprising observable and unobservable entities and properties thereof, exists independent of human cognitive activity. Science provides a literally true account of this world, which entails belief that the best scientific theories are true.
Common sense	If it "looks like a duck, walks like a duck, and quacks like a duck, it's probably a duck."
Empirical	Cause-and-effect relationships exist between real entities and phenomena, and offer mechanistic explanations of the natural world.
Explanationist	Our best scientific theories give us knowledge about unobservable entities, processes, and events. No miracles account for this.
Relative	As with inference to the best explanation, among a number of scientific theories, the most empirically successful is closer to the truth than its competitors.
Entity (object)	Entities exist and are mind-independent, but the truth of theories in which they are found cannot be determined.
Structural[b]	Everything that exists depends on the existence of structures, mathematical or theoretical, and structures depend on nothing else for their existence, i.e., there are no unobservable objects *per se*.
Semi[c]	Combines aspects of entity and structural realism to establishes principles allowing one to identify parts of theories likely to be retained as theories change.
Selective[d]	Not all the propositions of an empirically successful theory should be regarded as (approximately) true but only those elements that are essential for its success.
Metaphysical	Theoretical terms refer to *something* in the world.
epistemic[e]	Some of the claims about unobservable entities made by empirically successful scientific theories are at least approximately true, close to the truth, or verisimilar. They provide knowledge about the unobservable structure of the world.
Semantic	Scientific theories are true if the conditions (observable and unobservable entities and behaviors) under which they operate are true, which can be so only if the conditions are real.
Internal	The correspondence of objects with the world is true within an internal representation of the world (theory). There is no mind-independent perspective from which ultimate judgments of truth can be made.
Perspectival	States of affairs about the world are perspective-independent; whereas our scientific knowledge claims about these states of affairs are perspective-dependent.
Minimal	Theories, even false ones, that better "latch onto unobservable reality," move science forward.

[a] Psillos [559] characterizes realism in largely metaphysical terms as the joint holding of the semantic, metaphysical, and epistemic theses.
[b] Many subtypes of structural realism have been proposed, including syntactic, epistemic, ontic, causal, mathematical, nomic, and informational. All affirm an epistemic commitment to some kind of structure, rather than entities *per se*.
[c] A similar, but distinct, form of semirealism, "selective semirealism," has been proposed by Nanay [560] Its principle is "... there is always a fact of the matter about whether the singular statements science gives us are literally true, but there is no fact of the matter about whether nonsingular statements are literally true." They're not.
[d] Definition of selective realism from Peters [561].
[e] Adapted from Dellsén [562] and Morganti [563].

When considered in this way, defining realism as "the opposite of anti-realism" may be all that one can hope to achieve in practice. So, if we can't define realism, and it may or may not be of any importance in our scientific practice, why discuss it? There are at least two responses to this question, one philosophical and one practical. The philosophical response is actually no response at all, but rather a question itself. Why seek definitions and an understanding of the usage of *any* terms? Why define milk? Abstraction? Gravity? Why does the Oxford English Dictionary have definitions for >600,000 words and the Korean dictionary have >1.1 million entries? People must think there is some importance for these words beyond simply creating a *magnum opus* to entertain and titillate logophiles, philologists, lexicographers, and linguists, . . . and they would be right. Words are necessary and useful, have meaning, convey information, represent concepts and abstractions, and have intrinsic implications, all of which are of indispensable practical value. I am not an expert in any of the aforementioned fields, or in semantics, but if I seek to understand and manipulate the natural world – and be justifiably motivated to achieve these goals – I need the *concept* of realism, its implications, and à la Fine, the "metaphysical doctrine" it provides.

Waking and dreaming are wonderful, simple, common-sense, experiential metaphors for realism and anti-realism, respectively. When we dream, we experience a reality that often is indistinguishable from the one we experience upon waking. It is a reality we sometimes wish were not so, and at other times, was so. When one has a particularly pleasurable dream, one is likely to rue awakening. Here, the individual has an exceptionally clear intuitive understanding of reality and nonreality. The converse also is true. When one has a nightmare, one is grateful that reality was, in fact, not real. No one need know philosophy, science, or anything else for that matter, to distinguish reality from non-reality. Scientists have the same types of experiences when they do *gedankenexperiments*. They know they are dreaming, but they hope their dreams come true – that they can become real. There are few things more exciting, stimulating, and rewarding to the scientist than conceiving a hypothesis (nonreality) and then showing that its predictions are observable experimentally (reality). It is the concept of realism that drives such theory creation and testing.

Theory (imagination) and experiment (reality) are inextricably linked. Scientists imagine a world very much like their own, one in which they can establish initial conditions for objects, laws, etc., and then manipulate the world in different ways to see what happens.[224] Scientific progress comes from bouncing back and forth between theory and experiment. If a concordance between theory and experiment is found, most scientists would argue their thoughts and data reflect common-sense reality, at least as it is commonly understood.[225] In practice, scientists assume their experiments reflect reality – they are scientific realists – and thus have the potential to produce data of practical value, otherwise experimentation

---

[224] Note the doing an experiment may not require the scientist to do anything but observe existing phenomena. Examples are determining the periodicity of eruptions of the Old Faithful geyser in Yellowstone National Park or the gestation period of a mouse.

[225] Whether they do or do not is an active area of investigation by philosophers of science. Note that even if *realists*, who believe in the correspondence between observation and reality, are correct, the practical questions of whether the experimental results do reflect reality, or if a mechanism exists to determine their truth, can be difficult to answer.

Realists: "No miracles"	Anti-realists: "Pessimistic meta-induction"
"The positive argument for realism is that it is the only philosophy that doesn't make the success of science a miracle."	*"[T]he history of science, far from confirming scientific realism, decisively confutes several extant versions of avowedly 'naturalistic' forms of scientific realism."*
—Hilary Putman (1975)	—Larry Laudan (1981)[226]

Figure 5.29 The creeds of realism and anti-realism.

would be pointless. If the data were unreal, then any use or implications of them would be illogical in the natural world. And if one is not going to believe their own data, why would one do experiments that produce them? Consideration of these questions is important for scientists and philosophers of science, but it also has factionalized many into two factions – *realist* and *anti-realist* – each with their own creed (Fig. 5.29).[227]

5.7.2 Realism in Scientific Practice

The Taj Mahal provides an illustrative example of key elements of *common sense* realism. If one were to stand in front of it, one would not only see it, but one would get a sense of its Taj Mahal-ness – its color, height, extent, construction, etc. One could deepen their understanding by touching, smelling, or kicking it. Importantly, if one were a realist and never saw the Taj Mahal again, or even in the first place, they would have no reason to believe that it didn't continue to exist as it was when they stood in front of it or when they read about it (not considering changes due to aging or unexpected catastrophic events). Common sense, an important part of scientific realism, tells the realist the Taj Mahal is "real." Put colloquially, "if looks like a duck, walks like a duck, and quacks like a duck, it's probably[228] a duck."

Scientists want to understand the natural world and they do so initially by seeking explanations for empirical observations, but what if such observations are impossible? What if there exist objects and properties the scientists cannot perceive through their own senses, i.e., "unobservables?" Are these real or unreal? For example, ice and steam are explained as states of water formed at low and high temperature, respectively. We can see, smell, taste, and touch them both. Contrast this with electrons. We can't see them directly, don't know their color, nor do we know from what material they are made. They are theoretical

---

[226] Laudan uses the term "naturalistic" to indicate forms of scientific realism that exclude supernatural or spiritual components.

[227] Instead of characterizing the realism debate as one between creeds, Hacking [566] suggests that it is more attitudinal than doctrinal, i.e., it really is a debate between different movements.

[228] Note that "probably" is a nebulous term that could reflect a probability $p \sim 1$ that the object is indeed a duck, or $0.5 < p < 1$, in which case one's certainty might be tenuous.

entities postulated in 1897 by J. J. Thomson, in his "electrified-particle theory," to explain the behavior of cathode rays [567].[229] Quoting from that famous publication:

As the cathode rays carry a charge of negative electricity, are deflected by an electrostatic force as if they were negatively electrified, and are acted on by a magnetic force in just the way in which this force would act on a negatively electrified body moving along the path of these rays, I can see no escape from the conclusion that they are charges of negative electricity carried by particles of matter.

In this case, Thomson experimentally revealed *properties* of "particles of matter," but he did not know they *were* matter or even if such discrete entities existed. In fact, at that time, the prevailing theory of electricity involved an ether in which "electron" signified "a pure quantity of electricity, free of matter, its only inertia electromagnetic, and probably consisting of a strain in . . . an all pervading . . . ether."[230] Many thought Thomson was joking when he postulated the existence of an unobservable, discrete, material entity carrying a negative charge.

The existence of the electron, and other unobservable subatomic particles or forces, are looked upon as "real" by realists, who view causation from the viewpoint of direct cause-and-effect relationships between the entities they consider real and the phenomena for which these entities are responsible. If the phenomena are unobservable, for example, with theoretical systems and the theoretical entities operating within them, then the empirical realist cannot believe those systems and entities exist. The empirical viewpoint of causation is distinct from that of Humeans, who reject a direct relationship between entities and phenomena and instead see causation as merely a regular "conjunction of observable events"[231] as, for example, when the sun shines and heat always ensues.

A priori, establishing cause-and-effect relationships is important, if for no other reason then they are philosophically interesting and practically helpful. For the empirical realist, these relationships *explain*. They are the empirical bases for natural phenomena, even if the entities causing the phenomena are theoretical/unobservable [562]. Note that the empirical realist can accept the existence of theoretical/unobservable entities *even if* predictions cannot be made from them [574]. The empiricist perspective also is a key principle for *relative* realists, who, like those using inference to the best explanation, make empirical success the metric for deciding which theory among many is closest to the truth.[232]

It has been said that *"realism and explanation go hand in hand"* [575]. Lipton has suggested that causes and explanations also go hand in hand [348] because he considers the search for causes also to be a search for explanations. This is where the *explanationist* conception of realism is involved. It is more natural, according to Lipton, to use explanationist

---

[229] Thomson was awarded the Nobel Prize in Physics in 1906 "in recognition of the great merits of his theoretical and experimental investigations on the conduction of electricity by gases." At that time, and until 1913, he referred to the entities carrying negative charge as "corpuscles," although the Irish physicist G. F. Fitzgerald had suggested the use of the term "electron" in 1897 in a commentary accompanying Thomson's paper. Fitzgerald related corpuscles to electrons following the earlier work of George Johnstone Stoney, who had originally characterized the "units of charge found in experiments that passed electrical current through chemicals" and termed them "electrines," a term later changed to "electrons" [568–570].

[230] Quotation from Romer [571], p. 156. In essence, the electron *per se* was not an object, but simply a disturbance in an existing ether, the composition of which was unclear (c.f. Descartes's plenums [Section 4.4.3]). For an account of this ether, see Stoney [572].

[231] Quotation from Slaney [573].

[232] *Truth, truth, approximate truth*, and *verisimilious truth* are discussed in the next section.

thinking to ask what a causal theory explains than to infer as to what *would* account for observed effects and what effects the causes *would* have. In a sense, this is putting the cart before the horse or favoring accommodation over prediction. As opposed to trying to accommodate one's search with the limited number of causes extant, asking if something explains something else allows one to posit a huge variety of causes in an unbiased fashion and then select those that are most likely to account for the observed effects. The explanationist therefore asks not *"what* causes *p"* but *"why p."* For example, if one asks "what keeps the planets from launching off into space?" the answer would be gravity – end of story. But if one asks "why?" the entire worlds of physics (e.g., Newtonian mechanics) and cosmology open up and all knowledge, theory, and scientific practice of these worlds is revealed. This leads to deeper understanding and forms a basis for new theories, novel predictions, and practices.

Continuing with our *familienämhnlichkeit* metaphor, a nonidentical twin of empirical realism is *entity realism*, a concept associated with Hacking, Cartwright and Giere [576–578]. Entity realism is similar to other types of realism in the belief that entities are real, but as championed by Hacking, it can be divided into two parts: one for theories and the other for entities. The latter is distinguished from the former because, according to Hacking, experimentation, the physical performance of experiments on entities, is largely disconnected from theory *per se*. Entities are the key players here, not the truth of theories. The distinction between the two is based on the question that underlays each, viz.,

1. Are theories true, true-or-false, candidates for truth, or aim at the truth?
2. Do entities exist?

Entity realism differs from empirical realism in its agnosticism about theory truth, i.e., whether a theory is or is not true does not impact one's belief in the *entities* associated with the theory. In the quotation below, Hacking expounds on theories.[233] He holds:

...[theories appear] to be possible representations that are not, and perhaps could not be, literally and exactly true. But I am realistic about certain unobservable entities. Electrons do exist, I argue, even though we may be unable to give true descriptions of them over and above a purely phenomenological level. ...[A]n experimental argument is [not] the only viable argument for scientific realism about unobserved entities. I said only that it is the most compelling one, and perhaps the only compelling one.

For Hacking *et al.*, the ability to manipulate an entity, such as an electron, provides evidence of existence, not necessarily the theoretical aspects of its form and properties. Hacking's famous statement regarding entity realism thus was *"if you can spray them then they're real."*[234] He refers here to experiments in which the charge on a niobium ball could be decreased by bombarding (spraying) it with electrons. McMullin [580] echoed this idea when he paraphrased the key concept of Hacking's entity realism. "Intervening makes it

[233] Quotation from Hacking [579], p. 560.
[234] See Hacking [576], p. 23.

possible to know *that* the electron is, even though there is no similar assurance as to *what* it is."[235]

Structural realists certainly would agree with McMullin's characterization of realism as they believe there are no *whats* in the natural world, only *thats*. Unobservable elements of theories are characterized by their properties and relationships with other unobservable elements within a polynodic structure, or through a set of coherent mathematically defined relationships, yet with no explicitly definable form of their own. The origin of this idea may be found in Poincaré's discussion of the structure and dynamics of a four-dimensional world.[236] Worrall extended this concept to the question of realism in an effort to defend it against the pessimistic meta-induction argument of anti-realists [584]. Worrall's stated goal was to create a concept that gave realists and anti-realists "the best of both worlds." It *would* embody the pessimistic meta-induction but also *"some sort of realist attitude towards presently accepted theories in physics and elsewhere."* Structural realism was the concept he posited.

A common example of how structural realism accounts for the arguments from pessimistic meta-induction *and* scientific realism is that of the properties of light. Until the early nineteenth century, Newton's corpuscular theory of light was the accepted explanation. However, in 1817, Augustin Fresnel [585] reported that the behavior of light could not be explained through a particulate theory, but instead, was explained as a wave propagating through an elastic solid "aether" (see footnote 230, page 265). Fresnel developed mathematical equations (Fresnel's Laws) characterizing this behavior and used them successfully to predict the intensity of light refracted or reflected at the interface between two media and the hitherto unobserved phenomenon of the "white spot," a point of light at the center of a circular object's shadow.[237]

He argued that light moved as a wave in a mechanical, elastic medium with two components oscillating in a transverse fashion to each other and to the direction of propagation. Fresnel's wave theory, and its associated equations, explained diffraction descriptively and mathematically, a discovery for which the French Academy of Sciences awarded him the Grand Prix award in 1819. Fresnel's theory was the predominant optical theory for almost half a century, until James Clerk Maxwell showed that light was an electromagnetic wave [587]. Maxwell explained that:

...light itself ... is an electromagnetic disturbance in the form of waves propagated through the electromagnetic field according to electromagnetic laws. ... light consists in the transverse undulations of the same medium which is the cause of electric and magnetic phenomena.

---

[235] Quotation from McMullin [580], p. 63. For additional discussion, see Chang [564], Chakravartty [581], and Shaw [582].

[236] See Poincaré [583], pp. 80–81.

[237] Fresnel's equations also were used to derive the Brewster angle $\theta_B = \arctan\left(\frac{n_2}{n_1}\right)$, the angle at which polarized light will not be reflected at the interface of two media, the first and second media having refractive indices of n_1 and n_2, respectively. Brewster's original account of this property, read at The Royal Society in 1815 [586], was found in proposition XIX. It read "When a polarised ray is incident at any angle upon a transparent body, in a plane at right angles to the plane of its primitive polarisation, a portion of the ray will lose its property of being reflected, and will entirely penetrate the transparent body. This portion of light, which has lost its reflexibility, increases as the angle of incidence approaches to the polarising angle, when it becomes a maximum." For more information on the "white spot," see https://en.wikipedia.org/wiki/Arago_spot.

Maxwell's fundamental contribution to the advance of physics and optics was making the connection between light and electromagnetism, a connection revealing the "two components" of Fresnel to be electric and magnetic fields. Maxwell's theory showed that Fresnel's was flawed. Anti-realists have made this episode in scientific history the poster child for their pessimistic meta-induction campaign against realists, who themselves argue that the fact that Fresnel's laws remain valid and useful today cannot be a miracle.

As with all things philosophical, the truth, (if there is one) actually is not discrete, but rather a continuum of opinions whose extremes are represented by the pessimistic meta-inductionists and the no-miraculists. Arguments in the PoS, and within science itself, often evolve toward consensus (again, if one is possible) after the implacable positions of each camp have been propounded. Structural realism, according to Worrall [584], can account for elements of opinions that are polar opposites, and indeed, give realists and anti-realists the best of both worlds. It does so by arguing two key things: (1) a universalist realist view of science is not required to believe in the reality of a given phenomenon or theory, which enables realists to accept the fact that some, but not all, currently successful theories later may be proven wrong, which in turn mollifies at least some anti-realists; and (2) although a universalist realist perspective now cannot be made rationally, Fresnel's theory was not erased, only certain of its elements required modification, whereas others were carried over to Maxwell intact. What was left intact was the *structure* of Fresnel's theory. An "aether" was postulated in both theories. For Fresnel, it was "luminiferous" and for Maxwell it was "electromagnetic."

The existence of a medium in which light could travel thus was a structural feature of both theories, as was light's wave-like properties, transverse oscillations, and the mathematical formulation of its optical properties. Although Fresnel answered the question of how light behaved, he provided an incorrect answer to the question of *what* light was, a question Maxwell answered. Nevertheless, the answer to the latter question, a matter of matter, was irrelevant to the importance and practical use of Fresnel's and Maxwell's equations.[238] The supplanting of Fresnel's theory by Maxwell's was not a radical, Kuhnian-like paradigm shift, but rather only a change of aethers [589], and it was not until Einstein, in 1905, that material ethers disappeared entirely, replaced by electromagnetic fields. Of course, even then, the die-hard realist could argue that material ethers still existed; they were just matterless.[239]

As we have seen, structural realists seeks to circumvent the problem of pessimistic meta-induction by choosing which elements of a theory to consider real, namely structural. *Semi* and *Selective* realists do the same thing, but in different ways. Semirealists take a futuristic approach by identifying elements of current theories they feel will be retained in future theories – a kind of *optimistic* meta-induction. Selectivists only keep those (approximately) true propositions from empirically successful theories they believe are essential for their success.

---

[238] In fact, Fresnel's equation can be derived from Maxwell's equations [588].

[239] Space might be an example of this. Of course, even spaces in space, i.e., between galaxies, contain atoms, just at a density of <1 atom/m^3. Maybe space is where you find it and one cannot define it any more than that, or maybe space is equivalent to the empty set in set theory, i.e., the set containing no elements whatsoever.

Semirealists combine instantial causal features of the natural world with structure, i.e., relationships among the properties thus revealed. Entities believed real by entity realists based on empirical (experimental) studies are organized according to the mathematical or theoretical structures of structural realists – another kind of "best of both worlds" within the scientific realism universe.[240] In discussing fundamental aspects of semirealism [591], Chakravartty opined:

The most promising versions of realism, I thought – ones that genuinely grapple with skeptical arguments – are ones that associate epistemic warrant with *aspects* of theories as opposed to theories *simpliciter*. Entity realists limit belief to the existence of certain entities, based on the putatively demonstrable effectiveness of our causal knowledge of them. Structural realists limit belief to structural aspects of the world described by mathematical or other theoretical relations. (The epistemic form holds that this is all we can know, and the ontic form that this is all there is to know, because the world is exhaustively structural in nature.) It seemed to me that properly construed, these views come together: our best bets for knowledge of the world are supported, in the first instance, by our causal interactions with it, and the properties we detect in this way are best understood structurally, that is, in terms of their relations to other properties. That is the heart of semirealism . . . .

Close cousins of semirealists are selective scientific realists. They believe, as do all realists, that things exist mind-independently and, given enough experimental evidence, that unobservable things are real. Like semirealists, they also excise elements of theories they believe are nonessential. What is essential depends on one's definitions of the terms. Some have used practical factors as descriptors of these terms. Thus, essential elements of theories are those that are required to make a theory work in practice. Unfortunately, these distinctions may be as imprecise as those associated with essentiality. So, another way of looking at selectability involves using realism itself as a determining factor. There exists warrant for believing that working elements are real and warrant depends on the identification of elements (properties) that have remained critical parts of theories as time passes and theories are modified, sometimes significantly. This temporal element of semirealism allows the realist to rebut arguments of the pessimistic meta-induction, specifically, because over time, working elements are invariably linked to successful theories. Of course, if the semirealist can cherry-pick elements using the historical record, i.e., use hindsight, then antirealists would argue that it is difficult (impossible?) to justify a realist perspective at all.

This brings us to another question, namely whether a working element must remain true or approximately true to be considered real. Our prior discussion of light is a case in point. Here, a particulate, luminiferous ether was an obligatory component of the theory for many years. However, later experiments showed this ether was not real. It did not exist. Thus, pessimism should have been the prevailing perspective at the time of the original induction, one that later failed to meet the test of time. The semirealist would disagree, because the *property* of an ether remained an element of subsequent theories.

---

[240] In fact, Chakravartty has opined this is "the best approach for knowledge acquisition" [590].

Whether a particular embodiment of an ether was or was not real thus becomes irrelevant. Light, whatever it was, existed in something, even if one could not be sure of what that something was.

Ghins [592] has proposed four requirements for the selectivist to believe in the reality of an object.

1. Objects must be perceptible, or if not, at least one property common to observables must be ascertainable, for example, through instrumental means.
2. Object properties must be quantifiable.
3. Object properties are causal. They explain what is observed (directly or instrumentally).
4. Orthogonal experimental methods for revealing object properties yield concordant results.

These requirements are shared with observable objects, and it is this parallelism upon which belief rests. Whether theoretical objects exist is an entirely different question. In early work [593],[241] Ghins argues

> ...existence and truth assertions concerning familiar objects are warranted if they satisfy what we call the criteria of presence and invariance.

"Presence" and "invariance" are distilled from the four requirements above and mean what they imply, viz., an object must be discernible through sensory or instrumental means and its features are invariant with changing conditions.

Ghins's aim was to argue for a *"selective scientific realism"* that "extended" common sense to theoretical physics. This is exemplified by consideration of the truth of theoretical causative entities explaining various physical phenomena, for example, the electron. An important element of Ghins's selective realism is its fallibility, which upon reflection, is an inherent part of all types of realism, regardless of whether the most strident advocates of a pan-realist perspective are correct or not. Scientific truth, even so-called approximate truth, is temporally malleable because science is self-correcting. Mistakes are recognized and theories are revised or rejected. One could argue that these facts support an instantial type of realism, and this is, in fact, the philosophy to which most scientists subscribe. All science begins with belief in the reality of *something*. Scientists do real work, produce real data, and use these data to create all of the real human-made objects and processes on Earth, and in our accessible space (cosmos). Fallibilism is the underlying concept that controls what scientific realists select as real, or as stated by Ghins, selectivity comes only ...

> ...[t]o the extent that only some theoretical entities may be asserted to exist and some theoretical statements may be claimed to be true.

---

[241] In later work [592], Ghins suggests that selectivists remain agnostic regarding this question.

5.7.3 Metaphysics as Mind-Dependent Realism

In the beginning, there was nothing, which exploded.

—*Terry Pratchett [594]*

In the preceding discussion, we have used terms such as mind-independent, theoretical, unobservable, truth, structure, causation, property, etc. We know what these terms mean because we've been taught the English language. However, for most, "know" comes from reading a definition in a dictionary. We know *what* the word is in a literal sense, but we may not know *how* and *why* it is what it is. How did this word come about? What observational or thought processes were involved? Why was it needed? Of what value was the word at the time of its creation? Has the meaning of the word changed over time or has it remained constant, and why? If science is to provide the most accurate understanding of the natural world, it only can do so if the meanings of the terms it uses are universal. Is there a correspondence between the term "term" and the entity associated with it, and if so, how is correspondence defined? Questions such as these are the purview of metaphysics,[242] which seeks to determine *what is there*, *what is it like*, and *what is the relationship between the world and its observer*.[243] Such questions, and others, may be subsumed under the most fundamental metaphysical question of all, what is being *qua* being?.

Metaphysical realism, with respect to the natural world, has been characterized as a claim the world is real because *"(a) it exists and has properties (b) it can be referred to and (c) it can be known."*[244] The world *per se* is mind- and language-independent and comprises observable and unobservable entities, their properties, and their interrelationships (structure). Putnam [596] initially characterized metaphysical realism as a perspective in which

...the world consists of some fixed totality of mind-independent objects. There is exactly one true and complete description of 'the way the world is'. Truth involves some sort of correspondence relation between words or thought-signs and external things and sets of things. ... its favorite point of view is a God's Eye point of view.

After substantial criticism by philosophers of science defending their own versions of metaphysical realism, Putnam softened his position by saying *"the phrase 'metaphysical realism' ... refers to a broad family of positions, and not just to the* <u>one</u> *position I used to refer to"* [597]. This broad family has been likened to cutting cookies from the same dough or mapping *"different projections of the earth on the planisphere."* [595][245]

The metaphysicist can believe in the ontology of the natural world without, at the same time, thinking this ontology is justified through epistemic means. Truth *per se* thus is not achievable. What can be achieved is the belief that our best scientific theories are approximately true, close to the truth, or verisimilar, and that the entities upon which these theories are based really exist, i.e., there is a direct correspondence between actual entities and the

---

[242] In subsequent usage, "metaphysics" will be understood to be the metaphysics of science.

[243] Attempting to answer these questions could lead to an infinite regress, so the discussion here will be restricted to *metaphysical realism*, a theory of realism in itself.

[244] Quotation from Alai [595].

[245] For an explanation of planispheres, see https://en.wikipedia.org/wiki/Planisphere.

terms that are used to signify them. A simple example would be the term "sun," which in routine astronomical usage signifies an object at the center of our (solar) system, along with its properties and its physical relationships with other entities in our system. The existence of the sun does not depend on our thoughts, observations, or experience. It is mind-independent. It would still be there even if we could never see it.[246] For the sun and many other entities, this does not mean that knowledge of the sun's existence would be impossible. By observing the movements of the planets, one (especially if they were Newton or Kepler) *could* explain them by postulating the existence of a massive body that affected their trajectories through some means (gravity). However, the reality of the sun would remain a product of belief in a unitary existential system, but one that could be studied and understood scientifically.

Some think metaphysical aspects of scientific realism constitute a philosophical concept independent from scientific realism as a whole, i.e., metaphysicists can ruminate *ad infinitum* about the world *qua* world, but those ruminations contribute nothing to naturalism because metaphysics is literally "beyond the things of nature." Some argue that scientific realism includes some form of metaphysical realism [595]. This is reflected in phrases such as "... and theoretical entities exist mind-independently," or "... and theories are true in the correspondence sense").[247] If one considers such thought-dependent processes as imagination, postulation, interpretation, analysis, extrapolation, etc., whether philosophers of science argue that scientific realism does or does not incorporate metaphysics does not change the fact that metaphysical realism is an obligatory component of science practice. This viewpoint is supported by the fact, with the exception of deduction, which is strictly semantic, scientists must *think* about what they perceive if they are to glean anything of practical (read "real") use from their observational and experimental efforts. Just as eyesight requires both light detection by the retina and subsequent processing in the visual cortex, scientists observe and experiment to produce data that then is interpreted in the brain. Interpretation is not a normative process. It incorporates not only so-called "facts" but also the prevailing scientific dogma and the idiosyncratic thought processes and biases of the scientist. Thus, it is the mind-dependent processing of mind-independent information that cognizes the existence and characterization of the world, its components, and their behavior. The internal processing of information and its relationship to reality, realism, and metaphysics, what Putnam early in his career termed "internal realism," was a central component of his philosophy [596].

...it is characteristic of this view to hold that what objects does the world consist of? is a question that it only makes sense to ask within a theory or description. Many 'internalist' philosophers, though not all, hold further that there is more than one 'true' theory or description of the world. 'Truth', in an internalist view, is some sort of (idealized) rational acceptability – some sort of ideal coherence of our beliefs with each other and with our experiences as those experiences are themselves

---

[246] O'Hear has made a similar point. (See O'Hear [598], p. 6.) He writes "The truths science attempts to reveal about atoms and the solar system and even about microbes and bacteria would still be true even if human beings had never existed."
[247] Quotations from Alai [595].

<u>represented in our belief system</u> – and not correspondence with mind-independent or discourse-independent 'states of affairs'.[248]

Putnam created a furor among establishment philosophers of science when, in characterizing scientific realism, he referred to it as "internal" instead of "scientific." Many (all?) thought Putnam, a staunch realist, had repudiated scientific realism, but in actuality he had not (a fact that was not made clear for more than a decade) [597]. Putnam later clarified the relation of scientific to metaphysical realism when, in a personal communication to Sanjit Chakraborty [599], he said

For the internal realist, the so-called external world – the world outside my brain and body – is itself mind dependent. For a real realist, it isn't.

What is important here is the idea that internal, metaphysical, and scientific realism *simpliciter* all have important mind-dependent aspects. These aspects also are central to more relativistic and weaker types of realism designed to counter the pessimistic meta-induction. Two such types are *perspectival* and *minimal* realism. Each attempts a reconciliation between the mind-independent world of the classical realist and the mind-dependent (relativistic, epistemic) world of the perspectivist. According to Michela Massimi [600], the perspectivist believes science produces a true picture of the world, but still must recognize the caveats involved,[249] namely

…the boundaries of instruments, technologies, theories and model-building [epistemic considerations] that are the products of particular scientific communities at particular historical times, in particular social and cultural contexts [relativistic considerations].

The "boundaries" of Massimi are akin to the "symbolic representation of their environment" of Putnam. When scientists speak about the world, and depending on their particular beliefs about realism, their words may (for perspectivists) or may not (for anti-realists) have a correspondence relationship with the observable or unobservable entities to which the words refer. Perspectivists can assert their theories are true (or approximately true, verisimilar, etc.), at a particular time, and in the context of that time, without fear they may be falsified in the future, because in that future, the context within which those theories operate would be different, as would the scientist's perspective. Future perspectivists could agree with their anti-realist colleagues that a prior theory was indeed false, but not for the same underlying reasons – the anti-realist because they believed all along that the original theory "could not be trusted," and the perspectivist because of epistemic reasons. It may appear the perspectivist is changing the rules of the game to suit their own beliefs, time, and context, but this appearance is misleading because the perspectivist and anti-realist are arguing different subjects. The anti-realist is immersed in the philosophical component of science. The perspectivist lives in a wider and more practical milieu, one in which philosophy *per se* is much less important than understanding why and how scientists conceive

---

[248] Underlined words or phrases were emphasized in the original text.
[249] Quotation taken from Falk [601].

of, use, modify, or reject theories. In this latter sense, perspectivism shares certain features with "*instrumentalism*," which, as defined by Stanford [416],

... is the simple idea that conceptual resources like scientific theories can be useful guides to prediction, to action, and to further investigation without being literally or even approximately true, in just the way that we now think Newtonian mechanics was both radically false and nonetheless an extremely useful instrument.

Instrumentalism overcomes the pessimistic meta-induction because its main tenet has nothing to do with truth *per se*. The metaphor of catalysis would be a useful descriptor of instrumentalism. Catalysis is the process by which chemical reactions that normally would precede slowly, or not within one's lifespan, can be accelerated by orders of magnitude by a substance (protein, metal, chemical), the catalyst. Chemical reactions occur because the products of a reaction have lower energy than the starting materials – a thermodynamic principle. However, even though a reaction may have favorable thermodynamics, the rate at which the starting materials are converted into products may be exceedingly slow – an issue of kinetics. Catalysts increase reaction rates by huge amounts, making them possible within seconds, minutes, or hours, instead of millennia, but do not affect the thermodynamics. If one considers the attainment of truth about the natural world in the same way as a chemical reaction, theories would correspond to catalysts. i.e., they accelerate the rate at which empirical (not absolute) truths are revealed, but have no effect on the observables from which the truths are derived.

Saatsi has proposed a relatively new form of realism, *minimal realism*, that is stronger than instrumentalism but weaker than classical realism. It, too, catalyzes knowledge acquisition and is agnostic about truth. It seeks only to identify theories that are "closer to the truth" than others. Closeness to truth is not theoretical. It is evaluated by consideration of empirical adequacy, i.e., how accurately a particular theory aligns with what is actually known about the world. Saatsi terms this "*latching better onto unobservable reality*" [602]. He exemplifies this as follows.

Take the commonplace notion that theories – or theoretical representations, or models – can, in a given respect, provide better or worse representations of reality. One way in which a theory T' can be a better representation of reality than T, is with respect to observable phenomena, by virtue of being more empirically adequate. This sense of representational improvement in science is commonplace and acknowledged by realists and anti-realists alike. But it can furthermore be the case that the boost in empirical adequacy from T to T' is by virtue [Author emphasis] of T' being a better representation of the unobservable features of the world that lie behind the relevant empirical phenomena. If that is the case, then we can say that T' latches better onto unobservable reality than T – that is, T' latches onto unobservable reality in respects relevant to the improvement in empirical adequacy. It is this sense of latching onto unobservable reality, this sense of theoretical progress, that realists are minimally committed to.

Note that the minimalist is able to apply the "latching better onto" metric to *any* theories, including those that are patently false – as long as one latches better than the others. In this scenario, neither theory would contribute to knowledge accumulation, formally, but they *would* help bring the scientist closer to it. Accumulation of knowledge also corresponds to accumulation of truth, for if our knowledge were false, we would know nothing. It has been argued then, for every example of historical theory failure provided by anti-realists, the same example could be interpreted by realists as "getting closer to the truth" or "increasing verisimilitude."[250]

"Getting closer" includes theories that have been proven to be *entirely* false. This is true because, as with exploring, every wrong turn helps the explorer know *not* to walk down that path. Extrapolating, this means that progress is possible even if *all* prior theories were false, because eventually one would expect, or at least, hope, to find a suitable one. Therefore, by focusing their historical lens on the *progress of science* and not failure or success, realists can turn the pessimistic meta-induction into an "optimistic meta-induction" [604]. Edison espoused the same strategy in his famous response to a reporter's question of how he felt after failing 1,000 times to invent the light bulb.[251]

I didn't fail 1,000 times. The light bulb was an invention with 1,000 steps.

5.7.4 Anti-realism

Strict realism requires things to be real and theories to be true. Strict anti-realism requires the opposite.[252] Herein lies, in my opinion, the most fundamental question to be answered about the debate between the two: Why must each side of the debate argue in absolutes? If I say "there is a God," and you say "God doesn't exist," how am I to convince you I am correct, and vice versa? If I am a staunch Trump supporter and you are a staunch Biden supporter, how can we even begin a conversation if our positions are entrenched? The answer to both questions is "we can't." In this scenario, a debate of any kind would be useless, which begs the question "why is there a debate about realism and anti-realism if it would accomplish nothing?" I think Chang was right in one sense,[253] namely establishing the absolute truth of the natural world is impossible, so why should we try? I agree. However, I strongly disagree about *not* trying, *if the debate does not involve absolutist participants.* If subcategories of realism and anti-realism are argued, common ground may be found. Many of the scientific realism subtypes we have discussed strive to accomplish just that, creating a concept of realism satisfying both sides, i.e., a best of both world realism.

So, is the realism–anti-realism debate a pseudoissue?[254] Answers to this question are myriad. So, not only are philosophers of science debating the realism–anti-realism *issue,* they're also debating the *debate* itself! For the practicing scientist, whether the debate is actually a pseudoissue or not is irrelevant because, as discussed previously, it is understood by scientists that what they are doing is "real," … and that is enough. This does not mean,

---

[250] See Niiniluoto [603], Chapter 8.2.

[251] This quotation is attributed to Edison, but whether this is the actual wording is unclear.

[252] I restrict the discussion here to strict anti-realism to most easily contrast it with scientific realism writ large. Note that anti-realists have rebuttals to every type of scientific realism found in Table 5.10 and, like realism itself, many subtypes of anti-realism, including nominalism, idealism, phenomenalism, conventionalism, relativism, radical empiricism, and skepticism. Each presents challenges to realists, but in different ways. For a perspective on anti-realist arguments, see Park [605].

[253] See page 261.

[254] See Section 5.7.1.

by definition, all scientists are or should be strict realists. One can be both a realist and an anti-realist – a selective scientific realist – " ... *[t]o the extent that only some theoretical entities may be asserted to exist and some theoretical statements may be claimed to be true.*"[255]

One way of considering the debate is as realists operating primarily in the *now*, thinking of the future only with respect to theory prediction and validation, whereas anti-realists operating primarily in the future, expecting that current theory predictions and validations will be false and invalid, respectively. This is a classic "apples and oranges" situation in which realists argue *empirically* and anti-realists argue *historically*. Howard Sankey has approached the issue in a less metaphorical and more logical manner [606]. In the quotation below, he asks how the anti-realist can explain the realist belief that the success of our best scientific theories is due to their truth content and to the direct correspondence of terms in a theory to the actual entities in the natural world.

... an anti-realist denies the realist's claims about truth and reference. The ideal theory is neither true nor approximately true. Its terms fail to refer to any real thing. None of the entities postulated by the theory exist. ... Such an anti-realist is entirely without the resources to explain ideal methodological success. If a theory fails not only to be true but even approximately true, and none of its terms refer to any real entities, then the success of such theory is nothing short of a miracle.

This brings us to the anti-realists bulwark against realism, the *pessimistic meta-induction*.[256] According to this argument, the history of science is almost entirely a history of theories that were judged empirically successful in their day, only to be shown false later. Therefore, no reason exists to think that currently accepted theories are any different [607]. The popularization of the concept of pessimistic meta-induction often is attributed to Larry Laudan [607].[257] To bolster his case, Laudan discussed 13 different examples of theories once thought to be true being invalidated, in whole or in part, in the future. These included the phlogiston theory of chemistry, the caloric theory of heat, the electromagnetic aether, and theories of spontaneous generation. Realist opponents of Laudan argued that many of these examples were cherry-picked and thus did not reflect the true frequency of theory invalidation. However, many were absolutely true. It thus was clear, to Laudan and his proponents, that realism was a fantasy. The problem with the pessimistic meta-induction is that it focuses only on later invalidation of currently successful theories. It does not ask the simple question "What percentage of the most successful theories are later proven wrong?"

When we talk of current success and later failure, or of current validation versus later falsification, we again enter the realm of the absolute, a realm that is equally dangerous for realists and anti-realists alike. All successful theories will have caveats or *ceteris paribus* qualifications, which means realists must admit their theories are not absolutely true. This is one reason that terms like "approximately" and "verisimilar" are used as truth qualifiers.

---

[255] Quotation from Ghins [593].
[256] The term meta-induction is used because induction plays a major role in theory conception.
[257] Interestingly, neither "pessimism" nor "meta-induction" nor "induction" were mentioned in that work! Hilary Putnam defined the term "meta-induction" in 1978 [608].

On the other hand, it also is inappropriate for anti-realists to use those caveats and qualifications to argue unreality. This would be equivalent to arguing that Tiger Woods is not the golfer people think because he bogeyed one hole. We must look at theories in their entirety and judge their truth content accordingly. Problem – is there a normative procedure for judging? Answer – no, so where does that leave us? As you might expect, *somewhere* in the open interval $(0,1)$. Exactly where is impossible to determine, but what we can do is consider in a bit more depth what we mean by success and failure.

It is not unusual for theories to be modified as more data accumulate. These data may result in modifications of some of the theory's core propositions, auxiliary conditions, and their associated terms. However, the original theory is not discarded in its entirety and many of its theoretical terms, and its structure, are maintained. One example of this comes from the Alzheimer's disease (AD) field. For decades, the "amyloid cascade hypothesis" was the predominant theory of AD causation [609]. The theory postulated that AD developed after a cascade of events initiated by the formation of amyloid plaques (protein deposits) in the brain, leading to the neuronal injury and death responsible for the clinical signs of AD. These plaques were formed by the deposition of large amounts of fibrils of the amyloid β-protein ($A\beta$). This fibrillization process involved the self-association (aggregation) of $A\beta$ protein monomers into successively larger structures. Initial work suggested that neuronal toxicity required fibril formation. However, as theory testing and use proceeded, smaller assemblies also were found to be toxic. These toxic assemblies could be as small as dimers. Increasing evidence for the primacy of oligomers (protein aggregates containing two or more monomers, but smaller than fibrils) in AD etiology then led to a modification of the original theory, from the fibril-centric amyloid cascade hypothesis to an "oligomer cascade hypothesis" [610].

To the strict anti-realist, this change would interpreted as a failure of the original hypothesis and a vindication of the anti-realist position. To strict realists, this vindication "threw out the baby with the bath water." Its only justification was the replacement of the term "amyloid" by "oligomer," whereas the structure of the theory remained the same, both in terms of the existence of a cascade and of its neuropathological effects. Depending on your view of realism, failure and modification were the outcomes. From an unbiased perspective, each conclusion was wrong. Anti-realists were wrong because the essential structure of the theory was unchanged. Realists were wrong because, in practical terms, assigning etiologic primacy to oligomers instead of fibrils was a huge change.

Kuhnian paradigm shifts[258] can be considered in the same light. A paradigm, á la Kuhn, is a type of scientific *zeitgeist* that encompasses the manner in which scientific issues are perceived in a metaphysical sense and studied (see Section 6.3.1). Study includes selection of scientific questions and the strategies, methodology, instrumentation, and interpretive means used to answer them. Paradigm shifts are characterized by almost complete rejection of a current paradigm in favor of another. In such shifts, the baby *must* be thrown out with

---

[258] See Section 6.1 for a discussion of the social dimension in science and its relationship to paradigm change.

the bathwater because the shifts are so large the prior theory and its successor are incommensurable. Scientists committed to one theory or the other are unable to communicate
because they no longer have a common language. Even the suggestion the new paradigm's
foundation was the old paradigm, or the new paradigm gets closer to the truth than the
old, are not applicable. An example is the rejection of Newtonian mechanics in favor of
quantum mechanics. A more recent example is the discovery of proteinaceous infectious
particles termed "prions." This discovery destroyed the long-held dogma of microbiology
that all infectious organisms contained DNA or RNA to encode proteins necessary for
infection and propagation. An entirely new world of protein-only replication and disease
was created.[259]

Paradigm shifts provide strong support for an anti-realist science because not only are
successful theories proven false, but the scientific milieus in which they once operated
are destroyed. Scientists, in essence, move from one form of existence to another, leaving
their lives, as they knew them, behind. Even realists are forced to agree that their initial
confidence and belief in the natural world of the past was unwarranted. But wait – was
it? When the Ptolemaic heavens were superseded by the Copernican heavens, did all of
the entities thought to exist – be real – disappear? No, they did not. They simply were
organized in a different geometry than before. Did the form or relationships among the
planets change. No, they did not. The planets of the Ptolemaic system were carried over *en
masse* to the Copernican system, making the conclusion that the new paradigm was built
upon the old inescapable.

This discussion shows that some ostensibly inviolate tenets of realism and anti-realism
actually are not. Therefore, instead of trying to create best of both worlds realism sub-types
that are applicable both to realists and anti-realists, why not simply grant that both the "no
miracles" argument and the "pessimistic meta-induction" are reasonable and demonstrable, often simultaneously. How can this be? Because "reasonable" and "demonstrable" are
context-dependent, much as the "glass is half full" or "the glass is half empty" metaphor
describes precisely the same entity and its properties, but from different perspectives.
Where does this leave us? Well, as I'm sure such questions may be troubling to some,
and philosophically exciting to others, we find ourselves in the same quantum state as
Schrödinger's metaphorical cat, apparently being a strict realist and a strict anti-realist
simultaneously. Our interaction with the world will determine how our wave function
collapses.

5.7.5 Knowledge and Understanding

> We used to think that if we knew one, we knew two, because one
> and one are two. We are finding that we must learn a great deal more
> about 'and'.
>
> —*Arthur Eddington*

---

[259] See Section 5.3 for a discussion of prions and paradigm change.

As we discussed in Chapter 2, there seem to be as many definitions of science as there are definers of that term. However, be that as it may, it is unlikely anyone would disagree with the characterization of science as "wanting answers to questions about the natural world." How do we achieve that? Like Aristotle, we learn as much as we can about the world, and in particular, those portions producing our questions. We execute the scientific method. We use the knowledge thus obtained to satisfy our curiosity. But is our knowledge true? Is it real? We have addressed this question philosophically and metaphysically in the section above. What we have not discussed is a more fundamental question, one upon which our understanding of the world depends – what *is* knowledge? Do we really *know*, and if so, how?

I must first define the terms "understand" and "knowledge," as I use them here. These terms are abstractions, and thus it is difficult, if not impossible, to provide absolute definitions. In some senses, the two terms are mutually dependent, and thus related. To understand something, we must have at least some *a priori* knowledge, yet knowledge itself comes through understanding. I distinguish the two by considering *understanding* to be more conceptual and *knowledge* to be more propositional. Understanding can be defined as "the power of comprehending, especially the capacity to apprehend general relations of particulars"[260] or the "ability to grasp the full meaning of knowledge" or "the ability to infer."[261] In essence, one creates a whole that is greater than the sum of its parts. This type of understanding was advocated by Aristotle, whose "...conception of *epistēmē* [Greek; understanding, knowledge, explanation] [was] his insistence that it involve a grasp not just of a single isolated proposition, but of the whole causal and inferential network of propositions that [lay] behind it."[262] Saint Anselm (c. 1033–1109), who served as the Archbishop of Canterbury from 1093–1109, expressed a similar ideal in a dialogue concerning the devil.

...to elucidate the truth of the matter ...it is necessary that you not be content to understand merely one at a time those things...; rather, you must gather them all together in your mind as if in one view.[263]

To *understand* a system, be it biological, physical, computational, or mathematical, one must be know its components, establish their dynamic interrelationships, appreciate how alteration of individual or multiple components affects system behavior, consider the possibility that some unrecognized components of the system may exist, and be aware that some assumptions used as bases for understanding the system may be wrong. One then has a *conceptual understanding* of the system and its behavior.

Knowledge is more difficult to define and its study comprises an entire field of philosophy, epistemology. Plato, Aristotle's student, in the dialogue *Theaetetus*, has Socrates

---

[260] Merriam-Webster dictionary www.merriam-webster.com/dictionary/understanding.
[261] Wiktionary https://en.wiktionary.org/wiki/understanding.
[262] Quotation from Pasnau [13], p. 6.
[263] Quotation from Saint Anselm [611], in his book *Du Casu Diaboli*, Fall of the Devil, Chapter 12, p. 235 (trans.), written near the turn of the twelfth century.

consider the question "what is knowledge?" Here, Socrates serially discusses three possible definitions.

1. Knowledge is perception.
2. Knowledge is true opinion.
3. (Knowledge is) true opinion, combined with reason.

The strongest definition, (3), is argued by Socrates as follows.

> When, therefore, any one forms the true opinion of anything without rational explanation, you may say that his mind is truly exercised, but has no knowledge; for he who cannot give and receive a reason for a thing, has no knowledge of that thing; but when he adds rational explanation, then, he is perfected in knowledge . . . .

Knowledge may be broken down into three types, *knowledge-how*, *implicit*, and *knowledge-that*. In Chapter 3, we briefly discussed *knowledge-how* in the context of tacit knowledge, the type of knowledge that comes from experience and cannot be taught *pre facto*, for example, riding a bicycle or knowing what someone looks like. *Implicit knowledge* would also include bicycle riding, as well as a preschool children or adults speaking in a grammatically correct manner. Both know what is correct, but the children and many adults (unfortunately), cannot tell you precisely why.[264] Although different with respect to knowledge, the terms tacit and implicit often are used interchangeably.[265] The third category of knowledge, *knowledge-that*, also referred to as *propositional* or *explicit*, is our focus here. This type of knowledge can be explained explicitly verbally and may be true or false. It is the type of knowledge the scientific method purportedly produces, thus the question of truth or falsity is paramount. How do I *know that* the sun will rise in the East each day, cathode rays consist of negatively charged particles, or Avogadro's number is $6.02214076 \times 10^{23}$?

Analysis of knowledge has been a vibrant area of research in the last century, and, as with most things, different people have different definitions of it. However, "justified true belief (JTB)," a notion that can be traced back to Socrates's definition of knowledge, predominates. Here, belief is warranted only if sufficient justification can be made. This is embodied in the so-called "standard tripartite" logical explanation of knowledge.

A subject S knows that a proposition P is true if and only if:

1. P is true, and
2. S believes that P is true, and
3. S is justified in believing that P is true.

(1) is a prerequisite for any account of knowledge. If P is false in a mind-independent sense, knowledge is impossible, other than the knowledge that P is false. Similarly, it is possible, and quite common, to believe something is true when it isn't (2), but this state precludes one having knowledge. To have knowledge, P must be true *and* one must believe it to be

---

[264] This is especially likely to occur when those for whom English is a second language ask native speakers "why do you say it this way?" The answers: "it just is" or "it sounds right."

[265] The Cambridge Dictionary, for example, cross-references the terms (see https://dictionary.cambridge.org/us/dictionary/english/tacit-knowledge).

so. As Ludwig Wittgenstein has pointed out, one cannot have knowledge and at the same time say, *"I know it, but it isn't so."*[266] The crux of JTB is (3) – the warrant. One must justify their belief. The issue for scientists then is whether their beliefs *can* be justified. If so, science happily and productively marches on. If not, scientists have deluded themselves and the public.

There is problem with JTB, however, and that is (1). We certainly can stipulate to the truth of *P* semantically. But that does not mean, in reality, that *P* is true because (1) requires absolute truth, i.e., Truth, which is not achievable. Thus, in practice, the best we can do is truth. Truth must be ignored and truth embraced, which limits us, literally, to justified belief. This is the currency of science, what it uses to buy understanding of the natural world ... and that is enough, because true or not, the entities, their properties, and the structure of the world that we use operationally as true, have made the modern world. On this view, *truth* is enough. Nevertheless, if one begins to stray toward the absolute from the relative, it would be wise to heed the advice of Thomas Jefferson.

Ignorance is preferable to error; and he is less remote from the truth who believes nothing, than he who believes what is wrong.

What justifications for their beliefs do scientists invoke? The scientific method and its controlled experiments are the mainstay of scientific and clinical practice. If differences in results are obtained between groups, statistical methods are used to determine if they are significant. In doing so, one also determines which differences are insignificant, which is an important result as well, especially if one is determining whether, for example, a new drug is toxic in humans. The most used metric for significance is p-value, the probability that an observed difference is due to chance, i.e., is not consistent with the null hypothesis.[267] Most scientists *believe* $p < 0.05$ indicates significance and that the smaller the p-value the greater should be their confidence in this belief.[268]

The agreement of experiment with theory, and vice versa, also provide justification, as does the amount of corroboration. Corroboration can take multiple forms. It can be the number of times identical experiments are performed in the same laboratory, and by the same person, and yield identical results (within experimental error). Such repetition is obligatory if results of experiments are to be published in reputable scientific journals. Corroboration also takes the form of intra- and inter-laboratory agreement. Different researchers in the same lab should obtain the same results and different laboratories, or clinical trials, should be able to replicate these results. A failure to replicate decreases belief. It may not falsify a theory, but it demands that additional theory testing be done to provide additional corroboration or falsification.

Should scientists be comfortable with these justifications? This depends on what we mean by "comfortable," but for our purposes here, we will consider it means "generally

---

[266] This is an example of "Moore's Paradox," a logical paradox exemplified by the proposition "P and (I believe) not P."
[267] The null hypothesis, H_0, is that two experimental groups are not different.
[268] Whether this is warranted or not is a subject of vigorous debate among statisticians [612], from whom scientists and clinicians must take their lead (see Section 5.7.6).

accepted by the scientific community." We now can assess whether such acceptance is warranted and thus whether JTB itself is justified. Our assessment will focus on three critical parts of justification – statistics, determination, and replicability. But before we begin, we must discuss a particularly thorny, fundamental caveat of the entire JTB enterprise, the *"Gettier Problem,"* which was posed by Edmund Gettier in 1963 [613]. Gettier argued that JTB (see page 280)

...is false in that the conditions stated therein do not constitute a <u>sufficient</u> condition for the truth of the proposition that *S* knows that *P*. ...I shall begin by noting two points. First, in that sense of 'justified' in which *S*'s being justified in believing *P* is a necessary condition of *S*'s knowing that *P*, it is possible for a person to be justified in believing a proposition that is in fact false.

In his short, 2 1/2 page paper, Gettier presented two cases that satisfied the conditions of JTB, yet provided no knowledge. These cases were compelling and appeared to falsify the concept of JTB, a fundamental epistemologic tenet of knowledge. With no JTB, how could knowledge now be understood? The logic of the Gettier cases was ironclad, so much so that it has been said *"no epistemologist since Gettier has seriously and successfully defended the traditional view [of knowledge]"* [614], although many have tried.

I present below Gettier's *Case I*. I do so because it is a wonderful example of what can be discovered if one considers and challenges so-called "givens." A good scientist should do exactly that, not only with givens, but with assumptions in particular.

Suppose that Smith and Jones have applied for a certain job. And suppose that Smith has strong evidence for the following conjunctive proposition:
  (d) Jones is the man who will get the job, and Jones has ten coins in his pocket.
Smith's evidence for (d) might be that the president of the company assured him that Jones would in the end be selected, and that he, Smith, had counted the coins in Jones's pocket ten minutes ago. Proposition (d) entails:
  (e) The man who will get the job has ten coins in his pocket.
Let us suppose that Smith sees the entailment from (d) to (e), and accepts (e) on the grounds of (d), for which he has strong evidence. In this case, Smith is clearly justified in believing that (e) is true.
  But imagine, further, that unknown to Smith, he himself, not Jones, will get the job. And, also, unknown to Smith, he himself has ten coins in his pocket. Proposition (e) is then true, though proposition (d), from which Smith inferred (e), is false. In our example, then, all of the following are true: (i) (e) is true, (ii) Smith believes that (e) is true, and (iii) Smith is justified in believing that (e) is true. But it is equally clear that Smith does not know that (e) is true; for (e) is true in virtue of the number of coins in Smith's pocket, while Smith does not know how many coins are in Smith's pocket, and bases his belief in (e) on a count of the coins in Jones's pocket, whom he falsely believes to be the man who will get the job.

Intriguing, isn't it?[269] But does it mean all is lost and there can be no unassailable knowledge. In a philosophical sense, the answer is "yes," because we can never know if we indeed

---

[269] Note that there have been many weaker definitions of knowledge, including "knowledge tracks the truth," "knowledge cannot be obtained through a defect, flaw, or failure," or "knowledge requires that the evidence for the belief necessitates its truth."

know. However, science does not require, and cannot expect, unassailability. For scientists, JTB, as it has been understood since Plato, is good enough. In a seminal paper, "The inescapability of Gettier problems," published ~30 years after Gettier's original paper, Zagzebski [615] qualified the justification required by condition (3) of JTB.

Almost every contemporary theory of justification or warrant aims only to give the conditions for putting the believer in the best position for getting the truth. The best position is assumed to be very good, but imperfect, for such is life. Properly functioning faculties need not be working perfectly, but only well enough; reliable belief-producing mechanisms need not be perfectly reliable, only reliable enough; evidence for a belief need not support it conclusively, but only well enough; and so on. As long as the truth is never assured by the conditions which make the state justified, there will be situations in which a false belief is justified. I have argued that with this common, in fact, almost universal assumption, Gettier cases will never go away.

Zagzegski did provide an "out." He accepts the Gettier problem and its implication, viz., JTB cannot be considered knowledge, but he adds a third component to JTB – *luck*. Luck can allow us to find the truth for the wrong reason, but truth is reached nevertheless. We *were* right, but only because we also were lucky. It is important for philosophers of science, logicians, epistemologists, realists, and anti-realists alike to be aware of the Gettier problem. It *should* "never go away," because it reminds us of ours, and science's, limitations.[270] In this digital age, new types of Gettier problems have emerged, including computer-generated imagery (CGI). When watching television, one often sees what is not there, viz. CGI, and it is impossible to know this at that instant!

5.7.6 P-Values (?) and Statistical Significance (!)

When we look at colors, telling the difference between black and white is simple, and obvious, even for the color blind. When we design digital computers, the difference between 1 and 0 also is simple, and obvious (with the exception of quantum computers!). When we compare summer and winter temperatures, the differences are clear. We have no need of special cognitive abilities or preternatural sensations. In live/dead assays, be they of cells or animals, the results are unambiguous. When we consider extremes or polar opposites such as these, interpretation of observations or results is trivial. The problem is scientific research rarely involves extremes. It most often involves *subtle* aspects of, or effects on, entities, their properties, and systems.

If one thinks about the examples above in greater depth, they may find things are not so simple, obvious, clear, or unambiguous as originally thought. What if we're looking at shades of gray instead of black and white? What if instead of using digital computers we used analogue computers? How would we differentiate "on" (1) from "off" (0)? Would someone living in Antarctica (yearly temperature variation from $-22°F$ to $-112°F$)? have the same definition of "cold" or "hot" as someone living in Singapore (yearly temperature

---

[270] Zagzebski [616] opined *"... that Gettier problems are inescapable for virtually every analysis of knowledge which at least maintains that knowledge is true belief plus something else."*

variation from 76°F to 89°F)? What about assays of cells? What if one is measuring, for example, the effect of a toxic agent on electrical potential across a neuron's cell membrane? Is a change from the normal -70 mV to -68 mV meaningful?

We are taught that the way to determine if something is significant is through statistics. I doubt if any scientist has not been taught or is not aware of p-values, and in particular, $p < 0.05$, the standard "bright-line"[271] for determination of statistical significance. As commonly understood, the p-value is the probability of getting results at least as extreme as the ones you observed, given that the null hypothesis (H_0) is correct. So, for $p < 0.05$, there is less than a 5% chance an observed difference would be due to chance and thus, at least until the recent past, one would find this difference "significant."

I grew up with this notion. I used it in my scientific publications. I taught it to my students. It was demanded of me by journal reviewers and editors, ... and I demanded it of authors when I was an editor/reviewer myself. Therefore, in digging a bit deeper into the topic, I was surprised that my (and most others') notion of p-values and statistical significance were wrong![272] This meant that some of the assumptions about the reality/truth of my own data had to be reexamined, at least for p-values near 0.05. My work has been on the basic science of neurodegenerative diseases, especially its biophysics. If my "statistically significant" results turn out to be insignificant – *and wrong* – it may affect a large number of basic scientists worldwide who have based at least part of their own work on the assumption that mine was correct. This would cause a huge waste of time and resources as scientists, basically, "barked up the wrong tree." But if my work were translated into clinical medicine, for example, in the development and use of diagnostics, prognostics, or therapeutic agents, and it was wrong, the best that could be hoped was that erroneous results or nonefficacious treatments occur, not side-effects or death. More importantly, patients might have been given a useless treatment instead of one that worked.

Where did I, and science in general, go wrong and how can it be fixed? At the most fundamental level, the mistake was not digging more deeply into what a p-value was, how it was developed, and for what. In 1885, Edgeworth [617] defined statistics as *"the science of Means in general (including physical observations)"* and sought a method *"To find how far the difference between any proposed Means is accidental or indicative of a law?"* Note that "significance" *per se* does not occur in this aim. For Edgeworth, "significance" was not associated with some absolute property or metric, but meant "indicative of," i.e., there was *"... a sufficient ground for pursuing the investigation further."*

The methods of attacking [this problem] which will be here considered are those of the more formal kind, those which are afforded by the pure Calculus of Probabilities, as distinguished from Inductive Logic in general. ... it must be remembered that it is difficult to establish a scientific frontier between the doctrine of Chance and Induction.

---

[271] "Bright-lines," according to the Oxford English Dictionary (see www.oed.com/view/Entry/79457796? redirectedFrom=bright-line+rule#eid1264926830), consist of, relate to, or involve a clear, objective, or statutory distinction, as bright line rule, bright line test, etc.

[272] Descartes apparently was right – everything I learned from my esteemed teachers was wrong.

Note the clear distinction between significance in a formal mathematical sense and in an inductive sense. In Edgeworth's milieu, scientific questions were reduced to pure mathematics, divorced from nature, whereas science, rather than being mathematical, was an inferential process, be it by induction, abduction, or deduction. The two former processes had no bright-line constituents and the latter represented pure logic, *not* nature. Two decades later, Boring made a strenuous effort to distinguish "mathematical vs. scientific significance" [618].[273] He too sought a distinction between mathematics and scientific intuition.

It appears that the apparent inconsistency between scientific intuition and mathematical result is not due to the unreliability of professional opinion, but to the fact that scientific generalization is a broader question than mathematical description. . . . A knowledge of the 'probability that a difference is not due to chance' is distinctly worth while on the descriptive [Author emphasis] side; but this measure of significance does not necessarily apply to the general class for which a sample stands. . . . It is for this reason that mathematical measures of difference are apt to be too high and may need to be discounted in arriving at a scientific conclusion. The case is one of many where statistical ability, divorced from a scientific intimacy with the fundamental observations, leads nowhere.

This message, unfortunately, was ignored by Sir Ronald Fisher, who has been described as "*. . . a genius who almost single-handedly created the foundation for modern statistical science without detailed study of his predecessors*" [619]. The underlined phrase, meant to laud Fisher's innate intelligence and creativity, although unintended by the its author, also reflects the fact that some of Fisher's most important work was built upon William Sealy Gosset's (pen name "Student") development of the (Student's) t distribution[274] [620], which is similar to a normal distribution but has fatter tails (positive kurtosis).[275] Student's t distribution is a mainstay of statistical analysis of experimental data. Gosset considered the probabilities determined through the t distribution to be of substantial practical use, but not as a metric for confidence in an hypothesis (e.g., the null hypothesis), but to make Arthur Guinness, Son & Co. Ltd., the distiller, more profitable.[276] A failure to do so might have big economic consequence; it might mean the thriving or floundering of the brewery. Significance for Gosset had much more to do with substantive importance (real life economic issues) than with statistical importance (purely academic). In fact, in reporting results to the company, Gosset suggested

---

[273] This actually was the title of the paper.

[274] Student's definition of the equation for the t distribution stated that *"The equation is determined of the curve representing the frequency distribution of a quantity z, which is obtained by dividing the distance between the mean of a sample and the mean of the population by the standard deviation of the sample."*

[275] The t distribution is especially useful in situations in which data are limited, for example, as with the relatively small sample sizes encountered in most experimental work, because, relative to a normal distribution, it better reflects the likelihood that a randomly selected sample will be far from the mean.

[276] Fisher may have been more influential in the world of statistics than Gosset, but Gosset's work was placed on an equal footing with that of Fisher at the 1936 meeting of the Industrial and Agricultural Research Section of the Royal Statistical Society, when it was stated *"Mr. Gosset could be regarded as the father of modern statistical methods (which had themselves been most fully developed in agriculture), for the first test of significance applicable to small samples, now known as the t test, was due to him."* Gosset worked his entire life at Arthur Guinness, Son & Co. Ltd., the brewer, where he analyzed voluminous and varied data related to the production of beer. The work often involved small test plots of land on which grains were grown under varying conditions. Gosset had to find a method to determine which plot conditions were significantly better than others, but this was difficult due to the low sample sizes with which he was faced. It was this need that led to Gosset's development of the t distribution.

...that questions concerning 'the degree of probability to be accepted as proving various propositions' should be referred to a mathematician.

It was Fisher who played the role of the mathematician and who championed the bright line of 5% probability ($p < 0.05$) to determine if differences of means were "statistically significant" [621]. Fisher's use of the term "statistical significance" in his seminal treatise *"Statistical Methods for Research Workers"* [622] heralded the term's wide acceptance and use (though generally incorrectly).

It is usual and convenient for experimenters to take 5 per cent, as a standard level of significance, in the sense that they are prepared to ignore all results which fail to reach this standard, and, by this means, to eliminate from further discussion the greater part of the fluctuations which chance causes have introduced into their experimental results.

Thus began what has been called the *"dichotomization of evidence"* [623] and *"dichotomization of the p-scale"* [624]. In practice, because of Fisher's reputation and prominence, $p < 0.05$ became not just a suggested, arbitrary significance level at which one might have confidence in an hypothesis, but an inviolate determiner of a bright line of demarcation between significant and insignificant – *"And so the tool ... become the tyrant"* [625]. In historical context, establishing that a p-value of 0.05 meant a particular difference in means was "significant" had tremendous value. Fisher argued for this value, and it was accepted, for at least three reasons: (1) arbitrary bright lines with p-values of 0.015, 0.0325, and 0.07 already had been suggested by Edgeworth; (2) Fisher had just devised the method of analysis of variants (ANOVA), a whole new area of statistics, and few were familiar with the method;[277] and (3) the novelty of ANOVA required that some standard be set so that everyone using the method was consistent in their interpretation of its results.

Dichotomization has been recognized increasingly by statisticians and scientists as a fundamental problem for science, in general, and for clinical medicine. Put simply, dichotomization gives scientists and clinicians confidence in the significance or insignificance of their work when it is unwarranted. In a recent *Nature* Comment [626], one supported by >854 other experts in statistical modeling from >52 countries [626], Amrhein, Greenland, and McShane argue that *"the entire concept of statistical significance [should] be abandoned"* and the use of p-values *"to decide whether a result refutes or supports a scientific hypothesis"* should be stopped. They argue that p-values interpreted as significant often are biased upward in magnitude (i.e., p-values are deflated) while p-values interpreted as insignificant are biased downward in magnitude (p-values are

---

[277] Joan Fisher Box, Fisher's daughter (and the wife of another prominent statistician, George E. P. Box), characterized the field of mathematical statistics at this time as follows (see Box [541], p. 62): *"It is difficult now to imagine the field of mathematical statistics as it existed in the first decades of the twentieth century. By modern standards the terms of discourse appear crude and archaic and the discussion extremely confused. Admittedly, there existed sound bases in the theory already developed, if they could be distinguished. The whole field was like an unexplored archeological site, its structure hardly perceptible above the accretions of rubble, its treasures scattered through the literature; to assess its worth, one would need to dig and sift the soil beneath a modern home ...."* [541]

inflated). They also point out that experiment-to-experiment and study-to-study variation can be large enough to produce p-values of <0.01 and >0.30.[278]

Amrhein *et al.* suggest changing the word "confidence" in "confidence intervals" to "compatibility." This takes scientists out of the quantitative world of pure statistics and places them a more holistic, and realistic, scientific world that considers statistical analyses *in context,* i.e., with knowledge of the many factors that affect the behaviors of systems and a common sense approach to data interpretation. Data then can be described as "compatible" or "incompatible" with a particular hypothesis or theory. But what metric allows the scientist to distinguish the two? *How* compatible (or incompatible) are the data? Here again we come back to the need for a metric, but finding one is just as problematic as the search for a bright line. Here is where common sense can help. It demands compatibility or incompatibility when a p-value is small, but when it is not, we have a third type of "significance" characterization, namely "further experimentation will be necessary to reach a conclusion."

Goodman [627] has addressed the p-value dichotomy crisis by suggesting the inclusion in manuscripts of a "confidence index," an *estimation* of the likelihood the results of a study are true, in addition to other metrics, for example, p-values and confidence intervals. "Likelihood" certainly sounds Bayesian (in the sense that previous evidence is considered), and Goodman includes such methods, as well as investigator judgment and consideration of study limitations, in his formula for such an index. This is a reasonable idea, but it is not without the usual caveats related to metrology, including "how does one *justify* the method of measurement used?" There is no absolute answer to this question, yet the idea of including a more descriptive and less quantitative measure of significance in one's papers has merit.

Interpretation of confidence limits frequently leads investigators to the conclusion that no effect was observed in their experiments (the null hypothesis holds), when this may not be the case. Figure 5.30 illustrates this point for two point estimates of effect. The two points are identical, but one is associated with a large confidence interval (high p-value) and the other is not (low p-value), which leads to the unwarranted conclusions of confirmation or disconfirmation, respectively, of the null hypothesis. A recent meta-analysis of 791 published papers corroborates this fact [628], as 51% of null hypothesis "confirmations" incorrectly "proved" no effect. These confirmations can be particularly dangerous in clinical medicine, for example, when the null hypothesis is confirmed but the confidence interval includes segments of authentic increased risk that are ignored – a situation that would result in some patients not receiving therapy when they should be. Human lives generally are not at stake in preclinical research, although animal lives certainly are. Thus, errant null hypothesis conclusions may result in substantial waste of time, resources, and unnecessary use of animals. It also is true that confidence intervals from carefully considered, rigorously performed, and critically interpreted experiments in which great confidence exists can be identical to those from questionable studies [627].

---

[278] The example they give is researchers conducting two perfect replication studies of some genuine effect, each with 80% power (chance) of achieving $P < 0.05$.

BEWARE FALSE CONCLUSIONS

Studies currently dubbed 'statistically significant' and 'statistically non-significant' need not be contradictory, and such designations might cause genuine effects to be dismissed.

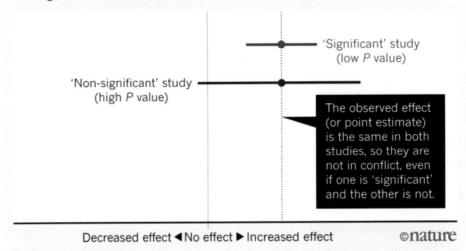

Figure 5.30 Proof or disproof of the null hypothesis? (Figure taken from Amrhein [626], with permission.)

Where does this leave us? Hurlbert *et al.* [624] made two proposals the American Statistical Association [629] and others [630, 631] have endorsed:[279] "...(1) that in research articles all use of the phrase 'statistically significant' and closely related terms ('non-significant' 'significant at $p = 0.xxx$' 'marginally significant', etc.) be disallowed on the solid grounds long existing in the literature;" and (2) "that direct formal requests be made to the editors and editorial boards of journals to modify their instructions to authors to include a disallowance of manuscripts that do not adhere to the above proscription." Such instructions might be akin to those below.

There is now wide agreement among many statisticians who have studied the issue that for reporting of statistical tests yielding p-values it is illogical and inappropriate to dichotomize the p-scale and describe results as "significant" and "nonsignificant." Authors are strongly discouraged from continuing this never justified practice that originated from confusions in the early history of modern statistics.

There is, indeed, "wide agreement" among statisticians. However, before addressing the statistical issues involved, I note that there is disagreement about how wide "wide agreement" actually is. For example, Hardwicke and Ioannidis have challenged the accuracy of a *Nature* perspective, which was signed by 854 scientists, suggesting "abandonment of the

---

[279] Reasonable questions or concerns about the ASA recommendations have been raised [612, 632, 633], but they do not change the value of the basic argument that p-values and statistical significance are used improperly to make conclusions about experiments, hypotheses, and theories.

notion of 'statistical significance' across science" [626]. They surveyed 248 signatories on the perspective "to understand how and why they signed ... and their scientific perspectives on various aspects of abandoning statistical significance."[280] Their responses showed that the apparent unanimity of opinion about eliminating p-values disappeared when certain potentialities were discussed. For example, 2/3 of respondents had concerns about increases in Type 1 (false positive) and Type 2 (false negative) errors.

Deborah Mayo, a prominent philosopher of science specializing in error analysis, and the recipient of the 1998 Lakatos Award for her outstanding contribution to the PoS, argues that throwing out p-values and statistical significance is akin to throwing out the baby with the bath water [634]. She points out that doing so would leave agencies, such as the FDA, in the position of having no standard for the acceptance or rejection of clinical trial results. The EPA would face the same problem, and public policy decisions could be compromised. For statisticians themselves, it would be difficult to determine the value of statistical models without a frame of reference involving significance. Mayo argues strongly for a focus on *error analysis* instead of p-values and statistical significance *per se*. An excellent exposition of her stance on error analysis and the so-called "statistics wars" may be found in her recent book *"Statistical Inference as Severe Testing: How to Get Beyond the Statistics Wars"* [635]. In personal correspondence with the author, Mayo summarized her view.

Practicing scientists ... aren't in need of a wake-up call to stop worshiping P-values, but rather a wake-up call as to the vital importance of error control in order to distinguish well from poorly tested claims. For this it would be good if they actually understood P-values.

Proponents of retiring p-values and statistical significance argue that it would result in a decrease in p-hacking and data dredging. However, Ioannidis argues that the prohibition of using p-values "may give bias a free pass" [636]. Mayo agrees [612].

In a world without thresholds, it would be harder to hold accountable those who fail to meet a pre-designated threshold by dint of data dredging. Forgoing predesignated thresholds obstructs error control.

Significance and confidence are concepts independent of truth. Although an inference may be made based on what might be considered strong statistical support, it still may be wrong. Recognizing and avoiding this problem also requires statistical analysis, and criteria, for decision-making. If a binary (yes/no) decision is required, there must be a bright line or equivalent. Imagine for a moment that $p < 0.05$ (or any other bright line) was prohibited. Digital decisions, i.e., 1 or 0, then would have to be made based on probabilities (analog). This would still require dichotomization of the probability distribution into "yes" and "no" regions separated by ... a bright line. How do we avoid this dilemma?

In an editorial in *The American Statistician* entitled *Moving to a World Beyond "p < 0.05,"* Wasserstein *et al.* [625] provide a variety of suggestions, as do 43 of those who signed on to the perspective of Amrhein. The suggestions very widely. They include decreasing pressure to discuss results with absolute certainty so as to further one's scientific

---

[280] The survey instrument (available here: https://osf.io/zkwgf/) consisted of 10 multiple choice questions presented on a single web page.

status, get funding, or be promoted; explaining *why* use of p-values was necessary; following the guidelines of the *"ASA Statement on Statistical Significance and P-Values"* [629]; working with experts to ensure that appropriate statistical measures are used; using the "the p-value as a heuristic," not deterministically; and countenancing uncertainty in all statistical conclusions by seeking ways to quantify, visualize, and interpret the potential for error.

A not uncommon thread running through most of the suggestions is the importance of nonstatistical factors in the interpretation of data. In essence, we expand our inferential space outside of pure statistics. We become more holistic. Amrhein [626] suggests the following.

Whatever the statistics show, it is fine to suggest reasons for your results, but discuss a range of potential explanations, not just favoured ones. Inferences should be scientific, and that goes far beyond the merely statistical. Factors such as background evidence, study design, data quality and understanding of underlying mechanisms are often more important than statistical measures such as P values or intervals.

Amrhein, Trafimow and Greenland [626] have a similar sentiment.

Rather than focusing our study reports on uncertain conclusions, we should thus focus on describing accurately how the study was conducted, what problems occurred, what data were obtained, what analysis methods were used and why, and what output those methods produced.

My own suggestion is two simple aphorisms.

Be quantitative in statistical analyses and qualitative in their discussion.
    Be statistical in analysis and scientific in explanation.

The point here, and this is one often misunderstood by scientists and statisticians alike, is statistical inference $\neq$ scientific inference. Each has its own processes and outcomes. Statisticians deal with populations, be they groups of humans or groups of numbers. They ask "Are these two groups different or the same with respect to a particular quality? If different, how different? If alike, how alike? How likely is it that phenomenon x occurred spontaneously, or for a particular reason?" In the past, how likely was x expected to occur? What is the probability that x will be observed now? You will notice that none of these questions have anything to do with science *per se*. Statisticians are not asking if one medicine is better than the other, whether it will rain tomorrow, or if horse #6 will win the Kentucky Derby. Medicine, rain, and horse are entities that might just as well have been called *bagel, lox,* and *cream cheese*. Statisticians deal with "groups" that are abstract mathematical concepts.

In a sense, statistics relates to science as deduction does to inference. Deduction is a process divorced from the natural world. It is a world of logical analysis without naturalistic factors. The conclusion that my toe is a bus driver, as bizarre as that may sound, *could* be concluded according to the rules of logic, if the propositions involved were appropriately framed. The scientist or clinician, by contrast, is not particularly interested in the theoretical machinations of the statistician. Theirs is the world of cause and effect. If a physician chooses to treat a patient with drug x instead of drug y, will the outcome be better? Should I pack rain gear for my trip? How much am I willing to bet on horse #6? Pure statistical analysis is a tool for making good decisions – not an end in itself – and requires a decidedly nonstatistical perspective, for example, background knowledge, proper study design, high quality data, and a clear practical goal. "I choose drug *y* because I think it's *better*." "I pack my raincoat because I don't want to get *wet*." "I bet on horse #6 because I want to win *money*." The goal of the scientist is not deterministic,[281] it is a holistic understanding of a entire system.

The exclamation point and the question mark in the title of this section were included to emphasize the need for tests of statistical significance (!) to do science and to ask if they are being used properly (?). In my opinion, retiring p-values and statistical significance is an overreaction to the problem of data analysis. There *is* a problem, but it lies less with statistics *per se* and much more with how the scientist understands and uses statistics. A quotation from Shakespeare's play "Julius Caesar "[282] is as applicable today as it was in 44 BC, when the play was set.

The fault, dear Brutus, is not in our stars / But in ourselves . . . .

A solution to the statistics wars is better education[283] in pure and applied statistics, better mentoring, better study design, and earlier and closer collaboration among scientists/clinicians and biostatisticians [638]. Localio *et al.* emphasized the last point in a recent editorial [639].

. . . 'For objective causal inference, design trumps analysis.' Optimal designs not only improve statistical power and precision, they minimize bias and often simplify the analysis. Consulting a statistician only for the analysis or to bless the sample size (or make an inadequate sample size appear legitimate) represents too little statistical science, far too late.

Many, including myself, have difficulty appreciating the nuances of biostatistical analysis. It is important to argue the philosophy and use of biostatistics, but what also is needed are simple recommendations that practicing scientists and clinicians can follow. In a 2019 editorial in The American Statistician, Wasserstein *et al.* [625] provided five recommendations that every scientist would do well to follow.[284]

1. Don't base your conclusions solely on whether an association or effect was found to be "statistically significant" (i.e., the p-value passed some arbitrary threshold such as $p < 0.05$).
2. Don't believe that an association or effect exists just because it was statistically significant.
3. Don't believe that an association or effect is absent just because it was not statistically significant.

---

[281] However, in some fields, for example, physics, this is precisely what is wanted.
[282] Act 1, Scene 2.
[283] Szucs and Ioannidis [637] suggest that the *". . . current statistics lite educational approach for students that has sustained the widespread, spurious use of null hypothesis significance testing (NHST) should be phased out."*
[284] For examples of new journal requirements for statistical analysis, see Nature
www.nature.com/collections/qghhqm/pointsofsignificance, Journal of Biological Chemistry
http://jbcresources.asbmb.org/collecting-and-presenting-data, or Science
www.sciencemag.org/authors/science-journals-editorial-policies.

4. Don't believe that your p-value gives the probability that chance alone produced the observed association or effect, or the probability that your test hypothesis is true.
5. Don't conclude anything of scientific or practical importance based on statistical significance (or lack thereof).

5.7.7 The "Replicability Crisis"

Nearly 60 years ago, Dr. Bernard K. Forscher published an amusing and remarkably prescient story, *"Chaos in the brickyard"* [640] (see Appendix B), that followed brickmakers, bricklayers, and the edifices they built as increased demand for edifices forced them to balance quality with productivity. In the end, quality was sacrificed for productivity and their edifices collapsed. This story is an allegory for how the edifice of science weakens due to nonscientific (e.g., social, institutional, and political) factors, including the training of excessive numbers of new scientists that overwhelm the number of available positions, the pressure to "publish or perish," career advancement, and gaining prestige in the scientific community [641–644]. It also foreshadowed the current crisis of irreproducibility[285] in science [645–649], a crisis that threatens the foundations of our knowledge of the natural world.

Some scientists (natural and social), clinicians, and statisticians 60 years ago *were* aware of the severe consequences of this problem. However, it was not until 2011 that awareness of the widespread misunderstanding and misuse of measures of statistical significance, and the consequent inappropriate acceptance or rejection of hypotheses, became inescapable. It began in psychology, and in contrast to Einstein's *annus mirabilis*, produced an *annus horribilis*[286] for psychologists and sociologists, a time during which many examples of poor science emerged [650–655]. One of the most visible and egregious examples was the claim that humans could manifest Extrasensory Perception (ESP or "psi") [656]. In a paper entitled "Feeling the Future: Experimental Evidence for Anomalous Retroactive Influences on Cognition and Affect," Bem reported results of tests of ESP in "9 experiments involving more than 1,000 participants." He concluded that 8 out of 9 experiments yielded "statistically significant" results indicating possession of ESP by participants. Common sense tells us that the claim is absurd, and Bem himself, in preparing for expected controversy, ends his paper with a paragraph entitled "On believing impossible things," in which he quoted the White Queen from Alice in Wonderland, who said, *". . . sometimes I've believed as many as six impossible things before breakfast."* Although meant to be humorous and thought provoking, I think all would agree that the White Queen's experiences do not corroborate Bem's findings and, in fact, lead the reader to consider Bem's experiments fantasies themselves. Controversy did, in fact, come, and from many sources, but for statisticians it was statistical in nature, not common sensical, doxastic, or hysterical [657, 658]. Wagenmakers *et al.* [659] may have summed up the consensus from the

---

[285] It is the *lack* of reproducibility that is the crisis. In discussing this, I will use replicability and reproducibility as synonyms.
[286] Horrible year.

substantial literature debunking Bem when they categorized his conclusions as "anecdotal," having concluded that

...Bem's p values do not indicate evidence in favor of precognition; instead, they indicate that experimental psychologists need to change the way they conduct their experiments and analyze their data.

Although Bem's ESP experiments are a particularly good example of statistical ignorance and misuse, they also have the quality of being irreproducible, as had all the prior experiments in the field. Similar difficulties have been reported by pharmaceutical companies trying, and failing, to replicate results of preclinical studies [648]. In the quotation below from Diaconis [660] twenty years earlier, if one replaces the word "skeptics" with "statisticians," his advice is identical to that of current biostatisticians.

...studies on psi...were 'crucially flawed.... Since the field has so far failed to produce a replicable phenomena, it seems to me that any trial that asks us to take its findings seriously should include full participation by qualified skeptics' (p. 386). ...[If Bem] is wrong, it offers a truly alarming massive case study of how statistics can mislead and be misused.

Like the five "don'ts" specified above, Wagenmakers *et al.* [659] also suggested ways to protect oneself from statistical misrepresentation and fallacy. They included avoiding "fishing expeditions," not "[torturing] the data until they confess," using multiple statistical methods (e.g., p-values, Bayes factors[287]), making all raw data publicly available before publication of the results, and enlisting the aid of skeptics in experimental planning and data review. The last suggestion, in essence, recapitulates the idea that the best way to test an hypothesis is to try to prove it wrong, i.e., falsify it (see page 252).

Are we stuck with p-values? No. Just as confidence in an hypothesis increases with corroboration, confidence in a statistical inference can increase if the data are analyzed using multiple methods. A good example is combining frequentist with Bayesian approaches, which can operate on the same data but generally answer different, but related, questions. The frequentist approach, for example, NHST is deterministic. It seeks a number (p-value) that provides the scientist with the likelihood their hypothesis (H_1) is true. The Bayesian approach, instead, seeks only to determine if one hypothesis (e.g., H_1) is to be believed over another (H_0).[288] Absolute truth is irrelevant. The frequentist analyzes the data they produce, without consideration of prior results. The Bayesian incorporates their *prior* belief in an hypothesis into their current data analysis. The Bayesian can always adjust their estimations (Bayesian conditioning) in the light of new data, whereas the frequentist cannot (at least for the data from one experiment). Frequentists assume a distribution of experimental data (where n is finite) are similar or identical to the distribution from all data (where n is infinite). The odds of this being true cannot be determined, but it is reasonable to suggest they are low. The frequentist can always be wrong. The Bayesian can never be wrong in the

---

[287] Wagenmakers own statistical analysis of the Bem data showed "[o]ut of the 10 critical tests, only one yields 'substantial' evidence for H_1, whereas three yield 'substantial' evidence in favor of H_0. The results of the remaining six tests provide evidence that is only 'anecdotal' or 'worth no more than a bare mention' " (quoting Jeffreys [661]).

[288] In practice, Bayesians would calculate a Bayes factor, the ratio of likelihoods that each hypothesis was true. The larger this factor, the greater would be the probability that the evidence supports one hypothesis over the other.

Figure 5.31 Same image, different conclusions!. (Credit: Bryan Ridgley (March 17, 2016). www.pinterest.com/pin/711216966123480421/)

pure statistical sense, but *inferences* from Bayesian analysis can themselves be wrong. Two "camps" among scientists and statisticians exist, one for each form of statistical analysis, but it also is true that each camp can come to the same conclusions (although to subtly different questions) about particular situations and data sets.[289] Note, in addition to ANOVA (analysis of variance) and Bayesian statistical approaches, many others exist and may be appropriate for a particular data set.

The inability to replicate someone's experiment does not necessarily mean the experiment was fraudulent, poorly designed, misinterpreted, etc. It does mean there could be value in more deeply examining the experiments, methods used, resulting data, statistical approaches, and interpretations, for whatever the outcome, science gets closer to truth. It is possible, even for experts, to come to different conclusions analyzing the same data (Fig. 5.31) [662, 663].

Replication is difficult...period. Why? It can be expensive and time-consuming; impossible due to lack of starting materials, subjects, or precise information about the materials and experimental methods used in the original experiments; difficult due to the new experiment's own methodological and analytical variability; or of little or no value to one's scientific corpus. Scientists must produce original data to be successful. They get no credit for reinventing the wheel. They may, like Ioannidis *et al.*, act as curators of scientific knowledge, disallowing or warning against the inclusion of poor science. Scientists may

---

[289] It should be noted that NHST can yield less accurate inferences with large biomedical data sets. By contrast, in physics and other exact sciences, NHST *can* be used effectively because the experiments are highly controlled and the results have great precision [637].

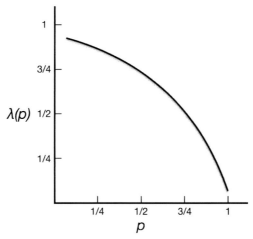

Figure 5.32 Speed vs. replicability. An illustration of the inverse relationship between the speed of scientific research and the probability that the research is replicable. (The value $\lambda(p)$ represents the scientist's expected speed if the desired reproducibility $p \in [0, 1]$ *ex ante.*)

(should) heed these warnings and improve their rigor, but they themselves are highly unlikely to spend any time checking the work of others, let alone attempting to repeat it.

Some of the biggest impediments to replicability are not scientific in nature. They are social. "Publish or perish" is a common phrase that turns out to be true. To a large degree, scientists are evaluated by the scientific community, and especially by grant review panels and appointments/promotion committees, on their productivity. Thus, there is tremendous pressure on scientists to publish many, but not always high quality, papers. As in Forscher's fable, increased productivity may come with decreased replicability (Fig. 5.32).

Reproducing poorly designed, executed, or interpreted experiments is difficult in an absolute sense, but a larger concern is the uselessness and wasted effort trying to replicate something that was fallacious from the start. The peer review process, a pillar of science, can contribute to the problem by not being the barrier to bad science it should. Scientific societies themselves can perpetuate bad research, or discount quality research, when different schools of thought compete against each other for preeminence. One may "win" over others, but not for purely scientific reasons. Scientists have opinions, and often cling to them for reasons of ego, pride, belief (but not fact), stubbornness, or resistance to change, all of which can affect the acceptance or rejection of results or theories. A good example of this was the strife between the "βaptists" and "τists" in the Alzheimer's disease field [664]. βaptists believed that the disease was caused by a protein, amyloid β-protein, that fills the brain in Alzheimer's disease patients. τists, by contrast, argued that tau, a protein that accumulates inside neurons, was the culprit. As this controversy raged, scientists and clinicians worked hard to support their positions instead of searching for the truth in an unbiased manner. It turns out that *both* proteins, as well as others, are involved in disease pathogenesis, just at different stages and through different mechanisms, so both sides were right *and* wrong.

The replication crisis, not surprisingly, affects fields with the greatest intrinsic variability, for example, psychology and medicine [654, 665, 666]. Psychologists and clinicians frequently try to prove, instead of test, a hypothesis (Fig. 5.33). One of the most disappointing investigations of replicability was that of "The Reproducibility Project: Cancer Biology," a project that studied replicability of preclinical research in cancer biology [667].[290] The project selected studies published in high-impact journals, including *Science, Nature, and Cell.* Replication of 50 experiments from 23 papers was attempted. Only 40% of the studies reporting positive effects could be replicated. More studies would have been assessed if their authors had responded to inquiries from the project (only 1/3 did) or provided useful information (only 40% did). Of those studies that could be replicated, only 15% of the effect size reported in the original studies was observed. It should be noted that some studies, whose replication was attempted by others, did confirm findings of the original studies.

As one moves to ever more precise fields of natural science, for example, to physics, computer science, and space sciences, replication becomes easier because of low intrinsic variation in these systems and their methods. Decreased system error means increased ability to correctly infer differences among test and control data sets, which changes the direction of scientific studies away from hypothesis testing *per se* (accuracy) toward improved precision. A good example is the Gran Sasso experiment (see page 253). All agreed that these particles traveled at or near the speed of light, c. The question was whether neutrinos moved faster. The high precision of the experiment (an uncertainty of $5.3 \times 10^{-6}\%$) enabled scientists to conclude that neutrinos did not.[291]

As replicable as particle physics experiments may be, some are not, and not because of failures in experimental design, intrinsic variability, execution, or analysis, but rather because replication would require an extraordinary and impractical amount of resources. Studies of the lifetimes of "charmed" particles is illustrative. Construction of the equipment, data collection, examination of 2.5 million images, culling 47 usable events from a total of 300,000, and then computationally determining the lifetimes of the particles in those events cost >$10 million and required 280 person-years. There were 99 co-authors on the paper reporting the results [669], including experimentalists, theoreticians, engineers, computer scientists, *et al.* No single person, laboratory, research center, or country could have done the experiment on its own [670]. However, what may be done in the future is a second (nonidentical) experiment, one using refinements in equipment and analytical methods.

Human clinical trials, by contrast, often collect data as percentages, which would limit the uncertainty to $1 \times 10^{-3}\%$, $\approx$2,000-fold higher. However, many trials have even higher uncertainty. Recent clinical trials of the Alzheimer's disease drug aducanumab, magnetic

---

[290] See https://elifesciences.org/collections/9b1e83d1/reproducibility-project-cancer-biology.

[291] Later studies [668] revealed neutrinos do have mass ($< 0.8eVc^{-2}$, which is >600,000 times less that of the electron) and therefore could *not* travel at or above c, i.e., without contravening relativity.

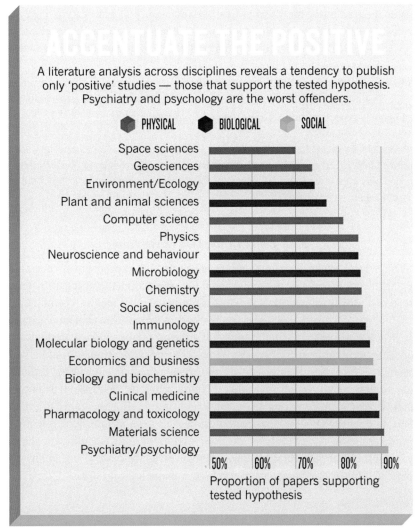

Figure 5.33 Don't test it, prove it!? An analysis of the scientific literature reveals that 70% or more of publications report efforts to *prove*, instead of *test*, one's hypothesis. (Figure from Yong [654], used with permission.)

resonance imaging (MRI) of brain blood volumes, Kaplan–Meier survival time analysis, and disease grading had uncertainties ~60,000-, ~200,000-, ~500,000-, and ~1 million-fold higher. It thus is not surprising that replication proves increasingly difficult as one moves toward biology and medicine. Even if high precision were obtainable in human clinical trials or psychology experiments, replication could be impossible, by definition, because one would have to use the *same* patients in a new study, which would be unlikely and certainly unethical. Therefore, what is done in these situations is to do a new study that is as similar to the old one as possible, but uses substantially increased numbers of patients.

This gives the trial more statistical power (the probability of not making Type II errors) and typically leads to the biggest *n* "winning," i.e., being the most believable trial. New studies may also be done with revised protocols designed to eliminate known weaknesses in an original experimental plan.

In closing, I note that in 1905, a century before the emergence of the current replication crisis, Poincaré [583] already had recognized the uncertainty of experimentation and warned against expecting future replications of experiments extant.

...every fact observed enables us to predict a large number of others; only, we ought not to forget that the first alone is certain, and that all the others are merely probable. However solidly founded a prediction may appear to us, we are never *absolutely* sure that experiment will not prove it to be baseless if we set to work to verify it.

5.7.8 On (T/t)ruth: God versus Hume

In practice, when a scientific community produces enough experiments[292] corroborating a theory, that theory is believed to be true and it becomes axiomatic in scientific research. Is it rational to do so? If so, why? If not, why? Is it rational in all realms (cosmology, particle physics, biology, chemistry, sociology, public policy, geopolitics, economics, philosophy) or only in some and only at specific times? If a theory is true, and we are logical, it must be rational to accept it as such. However, how does one define truth? Would it be rational to accept a specific definition of truth or must we be pluralistic and accept different types of truth, be they relative (truth), absolute (Truth), religious, or existential (however you define it)? If we cannot determine the definition of truth itself, how are we to evaluate it as it pertains to science? Don't all these questions lead us to an infinite regress that precludes us from determining if *any* theories are true? *Ceteris paribus*, should we even care if a theory is or is not true? Sir Peter Medawar, who received the Nobel Prize in Physiology or Medicine in 1960 for his discovery of acquired immunological tolerance, thought maybe not [671].

I cannot give any scientist of any age better advice than this: the intensity of the conviction that a hypothesis is true has no bearing on whether it is true or not. The importance of the strength of our conviction is only to provide a proportionately strong incentive to find out if the hypothesis will stand up to critical evaluation.

—Sir Peter Medawar

Medawar's advice is practical in nature. It seeks to guide scientists in asking and answering questions, regardless of their truth value. Four decades after Medawar, John Ioannidis [672] argued not only that determining the truthfulness of results *did* matter, but unfortunately, if one did so, they would be sorely disappointed, especially if the results were from clinical medicine.

---

[292] Who decides what is enough?

...false findings may be the majority or even the vast majority of published research claims .... However, this should not be surprising. It can be proven that most claimed research findings are false.

Medawar and Iaonnidis dealt with real world issues in basic immunology and clinical medicine. One may believe what they will about truth, but the question for them was whether knowledge gained through research provided understanding that could be used in practice, not whether research results were true *per se*. Scientists and clinicians, and especially those embarking on careers in these areas, *do* search for the truth because they equate it with abilities to manipulate scientific systems, to make predictions, and to prevent, ameliorate, or cure disease. However, these are relatively new concepts of truth. Throughout history, philosophers, natural philosophers, and scientists often believed, or were forced to believe, that the world was perfect, as it was created by God.[293] Whether you are a theist, agnostic, or atheist is not relevant to this discussion. God's truth was considered absolute and inviolate, by definition. The quest of those investigating how the world worked was to understand what God had created, i.e., absolute truth about the world – and their tacit assumption was this was possible. Now comes David Hume, who, in addition to his monumental contributions to philosophy in general,[294] was one of the most influential interlocutors in discussions of, among other things, cause and effect, induction, and truth. Hume's perspective was quite different from that accorded to God. Hume introduces the notion of cause and effect in the section of *Treatise* entitled *"Of the connexion or association of ideas"* [I.I.IV].[295]

...two objects are connected by the relation of cause and effect, when the one produces a motion or any action in the other, but also when it has a power of producing it.

Hume uses billiard balls as examples of objects. These objects may have two properties, their intrinsic exhibition of characteristics perceivable sensorially and, in some cases, properties not directly detectable by the senses, one of which is force ("power"). We know, through sight and experience, that a moving billiard ball impacting a stationary ball will cause that ball to move. This is the cause-and-effect connection. What we don't know is what links cause to effect. Hume states the effect cannot come from the balls themselves, nor through nonempirical means (*"abstract reasoning or reflexion [sic]"*). What we may infer is the cause-and-effect linkage is movement, through which the force to move the stationary ball is imparted. Hume continues by arguing that cause and effect must occur contemporaneously. We wouldn't expect, now could we observe, the first ball hitting the second, remaining stationary, and then the second ball moving. Here now comes induction, which tells us that *any* moving ball hitting *any* stationary ball should cause the latter to move, but will it? Hume argues we have no reason to think so.

[293] The word "God" is used here in a metaphysical, not existential, sense.
[294] These include the three-volume work, *"A TREATISE OF Human Nature: BEING An Attempt to introduce the experimental Method of Reasoning into MORAL SUBJECTS"* (1739), referred to here as "Treatise," and *"ENQUIRIES CONCERNING THE HUMAN UNDERSTANDING AND CONCERNING THE PRINCIPLES OF MORALS"* (1748), referred to here as Enquiries.
[295] X.Y.Z refers to the book, part, and section number, respectively, from the 1896 L. A. Selby-Bigge version of Treatise.

...there can be no *demonstrative arguments to prove, that those instances, of which we have had no experience, resemble those, of which we have had experience* [I.III.VI].

In the quotation below, knowledge (theory) suffers the same fate, according to Hume, in that confidence in our knowledge can only be probabilistic. The greater the number n of instances during which the theory is corroborated, the greater the confidence in our knowledge will be, but $p(truth) = 1$ is not possible. Thus, not only can there be no *Truth*, there cannot even be abiding *truth* (however long that may be), because the search for it is corrupted by psychological (pride) and sociological (approbation of the scientific community) factors. To wit:

...all knowledge degenerates into probability; and this probability; is greater or less, according to our experience of the veracity or deceitfulness of our understanding, and according to the simplicity or intricacy of the question. There is no Algebraist nor Mathematician so expert in his science, as to place entire confidence in any truth immediately upon his discovery of it, or regard it as any thing, but a mere probability. Every time he runs over his proofs, his confidence encreases; but still more by the approbation of his friends; and is rais'd to its utmost perfection by the universal assent and applauses of the learned world. Now 'tis evident, that this gradual encrease of assurance is nothing but the addition of new probabilities, and is deriv'd from the constant union of causes and effects, according to past experience and observation. [I.IV.I]

If the pessimistic meta-induction is correct, how can we believe newly induced theories are true, even when they are first conceived, and if we can't, it would seem futile to even attempt to prove a theory true. Maybe it is. Chang [564] has suggested a way of escaping the conundrum of induction – simply change the meaning of "Truth" and stop worrying about the existence or reality of entities.

It is time to face the fact that we cannot know whether we have got the objective Truth about the World (even if such a formulation is meaningful). ... So we need to have a notion of truth, a usable one in the operable realm, to underpin our practice of making, assessing, and accepting various statements. I would like to propose a pragmatist coherence theory of truth, according to which a statement is true if (belief in) it is needed in a coherent epistemic activity.[296]

It is possible the pithiest précis and prescription for the scientist or clinician confronting the issue of truth is *"don't take yourself too seriously."* However, what may be the most eloquent and beautiful prescription comes from Robert Boyle.[297]

And I will freely confess to you on this occasion, that, for my part, in the prospect I have of the future advancement of humane knowledge, I think most of those virtuosi, that now live, must content themselves with the satisfaction of having employed their intellects on worthy objects, and of having industriously endeavoured, by promoting useful knowledge, to glorify GOD and serve Mankind. For, I presume, that our enlightned posterity will arrive at such attainments, that the discoveries and performances, upon which the present age most values itself, will appear so easy, or so inconsiderable

---

[296] For Chang [673], epistemic activities are *"mental or physical actions contributing to the production or improvement of knowledge in a particular way, and following some discernible rules (though the rules may be unarticulated)."* Examples include prediction, explanation, and counting.

[297] From Boyle, R. *The Natural History of Human Blood*, postscript to "An Appendix to the Memoirs," in Boyle [44] Volume 4: 758–759.

to them, that they will be tempted to wonder, that things to them so obvious, should lye so long concealed to us, or be so much prized by us; whom they will, perhaps, look upon with some kind of disdainful pity, unless they have either the equity to consider, as well the smallness of our helps, as that of our attainments; or the generous gratitude to remember the difficulties this age surmounted, in breaking the ice, and smoothing the way for them, and thereby contributing to those advantages, that have enabled them so much to surpass us. And since I scruple not to say this of those shining wits and happy inquirers, that illustrate and enoble this learned age, I hope you will not think, that I, who own myself to be more fit to celebrate than rival them, would dissuade you from improving and surpassing the slight performances, that are, in this little tract, submitted to your judgment . . . .

6

Science as a Social Endeavor

... the scientific enterprise is carried out by a community of real human beings: a system of cognitively limited living organisms who are themselves embedded in and interacting with the very same nature that is the object of their cognitive striving. Moreover since the organisms in question are here members of a community, their involvement with the social world can be no more neglected than their involvement with the natural world In particular, such social involvements must inevitably produce impure or 'sullied' motivation – desire for fame and fortune for example – going far beyond the mere desire for truth beloved of the tradition. Finally, since we no longer conceive scientists as mere rational minds aiming to store up or accumulate more and more true descriptions of nature, we need no longer conceive the product of scientific activity as simply a theory or system of beliefs. Rather, scientific activity constitutes a complex social practice consisting only in part of accepted statements.

—Quotation from Friedman [674], p. 381, reviewing Philip Kitcher's book, "The Advancement of Science – Science without Legend, Objectivity without Illusions." Oxford University Press, 1996.

6.1 Science and the Public

The impact of science on the public is incontrovertible, but the impact of the public on science is as well. Robert Merton, in his early construction of the field sociology of science in 1938 [675], was well aware of this when he opined about the public and science.[1]

The increasing comforts and conveniences deriving from technology and ultimately from science invite the social support of scientific research. Readiness to accept the authority of science rests, to a considerable extent, upon its daily demonstration of power. Were it not for such indirect demonstrations, the continued social support of that science which is intellectually incomprehensible to the public would hardly be nourished on faith alone.

Merton's prescient words have, unfortunately, become all too true in the current climate of ideological and faith-based perspectives on science and its uses. However, the problem is

---

[1] See Merton [675], pp. 328 and 332.

even more fundamental – the public, for whom science often *is* incomprehensible, must depend on various "interpreters" to make sense of it. What if the interpreters are not unbiased scientists?

With the increasing complexity of scientific research, a long program of rigorous training is necessary to test or even to understand the new scientific findings. ... There results an increasing gap between the scientist and the laity. The layman must take on faith the publicized statements about relativity or quanta or other such esoteric subjects. ... Popularized and frequently garbled versions of the new science stress those theories which seem to run counter to common sense. To the 'public mind,' science and esoteric terminology become indissolubly linked. ... presumably scientific pronouncements ... are for the uninstructed laity of the same order as announcements concerning an expanding universe or wave mechanics. In both instances, the laity is in no position to understand these conceptions or to check their scientific validity and in both instances they may not be consistent with common sense. If anything, ... myths[2] ... will seem more plausible and are certainly more comprehensible to the general public than accredited scientific theories, since they are closer to commonsense experience and cultural bias.

The term "faith," as used above, is not restricted to religious faith. It incorporates any non-scientific perspectives of the natural world based on faith in something *other than science*, e.g., a trusted friend, an admired political figure, the news media, nonscientific internet sites, and social media, all of which may have their own agendas and intrinsic biases. This does not mean all these sources of information are wrong *a priori*. Each *can* "get it right," ... at times.

The conduct of modern science is impossible without public support. Put simply, if the public wants research to be done in a specific area or in a specific way, it can, but if not, it can't. What the public wants varies among individuals, social groupings, governments, states, regions, and countries, as do the processes through which opinions are formed. What are these processes and how *do* people decide whether they will support science, in general, and different research areas, in particular? There are many ways to formulate an opinion and decide whether and how to act on it (maybe as many as the number of people involved). However, the one constant is possession of knowledge. This is the metaphorical slab of granite that the sculptor shapes as they wish and is a prerequisite for opinion formation and decision-making. Ideally, one hopes that each sculptor starts with an identical slab, i.e., all possess the same knowledge. Unfortunately, we don't live in an ideal world. We live in a world in which each sculptor's slab (knowledge base) differs, and each sculptor sculpts (interprets knowledge) in their own way. This means one can expect to eventually see many different kinds of finished work (decisions).

But science is not art and its product, knowledge of the natural world, comes from rigorous experimentation and logic, not through idiosyncratic processes.[3] Scientists understand science and thus are able to assimilate information. Most laypeople acquire knowledge

---

[2] The actual quotation was "the myths of totalitarian theorists." Merton is discussing the impact of government-mandated societal norms on science (truth). However, if one substitutes "ideological" or "faith-based" for the word "totalitarian" in the original quotation, the same ideas hold true.

[3] This is not strictly true. Idiosyncrasies of scientists do affect science, but these are accounted for through the intrinsic error-checking mechanisms of science.

Table 6.1 *Issues about which truth and opinion collide.*

Subject	Scientific	Nonscientific
Climate change	Climatology	Climate deniers
The sun and planets	Astronomy	Astrology
Life on Earth	Evolutionary biology	"Creation science"
Medical care	Physicians	Quacks
Autism etiology	Immunology/Neurogenetics	Anti-vaxxers
Genetically modified organisms	Molecular biology	Idealogues[a] and fear-mongers

[a]Organizations doing admirable and important work, such as Greenpeace [676], sometimes value their ideologies over scientific research and thus cherry-pick data favorable to their cause. Organizations such as the Non-GMO Project [677], whose stated purpose is to eliminate all GMO crops and organisms, are political movements who seek to achieve their goals by misrepresenting science or by citing nonscientific sources of information.

only *after* its translation into a form understandable to them. We would hope the scientists who did the work could be the translators themselves, but sometimes this is not the case (scientists often have a hard time translating into "English"). Instead, most people learn about science through the media (newspapers, radio, television, streaming services, social media sites, the Internet, etc.), political, topic-focused, or religious organizations, or simply by word-of-mouth (especially from celebrities and political figures).

How *does* scientific knowledge, as published in the scientific literature or as thought of by the scientist, compare with the impressions given the public by the media? How is this knowledge parsed to make the science comprehensible? What is lost and what is gained? How does the knowledge imparted affect public awareness, perception, and opinion of issues such as stem cell research, end-of-life questions (e.g., what is brain death? – the old definitions are changing), health-care costs, nuclear power, climate change (global warming), genetically modified organisms (GMOs), genetic testing, etc.? And, most fundamentally, how is the public to distinguish truth from opinion and science from nonscience (e.g., pseudoscience)?

Table 6.1 lists subjects about which truth and opinion often differ dramatically. Note that the categorical assignments of subjects to "nonscientific" does not mean these subjects are devoid of all scientific knowledge. For example, "anti-vaxxers" often cherry-pick reliable scientific data to support their opinions (Table 6.1).[4] Figure 6.2 provides a more general sense of the inferential and epistemological processes through which scientific and nonscientific "facts" are produced. It is a good example of the lack of rigor (e.g., systematicity) in nonscientific processes.

The demarcation between truth and opinion is particularly important with respect to the education of our citizens. We are taught as children how our universe came to be,

---

[4] Figure 6.2 is an amusing, but often all too true, illustration of the differences in perspective among scientists and creationists. Contrasting creationists with scientists is a particularly stark means of illustrating these differences, but creationism is but one of many types of nonscientific perspectives of the natural world. I also note that some ostensibly scientific claims may be ridiculous.

how it is organized, and how life developed on Earth. What we are taught is based on the best scientific evidence available. One of the most oft-cited and illustrative examples of the difference between scientific and nonscientific was the question of the organization of the sun and planets. It is understandable that human nature would lead ancient observers of the heavenly bodies to first consider the Earth the center of the universe. This geocentric notion *was* scientific, as it was based on empirical observation and theory creation. Geocentricity was made part of the Bible[5] and was an explicit component of the doctrine of the Roman Catholic church. What was nonscientific was the adherence to this idea in the light of compelling observational evidence for heliocentricity. It was heretical to dispute geocentricity because it was ". . . explicitly contrary to Holy Scripture."[6] The belief in scientific theories based on religious doctrine is nonscientific, by definition. However, the refusal to *evaluate* counter claims based on scientific evidence also is unscientific. The search for truth by scientists requires the constant challenging and testing of prevailing hypotheses and theories, and the development and incorporation of new theories that better explain nature than do the old ones.

A similar, but more modern, example of the science/nonscience dichotomy is the question of the origin of human beings. Scientific studies have provided a wealth of information concerning the history of the Earth and the emergence and development of species. Religious studies have provided different perspectives on these processes, including the perspective that the creation of the Earth and the life on it were acts of God ("creation science").[7] Current controversies regarding the incorporation of "creation science," an oxymoron, into school curricula reflect the desire of some to introduce religious beliefs into an educational system whose foundation is built upon facts. One certainly is entitled to have their own opinions, but unless these opinions are based on facts, they are intrinsically unscientific. Figures 6.1 and 6.2 provide a general sense of the inferential and epistemological processes through which scientific and nonscientific "facts" are produced. It is a good example of the lack of rigor (e.g., systematicity) in nonscientific processes.

Unfortunately, much of the information accessible by the public is pseudoscientific or anecdotal. Some in the media have recognized this and even written articles about the almost daily new "cures" that have been found (none of which actually are) [679]. The inability to differentiate science from nonscience has contributed to the public's mistaken belief that childhood immunization with MMR (measles–mumps–rubella) vaccines or vaccines containing thimerosal, leads to autism, a claim thoroughly refuted by countless scientific studies [680, 681]. Nevertheless, many parents have refused common childhood immunizations for their children [682–684]. The result of this "anti-vaxxer" movement is dangerous increases in the number of susceptible people in the population, which has led to the reemergence of diseases that previously had largely been eradicated, e.g., measles. Deadly outbreaks have been reported around the world, including in the United States,

---

[5] See 1 Chronicles 16:30.
[6] See the written decision of the Inquisition with respect to the alleged heresy of Galileo Galilei [678].
[7] It is ironic that the word *science* has been transmogrified in an effort to legitimize the nonscientific.

Figure 6.1 Differing perspectives of life on Earth. (Credit: Creationist Method by John Trever, *The Albuquerque Journal*.)

Europe, and Asia [685–688]. In addition, those not immunized against common childhood disease may, as adults, suffer far worse symptoms, or even death.

A particularly amusing, though unfortunate, example of how the media may be unable to distinguish science from pseudoscience, and therefore creates unwarranted public attention to nonscientific pronouncements, was the reporting by ABC News, the Associated Press, the Washington Post, and others that "the first independent scientific review of [UFOs] . . . in almost 30 years" concluded that "the physical evidence in some UFO sightings merits further serious scientific review" [689]. The "scientific review" was conducted by "scientists" in a society, the *Society for Scientific Exploration*, whose journal seeks to provide "a critical forum for sharing original research into conventional and unconventional topics," but which in reality publishes articles on telepathy, extrasensory perception, communicating with the spirit world, and, of course, UFOs.[8] The magazine *Time* did correctly characterize this society as "fringe science" [690].

The Internet is the spawning ground for the unprecedented promulgation of falsities, especially about politics and science. A particularly informative investigative report has been published by the organization Avaaz in 2021 [691]. They focused on Facebook and how its platform disseminated false news that had massive effects on the US presidential election of 2020. The executive summary of this report placed particular emphasis on four aspects of Facebook's behavior.

---

[8] The reader may be interested to know that I became aware of this journal when reading what I thought was a very interesting essay discussing the demarcation problem. I thought, initially, that a journal with this name likely is a reputable, scientifically rigorous, publication, as would most people. However, as I read the essay, I encountered numerous "radical challenges to mainstream views" [quoting from the essay] that I thought were a bit too radical, so I investigated the journal further. When I saw articles on "Communication with the Spirit World of God" and "ESP and Parapsychology in Everyday Life," and I did more research on the society itself, I realized that this society and its journal were pseudoscientific.

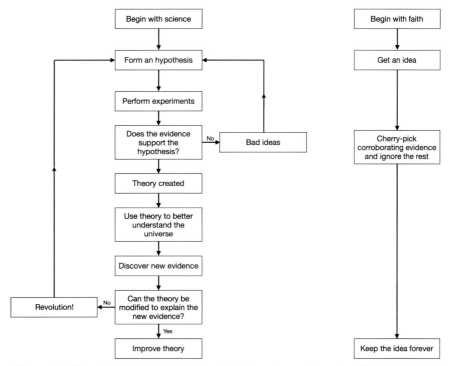

Figure 6.2 Scientific method vs. faith method. Based on a figure by Wellington Grey www.flickr.com/photos/dullhunk/2182220033.

1. Facebook could have prevented 10.1 billion estimated views for top-performing pages that repeatedly shared misinformation.
2. Avaaz identified 267 pages and groups (for example, "Stop the Steal," Boogaloo, QAnon, and militia-aligned) with a combined following of 32 million spreading violence-glorifying content and promoting imagery and conspiracy theories related to these movements.
3. Nearly 100 million voters saw sites falsely alleging voter fraud.
4. The top 100 most popular false or misleading stories were viewed an estimated 162 million times.

Facebook also posted misinformation about the COVID-19 pandemic. In fact, Avaaz deemed Facebook an "epicentre of coronavirus misinformation" [692]. Loomba [693] suggests that COVID-19 misinformation is one reason people may be hesitant to be vaccinated, just as with MMR vaccination.

Misinformation can produce lasting damage. It has been reported an astounding 35% of registered voters, a year *after* the election, said the 2020 presidential election should "probably" or "definitely" be overturned [694]. Recently, Vosoughi *et al.* published results of an in-depth statistical study of news stories "tweeted" from 2006–2017. The data comprised ≈126,000 stories tweeted by ≈3 million people >4.5 million times. Their findings?

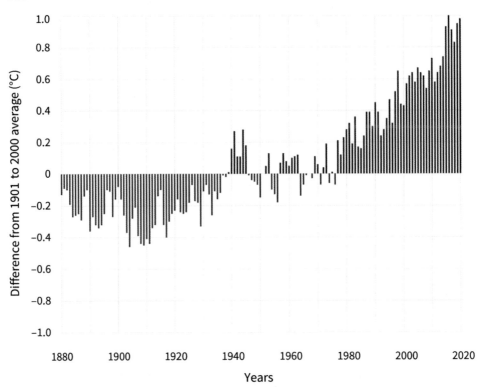

Figure 6.3 Global warming. Yearly global average surface temperatures from 1880 to 2020.
Blue bars indicate cooler-than-average years; red bars show warmer-than-average years.
Source: NOAA Climate.gov graph, based on data from the National Centers for
Environmental Information.

Falsehoods spread much more rapidly and broadly than the truth. They were of the order of
twice as likely to be retweeted. Psychological factors contributed to this. People's perceived
falsehoods as more "novel" and novelty was more exciting and interesting than truth.
False stories inspired fear, disgust, and surprise, whereas true stories inspired anticipation,
sadness, joy, and trust. Fear certainly drives propagation of falsehoods more than joy.

Falsehoods about the theory of global warming are rampant, a subject that has been
exceptionally controversial (it shouldn't be). It is not an exaggeration to say global warming[9] is a serious threat to life on Earth [695]. A histogram of global temperature differences
from 1880–2020 demonstrates this clearly (Fig. 6.3). One does not need to be a scientist to
immediately recognize a significant, progressive, monotonic increase in temperature that
began after the advent of global industrialization in the early twentieth century. Those arguing that this phenomenon does not exist have no scientific bases to support their opinions.
Ignorance, in fact, is now being used by doubt-mongers (anti-vaxxers, global warming

---

[9] The United Nations Framework Convention on Climate Change defines climate change (global warming) to mean "a change
of climate which is attributed directly or indirectly to human activity that alters the composition of the global atmosphere and
which is in addition to natural climate variability observed over comparable time periods."

deniers) to discredit science [696]. They do so by socially constructing ignorance through the strategic and systematic expression of suspicion about scientific truths.[10]

Misinformation about astronomy,[11] cosmology, or evolution does a huge intellectual disservice to the public. Medical misinformation does actual physical damage to people. Medical quacks dispensing advice and hawking cures for illnesses have existed for as long as has humanity. Patients with life-threatening diseases often turn to these people in the hope of miracle cures, which don't exist. Nevertheless, taking advantage of the inability of people to distinguish science from nonscience, quacks propound preposterous ideas to further their own beliefs or make money through sales of useless or injurious products and books. Most of these snake oil salesmen have little or no expertise to warrant belief in their claims. More insidious is the fact that physicians may also participate in this unethical, immoral exercise. In the field of Alzheimer's disease, quacks and pseudo-science are rampant. Invasive, bizarre, and expensive surgical procedures [698] and treatments [699, 700], as well as a panoply of nutritional and life style changes, have been suggested to prevent or cure the disease. No scientific evidence shows that any of these approaches prevent Alzheimer's disease or can improve the disease state in those already affected [701].

The special importance of science has not been overlooked by businesses and their advertisers, who preface their product information with references to science such as "results of scientific studies," "by a leading laboratory," or "doctor X says" in an effort to bolster the value of their products in the eyes of the public. This process is not scientific because the claims promulgated rarely meet accepted standards of scientific practice and the people making them are paid well for their endorsements, not their scientific acumen or accuracy. Marketing and sales divisions of companies are not in the science business, they're in the profit business. The public is misled because it is rarely equipped to determine the validity of the claims. People may think if the product was produced and tested in a laboratory or was vouched for by a physician, then it must do what it says. This may be true, *if a person can demark a pseudo-scientific claim from a scientific one.*

Two of the best examples of nonscientific argument and its impact on society are public perceptions of cosmetics and genetically engineered foods. Each is an extreme example of the unethical, although unfortunately, legal misrepresentation of science to sell products. In the cosmetics industry, scientific words, innocuous, but impressive sounding ingredients, and vague and misleading neologisms are used to attract buyers because they are thought to carry great weight and legitimacy. In the food industry, companies prominently inform the public that their products are "non-GMO," i.e., are not made using genetically modified organisms. These companies seek to take advantage of the common public misperception

---

[10] "Agnotology" is the study of ignorance. In our case, we are particularly concerned with the strategy of doubt-mongers and businesses to sow "culturally induced ignorance or doubt, typically to sell a product or win favour, particularly through the publication of inaccurate or misleading scientific data."

[11] It is interesting, but not surprising, that the same type of pseudo-scientific tripe was expounded by astrologers in the eleventh century. However, what is *very surprising* is the fact that even then, some natural philosophers, such as the famous Islamic scholar al-Biruni, condemned this pseudoscience. Referring to astrologers, al-Biruni wrote "...[they] would stamp the sciences as atheistic, and would proclaim that they lead people astray to make ignoramuses, like him, hate the sciences. For this will help him conceal his ignorance, and to open the door for the complete destruction of both of sciences and scientists" (see Chapter 6, *al Biruni-On the Importance of the Sciences*, in Levi [697].

that the process of genetically engineering foods is the equivalent of that used to create Frankenstein, and with even more dire consequences.

If one wants to improve their appearance, the skin often is the first place they start. Who wouldn't want to buy a product containing nano or micro ingredients, peptides, bioactive substances, antioxidants, biomolecules, or bio-stimulating cream with microlift? Luckily, "[t]hanks to the *groundbreaking research of scientists* at Estée Lauder," you can buy "a daily moisturizer with some of the most *advanced protection technology*." Unfortunately, these claims are unscientific because they provide no evidence that the research is ground-breaking, let alone scientifically rigorous. If you need more collagen, "*advanced Shiseido technology* and ingredients including Super Bio Hyaluronic Acid N for intense moisture and Hydroxyproline" certainly can help you. The problem is that healthy people don't need more collagen and you can't get more by pouring *Super Bio Hyaluronic Acid N*[12] or the amino acid hydroxyproline onto your skin, other than maybe making it sticky. Dior may be the most technologically advanced skin care company as they are "targeting the 'heart' of the skin: *stem cells*." Problem – the skin has no stem cells, therefore the claim has no scientific foundation. The advertising of Estée Lauder, Shiseido, and Dior is not unique in the industry, but it is reflective of the vagueness, omissions, and falsehoods that characterize 86% of cosmetics ads invoking science as a buying incentive [702, 703].

Genetic engineering has benefited humankind in numerous areas, including animal husbandry, agronomy, molecular biology, microbiology, genetics, immunology, and medicine. Put simply, genetic engineering is the laboratory modification of DNA sequences. Modification of sequences within germ cells (sperm or eggs) enables creation of new organisms with desired characteristics. These genetically modified organisms (GMOs) are defined by the World Health Organization as, "Organisms (i.e. plants, animals, or microorganisms) in which the genetic material (DNA) has been altered in a way that does not occur naturally by mating and/or natural recombination" [704]. Actually, creating new organisms with desired characteristics is nothing new. Creating animals or plants with desired characteristics has been done for thousands of years.

GMOs have indeed extended the limit of what "feeble man can do." In doing so, many aspects of our lives have been revolutionized, including the ability to substantially increase food crop yields, create new foods that contain higher levels of natural vitamins (A, C, and E) and maintain their freshness longer, or use bacteria, plants, or animals to produce insulin, clotting factors (IX and VIII), human growth hormones for those affected by dwarfism, tissue plasminogen activator (tPA) to treat strokes, and calcitonin to treat osteoporosis [705, 706]. Potential problems do exist with *some* GMOs, including the potential immunogenicity (the induction of immune responses) of the nonhuman proteins produced, unintended gene transfer to native plants, bacteria, or animals, and direct toxicity [706, 707]. Scientists have documented the benefits and the dangers of GMOs on

---

[12] This is an unspecified compound that does not exist in nature and isn't even in the same chemical class as collagen. Hyaluronic acid is a glycosaminoglycan (sugar) whereas collagen is a protein.

a case-by-case basis, arriving at the conclusion that most GMOs are as safe, or safer,[13] than natural products. The National Academy of Sciences reported ". . . there were no differences in terms of a higher risk to human health between foods made from GE crops and those made from conventionally-bred crops" [709]. An especially broad meta-analysis (1783 original research papers, reviews, relevant opinions, and reports) of the literature addressing major issues in the debate on GE crops found no reports of "significant hazards" [710]. However, nonscientific consideration of the data extant on GMOs, including reliance on scientifically flawed studies, selective review of the scientific literature, or reliance on "folk biology" and intuition [711], has been used politically to impede progress in this area. A food industry analyst, Phil Lempert, may have stated most succinctly why certain issues become so muddled by various social and political factions when he said, regarding the National Academy of Sciences report, "It's an emotional issue, it's not a science issue" [712].

If one applies these criteria to the issues shown in Table 6.1, it becomes clear why the scientific and nonscientific aspects of each issue were categorized as they were. For each of the first five issues, beliefs, opinions, specious arguments, and illogic, among others, categorize the nonscientific from the scientific. Anti-vaxxers are particularly guilty of ignoring overwhelming statistical evidence showing no connection between childhood vaccination and autism in favor of believing anecdotal accounts of contemporaneous immunization and autism diagnosis. The issue of genetically modified organisms is a particularly good example of how persuasive argumentation can trump critical evaluation, especially when the arguments come for an ostensibly reliable and reputable organization such as Greenpeace.

Science not only informs public policy issues, but has an important role in the legal system. Legal decisions often depend on scientific information. The legal process, unlike the scientific process, does not make an unbiased search for truth preeminent. Its paradigmatic principle is that vigorous argumentation among defense and plaintiff attorneys will reveal the truth. Unfortunately, this process actually results in decisions based on how well an attorney presents their case, not on its merits. Expert scientific testimony often is used to buttress a position. However, attorneys choose scientists as expert witnesses not because of their scientific acumen or impartiality *per se* but because they can help them win their cases. This process often pits scientists expressing a view consistent with that of the global scientific community with those who represent a view that is not supportable scientifically or that relies on what is called "bad science," i.e., results obtained in studies not using the scientific method or its logical underpinnings correctly.

Science may play a definitive role in the court room, especially in the United States, where judges have drawn on philosophy of science in deciding when to confer special

---

[13] An interesting study by Bruce Ames [708] showed almost three decades ago that only 0.01% of the total weight of pesticides consumed by Americans in their food came from unnatural pesticides (those chemically synthesized and applied to crops). 99.99% of the pesticides were naturally produced by the plants themselves! The advent of GMOs likely has decreased this number, but the Ames study emphasizes a number of points: (1) superficial understanding of words like "pesticide" may lead the public to think – unscientifically – that these useful chemicals are toxic to humans by definition (they are not) and (2) the proteins produced by the new genes inserted into GMOs are remarkably similar, or identical, to those normally produced.

status to scientific expert testimony. A precedent-setting case was Daubert vs Merrell Dow Pharmaceuticals (92102, 509 U.S. 579, 1993). In this case, the Supreme Court argued in its 1993 ruling that trial judges must ensure that expert testimony is reliable, and that in doing this the court must look at the expert's methodology to determine whether the proffered evidence is actually scientific knowledge.[14] As reported by Abboud [715],

The court defined the criteria by which scientific knowledge – which for them included a least theories based on evidence, expert testimony from scientists, and scientific techniques – could be introduced and used in court cases as evidence. The Daubert Standard states that the judge of a case is responsible for determining what claims are admissible as scientific knowledge and as evidence in the case. The admissibility should be determined by the falsifiability of the claims, by whether or not they had passed peer reviewed, by the general scientific acceptance of the claims, and for techniques, by their error rates of the techniques. Daubert v. Merrell Dow Pharmaceuticals, Inc. set a landmark precedent in the US judicial system and influenced most subsequent legal cases that appealed to science to establish facts in trials.

We have cited, *en passant*, other areas in which science may have a particularly high impact on public policy, specifically, and on the public, in general. However, seeking truth about the natural world and deciding how to live in it are two independent things. Bauer has argued strongly for this ideal.[15]

[It is fallacious to think] that science can and should decide questions of public policy. ... Any question of 'ought' or 'should' is by definition trans-scientific[16] and has no answer within science, no matter how many technicalities happen to be entangled in it.

Until the mid- to late twentieth century, scientists had largely striven to divorce themselves from nonscientific issues.[17] This has become increasingly difficult in the light of public challenges to science's value and validity. Scientists have had to get involved in public debate to ensure that scientific knowledge was being proffered accurately and used appropriately by those whose responsibility was to make policy decisions. Panels, committees, organizations, etc. usually make such decisions, not individuals, and this means that some type if consensus ("group think") must be reached, which, in turn, means that social dynamics and not pure facts and logic may guide decisions. How is the public to be protected from decisions made by panels such as that shown in Fig. 6.4? Are there principles that could be employed by such panels to mitigate the effects of group think? The answer is "yes," and it is the "precautionary principle."

The precautionary principle, in essence, is "err on the side of caution," or as demanded in the Hippocratic Oath, "do no harm." This is particularly important when science cannot provide iron-clad truth, which is the case in the overwhelming majority of cases. The

---

[14] Haack [713] and Foster & Hubner [714] have argued by equating the question of whether a piece of testimony is reliable with the question whether it is scientific (as indicated by a special methodology), the court was producing an inconsistent mixture of the philosophies of Popper and Hempel, and this led to considerable confusion in subsequent case rulings that drew on the Daubert case.

[15] See Bauer [36], p. 123.

[16] Questions put to science that have no scientific answer are 'trans-scientific,' according to Weinberg [91].

[17] Historically, this has not been strictly true, as exemplified by the conflict about heliocentricity between astronomers and the Catholic Church.

"Everyone in favor raise your hand!"

Figure 6.4 "The danger of group think." Although humorous, this scenario is common, and dangerous. (Credit: Niven and Vasshus [770].)

principle is thought to have first been propounded in 1970 in the context of protecting the environment from damage, especially "human-made." It is useful as a reminder that science has the potential to produce knowledge that could be used improperly, unethically, or immorally, and thus scientists and the public should consider this potential in any deliberations. In practice, scientific evidence can never show that damage will *not* occur, thus stringent application of this principle is illogical and potentially damaging on its own. Cass Sunstein (Karl N. Llewelyn Distinguished Service Professor of Jurisprudence, Law School and Department of Political Science, University of Chicago), echoed these concerns [716] *and* acknowledged the value of the principle.

In its strongest and most distinctive forms, the principle imposes a burden of proof on those who create potential risks, and it requires regulation of activities even if it cannot be shown that those activities are likely to produce significant harms. Taken in this strong form, the precautionary principle should be rejected, not because it leads in bad directions, but because it leads in no direction at all. The principle is literally paralyzing – forbidding inaction, stringent regulation, and everything in between. ... The salutary moral and political goals of the precautionary principle should be promoted through other, more effective methods.

6.2 Science and the Scientist

6.2.1 Individuals and Teams

Historically, scientific experiments were performed by the individual. For the ancients, their "laboratory" was the world itself, and the instruments were their five senses. As science developed and instruments and techniques were invented, experiments increasingly were done in laboratories, be they just designated spaces or small rooms. If these sites, and the natural philosophers and scientists working there, were in distant cities, they were, for all intents and purposes, isolated from one another because of the difficulty of travel. Walking, crude wheeled vehicles, or riding an animal limited the ability to share ideas and results by direct interactions. Travel by boat was relatively rapid, but could not be used for many locations inland. Assuming travel was even possible; cost, physical comfort and safety, and time usually precluded it. Assuming one *could* travel, few except the wealthy could afford to do so and few had the fortitude or time. For example, during the Roman Empire, one-way journeys from Rome to Athens or London have been estimated to have taken one or two months, respectively [717]. Dissemination of information thus remained slow, as sharing "lab notebooks," be they papyrus scrolls, parchment, bamboo, or other materials (e.g., tablets constructed with wax, stone, clay, wood, or cloth), was unfeasible.

Considering these impediments to collaborative work, it is not surprising that science was done primarily by individuals. If "collaborative work" is identified with multi-author scientific papers, then it was not until the mid-seventeenth century, with the advent of scientific journals in France (*Journal des sçavans*),[18] and then England (*Philosophical Transactions*),[19] that it was evidenced. Travel and communication had improved dramatically compared with early in the first millennium, but the desire to collaborate remained low, no doubt because of the pervasive fear of plagiarism and the attendant loss of recognition of one's original work.[20]

Modern science is rarely a solitary exercise, and learning science is rarely done in solitude. The complexity of the natural world largely precludes solitary scientific practice. No one person knows everything and no one person can do everything. Answering many scientific questions requires a multi-disciplinary approach. This requires collaboration within laboratories, among laboratories, locally, nationally, and internationally. Collaborations are not the human equivalent of a multi-core computer cpu. The multi-core cpu distributes processor time among its cores based on rigid rules of time allocation and the intrinsic structure of the chip. These organizational and functional features are inviolate. Human multi-processing, i.e., collaboration, does not involve rigid rules but does involve both tangible physical interactions and intangible sequelae (e.g., joy, excitement, feelings of accomplishment, and camaraderie). Cross-stimulation between and among scientists is a particularly effective way in which novel ideas emerge.

One *can* be a scientist, and do science, in isolation. There are no legal, ethical, philosophical, or even practical barriers (in some cases) to doing so. Science practiced in isolation may satisfy and enrich the curiosity and intellect of the individual, but it is not science in

---

[18] Note that the first issue of this journal appeared two months before that of the *Philosophical Transactions*.

[19] The full title of the journal was *Philosophical Transactions: Giving some Accompt [sic] of the Present Undertakings, Studies and Labours of the Ingenious in Many Considerable Parts of the World.*

[20] Zuckerman and Merton [718], in a Citation Classic (a bibliometric concept introduced by Eugene Garfield in 1977 (see [719]) to designate highly cited papers in a scientific discipline), suggest that "... *intent upon safeguarding their intellectual property, many men of science still set a premium upon secrecy (as is evident in their correspondence with close associates). They maintained an attitude and continued a practice of (at least, temporary) secrecy.*" At the turn of the nineteenth century, the majority of collaborative research reports appeared in France. Only then did collaborative research became more common in England (and Germany). It has been suggested that it was the emergence, at that time, of science as a *profession* that fostered collaboration in the sense it is understood today [720].

the way in which it now is done, namely the acquisition and *dissemination* of knowledge about the natural world. As early as ≈3,000 years ago in Egypt, methods to care for people with injuries or tumors were committed to paper (papyrus).[21] This enabled Egyptian physicians and others to better understand the natural world, at least in a medical sense. If such knowledge had been generally available, the work would have had substantially more impact on society. It also is true that the work itself might have been more informative and useful if other physicians and learned men[22] had contributed their own individual knowledge and experience to it, i.e., if the work was performed within a community of physicians. Five millennia later, the importance of community was emphasized tacitly when Anton van Leeuwenhoek, the seventeenth century scientist known as "the father of microscopy," said *"whenever I found out anything remarkable, I have thought it my duty to put down my discovery to paper, so that all ingenious people might be informed thereof."* A century later, Joseph Priestley, credited with the discovery of oxygen, echoed van Leeuwenhoek in a pronouncement that *"... whenever he discovered a new fact in sience [sic], he instantly proclaimed it to the world in order that other minds might be employed upon it besides his own."* Priestley and van Leeuwenhoek had argued for wide dissemination of knowledge within the society of scientists because they were convinced of the practical and theoretical benefits of doing so.[23]

Heesen, in writing about the incentive to share science [721], opines the social value of scientific work is highest when shared. He considers dissemination of scientific results so central to the scientific enterprise that, if not shared, they would not even *be* science.

... work that is not widely shared is not really scientific work. Insofar as science is essentially a social enterprise, representing the cumulative stock of human knowledge, work that other scientists do not know about and cannot build on is not science. The sharing of scientific work is thus a necessary condition not merely for the success of science but in an important sense for its very existence.

It has been argued, in fact, all scientific knowledge has social origins and that the interaction of scientific practices with social elements is necessary for the logical and cognitive structures of enquiry. What are the roles of individual scientists, groups, departments, institutions, and countries in the social network from which scientific knowledge emerges? Is science simply the sum of strictly scientific contributions by individuals, or do the various social groups contribute to a synergy, dysergy, both, or neither? Do the social groups involved affect how scientific questions are conceived, chosen, studied, and answered. Do these groups, or subsets thereof, determine whether knowledge is true, likely true, undetermined, likely false, or false? Are these determinations made through some sort of normative societal behavior? These questions fall under the rubric *sociology of science* (SoS).[24]

---

[21] See Section 4.1.

[22] The word "men" is accurate and appropriate here because it is likely there were few or no learned women in Egypt at the time.

[23] In their times, however, the epistemology of the scientific knowledge thus disseminated was not a major consideration.

[24] The field "sociology of science" is huge. It is impossible to discuss all its aspects here, or anywhere. Instead, the discussion will be limited to certain communities of scientists and how they function in their continuing efforts to understand the natural world.

SoS is a relatively new field. It was founded by Robert K. Merton, who, in 1949,[25] published the seminal work in this area, "Social Theory and Social Structure." In 1994, Merton was awarded the National Medal of Science by President Bill Clinton "For founding the sociology of science and for his pioneering contributions to the study of social life, especially the self-fulfilling prophecy and the unintended consequences of social action."[26] Merton also has been credited with coining, among others, the terms "role model," "self-fulfilling prophecy," and "unintended consequences," and with creation of the "focus group" method of investigating public opinion.

6.2.2 The Coin of the Realm: Recognition

> ...cognitive wealth in science is the changing stock of knowledge,
> while the socially based psychic income of scientists takes the form
> of pellets of peer recognition that aggregate into reputational wealth.[27]

We know in many areas of human endeavor, good works are rewarded in a rather practical manner, namely with money. Of course, there exist a vast array of endeavors of a more spiritual and less remunerative nature, but by in large, people work and get paid for it. What is the payoff for the successful scientist? Although senior scientists working at top tier universities, or private companies, can garner substantial salaries, I know of no one who originally became a scientist to make money. The rewards of science are more intellectual and psychological in nature, and include, among others: eponymy in discovery (Newton's laws, Le Châtelier's principle, Lorentz transformation, Schrödinger equation); eponymous methods (Euclidean geometry, Boolean algebra, Lord's procedure for hemorrhoids, Heimlich maneuver); recognition by scientists and the public (awards, prizes, lectureships, Professorial Chairs, election to national and international academies), and "being first." The Nobel prize–winning economist, Paul Samuelson, opined:

Not for us is the limelight and the applause [of the world outside ourselves]. ... In the long run, the economic scholar works for the only coin worth having – our own applause.[28]

What was true of economists in 1962 is also true for scientists today, who expect their work to be acknowledged within their scientific communities. I cannot emphasize enough the importance, appropriateness, and necessity of acknowledgment. In the quotation below, Merton emphasizes both the practical and ethical necessity of acknowledgment in a particularly profound way.[29] I include this lengthy quotation because I believe all students, budding and practicing scientists, and clinicians should understand it and act in concert with it.

---

[25] Revised and expanded editions of the book were published in 1957 and 1968.

[26] The National Medal of Science is the highest honor the United States bestows upon its scientist. Merton's peers in Sweden and England honored Merton with membership in the Royal Swedish Academy of Sciences (the first foreign sociologist to be so honored), and the British Academy (England's most prestigious national academy for sociologists), respectively. It has been said that if there existed a Nobel Prize in sociology, it would have been awarded to Merton. Ironically, Robert Merton did win the Nobel Prize in Economics in 1997, but it was not Robert K. Merton ("Mr. Sociology"), but rather his son, Robert C. Merton.

[27] Merton [341], p. 331.

[28] Samuelson [722], p. 18.

[29] See Merton [341], p. 335.

[T]he failure to cite the original text that one has quoted at length or drawn upon becomes socially defined as theft, as intellectual larceny or, as it is better known since at least the seventeenth century, as plagiary. ... the bibliographic note, the reference to a source, is not merely a grace note, affixed by way of erudite ornamentation. ... The reference serves both instrumental and symbolic functions in the transmission and enlargement of knowledge. Instrumentally, it tells us of work we may not have known before, some of which may hold further interest for us; symbolically, it registers in the enduring archives the intellectual property of the acknowledged source by providing a pellet of peer recognition of the knowledge claim, accepted or expressly rejected, that was made in that source.

We also need to be aware of what Merton has called " 'obliteration by incorporation,' ... the obliteration of the sources of ideas, methods, or findings by their being anonymously incorporated in current canonical knowledge" [341, 723, 724].

Must we cite everything relevant to the work presented in our manuscripts, seminars, poster presentations, or other forms of scientific communication? What is relevant? Are there things that we would consider self-evident, as stated in the *Declaration of Independence*, and thus not necessary to acknowledge or cite? If so, what are these things? Let us argue extremes and then increasingly constrain the argument until we reach a conclusion. At one extreme, we would need to cite a reference for every proposition made, either directly or implicitly, in one of our manuscripts. For example, if we used the term "H_2O," we would need to cite those publications that specified what this chemical is? How would we do this? Would we cite a textbook? Not likely, because general chemistry textbooks themselves would not have done so. If not, what should we cite? Publications that originally described water as having the chemical composition OH? Those that said water was H_nO_n? The first publication to suggest water was H_2O did so based on false assumptions. And how would we incorporate the fact that water is not *fully* H_2O, but rather exists in equilibrium with H^+, H_3O^+, and OH^-? Clearly, we cannot, and should not, cite everything.

At the opposite extreme, we would acknowledge no one. This is unacceptable, for obvious reasons, and from the sociological perspective, this just is not *"the way the game is played."* However, scientists *do* routinely omit – through anonymous incorporation – acknowledgments when the sources of ideas, methods, or findings are considered general knowledge or taken for granted [341, 723, 724]. For example, obliteration by incorporation *could* be used to justify not acknowledging the discoverers of the structure of water because water *is* a chemical principle that is part of canonical chemical knowledge.

Where does this leave us? Unless we are writing science fiction, we acknowledge something. Conversely, it is impractical, if not impossible, to cite everything – and why should we? The normative sociology of science dictates that facts understood by canon or consensus rarely must be accompanied by citations of their original discoverers. Nevertheless, we do need to make judgments about what facts, opinions, hypotheses, experiments, or other intellectual property or information should or should not be acknowledged. There must be an interval within our "citation continuum" that comprises those works of science that are subject to citation, at least within the disseminated wisdom of science (manuscripts, seminars, editorials, posters, reviews, chapters, books, commentaries, the media) for which these citations are relevant.

What rules govern the selection of citations within this interval? Formally, there are none. However, empirical standards do exist. I discuss here my own perception of and experience with these standards. When introducing one's work in a manuscript, it is customary to discuss briefly the scientific discoveries on which the work is based and the history leading one to do the work. Appropriate acknowledgment is required here. If one uses a technique not developed by oneself, appropriate acknowledgment of the source of the methodological details of the technique is required. Even if one does use their own methods, these either must be specified within the *Methods* section of the manuscript or included as citations of the paper in which the method was published. It is particularly important that a *Discussion* section compare and contrast the results reported with what is known already. This extant knowledge requires appropriate citation.

Merton has opined citations are not merely grace notes. However, I argue that sometimes the bibliographic note is, and should be, *precisely* a grace note. In fact, Merton himself states this explicitly when he writes *"Exploring the literature of a field of science becomes not only an instrumental practice, designed to learn from the past, but a commemorative practice, designed to pay homage to those who have prepared the way for one's work."* Those who have prepared the way, in particular, the great scientists of the past remain in danger of not being cited properly because of obliteration by incorporation. Citation of their work eliminates this last possibility, and rightly so.

In addition to providing the scientist of today with an opportunity to honor historically eminent scientists and their contributions to science, the action of citing prior great works involves (and certainly should involve) reading the original literature and acquainting oneself with what really was actually written, not with what was paraphrased or summarized in a textbook. This is a tremendously valuable learning experience for the scientist, and especially for students. It also is fun and stimulating. It links the scientist, through historical investigation, to the originators of the ideas that guide research today.

Two examples illustrate this point. The first is found in Bitan *et al.* [725], a publication describing the behavior of Aβ, a key peptide linked to the pathogenesis of Alzheimer's disease (AD). Aβ exists in a complex equilibrium among its monomer form, oligomer forms (varying in size from $2 - n$, where n can be quite large [>50]), and fibrils (polymeric forms of Aβ). Removing specific forms of Aβ from a mixture to study their characteristics perturbs the equilibrium, which then is reestablished according to Le Châtelier's Principle. It *is* appropriate to acknowledge Le Châtelier's Principle in discussions of the Aβ system and I did so by citing his original publication of 1884 in the French journal *Comptes Rendus Académie des Sciences* [726].

We commented in our manuscript, "if oligomers were in rapid equilibrium with monomers or dimers, once the monomers and dimers in the Aβ preparation had passed through the [filter] membrane, the remaining oligomers would quickly dissociate (according to LeChâtelier's principle ...), leading to passage of the majority of the peptide through the membrane." We could have chosen to say simply that once monomers and dimers were removed, equilibrium would be reestablished by oligomer dissociation, but what evidence could we provide to support this contention? Should we have been content in a supposition that every reader would know this would happen? *Does* everyone know this would happen? Once might think that readers of the *Journal of Biological Chemistry*, in which our paper was published, would have remembered those portions of their undergraduate physical chemistry courses dealing with chemical equilibria. However, many would not, and they simply might accept our statement as fact – a very dangerous behavior for a good scientist. Instead, we made no such assumptions for the readers, *nor for ourselves*, but rather supported our statement with the *original* literature on the subject.

Serendipitously, the University of California library system owns the journal, *Comptes Rendus Académie des Sciences*, so I was able to hold volume 99 of the journal in my hands and read Le Châtelier's article in the original French. I did not feel that I was "paying homage," but rather that I was connected – physically, intellectually, and existentially – to Le Châtelier and to his time. It was a thrilling experience that cannot be described well verbally but may only be expressed in emotional terms that are decidedly nonverbal. As an ancient Buddhist proverb says, "I can show you the moon by pointing to it with my finger, but you cannot experience the moon by looking at my finger, you must see the moon for yourself."

In the second example, from Lomakin *et al.* [727], we used a technique termed quasielastic light scattering spectroscopy (QLS) to study the motion of Aβ peptides in salt solutions during aggregation. It is possible to determine how large aggregates are by measuring their rate of diffusion D and converting D into a quantity, the hydrodynamic radius R_h, which represents the size of an ideal sphere with an identical D. The Stokes–Einstein equation [728] is used for this purpose. It is common practice, however, not to cite this publication but rather simply to state that the Stokes–Einstein equation was used. Why should a reader assume that this use of the equation is proper? How would one know unless one could read the original publication? It is for this reason that the text reporting our use of QLS included not only the citation of one of Einstein's four papers published in the *anno mirabilis* of 1905, *Über die von der molekularkinetischen Theorie der Wärme geforderte Bewegung von in ruhenden Flüssigkeiten suspendierten Teilchen. Annalen Der Physik* [728], but the equation itself.[30]

6.2.3 The Priority Rule: "Be First."

If it ain't published, it ain't done.

When my students would rather do experiments than write papers, I utter the words above, which, for all intents and purposes, are true. Why? Because the results of scientific studies must be available to the scientific community if they are to contribute anything at all to the quest for knowledge. As discussed with respect to physicians in ancient Egypt, and to

---

[30] For a spherical particle, the relation between its radius R_h and its diffusion coefficient D is given by the Stokes–Einstein equation:

$$R_h = \frac{k_B T}{6\pi \eta D}. \tag{6.1}$$

Here k_B is the Boltzmann constant, T is the absolute temperature, and η is solution viscosity. For nonspherical particles, the equation defines the apparent hydrodynamic radius of the particle.

those who work in isolation, the knowledge they produce may be useful to them personally, but it cannot be considered knowledge in the sense of something the generic *we*, i.e., our society, know because we aren't even aware of its existence. To advance one's career, it is critical to publish novel work, which means being first. Two or more investigators may do superb work studying a particular question, but the one who publishes their discovery first receives the credit in the vast majority of cases. It is this publication that likely will be cited by others in their own manuscripts, review articles, book chapters and books, or commentaries. The first publication of a particularly significant advance often is covered by the media (newspapers, magazines, radio, television, cable and streaming, and social). It is not uncommon for the scientist responsible for the work (usually the principal investigator/laboratory head) to be interviewed by the media. Dissemination of one's work has major benefits. Papers reporting similar or identical results that are published after the first report often are labeled "me too" papers, and their authors receive little recognition.[31]

In many different social groups, including scientific, public, political, economic, and commercial, the stature of the person/laboratory that was first is increased. In science, this is important because it can result in invitations to: (1) speak at scientific meetings, conferences, etc., and have the work presented included in later compendia, proceedings, or journal special issues; (2) write review articles, opinion pieces or perspectives, book chapters, or entire books; (3) lead or participate in college or graduate level courses on the investigator's work/subject; (4) speak to colleagues nationally and internationally at various universities, institutes, or research centers; (5) sit on national and international scientific review boards; or (6) present to, collaborate with, or consult for businesses involved in basic science and clinical/experimental medicine (e.g., pharmaceutical ("Big Pharma") or biotechnology companies). It is unfortunate but true that scientists must sell themselves to be successful, and advertising (recognition) helps do that. As antithetical to the scientific temperament as it may be, one must "get their name out there" if they are to maximize their chances for career advancement and success.

Public stature increases the opportunities for further exposure through speaking engagements for the lay public or service on advisory boards of various organizations throughout the world. In addition, much in the same way that public opinion is thought to bias jurors, awareness (lay and scientific) of a scientist's stature may increase their likelihood of winning funding, which is a highly competitive process. This is because reviewers may favor applications from well-known and successful laboratories instead of evaluating the applications solely through objective, un-biased scientific principles. Implicit (or explicit) bias also may work to one's benefit, or detriment, more indirectly, e.g., by affecting public policy decisions and allocation of funds to particular types of research. For example, the stature of Dr. Anthony Fauci, Director, Institute of Allergy and Infectious Diseases (NIAID-NIH),

---

[31] This issue is especially important when submitting grant applications. If a study section or review committee thinks an application proposes studies that appear duplicative or highly similar to work extant, the application will not be considered novel and thus will fail to be funded.

as "the nation's top infectious disease expert,"[32] no doubt has increased confidence in his knowledge and evaluations of the COVID-19 pandemic, the direct effect of which, for most, is to believe what he says and expect the government to act on his advice.[33]

Do all who publish first enjoy these benefits? No. The rewards of being first depend on a variety of factors, the most important of which is the significance of the published work. Quantum leaps in knowledge that have, or are likely to have, profound impacts on science and society are particularly rewarding for the scientist, both scientifically (problem solved!) and socially (recognition). However, incremental advances also are appreciated, but to a lesser degree. This is the way science traditionally was thought to progress,[34] i.e., brick by brick, and the work of each bricklayer was recognized and appreciated. Scientists contributing incrementally to knowledge are the norm, not the exception, in science. These contributions are what makes most scientists successful.

Whether a finding is significant or not is easy to tell if the work has obvious strengths, applications, implications, or is a quantum advance. It is more difficult if the finding is not obviously ridiculous. Thus, some theories or discoveries are embraced and some are eschewed, with the bases for such acceptance or rejection often "conventional wisdom" and personal biases. Work that appears insignificant at the time could turn out to be monumentally important in the future – a kind of "reverse pessimistic induction," i.e., optimistic. Rarely, but not infrequently, the pessimistic meta-induction *is* validated when ostensibly important discoveries are found to be wrong (falsified) or useless.[35] The work of Ptolemy and Copernicus is an example of significant→insignificant (Ptolemy) and its inverse, insignificant → significant (Copernicus), although in this case it would be more accurate to talk about accuracy, acceptance, and their opposites. At the time of Ptolemy, and for 15 centuries more, his geocentric universe was the basis for countless successful and informative studies of the Earth, planets, and stars, not to mention the practical value for navigation at sea. It worked, and it worked well. However, eventually optimism about the theory was vitiated by Galileo's observations and Kepler's heliocentric model with ellipses. The new theory was intuitive, simple, and accurate.

One of the best examples of unappreciated work that later became monumentally important was that of Barbara McClintock, a plant cytogeneticist who, eventually, was recognized for revolutionizing our understanding of chromosome structure and its genes and for discovering "mobile genetic elements," an achievement for which she was awarded the 1983 Nobel Prize for Physiology or Medicine. I say eventually because, while she was doing the work that won her the prize, few thought it to be of value and many were hostile to, and dismissive of, McClintock. Only after technical advancements in molecular biology allowed others to corroborate McClintock's work did the naysayers realize McClintock had

---

[32] This moniker, from wherever it came, is inaccurate. There are many, many people with equal or superior knowledge to that of Dr. Fauci. Nevertheless, politically, if one decides, for whatever reasons, to accord this moniker to a chief presidential advisor on the COVID-19 pandemic, then the moniker stands.

[33] Whether this happened is debatable. What isn't debatable is the fact the government followed Dr. Fauci's advice and increased funding earmarked for COVID-19 research.

[34] Not so much anymore. See Section 6.3.

[35] Over time, one would predict four eventual outcomes for any finding: significant → significant, significant → insignificant, insignificant → significant, and insignificant→insignificant.

Table 6.2 *Battles over who should be credited with a discovery.*

Combatants	The discovery
Newton vs. Leibniz	Calculus
Priestley vs. Lavoisier vs. Scheele	Oxygen
Darwin vs. Wallace	Theory of evolution
Watson and Crick vs. Franklin[a]	DNA structure
Luc Montagnier vs. Robert Gallo	AIDS is caused by HIV

[a]The battle was not actually between Franklin and Watson & Crick, but rather between those favoring Watson & Crick and those who felt Franklin should have been recognized as the one who's data showed the double helical structure and should have received the Nobel prize along with Watson & Crick [732]. Franklin's untimely death in 1958 precluded her receiving the prize, which only is awarded to those living.

always been right and began rewarding her with the myriad prizes and honors she deserved. McKlintock's problems did not derive from her science *per se*, but rather from the social dynamic of science at the time. Because McKlintock's hypothesis clashed with the conventional wisdom, the society of scientists chose not to entertain the idea that the hypothesis was the best explanation for the data. McClintock was, in fact, first to propose the idea of "jumping genes," but she was not be recognized as such because of *a priori* rejection of the idea.

The fight to be recognized as first can become a protracted battle, one in which the history of a discovery may be manipulated by scientists and nonscientists to suit their own biases or because of their own ignorance. A number of well-known battles[36] are listed in Table 6.2. Among the causes of these battles were: (1) ignorance of the first publication of the discovery, either by competing scientists or the scientific community at large, so that the one publishing second receives credit; (2) independent discovery of the same thing, followed by one scientist being better at "selling" their discovery to scientists and the public; (3) differences in the number of citations of one discoverer's work compared to that of another, so that the more cited scientist becomes even more cited and the less cite scientist even less cited (until that person is largely forgotten); (4) intrinsic bias of the scientific community toward the most well known, but not necessarily the first, discoverer; and (5) legal determination of who has priority in discovery.[37]

[36] I do not include Ptolemy and Copernicus in this category because there was no contemporaneous battle between these two scientists (they lived in two different millennia). The contemporaneous battle was between scientific and religious explanations of the natural world.

[37] The battle between Montagnier and Gallo over discovery of the AIDS virus is such an example. This fight not only had implications for who might receive a Nobel prize, but it had huge financial implications because the discovery would be intellectual property that was patentable, which meant the patent holder could sell AIDS vaccines and tests worldwide. The battle ended in 1987, but only when then President Ronald Reagan and French Prime Minister Jacques Chirac announced the United States and France would acknowledge Montagnier and Gallo as co-discoverers of the AIDS virus and share patent rights for tests to screen blood for the presence of AIDS antibodies. However, notwithstanding this agreement, only Montagnier shared in the 2008 Nobel Prize for Physiology or Medicine for "discovery of human immunodeficiency virus." Gallo was snubbed, which created yet more controversy. We all will have to wait until 2058, when records of the machinations of the Nobel committee are released, to know why.

Today, with the hope of precluding fights over priority and of according credit appropriately, scientists and journal editors often work together to simultaneously publish two or more reports of the same discovery. This happens frequently because simultaneous discovery is quite common in science. In publications appearing after these initial reports, scientists generally will cite all the contemporaneous reports so as to give credit equally. However, for the reasons discussed above, even simultaneous publication cannot preclude favoritism towards one discoverer over another. In addition, although identical in principle, *simultaneous* discovery does not mean *identical* discovery. The methods and details of procedure by which the discovery was made rarely are the same. Data from one discoverer may be more convincing, thorough, rigorous, or greater in scope or amount than data from another. Scientists thus can make a determination of which report(s) are most convincing to them and only cite *those* reports in the future. Note that these determinations may differ among scientists because of their use of distinct evaluative metrics. These and other vagaries have substantial impact on determination of priority and citation choice.

Traditionally, priority in publishing was established by the date of acceptance of a manuscript, not when a paper or electronic version becomes available to the community. However, the need for scientists to publish in peer-reviewed journals is complicated by the often long (sometimes years) process of manuscript submission, review, revision, and finally, acceptance. This substantially retards scientific progress. It was because of this that, in 1991, a physicist at Los Alamos National Laboratory, Paul Ginsparg, created a server, "arXiv," on which new work could be shared *prior to publication*. Astronomers, mathematicians, and computer scientists were heavy users of arXiv. In fact, as of January 2022, two million papers had been shared on arXiv. In the past, work uploaded to arXiv was subsequently published in a peer-reviewed journal, but an increasing number of scientists now disseminate their work only through an arXiv so the "date of publication" becomes the "data of addition to an arXiv." Recently, other scientific communities have embraced the arXiv idea. Since 2016, 50 new arXivs have been created, ranging from psychology (PsyArXiv) and sociology (SocArXiv) to earth science (EarthArXiv), ecology (EcoEvoRxiv), chemistry (ChemRXiv), and biology (BioRXiv). One can predict confidently this trend will continue as scientific publishing continues a move from print-to-electronic and anonymous-to-open peer review.

Regardless of the mode of scientific publishing, will everyone who doesn't publish first be marginalized and forgotten? Aren't those who publish second, third, or later entitled to recognition for their efforts? In my experience, they *are* recognized and appreciated, but not to the degree of those who came first. One may describe this as "the rich getting richer and the poor getting poorer," i.e., recognition builds recognition and lack of recognition erodes recognition. Merton likened this behavior to that discussed by Jesus, as quoted by Matthew, and thus termed it "The Matthew Effect" [723].[38] Merton's inspiration for, and

---

[38] The relevant biblical passages are Matthew 13:12 http://biblehub.com/matthew/13-12.htm and Matthew 25:29 http://biblehub.com/matthew/25-29.htm. The term "Matthew Effect," as used by Merton, is a maladaptation of Jesus's words. The scientific effect to which Merton refers certainly was *not* exemplified in what Jesus suggested, which specifically had to do with the glory of God in the individual, viz. those who believed in that glory would be graced by God whereas those that did not would, in essence, be punished by God. Here, for consistency with the sociological and scientific literature, and for simplicity, I use this term as intended by Zuckerman and Merton.

study of, the Matthew Effect arose largely because of the work of Harriett Zuckerman, who, as a graduate student, interviewed 41 Nobel laureates in science to learn about "patterns of productivity, collaboration, and authorship" [730, 731].[39]

For unto everyone that hath shall be given, and he shall have abundance, but from him that hath not shall be taken away even that which he hath.

—The Bible, King James Version, Matthew 13:12

There can be many types of Matthew effects. As discussed above, they may involve popularity contests and citation frequencies, but they may also be operative with respect to recruitment of graduate students and fellows, giving talks, writing review articles, editing books, involvement in professional organizations, or funding. For example, in discussing papers published by different laboratories, it has been said that ". . . eminent scientists get disproportionately greater credit for their contributions to science while relatively unknown ones tend to get disproportionately little for their occasionally comparable contributions." Three Nobel laureates agreed. A physicist opined *"The world is peculiar in this matter of how it gives credit. It tends to give the credit to already famous people."*[40] A second said, *"In papers jointly published by scientists of markedly different rank and reputation, the man who's best known gets more credit, an inordinate amount of credit."* Unfortunate, but often true, is the comment, *"If my name was on a paper, people would remember it and not remember who else was involved."*[41]

With respect to the relative levels of recognition to be expected from work in my own laboratory, I warn students that it is likely *I* will get most (and sometimes all) the credit for their published work. I do this so students understand this unfair social convention *pre facto*, at least until it is eliminated. I also point out, in multi-author publications, it is best to be first author. This is the "priority rule for publications" and it has the same types of implications as does the priority rule for discovery. This is very important, and in very practical ways, for their career development, as their worth, in part, will be judged in the future by the answer to the question, "How many 'first-author' papers has this student published?"

What holds for the individual also holds for institutions. Institutions of high repute, e.g., Harvard, Cambridge, Oxford, and the MRC and NIH do themselves experience Matthew effects. They receive disproportionate shares of research support in the form of grants from government agencies, private companies, philanthropic organizations, and individual donors. They are able to recruit and retain the highest caliber researchers. They have their pick of the most promising students in the world. They are inordinately respected by the public and deferred to by the press. How much of this is deserved? Are there *any* other institutions that may be considered elite in terms of their public perception and actual quality, as judged by unbiased metrics? Is it possible for a scientist who trained at "lesser" institutions to be successful and respected? Certainly, these elites do have reason to be

---

[39] Merton acknowledged this when he wrote *"It is now [1973] belatedly evident to me that I drew upon the interview and other materials of the Zuckerman study to such an extent that, clearly, the paper should have appeared under joint authorship"* [341]. Interestingly, or ironically, Merton married Zuckerman in 1993.

[40] Zuckerman [730], as quoted by Merton [341], p. 318.

[41] Quotation from Zuckerman [732], as quoted by Merton [341], p. 319.

considered so, but for the individual scientist, it is their *own* work upon which they will be evaluated for industry jobs, academic positions, advancement, honors and awards, etc.

Science may be a special case in which "quality will out," i.e., the evaluation of the individual scientist or his or her laboratory occurs outside the space of institutional consideration or merit. We know this is true for a variety of reasons. Merit is not judged, and cannot be perpetuated, by institutional affiliation, but rather by the value of the research product, as judged initially by the "field," and then by the ability of the new knowledge revealed to stand the test of time. Researchers who are successful and respected in their fields can become commodities that are actively traded among universities and companies. Their recognition may result in lucrative offers to move from their supposed lower-tier institutions to top-tier institutions because of institutional desires for excellence (and maybe high ratings in U.S. News & World Report institutional rankings). This does have practical value for the individual scientist. For example, media reports of exciting results of investigations that usually start with "Dr. Jones has discovered ...," become "Harvard researcher Jones ...." This can add to the scientist's reputation.

The institutional "pedigree" of the investigator *is* important. The society of scientists likely will assume the person is a good scientist if the pedigree includes time spent at the top institutions in the world. This does not mean the person actually *is* good, but the pedigree becomes something that prospective employers will accept without additional evidence. Coming from a lower-tier institution does not preclude success, as success is judged to a large degree on publication quality (journal impact factors and scientific impact *per se*) and quantity. It also is judged on more intangible factors, including clarity of presentation at seminars, originality of thought, ability to "think on one's feet," ability to collaborate with others, personality, friendliness, attention to detail, work ethic, and many others.

6.2.4 Knowledge and Its Acquisition

In its infancy, sociologists of science concentrated on its structure, especially its institutions and organization [733], i.e., ostensibly nonscientific aspects of knowledge and its acquisition. However, beginning in the 1970s (no doubt due to the influence of Thomas Kuhn (see Section 6.3.1)), the remit of SoS expanded to include not only the *static* structure of science, but also its *dynamics* [734]. This included interactions of scientists with nonscientific elements of the social fabric, such as the public at large, politics, religion, the law, the economy, regulatory agencies, and other stakeholders, all of whom have substantial impact on whether, and in what manner, science is practiced and how knowledge provided by it will or will not be used. An important element of this dynamics is knowledge acquisition, which is the *raison d'être* of science. How is this done? What are the sociological bases of scientific ideas? Why do we have confidence in the verity of the knowledge? What epistemic analyses or principles do we apply to answer such questions? "Sociology of scientific knowledge" (SSK), a sub-field of SoS, seeks to provide answers to these questions [733].[42]

---

[42] Historically, two conceptions of SSK may be contrasted. The "weak program" viewed SSK as a means to understand the sociological underpinnings of failed or false theories. The so-called "Edinburgh School" of sociologists generalized the analysis to include *all* theories, regardless of their success. This conception of SSK, termed the "strong program," focused on psychological, social, and cultural factors influencing specific knowledge claims. For a more detailed discussion, see Longino [735].

We have defined knowledge, at least in part, as justified true belief. The focus was on the individual and how that person could "know." A person who takes responsibility for determining what *they* themselves consider knowledge, e.g., by doing their own epistemological work, practices what Hardwig [736] refers to as "epistemological individualism." We all do this every day, and have done so since infancy, even though we may not realize it. When we touch a hot surface, we know it will burn us. If we feel ill, we may seek medical attention, have tests performed, and then, on this basis, know what's wrong with us. If someone says the sunset is beautiful, we likely will look for ourselves so we will know whether it is or not. However, in science as in life, an epistemological individualism is often difficult, or impossible.

In Aristotle's time, philosophers interested in understanding the natural world were limited in their endeavors because of the crude technology available to them. Knowledge was limited. However, as the many mysteries of nature were solved, the amount of information produced was overwhelming. This information explosion precluded humans from knowing about every aspect of nature, or even of their own fields. Science and medicine became more and more specialized. For example, no longer could people simply be "biologists." They had to be defined by the minute area of nature they studied. Thus, they became physiologists, immunologists, primatologists, molecular biologists, etc., and within each specialty, they often were classified according to the animals used in their experiments. Drosophila geneticists were "fly people." Neuroanatomists, neurophysiologists, and specialists in organismal development working on the nematode *C. elegans* were "worm people." Physicians could no longer function as generalists who saw all types of patients. Sub-fields of medicine, such as urology, neurology, or orthopedics, had to be created, and within those sub-fields, more specialization occurred. In orthopedics, for example, physicians may specialize in hands, knees, shoulders, etc. In oncology, there are specialists in every type of cancer (of which there are >100). A neurologist may specialize in neurosurgery, neurooncology, neuroradiology, neuropathology, memory disorders, neuromuscular diseases, pain management, etc. Epistemological individualism no longer was possible. One could not figure most things out for themself, or even be capable of doing so. They had to accept as true what was written in textbooks or conveyed to them by mentors or colleagues. Reliance on experts for knowledge was unavoidable, which perforce meant that knowledge no longer was the sole purview of specialists. It now came, at least in part, from the *society* of scholars. Knowledge no longer sits with the individual, but rather with the community as a whole. One thus would say *"We* know *p"* instead of *"I* know *p."* In the quotation below, Hardwig [736] eloquently explains how this works (bolding is Hardwig's).

Scientists, researchers, and scholars are, sometimes at least, knowers, and all of these knowers stand on each other's shoulders in the way expressed by the formula: B knows that A knows that p. These knowers could not do their work without presupposing the validity of many other inquiries which they cannot (for reasons of competence as well as time) validate for themselves. ... It would, moreover, be impossible for anyone to get to the research front in, say, physics or psychology, if he relied only on the results of his own inquiry or insisted on assessing for himself the evidence behind all the beliefs he accepts in his field. Thus, if scientists, researchers, and scholars are knowers, ... **within the** structure of knowledge, ... the expert is an expert partially because he so often takes the role of the layman **within his own field**. ... [T]here is clearly a complex network of appeals to the authority of various experts, and the resulting knowledge could not have been achieved by any one person.

The complex network of which Hardwig speaks is illustrated nicely by the following sequence of logical propositions. Here D and E may be said to know p vicariously, i.e., through the community, but not as individuals.

A knows that m.
B knows that n.
C knows (1) that A knows that m, and (2) that if m, then o.
D knows (1) that B knows that n, (2) that C knows that o, and (3) that if n and o, then p.
E knows that D knows that p.
ad infinitum

Science works in precisely this way, as prior knowledge expands through the social network to gradually be accepted as truth, the foundation/shoulders upon which subsequent scientific investigation and discovery sit/stand. Newton's words, in a letter to Robert Hooke in 1676, often are used to illustrate this point [737].

What Des-Cartes [sic] did was a good step. You have added much several ways, & especially in taking the colours of thin plates into philosophical consideration. If I have seen further it is by standing on the sholders [sic] of Giants.

Newton was being modest, and he was not wrong. It *was* from on the shoulders of his predecessors that he saw further. I do not know if Newton actually jumped from one expert's pair of shoulders to another, allowing him to see farther than different experts with differing specialized knowledge. Practicing scientists/physicians do this routinely, but they also do this within each of their own scientific/medical social groups, not by jumping about, but by the equivalent of "crowd surfing."[43] The knowledge base of the crowd surfer depends upon the crowd and where the "waves" they produce take, and finally deposit, the surfer. Being a member of a group means you accede, at least in part, to the consensus of the group on matters of importance to it. You forfeit some of your own epistemic individuality. In one respect, everyone is "on the same page," which accelerates scientific progress in the circumscribed area defining each community. Accumulating knowledge *per se* is considered progress. If this knowledge is strongly corroborated in the future, one's belief in the veracity of that knowledge is increased. If knowledge extant later is falsified, this *also* constitutes progress. It is progress in the same sense as taking a wrong turn while driving, which eventually will lead to the realization that you're not going in the correct direction. The wrong turn has actually helped you to find the correct way. However, assuming that the collective wisdom of the group is true, and using this wisdom as a foundation of your own research, also can stifle discovery, perpetuate scientific misunderstanding, and lead to the ostracization of those who don't subscribe to the group's "bylaws" (not because the

[43] See https://en.wikipedia.org/wiki/Crowd_surfing.

ostracized individual's hypotheses are incorrect, but simply because they don't align with those of the group). Francis Crick put it this way:[44]

It is astonishing how one simple incorrect idea can envelop the subject in a dense fog. ...Only the gradual accumulation of experimental facts that appeared to contradict our base idea could jolt us out of our preconception.

Crick, in essence, is bemoaning the fact that scientists put their faith in the opinions of experts, and the "conventional wisdom," in their respective fields, both of which turned out to be wrong. This begs the foundational philosophical question of the rationale for believing experts? How could, and why should, scientists have confidence in their veracity? These questions have become particularly relevant and important in light of the COVID-19 pandemic, especially with respect to the widespread misunderstanding of the pandemic by the lay public. Goldman, in an aptly titled paper "Experts: Which ones should you trust? [740], and Hardwig [670], among others, have addressed the sociological components of these questions and come to the rather unsatisfying conclusion that there exist no *absolute* standards enabling one to answer such questions. Instead, at one extreme, confidence in experts is inescapably tautological – one trusts simply because one trusts – or as expounded by Burge [741] and Foley [742], respectively:

A person is entitled to accept as true something that is presented as true and that is intelligible to him, unless there are stronger reasons not to do so. The justificational force of the entitlement described by this justification is not constituted or enhanced by sense experiences or perceptual beliefs.

   [It is] reasonable for us to be influenced by others even when we have no special information indicating that they are reliable.

Burge has termed this type of confidence "fundamental authority," an *a priori* confidence independent of any other factors, or simply blind faith. A person entirely ignorant about a particular subject might accord absolute authority to a so-called expert. This is not entirely irrational, as how is one to judge knowledge claims with no bases for judgment? Similarly, one cannot evaluate the expertise of an "expert" without prior knowledge of why that person *should* be considered one. "Derivative authority," in contrast, *is* determined with the help of sense experiences, perceptual beliefs, extant knowledge, opinions of others, common sense, etc. Confidence, in this case, is relativistic because it is determined relative to a person's knowledge base, temperament, biases, political and religious beliefs, etc. Deciding whom to believe is equally relativistic. Kitcher [418] believes scientists practice a similar type of relativism, but terms it "calibration." Scientists compare their or others' knowledge of an issue to that of the expert and then decide whether, and how much, the expert should be believed. Opinion polls exemplify a calibrative basis for belief, i.e., one calibrates their own knowledge base, ideas, or actions relative to those of others. With respect to the COVID-19 pandemic, the confidence of the vast majority of Americans in

---

[44] Quotation from Crick [739], p. 140.

the expertise of Dr. Anthony Fauci was a powerful standard for calibration of one's own knowledge. It has lead people to accept, on faith, Fauci's pronouncements.[45]

6.2.5 The Sociology of Research Funding

Scientists also recognize the effects of SoS on scientific practice, but in quite practical terms. Unless a scientist is independently wealthy, grants must be won to pay for personnel, supplies, equipment, travel, publishing papers, etc. However, the funding process often is decidedly nonscientific. The "Golden Rule" – they who have the money make the rules – is how funding works for all sources, including governmental, private, and commercial. The determination of how a limited amount of money will be distributed almost always involves a combination of scientific and nonscientific (social) considerations. For example, if Bill Gates or Mark Zuckerberg have a favorite scientific question, they will pay scientists to work on it. The scientists must do what Gates and Zuckerberg want, not what they might want in an ideal world. The same holds true for the NIH, which is the primary funder of biomedical research in the United States, and which has the mandate of funding research in the public interest. Who determines what's in the "public interest?" It is not just scientists or the scientific community. It is the combination of scientists and nonscientists (politicians, boards of directors, private entities) that do so, and in the final analysis, the nonscientists have veto power.

At the NIH, grant applications only can be submitted to fund research determined to be in the public interest. Once submitted, an application is evaluated by a focused group of experts, a "study section." Study sections comprise a select group of scientists, usually senior scientists, that read grant proposals and discuss their scientific merit. The decision of what study section will review a given proposal depends on the aims of the application and the expertise resident in the different study sections impaneled by the NIH, of which there are almost 200. If the appropriate expertise does not exist in these regular study sections, *ad hoc* study sections may be created to ensure knowledgeable review of a proposal. Creation of specific study sections, their impaneling, and assignment of applications to study sections all are points in the review process at which decision-making has social components, because it is done by individuals within the NIH bureaucracy. The decisions of these individuals are affected by their personal interpretations and valuations of proposals, which may, or may not, reflect the prevailing wisdom of the scientific community and the NIH. Bias and nonscientific considerations thus are intrinsic to the entire process.

If an application is submitted by a scientist who is deemed to possess adequate knowledge and skills, and addresses a significant question, is technically feasible, rational, and rigorous, and will be done in an environment suitable for the proposed work, members of the study section likely will accord the application a good score.[46] The NIH and other

---

[45] For a marvelous treatment of SSK, I recommend Kitcher's "The Advancement of Science. Science without Legend, Objectivity without Illusions." Oxford University Press, 1996.

[46] Scores are numerical indicators of the perceived value of the work, much like a golfer's handicap is an estimate of the golfer's skill. Each member of the study section provides a score, all of which then are averaged. Scores then are ranked.

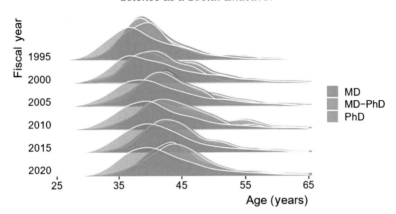

Figure 6.5 Age of principal investigators receiving NIH R01 support for the first time.

granting institutions have a limited amount of money, so only the highest ranked applications have a chance of being funded. At the NIH, in the recent past, one had to have a single digit percentile rank to have any hope of being funded. Pay lines vary depending on how much money Congress approves for the NIH. Nevertheless, only a minority of applications are funded through this highly competitive process, one in which social factors heavily influence decisions. The difficulty in being granted an R01 is illustrated by the fact, in 2020, the mean age at which a principal investigator with a Ph.D. degree received their *first* R01 was 43 (median of 44) (Fig. 6.5).[47] Those with M.D. or M.D./Ph.D. degrees were even older. This trend was the same for investigators identifying as men or women.

The penultimate review of grant applications is done by Institute and Center National Advisory Councils or Boards, of which there are 20, one for each NIH Institute or Center. Depending on which Council/Board is to be impaneled, the President of the United States, the Secretary of the Department of Health and Human Services (DHHS), the NIH Director and Deputy Director, and Institute or Center Directors make the appointments. Appointees are "chosen for their expertise, interest, or activity in matters related to health and disease" and comprise both scientific and public representatives. Among others, councils may comprise MDs, PhDs, MPHs, NGO (nongovernmental organization) directors, and pastors. Each Council/Board is tasked with advising institute directors "on policy matters and scientific opportunities, to recommend concurrence or nonconcurrence with the initial review groups' evaluations of applications for support of research and research training and to provide redress of real or apparent errors that may have occurred during initial review." The Institute/Center director makes final funding decisions based on staff and Council/Board advice. Social and political factors are intimately involved in all aspects of final review

---

[47] Data are from NIH Extramural Nexus (11/18/21) https://nexus.od.nih.gov/all/2021/11/18/long-term-trends-in-the-age-of-principal-investigators-supported-for-the-first-time-on-nih-r01-awards/.

because they were involved in *all* appointments, from the Secretary of DHHS down to Council/Board members, and affect how these appointees discharge their duties.[48]

In a perfect world, one would like peer review, and science in general, to be entirely objective. Unfortunately, this is not always the case because, in addition to social factors, a myriad of other elements are usually involved. Objectivity itself may be illusory. What then *is* objectivity? Can it be defined and practiced uniformly? According to Bauer [36], objectivity does exist, but it is socially determined – by consensus – and the only way to achieve consensus is by purposefully involving as many biased people as possible! Upon reflection, this idea *is* rational. If everyone is objective and unbiased, in an absolute sense, one could imagine a group of people, after reviewing the same information, immediately coming to consensus because each would have used the same approach for evaluation and the same weighting of each evaluative parameter. This is akin, metaphorically, to data being entered into a number of identical computers. The same data will invariably yield the same result. All applications will be evaluated in an identical manner, and although the outcome will vary with each, for any one application, the results will be identical. It thus would seem that seeking categorical elimination of all bias in evaluation was what was irrational, not its inclusion.

To assure that scientific knowledge is reliable ... requires that interactions among scientists be unconstrained and that scientists be as varied as possible in their biases. Science progresses through continual winnowing under consensually governed institutions. Objectivity comes into science because ideas and results are exposed to the criticism of people with disparate and conflicting and competing intellectual approaches and beliefs, personal biases, social goals, hidden agendas, and the rest, so that by and large-consensus among all of them can only be achieved when they are left no other option than to agree with each other, when the puzzle itself demands and allows it, when the players submit jointly to reality therapy. Scientific activity therefore becomes more efficient and more reliable the more it includes the whole range of human types-geographic, sexual, intellectual, emotional ....

There are problems with these scenarios. The first is *consensus* is distinct from *truth*. Because everyone agrees does not mean they are correct in their judgment (remember the original consensus regarding Ptolemy's geocentric heavens). All could be right, wrong, or a mixture of the two. Also, if opinions *are* biased, that does not mean all are wrong. Not all conspiracy theorists are wrong, and not all scientists, and even mathematicians, are always right. Biases cannot be completely avoided and, in many cases, nor should they. The questions thus become how, when, and how much should we allow our biases to affect our judgment? As you might expect from prior discussions in this book, there are no absolute answers, and to even provide cursory answers would be impossible in a chapter, in an entire book, or even in a lifetime. Judgments about grant applications are inextricably intertwined with bias in each reviewer. One must always be aware of this, sometimes to understand what appear to be erroneous conclusions and try to correct them, and sometimes, for one's own well-being, to accept contentious conclusions, say *"c'est la vie,"* and move on.

---

[48] One of the best, and most common, examples of the vagaries of the review process is illustrated by the fact that many applications considered inferior (and thus not funded) by one study section may, when resubmitted to another, "walk on water" (and thus be funded).

6.3 Scientific Progress: Paradigms, Programmes,[49] and Traditions

There are many sociological aspects affecting how we think about and practice science. It seems that the progress of science could be considered the never-ending proposition, corroboration, and refutation of theories. Thus, like aging, science would progress in a continuous, monotonic manner, not a halting, sporadic process with quantum leaps and bitter retreats. Scientists would have control over their own work, choose precisely what questions to study, come up with their ideas independently of others, be driven solely by curiosity about the natural world (largely unaffected by the nonscientific world), infer from and interpret their results in an unbiased and objective manner, and drive science forward through accumulation of the knowledge they, as individuals, had acquired. This was a continuous process, much like how Egyptian pyramids were built, i.e., by the systematic addition of individual stones. The bedrock of science, the "stones," comprised facts and logic produced by countless scientists employing the scientific method. Advances in science were facilitated because this bedrock was common to all working in their respective fields at specific points in time. Scientists focused on accepted questions and the methods to answer them. Hypothesis formulation and testing, theory choice and acceptance, determination of epistemological value, and even what was considered fact, i.e., a true representation of reality, became, or were tacitly accepted as, givens. However, historians and philosophers of science studying the progress of science through the ages have concluded that movements from *Era x* of science to *Era y* and beyond usually involve distinct mechanisms. Feyerabend expounded on a number of these mechanisms. He discussed how different mechanisms were, or might have been, involved in progress from the astronomy of Ptolemy/Aristotle to that of Copernicus/Galileo (Table 6.3). We focus here on three of these mechanisms, all of which were centers of discussion in the twentieth century: "paradigms," "research programmes," and "research traditions."

6.3.1 Thomas Kuhn: Paradigms

Until 1962, the notion that scientific progress was *not* driven solely by scientific questions and answers was almost heretical. Then came Thomas Kuhn, his book *"The structure of scientific revolutions"* [197], and a perspectival sea change. Kuhn argued that scientific progress was inextricably ensconced in, and affected by, society at large. One no longer could explain the advance of science as the result of pure, value-free, unbiased, logical, apolitical, and areligious activities practiced by scientists in isolation. Science was more holistic. It not only involved the aforementioned activities but was dependent on, and affected by, society, with all its uncertainty, illogic, relative ignorance, and doxastic perspectives, intransigence (especially among older scientists), cynical goals, and political and social agendas. Each discipline appeared to practice their own form of science, determined through the consensus of the members of that discipline. Kuhn introduced the word "paradigm" to describe this type of socially determined scientific practice. To

---

[49] Hereafter, the word "programme," without quotation marks, is synonymous with Lakatos's systems of scientific progress.

Table 6.3 *Mechanisms of scientific progress. A chronological listing of modes of characterizing scientific progress from Ptolemy/Aristotle to Copernicus/Galileo (and beyond), à la Feyerabend [298].*

Philosopher	Mechanism	Description
Religious adherents	Naive empiricism	The bible is our only source of knowledge about the natural world.
Ancient astronomers	Sophisticated empiricism	Revision of existing field in light of new evidence
Astronomers	Conventionalism	Simplification of sophisticated empiricism
Karl Popper	Falsificationism	Revision of the field's theories/principles when prior tenets are refuted
Thomas Kuhn	Paradigms	Unexplainable anomalies lead to scientific progress through paradigm replacement
Imre Lakatos	Research programmes	Sequence of theories with irrefutable hard cores and modifiable auxiliaries
Larry Laudan	Research traditions	Domain-specific assumptions about entities, processes, methodology, theory construction

Kuhn, paradigms were systems to solve scientific puzzles. Paradigms, though, were much more. They represented the *gestalt* of science at the time. Paradigm and *gestalt* thus were almost synonymous. Under the rubric of paradigm lay theories (characterized by their predictive value, simplicity, consistency, and plausibility), methods, general perspectives on science and its conduct, acceptable scientific questions and practices, credit (recognition) worthiness, grounds for ostracization, effort distribution, etc. According to Kuhn [197],

Paradigms should not be perceived ...as the product of the work of the individual but rather as the product of the work of a 'community of scientists.'

In studying the history of science, Kuhn posited that scientific progress occurred in two fundamentally different ways: (1) traditional incremental advances leading to new knowledge and revised theories ("normal science") and (2) quantum leaps in understanding that reject existing paradigms in favor of new ones ("revolutionary science"). Examples of normal science would be miniaturization of electronic components, development of vaccines for COVID-19, or improving the accuracy and precision of estimates of the age of the universe. Examples of revolutionary science would be the periods in which scientists came to accept heliocentricity, gravity, cell theory, relativity, and quantum mechanics. These latter discoveries revolutionized our understanding of the natural world, rejecting old paradigms and ushering in new ones.

The initiators of revolutionary science are anomalies or puzzles, i.e., things that cannot be explained by theories within the prevailing paradigm. No theory can explain every element of the dynamics of a system. No one expects it to and unexplained elements may be neglected or ignored completely if the community feels they are not integral to the theory.

However, some anomalies can be problematic because they come from elements in a theory upon which the theory itself was built. Initially, the scientific community may be aware of these anomalies, but continues to function within the existing paradigm. They do so because their paradigm has worked well in the past and continues to be useful. Eventually, as additional anomalies accumulate that cannot be explained by or continue to be inconsistent with theory, whether through continued experimentation and theorizing or the retirement or deaths of die-hard proponents of the old paradigm, the anomalies must be explained. It then is time to create new theories and new paradigms. As an example, it was anomaly that led Watson and Crick to the correct structure of DNA. Crick later commented:[50]

... we were acutely aware that something was wrong and were continually trying to find out what it was. It was this dissatisfaction with our ideas that made it possible for us to spot where they mistake was. If we had not been so conscientious in dwelling on these contradictions we should never have seen the answer.

Feynman, in a somewhat less eloquent manner, also opined on interpretation of facts and the implications of inconsistencies or contradictions in those facts.

... if your conclusions contradict common sense, then so much for common sense; if they conflict with received philosophical opinion, then too bad for received opinion; but if they deny the very facts of our experience, then you must consign your conclusions to the flames.

What happens as we watch our old theories and paradigms burn? With what do we replace them? How does science continue to function? Science is a self-correcting enterprise. Theories are constantly being tested. If they are corroborated, they are used. If not, then efforts are made to understand why. Sometimes, small tweaks are enough to make a theory consistent once more. At other times, auxiliary assumptions or theories, i.e., small additions or qualifiers of the main theory, may be useful. Often, however, tweaking theories fails or the theory itself is falsified. Science has found the error and now must correct it. During periods of revolutionary science, new theories are proposed by many. Most will be found wanting and science may be thrown, to some degree, into chaos. This is normal, productive, and necessary. Metaphorically, if one gets lost, they soon realize they have traveled in the wrong direction and they must try a new direction. There are many. Some eventually help the traveler get to their destination and many won't, but being lost (read "chaos") eventually ends when the right direction is found. From then on, this information will be what travelers use to reach their own destinations. If directions were paradigms, the necessity of rejecting the old paradigm in favor of a new, more consistent one, would be obvious.

I have characterized paradigms as more than just theories guiding a scientific community. As did Kuhn, I likened them to gestalts, and in doing so, incorporated sociological, psychological, economic, and technological elements in them, in addition to purely scientific ones. But these are not their only elements. Central to any paradigm is its lexicon, through which knowledge is disseminated, scientific communication enabled, and theories promulgated and discussed. Kuhn argued, in addition to replacement of the theoretical

---

[50] See Crick [739], p. 140.

foundations of a paradigm by those of another, that the *language* of science would change. A set of terms used previously to describe and explain a phenomenon now would be replaced by a second set, within which some old terms remained but many would be replaced. In addition, novel terms would need to be created to account for entirely new objects, mechanisms, or ideas. Even identical terms in each set might not be synonymous. Imagine a scientist adherent to the old paradigm talking with one adherent to the new. With respect to the same phenomenon or theory, one scientist might refer to it as "blix" while the other might use the term "schmeh." If new phenomena were integral to the new paradigm, blix and schmeh might refer to entirely different things. Scientists working in two different paradigms thus would have no common language or perspective enabling them to understand each other. Neither would always know about what the other was talking and, in this sense, the two paradigms and their component languages would be "incommensurable." Kuhn used the metaphor of the rabbit–duck image (see Fig. 5.6) to illustrate incommensurability.[51] In this case, although two groups of scientists might be looking at the same image, those working within one paradigm would see a rabbit while those working in another would see a duck. Rabbits and ducks would be incommensurable, just as are Newtonian and quantum mechanics.

We have discussed extremes here – normal and revolutionary science – but science does not progress only through one type of period or the other. In addition to the incremental advances commonly occurring in a particular field during normal science, larger changes may occur. They are not quantum in nature, but they force scientists within the field to revise at least some of their prior assumptions and knowledge. It is the magnitude of such revisions that communities of scientists use in deciding whether to label an advance revolutionary. However, the definition of magnitude can be highly subjective, which means that whether an advance is or is not considered revolutionary also is subjective, at least in part. As an example, few would argue with the statement "after Einstein we lived in a different world," but many would argue with the statement that "mRNA vaccines[52] are revolutionary."

Kuhn's discussion of incommensurability in *The Structure of Scientific Revolutions* put the term and the idea "on the map," as it were, but Paul Feyerabend also wrote about it the same year [744].[53] Feyerabend, instead of considering incommensurability in the context of paradigms, focused on its use as a descriptor of situations in which concepts

---

[51] Kuhn's concept of incommensurability did change over time. According to Sankey [743], *"Originally, incommensurability was a relation of methodological, observational and conceptual disparity between paradigms. Later Kuhn restricted the notion to the semantical sphere and assimilated it to the indeterminacy of translation. Recently he has developed an account of it as localized translation failure between subsets of terms employed by theories."*

[52] These are one type of vaccine used to immunize people against COVID-19 and its variants. The others comprise viral proteins or inactive virus particles.

[53] In his 1962 paper, Feyerabend explained incommensurability as follows: *... there exist pairs of theories, T and T', which overlap in a domain D' and which are incompatible (though experimentally indistinguishable) in this domain. Outside D', T has been confirmed, and it is also more coherent, more general, and less ad hoc than T'. The conceptual apparatus of T and T' is such that it is possible neither to define the primitive descriptive terms of T' on the basis of the primitive descriptive terms of T nor to establish correct empirical relations involving both these terms (correct, that is, from the point of view of T). This being the case, explanation of T' on the basis of T or reduction of T' to T is clearly impossible .... Altogether, the use of T will necessitate the elimination both of the conceptual apparatus of T' and of the laws of T'. The conceptual apparatus will have to be eliminated because its use involves principles ... which are inconsistent with the principles of T; and the laws will have to be eliminated because they are inconsistent with what follows from T for events inside D'.*

underlying fundamental theories could not be reduced to a common language, or, according to Oberheim and Hoyningen-Huene [745], when *"...the main concepts of one [theory] could neither be defined on the basis of the primitive descriptive terms of the other, nor related to them via a correct empirical statement."* For Feyerabend, like Kuhn, changes in fundamental theories lead to changes in meanings, which altered scientists' view of the world.[54]

Along with Merton, Kuhn may have been one of the most important sociologists of science in the twentieth century. This, however, does not mean that their views were instantly accepted and incorporated into the bedrock of SoS. Sir Karl Popper vehemently disagreed with Kuhn's paradigmatic view of the history of science. Remember that Popper had argued for what many might say was a binary decision tree regarding science and nonscience (pseudoscience). If a theory could be falsified, then it would be scientific. It not, then not. Kuhn's notions of cycles of normal science and revolutionary science were problematic for Popper, for this suggested there were three types of science practice, *viz.*, *revolutionary science*, *normal science*, and *pseudoscience*. Although he eventually agreed in principle that a form of "normal science" *was* practiced, he did not think much of it, as the quotation below indicates.

'Normal' science, in Kuhn's sense, exists. It is the activity of the non-revolutionary, or more precisely, the not-too-critical professional: of the science student who accepts the ruling dogma of the day; who does not wish to challenge it; and who accepts a new revolutionary theory only if almost everybody else is ready to accept it – if it becomes fashionable by a kind of bandwagon effect. ...In my view the 'normal' scientist, as Kuhn describes him, is a person one ought to be sorry for. ...He has learned a technique [i.e., operates within the confines of a paradigm] which can be applied without asking for the reason why (especially in quantum mechanics). As consequence, he has become what may be called an applied scientist, in contradistinction to what I should call a pure scientist. He is content to solve 'puzzles'.

There is much in each person's concepts of SoS that remains applicable to science in the twenty-first century, and some that are not. In particular, natural, social, and political scientists, psychologists, and economists, found Kuhn's "paradigms" compelling and applicable in their fields,[55] so much so that the word became an important part of their lexicons [747]. For example, psychologists Reese and Overton [748] echoed Kuhn when they stated

...[t]heories built upon different world views are logically independent and cannot be assimilated to each other. They reflect different ways of looking at the world and, as such, are incompatible in their implications. ...Different world views involve different understanding of what is knowledge and hence of the meaning of truth.

---

[54] Feyerabend, in his studies of the paradigmatic shift from geocentricity to heliocentricity, suggested multiple theories to explain it, including naive empiricism, sophisticated empiricism, conventionalism, falsification, crisis theory (Kuhn), or research programmes (Lakatos [see below]). See Feyerabend [298], pp. 45–46, for explanations of these theories.

[55] Kuhn and Lakatos strenuously disagreed about this applicability. They felt strongly that sociologists, political scientists, economists, *et al.* had misconstrued the meaning of paradigms and invoked them in fields in which they were not applicable. Lakatos even went so far as to characterize "these efforts as little more than 'phony corroborations' that yield 'pseudo-intellectual garbage'" [746].

Palermo [749] used Kuhn's ideas of revolutionary science and paradigm change to answer the question "is a scientific revolution taking place in psychology?" His answer was "yes."

...there seems little doubt that experimental psychology is ripe for a revolution – if it is not, in fact, currently in the midst of one. All of the historical and social elements which Kuhn has described as symptomatic of revolutions in other sciences appear to be present in psychology at the moment.

Many philosophers of science were not as sanguine about the value of Kuhn's ideas and offered many criticisms of them. For example, it was argued that if two paradigms were incommensurable, i.e., like apples and oranges, how could one determine if one paradigm actually was, or might become, better than another? How could one find a rational explanation for paradigm change. Opponents of Kuhn concluded one couldn't and that his interpretations of the dynamics of scientific progress yielded results that were irrational [747, 750]. One thing is certain, Kuhn's ideas have stimulated, and continue to stimulate, a torrent of discussions, criticisms, conferences, books, articles, and competing theories of scientific progress, one of which was Imre Lakatos's concept of "scientific research programmes."

6.3.2 Imre Lakatos: Research Programmes

One vehement critic of Kuhn was Imre Lakatos who, when discussing the theories of Popper and Kuhn, wrote:[56]

For Popper scientific change is rational or at least rationally reconstructible and falls within the realm of the logic of discovery. For Kuhn scientific change – from one 'paradigm' to another – is a mystical conversion which is not and cannot be governed by rules of reason and which falls totally within the realm of the (social) psychology of discovery. Scientific change is a kind of religious change – [a matter of mob psychology].

In the quotation below, Lakatos explains his own sense of scientific progress, and contrasts it with Popper's "falsificationism" and Kuhn's "conventionalism" (paradigms as conventions of the scientific community).[57] Lakatos's explanation makes clear that all three theories of major scientific progress do have commonalities.

According to my methodology the great scientific achievements are research programmes which can be evaluated in terms of progressive and degenerating problem shifts; and scientific revolutions consist of one research programme superseding (overtaking in progress) another. This methodology offers a new rational reconstruction of science. It is best presented by contrasting it with falsificationism and conventionalism, from both of which it borrows essential elements.

Some considered Kuhn to have promulgated a theory of scientific change with only two essential components, normal and revolutionary science, and no mechanism for moving

---

[56] See Lakatos [750], p. 93.
[57] See Lakatos [750], p. 110.

rationally from one to the other. Popper, in contrast, provided an absolute criterion for change – in whatever form – falsification. There were no compelling reasons to abandon one theory for another if the prior theory was corroborated and never falsified, but if the theory were falsified, it would have to be replaced by the theory that survived the fiercest forms of testing. In a sense, Popper thus also had a two-component theory, but, unlike Kuhn, with a normative procedure for creating change. Popper's perspective on science was more metaphysical than Kuhn's. Popper, as encompassed in the title of one of his most famous works, sought to define "The Logic of Scientific Discovery." Kuhn, in contrast, sought to understand science in practice, i.e., what science *was* rather than what it should be.

As these histrionics ebbed over the decades since the 1960s and thoughtful, objective consideration of Kuhn's paradigmatic program of scientific progress occurred, people realized that Kuhn's "structure" was not binary – he never argued it was – but rather was a continuum between two extremes, normal and revolutionary science, with increasing recognition of anomalies occurring in between.[58] However, even this recognition could not adequately answer the simple question of when the anomalies were considered so significant that an existing paradigm had to be abandoned. Here is where Lakatos's overblown "mob psychology" would be invoked, a psychology that was in fact simply the consensus of the scientific community, not of an emotional, irrational mob. Even from this point of view, Kuhn's ideas were felt to be untenable, as one then would have to determine what consensus meant and how it was reached – questions without absolute answers.

To reconcile the quasi-binary theories of scientific activity by Popper and Kuhn, Lakatos suggested that progress in science was mediated through "research programmes" [750]. Instead of invoking Kuhn's explanation of scientific progress as a shift in paradigm or Popper's idea of corroboration or falsification,[59] Lakatos considered progress to manifest as the replacement of an existing programme by another. The "new" programme could be a contemporaneous programme found through experimentation to be better than the older programme, or it could be novel. A key characteristic of programmes was their progressive nature. They moved forwards through experimentally mandated continuous modification or replacement of existing theories. The new theories had to be more explanatory, consistent, and predictive than the old, but depended on a set of initial assumptions or, as Gholson has termed them, "commitments" [747], that were held in common. These commitments constituted the "hard core"[60] of a programme, around which existed a "protective belt" of auxiliary hypotheses, assumptions, givens, etc., which could be eliminated, modified, or

---

[58] Remember Crick's explicit statement that it was, in essence, the eventual explanation of anomalous "incorrect idea[s]" or "preconception[s]" that led him and Watson to the structure of DNA.

[59] Falsification *could* be employed with programmes, but it was not the unitary falsification proposed by Popper. Falsification of a programme required the invalidation of its multiple (sequential) theories, not a single overarching one. Lakatos termed this "sophisticated methodological falsificationism."

[60] Lakatos used Newton, and his three laws of motion, to illustrate the hard core of a research programme (one that later would degenerate in the face of relativity and quantum mechanics).

supplemented to "save the theory," i.e., maintain its consistency with potentially falsifying observations.[61] As an example, Duhem, in discussing physics, wrote:

... [it] is not a deductive science whose propositions follow from principles evident a priori. It is a science whose origin lies with experience, and the principles of cosmology are nothing but hypotheses conceived with a view to saving the phenomena that experience has made known to us.

For Popper, however, a refutation of one theory would invalidate an entire programme. Lakatos disagreed.

It is not that we propose a theory and Nature may shout NO; rather, we propose a maze of theories, and nature may shout INCONSISTENT.[62]

Lakatos's maze was actually a continuous sequence of theories, $T_1, T_2, T_3, \ldots T_n$, in which anomalies in T_i were explained by T_{i+1}. Feyerabend, who was a colleague of both Popper and Lakatos at the London School of Economics (see Section 4.7.1), categorically rejected Lakatos's "relaxed" view of programmes, characterizing them as "epistemological anarchism" [298].[63]

The dynamics of research programmes was characterized by the initial creation of a programme, its subsequent development and maturation, and then its degeneration and eventual replacement. The term degeneration is applied to a programme when it ceases to be progressive, i.e., it can no longer adequately explain phenomena, becomes increasingly inconsistent with observations, or it loses its ability to foster new predictions. Note, however, that progressive programmes could continue to be of use, even in the presence of anomalies and even if scientists knew that the programme contained elements that were not true, at least until a superior programme emerged. Programme degeneration can occur in one or both of two domains, theoretical and empirical [750]. The former domain, in contrast with the latter, is more immune to degeneration because *ad hoc* revisions in a programme's protective belt may protect its central theories. Even if such protection appears insufficient, scientists still have to be convinced, a process that has decidedly social components (bias, community pressure, dogmatism, pride, avarice, respect, etc.) [747]. Once the community of scientists *is* convinced, there remains the question of which among many competing programmes extant becomes "the chosen one," and why.

Choosing a replacement program involves many of the same considerations as choosing a new theory. If a prospective programme appears to follow directly from the historical sequence of prior programmes, one may induce that it would be rationale for it to replace the degenerated programme. Deduction may also apply, to a degree, but it would be rare that the abstract tools of logic could be applied to programme choice. This would be

---

[61] To learn more about the origins of the idea of "saving theories," see Duhem's 1908 publication "ΣΩZEIN TA ΦAINOMENA: Essai sur la notion de théorie physique de Platon à Galilée" [751], which was translated into English in 1969 by Edmund Doland and Chaninah Maschler [752].

[62] See Lakatos [750], p. 45.

[63] Before Lakatos' death, he and Feyerabend had intended to coauthor a paper in which, according to Feyerabend, *"I was to attack the rationalist position, Imre was to restate and defend it, making mincemeat of me in the process."* For a further discussion of the Lakatos-Feyerabend debate, see Motterlini [753].

especially true for programmes in the biological sciences where there is such complexity that choosing propositions for deductive purposes is difficult if not impossible. Employing abductive principles to the problem of programme choice is what is most commonly done. This makes sense because when one considers Kuhnian scientific revolutions or Lakatosian programme changes, the movement from old to new cannot be incremental, which means that the elements upon which induction was based in the past will no longer be adequate for the purpose.

Research programmes are characterized by more than hard cores, auxiliary hypotheses, or progression and degeneration. As the title of Lakatos's seminal work, "The Methodology of Scientific Research Programmes" [291] implies, specific methodologies are associated with research programmes. According to Lakatos, these methodologies are "powerful problem-solving machines, which, with the help of sophisticated mathematical techniques, digest anomalies and even turn them into positive evidence."[64] In this sense, programmes are similar to paradigms, the functions of which have been said to "supply puzzles for scientists to solve and to provide the tools for their solution" [754]. However, Lakatos "methodology" is more implicit than explicit and more metaphysical than normative.

The Methodology of Research Programs might also have been entitled *"A Rational System for Scientific Progress,"* as its methodological component was more of an adumbration of programme principles rather than an exposition of the technical armamentarium of a particular group of scientists. These principles included "negative heuristics" and "positive heuristics" that helped scientists determine impermissible and permissible research methods and approaches, respectively. The former protected the hard core of the programme from potentially confounding information while the latter mandated research directions supportive of both the hard core and auxiliary hypotheses [755].

6.3.3 Larry Laudan: Research Traditions

As we have seen, philosophical study of the progress of science has produced a variety of perspectives, all of which have in common a dependence on sociological, as well as scientific, factors. Kuhn felt Popper's falsificationism was too rigid and Lakatos argued that Kuhn's concept of paradigms was too flexible. Larry Laudan had disagreements with all three, which, just as Lakatos had attempted with Popper and Kuhn, he sought to remedy with yet another concept of scientific progress, "research traditions" [756].

Laudan begins his exposition of research traditions stating *"[t]heories are inevitably involved in the solution of problems,"* and then poses the question how does one determine the adequacy of a theory for solving empirical and theoretical problems (puzzles)? He categorically rejects any notion of absolutism (a à Popper). Instead, he mandates the use of "comparative modalities," i.e., determining how one theory "stacks up" against others.

If you are a biologist, or any other scientist studying complex systems, you are acutely aware that the more general and all-encompassing is the theory, the less useful it is in

---

[64] See Lakatos [291], p. 4.

elucidating specific elements within it. For example, the theory that humans are conceived and develop from a single cell may be true, but it doesn't explain the necessity or mechanisms of oogenesis or spermatogenesis, where and how fertilization occurs, placental development and function, asymmetric cell division, etc. Theory reductionism is required to enable scientists to ask discrete, answerable questions and to develop the technology necessary to do so. The theory that fertilization occurs in the Fallopian tubes after sperm have transited the cervix and uterus is readily understandable, enables the design of experiments to test the theory, and enables predictions to be made about how the process may be affected by a variety of things, be they genetic, anatomical, or physiological. It thus is obvious that theory reductionism, by definition, requires something to reduce – the single cell theory of human reproduction in this case.

Let us now consider the situation at the time a particular theory is conceived and ask why it, and not other theories, prevailed. Laudan would answer *when compared* with other theories, this one was most adequate. How could this be determined? Would you ask the same questions you did in comparing fertilization theory to other theories of the same process? The answer clearly is "no," because, *at the time*, each theory comprised different elements, some of which were shared among theories and many that were not. In essence, you would be comparing apples with oranges, and in some cases, apples with bacteria. How do Laudan's traditions resolve this conundrum? Answer: create two classes of theories, "super theories," which are overarching theories within which are auxiliary or "mini-theories" of narrower breadth and more limited scope. Importantly, each class of theory would be evaluated using different criteria. Examples of super theories and mini-theories would be the theory of evolution and mechanisms of DNA mutation, respectively. Laudan argues that only through this approach can one develop a theory of scientific progress that is "historically sound and philosophically adequate." This point is debatable, and Popper, Kuhn, Lakatos, Feyerabend *et al.* did so vigorously, but for now, let us accept Laudan's premise and discuss the inadequacies in prior theories it purports to eliminate.

Laudan raises a number of issues with Kuhn's use of paradigms in the characterization of scientific progress. These issues arise because of Kuhn's apparent lack of appreciation of the contexts of paradigm evaluation and change, specifically whether they involve unitary theories or maxi-theories (akin to mini-theories and super theories). Further problems arise because of what Laudan argues are Kuhn's nebulous description and implicit nature of paradigms themselves. As an example, how can broad conceptual issues be considered if the evaluative questions focus only on elements of the whole, e.g., to what degree one paradigm explains anomalies better than another? How can one answer the question of whether the conceptual foundations of a paradigm are even reasonable? What is the relationship between a paradigm and its components? Does one determine or define the other? How can a paradigm evolve if its core components are protected from refutation or anomaly? How may a paradigm be understood and evaluated if its organization and dynamics are not stated explicitly but depend only on a limited set of exemplars such as "archetypal applications of a mathematical formulation to an experimental problem?"

Laudan also points out contradictions between the theory of paradigmatic change and how science actually is done. The example he gives is of two scientists ostensibly working within the same disciplinary matrix (network of commitments-conceptual, theoretical, instrumental, and metaphysical) yet having radically divergent views of scientific ontology and methodology. This example is akin to that of two people seeing the same glass of water and one saying "it's half full" while the other says "it's half empty."

Laudan considered Lakatos's research programmes to be a marked improvement on Kuhn's paradigms, yet with some of the same problems, as well as its own. Like paradigms, programmes were empirically oriented and thus failed to explain conceptual change *per se*. The hard cores of programmes also were protected from refutation and anomaly, which made them resistant to re-evaluation, modification, or substantial change. This rigidity meant that the sequence of theories characterizing programmes had to involve precursor-product relationships. In contradistinction to modern scientific practice, novel changes to, or complete replacement or elimination of, foundational theories were prohibited.

So what did Laudan do to create a theory of scientific progress without the flaws he saw in paradigms or programmes? He answered the question as follows.

Although there are numerous common elements between my model and those of Kuhn and Lakatos (and I readily concede a great debt to their pioneering work), there are a sufficiently large number of differences that I shall try to develop the notion of a research tradition more or less from scratch.

Laudan begins by pointing out common traits of research traditions, which I paraphrase here: (1) they are maxi-theories exemplified and partially constituted by constituent, often short-lived theories, which may be conceived contemporaneously or by succession; (2) like Kuhn's disciplinary matrix, each tradition can be distinguished from another through comparison of its metaphysical and methodological commitments (a là Gholson, see above); and (3) traditions may be continually reformulated over long periods of time, often with new formulations that contradict the old. He then provides a working definition of research traditions.

... a research tradition is a set of general assumptions about the entities and processes in a domain of study, and about the appropriate methods to be used for investigating the problems and constructing the theories in that domain.

One might argue that Laudan's definition of research traditions could equally well be applied to paradigms because they, like paradigms, specify permissible questions, as well as the empirical (experimental, observational, analytical, etc.) and theoretical approaches used to answer them. Traditions are, in essence, *"a set of ontological and methodological 'do's' and 'don'ts.'"* The ontological and methodological features of a tradition generally are interdependent, which would be expected if, as occurs daily in science practice, the methods available to a researcher determine which questions are potentially answerable, and the questions to be answered specify which methodologies must be available to do so.[65]

---

[65] It was just this latter need that propelled the development, 40 years ago of the "gene machines" that revolutionized science by enabling chemical synthesis of genes and sequencing of DNA. This, along with PCR (polymerase chain reaction; a method to produce enough DNA (for sequencing) from infinitesimally small amounts of starting material), is why the human genome and the genomes of many other species have been sequenced in their entirety. It is also why it seems that every detective show or murder mystery has "DNA" as a central element of a story.

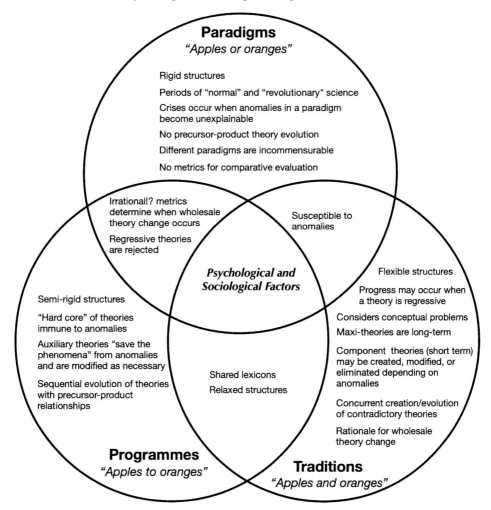

Figure 6.6 Ontological and empirical structures to solve scientific problems/puzzles. Venn diagram of theories of scientific practice and progress. Nonscientific factors, including sociological, psychological, political, or religious, have tremendous impact on the dynamics of the three different systems. Among these, sociological factors predominate. (See text for details.)

Kuhn, Lakatos, Laudan *et al.*, although presumably studying the same history, have come up with overlapping but distinct interpretations of science and its progress. As a practicing scientist, I see merit in each, and that significant commonalities, and differences, exist among their ontological and empirical frameworks (Fig. 6.6).

Given that one subscribes to the notions of research traditions, paradigms, and programmes, the most critical issue becomes the means by which scientists choose to replace one with another. As shown in Fig. 6.6, the means of moving from one paradigm or programme to another is a bit nebulous (irrational!?), something Laudan sought to avoid

through his conception of traditions and the use of modalities allowing rational evaluations and comparisons. These modalities included adequacy, progress, and appraisal. To
provide metrics[66] upon which decisions could be based, Laudan made two overarching
stipulations.

(1) [T]he solved problem – empirical or conceptual – is the basic unit of scientific progress; and
(2) the aim of science is to maximize the scope of solved empirical problems, while minimizing the
scope of anomalous and conceptual problems.

One might formulate these ideas in the equation

$$T_x = \frac{i_p P - i_a A - i_c C}{P + A + C},$$

where T_x is the normalized merit of tradition x; P, A, and C are the numbers of problems
solved, and the numbers of anomalies and conceptual problems remaining, respectively;
i_p, i_a, and i_c are constants reflecting the weighting of each variable, and $i_p + i_a + i_c = 1$.
Although simple, this method of tradition evaluation is superior to the *ad hoc* methods
applied to paradigms and programmes because it provides the means to weight (i_x) the
importance of each component term.

Earlier, I qualified the process of tradition choice by writing "at the time." This is important to note with respect to the relative merits of extant traditions as some of the most
important advances in science occurred because one or a few scientists disregarded this
quantitative comparative approach. They did so because they believed, implicitly, that a
technically disfavored tradition, *in the long run,* would supersede the originally favored tradition. The transition in traditions from geocentricity→heliocentricity is a good example,
as are the transitions from Newtonian→quantum mechanics, corpuscular→wave theory of
light, and nucleic acid only→"protein also" dogma of molecular biology (see page 122).

The replacement of the corpuscular theory of light by the wave theory is a particularly
good example of scientists ignoring anomalies (diffraction, interference) because the existing concept solved many prior problems in optics. It also is a good example of scientists
believing in a new concept, regardless of current wisdom, because they believed *it* might
now, or later, account for anomalies and explain prior problems better.[67]

Temporality is a critical part of Laudan's evaluation of the adequacy and progress of a
tradition. He considers two temporal modes, synchronic and diachronic. The former mode
evaluates the *current* adequacy of a tradition. The latter mode involves two questions: (1)
what has been the absolute progress of the tradition from time t_i to t_j, i.e., the difference
$T_i - T_j$? and (2) what has the rate of progress been between any two arbitrary times t_x
and t_y, i.e., $\frac{T_x - T_y}{t_x - t_y}$? Note that synchronic and diachronic appraisals may differ widely. For

---

[66] These metrics are "approximative," i.e., they can be used to quantitatively assess traditions, but they have varying levels of
precision.
[67] A passage from Aspect [757] discusses this particularly well. "*Despite its value, Huygens' wave model, described in full
detail in 1690 in his Treatise on Light, was ignored by most scientists for more than a century. If they had adopted the
corpuscular model, it was because of the tremendous prestige conferred upon Newton, who had managed to explain the
motions of planets through his law of universal gravitation. It took more than one century before Thomas Young, in England,
and Augustin Fresnel, in France, developed the wave model of light, as the only one able to account for the phenomena of
interference and diffraction … .*"

example, a brand new discovery would have high values for each, but if, after some period of time, only a few new problems were solved; synchronic progress would remain high but diachronic progress would be substantially lower. The converse may also be true, e.g., if substantial progress occurred at the end of some time interval, diachronic merit could be high while synchronic merit could be low.

Two important points are: (1) as with the corpuscular→wave transition and the effects of context (time) on its evaluation, context can substantially alter (sometimes by 180°) one's sense of things (examples we have already discussed, *but with different contexts*, include history, ignorance/knowledge, realism, learning, abduction, simplicity, explanation, importance, right/wrong, and reasonable/demonstrable; and (2) truth and falsity are irrelevant to the comparative evaluation of traditions because of the fundamental philosophical problems involved in doing so and because the metrics of comparison are akin to logical propositions (meaning the context of the determination of the merits of traditions is logical, not empirical *per se*) and thus have nothing to do with absolute truth or falsity.

Laudan continues by considering the problematic common assumptions that theory appraisal occurs within a single context and, within this context, appraisal depends only on empirical adequacy. As discussed above, history tells us that the "single context" argument is false and that empirical adequacy cannot account for all features of theories that are relevant to their appraisal. Clearly, there may be many contexts of appraisal, but Laudan explicates two: acceptance and pursuit. The context of acceptance is the one used daily by most practicing scientists and physicians. If one wants to prescribe a drug for a patient with high blood pressure, one implicitly *accepts* those parts of the operative tradition dealing with this situation and likely prescribes a diuretic. If one is a particle physicist, one *accepts* the value and necessity of particle accelerators. The rationale for such choices, according to Laudan, is that the tradition chosen exhibits the *"highest problem-solving adequacy."* Pursuit, in contrast, has a different rationale, one more dependent on potential diachronic metrics than on synchronic evaluation. A scientist may choose to work on a highly speculative hypothesis, one that has little merit,[68] because they believe it has the potential to answer more problems and explain more anomalies than any in existence *at the time*.

This does not mean the scientist must devote themself *only* to the novel hypothesis. Scientists often use the strategy of "multiple working hypotheses" (see Section 5.5.1), wherein they begin by not assuming any of the theories extant are true and holding their evaluation in abeyance until they accumulate sufficient experimental evidence to determine which hypothesis merits acceptance. What may be the best modern example of this type of pursuit was Stanley Prusiner's search for a theory of "slow virus diseases" (see page 167). Prusiner courageously and doggedly pursued an ostensibly outrageous theory that had no precedent, no prior experimental support, and no practical merit, but if true, would solve the puzzle of the etiology of slow virus diseases, something no theory at the time could do. A huge experimental corpus later proved his hypothesis was true, a feat recognized by

---

[68] That is, problem solving success, inductive support, level of falsifiability, or ability to make novel predictions.

his winning the Nobel Prize in 1998. Kuhn certainly would object to the notion of multiple working hypotheses because it would require the simultaneous existence of two or more paradigms. Lakatos would criticize the idea because it would mandate concurrent, distinct hard cores. Laudan [758], and Feyerabend, in contrast, argue *"the emergence of every new research tradition occurs under just such circumstances."*[69]

---

[69] This brings up the interesting and intriguing question of how these three titans of the philosophy of science could examine precisely the same history and develop three distinct models for scientific practice and progress.

7

Epilogue

...possession of knowledge and theoretical tools is insufficient to support the conduct of novel, significant science. Such science requires many intangible elements, including curiosity, creativity, initiative, energy, logic, enthusiasm, common sense, critical thinking, doggedness, patience, stamina, perspective, and thoughtfulness. One also must understand what I refer to as 'the scientific gestalt,' by which I mean the manner in which the component parts of science, including rote knowledge extant, scientific paradigms (both technical and intellectual), science history, psychological and social aspects of science, and philosophies of science, integrate together to produce a highly dynamic and highly complex system, 'science,' that must be appreciated intellectually as well as viscerally and intuitively.

Appendix A

A Simple Question with Profound Implications

We have discussed the passions of Max Delbrück (Section 3.3.2) and Leroy Hood (Section 3.3.3), each of whom pursued obviously ambitious, important, novel questions. However, ostensibly simple questions often can lead to profound and important insights, *if one considers the question deeply enough.* An example is a question asked by Ludwig Wittgenstein in 1945 (see below)[1] that, although not intended by Wittgenstein, was relevant to one of the great mysteries in neuroscience and neurology – how we initiate movement. Wittgenstein is considered by many to be the greatest philosopher of the twentieth century [759–761]. He was a student of Bertrand Russell (also one of the greatest philosophers of the twentieth century) at Cambridge from 1911–1913 where he was exposed to Russell's philosophy, *logical atomism.* This philosophy dealt with the way we see the world, in terms of logical propositions (sentences), their constituent parts (atoms), and the logical relations among them. Russell applied this philosophy to logic and mathematics. Wittgenstein focused on linguistics. Their efforts and those of others, including Alfred North Whitehead and G. E. Moore, would spawn a major branch of philosophy, *analytic philosophy,* the aim of which is to use logic to *"... attain conceptual clarity"* and to ensure *"that philosophy [is] consistent with the success of modern science."*[2]

Wittgenstein was particularly interested in the logical structure of language and how language was used, or more accurately, misused, in philosophy. He is known in particular for two works, *Tractatus Logico-Philosophicus,* published in 1922,[3] and *Philosophical Investigations (PI),* published posthumously in 1953. One of the concepts addressed in *PI* was *will.* What do we mean when we use the word "will?" How do we "attain conceptual clarity" with respect to this term and how are we to use it in a manner that is "consistent with ... science?" Wittgenstein used the simple action of raising his arm to address these questions. He asked:

---

[1] I thank Stuart Firestein for making me aware of this question.

[2] Quotations are from www.philosophybasics.com/movements_analytic.html. Note that *Analytic philosophy* has been defined in multiple ways (e.g., see www.britannica.com/topic/analytic-philosophy or Beany [762]) and also has been termed *linguistic philosophy.*

[3] *Tractatus Logico-Philosophicus* was originally published in 1921 in German as *Logisch-Philosophische Abhandlung.*

... 'When I raise my arm,' my arm rises. And now a problem emerges: what is left over if I subtract the fact that my arm rises from the fact that I raise my arm? ((Are the kinaesthetic sensations my willing?)) When I raise my arm, I don't usually *try* to raise it. [It just goes up.][4]

One interpretation of the meaning of this quotation is *"[a] bodily movement ['I raise my arm'] plus an existentially prior act of will equals an action ['my arm rises']"* [763]. What is this "existentially prior act?" Wittgenstein does not know, nor does he think he can know—"The world is independent of my will."[5] This question has been the subject of study for neurophysiologists and psychologists seeking to understand the physical bases of initiation of movement, free will (volition), and consciousness. The latter goals have tremendous implications in politics, and especially in the law [764], where *intent* and *action* are key principles determining responsibility but the concept of an "existential prior act" has not been considered.

What *has* science revealed about the existential prior act and how it fits into the sequence of intent/will→action? It turns out we know quite a lot [764], but we still do not have *the* answer [765]. If one observes a subject performing voluntary and involuntary movements and correlates these movements with measures of brain activity, one can determine a temporal sequence of events related to conscious will and subsequent movement. Elucidating neuronal pathways responsible for movement, for example, lifting one's arm, has been accomplished. Establishing a temporal relationship between this action and the desire to perform the action has been accomplished. Much progress also has been made in understanding what happens in the brain *before* either the will to move or the movement itself occurs. It has been found that electrical changes in the brain occur ≈0.5 s before movement occurs and that subjects' "first feeling the urge to act" is observed ≈0.2 s beforehand [766]. So, something (an existentially prior act?) is happening in the brain *before* we consciously *will* our arm to rise.[6] Libet argues that the volitional process thus is initiated unconsciously [769], which raises the question "did we move our arm because of our own free will or because something in our brain, over which we have no control, told us to?" Continued work in this area suggests that although the existential prior act may indeed be unconscious in nature, we may still consciously exert our will by stopping the action once we realize it has been "ordered" by our brain. It is the exercise of our own free will that may be the difference between the fact that my arm rises from the fact that I raise my arm. However, the fact that we have not *initiated* the action consciously suggests that Wittgenstein was at least partially correct when he wrote "the world is independent of my will." The entire chain of events linked to raising our arm thus can be schematized as follows.

---

[4] Quotation from Wittgenstein [368], §621–622.
[5] From Wittgenstein [20] §6.373.
[6] Not everyone agrees this is true. See Trevena and Miller [767] or Pockett and Purdy [768].

"Existential prior act" → I become aware of the process of raising my arm ("I raise my arm") → I allow the process to continue of my own free → "My arm rises"

We now have a new question: What initiates the existential prior act, that is, our unconscious desire to raise our arm? or, as Wittgenstein might ask, "what is left over if I subtract the fact that my arm rises from the fact that my arm hasn't risen? We don't know, and maybe this new question represents an unknowable unknown? I don't know!"

Appendix B

Chaos in the Brickyard

by Bernard K. Forscher [640]

Once upon a time, among the activities and occupations of man there was an activity called scientific research, and the performers of this activity were called scientists. In reality, however, these men were builders who constructed edifices, called explanations or laws, by assembling bricks, called facts. When the bricks were sound and were assembled properly, the edifice was useful and durable and brought pleasure, and sometimes reward, to the builder. If the bricks were faulty or if they were assembled badly, the edifice would crumble, and this kind of disaster could be very dangerous to innocent users of the edifice as well as to the builder who sometimes was destroyed by the collapse. Because the quality of the bricks was so important to the success of the edifice, and because bricks were so scarce, in those days the builders made their own bricks. The making of bricks was a difficult and expensive undertaking and the wise builder avoided waste by making only bricks of the shape and size necessary for the enterprise at hand. The builder was guided in this manufacture by a blueprint, called a theory or hypothesis.

It came to pass that builders realized that they were sorely hampered in their efforts by delays in obtaining bricks. Thus, there arose a new skilled trade known as brickmaking, called junior scientist to give the artisan proper pride in his work. This new arrangement was very efficient and the construction of edifices proceeded with great vigor. Sometimes, brick-makers became inspired and progressed to the status of builders. In spite of the separation of duties, bricks still were made with care and usually were produced only on order. Now and then, an enterprising brickmaker was able to foresee a demand and would prepare a stock of bricks ahead of time, but, in general, brickmaking was done on a custom basis because it still was a difficult and expensive process.

And then, it came to pass that a misunderstanding spread among the brickmakers (there are some who say that this misunderstanding developed as a result of careless training of a new generation of brickmakers). The brickmakers became obsessed with the making of bricks. When reminded that the ultimate goal was edifices, not bricks, they replied that if enough bricks were available, the builders would be able to select what was necessary and still continue to construct edifices. The flaws in this argument were not readily apparent and so, with the help of the citizens who were waiting to use the edifices yet to be built, amazing things happened. The expense of brickmaking became a minor factor because large sums of money were made available; the time and effort involved in brickmaking was reduced by ingenious automatic machinery; the ranks of the brickmakers were swelled by augmented training programs and intensive recruitment. It even was suggested that the production of a suitable number of bricks was equivalent to building an edifice and therefore should entitle the industrious brickmaker to assume the title of builder and, with the title, the authority.

And so it happened that the land became flooded with bricks. It became necessary to organize more and more storage places, called journals, and more and more elaborate systems of bookkeeping to record the inventory. In all of this, the brickmakers retained their pride and skill, and the bricks were of the very best quality. But production was ahead of demand and bricks no longer were made to order. The size and shape was now dictated by changing trends in fashion. In order to compete successfully with other brickmakers, production emphasized those types of brick that were easy to make and only rarely did an adventuresome brickmaker attempt a difficult or unusual design. The influence of tradition in production methods and in types of product became a dominating factor.

Unfortunately, the builders were almost destroyed. It became difficult to find the proper bricks for a task because one had to hunt among so many. It became difficult to find a suitable plot for construction of an edifice because the ground was covered with loose bricks. It became difficult to complete a useful edifice because, as soon as the foundations were discernible, they were buried under an avalanche of random bricks. And, saddest of all, sometimes no effort was made even to maintain the distinction between a pile of bricks and a true edifice.

References

[1] Purcell EM, Pound RV: A Nuclear Spin System at Negative Temperature. *Physical Review* 1951, **81**:279–280.

[2] Needham P: Determining Sameness of Substance. *The British Journal for the Philosophy of Science* 2017, **68**(4):953–979, http://dx.doi.org/10.1093/bjps/axv050.

[3] Firestein S: *Ignorance*. Oxford University Press 2012.

[4] Sills J: Why Science? Scientists Share Their Stories. *Science* 2017, **356**(6338): 590–592.

[5] Gauch HG: *Scientific Method in Brief*. Cambridge University Press 2012.

[6] Popper KR: *The Logic of Scientific Discovery*. Routledge 1959.

[7] How Our Dictionaries Are Created. In *Oxford English Dictionary*, 3rd edition, Oxford University Press 2014.

[8] Ross S: Scientist: The Story of a Word. *Annals of Science* 1962, **18**(2):65–85.

[9] Bacon F: *Novum Organum by Lord Bacon*. P. F. Collier & Son 1902.

[10] Markie P, Folescu M: Rationalism vs. Empiricism. In *Stanford Encyclopedia of Philosophy*. Edited by Zalta EN 2017, https://plato.stanford.edu/entries/rationalism-empiricism/#1.2.

[11] Atchison J: *Language Change: Progress or Decay*. Cambridge University Press 2001.

[12] Shapiro AE: Newton's "Experimental Philosophy." *Early Science and Medicine* 2004, **9**(3):185–217.

[13] Pasnau R: *After Certainty. A History of Our Epistemic Ideals and Illusions*. Oxford University Press 2017.

[14] Science. In *Oxford English Dictionary*, 3rd edition, Oxford University Press 2014, www.oed.com/view/Entry/172672?redirectedFrom=science#eid.

[15] Weaver W: Science and Complexity. *American Scientist* 1948, **36**: 536–544.

[16] Smart JJC: *Between Science and Philosophy: An Introduction to the Philosophy of Science*. Random House 1968.

[17] Nagel E: *The Structure of Science: Problems in the Logic of Scientific Explanation*. Harcourt, Brace & World 1961.

[18] Van Fraassen BC: *The Scientific Image*. Oxford University Press 1980.

[19] Elliott K, McKaughan D: Non-Epistemic Values and the Multiple Goals of Science. *Philosophy of Science* 2014, **81**:1–21.

[20] Wittgenstein L: *Tractatus Logico-Philosophicus*. Harcourt, Brace & Company, Inc. 1922.

[21] Hoyningen-Huene P: Systematicity: The Nature of Science. *Philosophia* 2008, **36**:167–180.

[22] Hoyningen-Huene P: *Systematicity: The Nature of Science*. Oxford University Press 2013.

[23] Roddenberry G: The Immunity Syndrome. *Star Trek. Episode 18. NBC.* January 19, 1968

[24] Nickles T: *The Problem of Demarcation*. University of Chicago Press 2013.

[25] Schindler S: *Theoretical Virtues in Science: Uncovering Reality through Theory*. Cambridge University Press 2018, www.cambridge.org/core/books/theoretical-virtues-in-science/BEECC887D80F03C8FFAC1EDBA0D30B05.

[26] Pigliucci M, Boudry M: *Why the Demarcation Problem Matters*. University of Chicago Press 2013:1–6.

[27] Alters BJ: Whose Nature of Science? *Journal of Research in Science Teaching* 1997, **34**:39–55, http://dx.doi.org/10.1002/(SICI)1098-2736(199701)34:1<39:: AID-TEA4>3.0.CO;2-P.

[28] Bormann K: The Interpretation of Parmenides by the Neoplatonist Simplicius. *The Monist* 1979, **62**:30–42.

[29] Preus A: *Historical Dictionary of Ancient Greek Philosophy*. Rowman & Littlefield, 2nd edition 2015.

[30] Mulligan K, Correia F: Facts. In *The Stanford Encyclopedia of Philosophy*, Spring 2013 edition. Edited by Zalta EN, Metaphysics Research Lab, Stanford University 2013.

[31] Wikipedia: Fact 2017, https://en.wikipedia.org/wiki/Fact.

[32] Eliot TS: Fragment of the Agon, in Sweeney Agonistes. In *The Complete Poems and Plays, 1909–1950*, Harcourt Brace & Company 1952: 80–81. [Unfinished poem].

[33] Sargent RM: *The Dissident Naturalist. Robert Boyle and the Philosophy of Experiment*. University of Chicago Press 1995.

[34] Franklin A: Review Essay: Experimental Questions. *Perspectives on Science* 1993, **1**:127–146.

[35] Pickering A: *Science as Practice and Culture*. University of Chicago Press 1992.

[36] Bauer HH: *Scientific Literacy and the Myth of the Scientific Method*. University of Illinois Press 1992.

[37] Sparkes A, et al. Towards Robot Scientists for Autonomous Scientific Discovery. *Automated Experimentation* 2010, **2**:1–1, www.ncbi.nlm.nih.gov/pubmed/20119518.

[38] Polanyi M: *Personal Knowledge*. University of Chicago Press 1958.

[39] Chang H: *Inventing Temperature: Measurement and Scientific Progress*. Oxford University Press 2004.

[40] Ireland J: *"Udana" and the "Itivuttaka": Two Classics from the Pali Canon*. Buddhist Publication Society 1997.

[41] Asquith PD, Kyburg HE (Eds): *Research in Philosophy of Science Bearing on Science Education*. Philosophy of Science Association 1979.

[42] Matthews MR: *Science Teaching: The Role of History and Philosophy of Science*. Routledge 1994.

[43] Santayana G: Reasons in Common Sense. In *The Life of Reason or The Phases of Human Progress, Volume 1*, Charles Scribner's Sons 1917.

[44] Boyle R: *The Works of the Honourable Robert Boyle. In Six Volumes. To Which Is Prefixed the Life of the Author*. J. and F. Rivington 1772.

[45] Bauerle C, et al.: *Vision and Change in Undergraduate Biology Education: A Call to Action*. American Association for the Advancement of Science.

[46] Massimi M: *Perspectivism*. Routledge Handbooks in Philosophy 2017:164–175.

[47] Chang H: Is Water H_2O? Evidence, Realism and Pluralism. In *Boston Series in the Philosophy of Science, Volume 293*. Edited by Cohen RS, Renn J, Gavroglu K, Springer Science + Business Media B.V. 2012.

[48] Cavendish H, et al. XXXVII. The Report of the Committee Appointed by the Royal Society to Consider of the Best Method of Adjusting the Fixed Points of Thermometers; and of the Precautions Necessary to Be Used in Making Experiments with Those Instruments. *Philosophical Transactions of the Royal Society of London* 1777, **67**:816–857, https://royalsocietypublishing.org/doi/abs/10.1098/rstl.1777.0038.

[49] Chang H: The Myth of the Boiling Point. *Science Progress* 2008, **91**(3):219–240, www.jstor.org/stable/43423228.

[50] Kenrick FB, Gilbert CS, Wismer KL: The Superheating of Liquids. *The Journal of Physical Chemistry* 1923, **28**(12):1297–1307, https://doi.org/10.1021/j150246a009.

[51] Soffar H: What Are the Importance of Atomic Clocks and How Do They Work 2018, www.online-sciences.com/technology/what-are-the-importance-of-the-atomic-clocks-and-how-do-they-work/.

[52] Grebing C, et al. Realization of a Timescale with an Accurate Optical Lattice Clock. *Optica* 2016, **3**(6):563–569, www.osapublishing.org/optica/abstract.cfm?URI=optica-3-6-563.

[53] Starr M: We Now Have Atomic Clocks so Precise, They Could Detect Space-Time Distortion 2018, www.sciencealert.com/scientists-have-developed-atomic-clocks-so-precise-they-could-detect-gravitational-waves.

[54] Blokker E, et al. The Chemical Bond: When Atom Size Instead of Electronegativity Difference Determines Trend in Bond Strength. *Chemistry – A European Journal* 2021, **27**(63):15616–15622, https://doi.org/10.1002/chem.202103544.

[55] Hayes W: Max Ludwig Henning Delbrück 1906–1981. *National Academy of Sciences* 1993.

[56] Delbrück M: The Arrow of Time – Beginning and End. *Engineering & Science* 1978.

[57] Judson HF: *The Eighth Day of Creation: Makers of the Revolution in Biology*. Simon and Shuster 1979.

[58] Delbrück M: Light and Life III. *Carlsberg Research Communications* 1976, **41**(6):299–309, http://dx.doi.org/10.1007/BF02906138.

[59] Bohr N: Light and Life. *Nature* 1933, **131**(3308):421–423, https://doi.org/10.1038/131421a0.

[60] McKaughan DJ: The Influence of Niels Bohr on Max Delbruck: Revisiting the Hopes Inspired by "Light and Life". *Isis* 2005, **96**(4):507–529, https://doi.org/10.1086/498591.

[61] Fischer EP, Lipson C: *Thinking About Science: Max Delbrück and the Origins of Molecular Biology*. W.W. Norton & Company 1988.

[62] Schrödinger E: *What Is Life? The Physical Aspect of the Living Cell*. Cambridge University Press 1967.

[63] Timoféeff-Ressovsky NW, Zimmer KG, Delbrück M: Über die Natur der Genmutation und der Genstruktur. *Nachrichten von der Gesellschaft der Wissenschaften zu Göttingen: mathematisch-physische Klasse, Fachgruppe VI: Biologie 1* 1935, **6**(13):189–245.

[64] Abir-Am P: The Discourse of Physical Power and Biological Knowledge in the 1930s: A Reappraisal of the Rockefeller Foundation's "Policy" in Molecular Biology. *Social Studies of Science* 1982, **12**(3):341–382, www.jstor.org/stable/284665.

[65] Harding C: Max Delbrück (1906–1981), http://resolver.caltech.edu/CaltechOH:OH _Delbruck_M.

[66] Ellis EL, Delbrück M: The Growth of Bacteriophage. *The Journal of General Physiology* 1939, **22**(3):365–384.

[67] Luria SE, Delbrück M: Mutations of Bacteria from Virus Sensitivity to Virus Resistance. *Genetics* 1943, **28**(6):491–511, www.genetics.org/content/28/6/491.

[68] Gad S: 1969, www.nobelprize.org/prizes/medicine/1969/ceremony-speech/.

[69] Kendrew JC: How Molecular Biology Started. *Scientific American* 1967, **216**(3):141–144, www.jstor.org/stable/24931441.

[70] Timmerman LD: *Hood: Trailblazer of the Genomics Age.* Bandera Press LLC 2017, https://books.google.com/books?id=bjOHDAEACAAJ.

[71] Hood LE: My Life and Adventures Integrating Biology and Technology 2002, www.kyotoprize.org/wp/wp-content/uploads/2016/02/18kA_lct_EN.pdf.

[72] Willyard C: New Human Gene Tally Reignites Debate. *Nature* 2018, **558**(7710):354, http://dx.doi.org/10.1038/d41586-018-05462-w.

[73] Salzberg SL: Open Questions: How Many Genes Do We Have? *BMC Biology* 2018, **16**:94–94, www.ncbi.nlm.nih.gov/pubmed/30124169.

[74] Dreyer WJ, Bennett JC: The Molecular Basis of Antibody Formation: A Paradox. *Proceedings of the National Academy of Sciences* 1965, **54**(3):864–869, www.pnas.org/content/54/3/864.

[75] Dr. Henry Huang, Plaintiff v. California Institute of Technology, et al. Defendant Case No. CV 03-1140 MRP 2004.

[76] Department of Molecular Biotechnology 2001, http://nick-lab.gs.washington .edu/mbt-web/.

[77] Ideker T, Galitski T, Hood L: A New Approach to Decoding Life: Systems Biology. *Annual Review of Genomics and Human Genetics* 2001, **2**:343–372, https://doi.org/10.1146/annurev.genom.2.1.343. [PMID: 11701654].

[78] Institute for Systems Biology 2000, https://systemsbiology.org/.

[79] Clough M: Teaching and Assessing the Nature of Science: How to Effectively Incorporate the Nature of Science in Your Classroom. *The Science Teacher* 2011, **78**:56–60.

[80] Clough MP: *History and Nature of Science in Science Education.* SensePublishers 2017:39–51, https://doi.org/10.1007/978-94-6300-749-8_3.

[81] Thomas L: *The Wonderful Mistake: Notes of a Biology Watcher: Incorporating the Lives of a Cell, and, the Medusa and the Snail.* Oxford University Press 1988.

[82] Merton RK: *Social Theory and Social Structure, Rev. ed.* Free Press 1957.

[83] Angli GH: *Exercitatio Anatomica de Motu Cordis et Sanguinis in Animalibus.* Charles C. Thomas 1928.

[84] Feynman R: *No Ordinary Genius: The Illustrated Richard Feynman.* W.W. Norton 1994.

[85] Brewster D: *Memoirs of the Life, Writings, and Discoveries of Sir Isaac Newton, Volume 2.* Thomas Constable and Co. 1855.

[86] Lynd RY: *The Pleasures of Ignorance.* Methuen 1928.

[87] Department of Defense News Briefing – Secretary Rumsfeld and Gen. Myers https://archive.defense.gov/Transcripts/Transcript.aspx?TranscriptID=2636.

[88] de Magalhães JP, Wang J: The Fog of Genetics: Known Unknowns and Unknown Unknowns in the Genetics of Complex Traits and Diseases. *bioRxiv* 2019, www.biorxiv.org/content/early/2019/02/18/553685.

[89] Johnson S: *Farsighted. How We Make the Decisions That Matter the Most.* Riverhead Books 2018.

[90] Valentin J (Ed): The 2007 Recommendations of the International Commission on Radiological Protection. *Annals of the ICRP* 2007, **37**(2–4):1–332.

[91] Weinberg AM: Science and Trans-Science. *Minerva* 1972, **10**(2):209–222.

[92] Geer, Jr DE: Unknowable Unknowns. *IEEE Security Privacy* 2019, **17**(2):80–79.

[93] Haldane JBS: *Possible Worlds: And Other Essays*. Chatto and Windus 1927.

[94] Žižek S: Philosophy, the "Unknown Knowns," and the Public Use of Reason. *Topoi* 2006, **25**:137–142, https://doi.org/10.1007/s11245-006-0021-2.

[95] Gross M, McGoey L: *Introduction*. Routledge 2015.

[96] Gross M, McGoey L: Introduction. In *Routledge International Handbook of Ignorance Studies*. Edited by Gross M, McGoey L, Routledge 2015:1–14.

[97] Duncan R, Weston-Smith M (Eds): *The Encyclopaedia of Ignorance: Everything You Ever Wanted to Know About the Unknown*. Pergamon Press 1977.

[98] Little B: From the Editor. *The Journal of Product Innovation Management* 1985, **2**(3):131–133.

[99] Smithson M: Ignorance and Science: Dilemmas, Perspectives, and Prospects. *Knowledge* 1993, **15**(2):133–156, https://doi.org/10.1177/107554709301500202.

[100] Smithson M: *Ignorance and Uncertainty: Emerging Paradigms*. Cognitive Science. Springer-Verlag Publishing 1989.

[101] Kassar NE: What Ignorance Really Is. Examining the Foundations of Epistemology of Ignorance. *Social Epistemology* 2018, **32**(5):300–310, https://doi.org/10.1080/02691728.2018.1518498.

[102] LeMorvan P: On Ignorance: A Reply to Peels. *Philosophia* 2011, **39**(2):335–344, https://doi.org/10.1007/s11406-010-9292-3.

[103] Zimmerman MJ: *Living with Uncertainty: The Moral Significance of Ignorance*. Cambridge University Press 2008.

[104] Medina J: *The Epistemology of Resistance*. Oxford University Press 2013.

[105] Alcoff LM: *Epistemologies of Ignorance. Three Types*. State University of New York Press 2007.

[106] Tuana N: The Speculum of Ignorance: The Women's Health Movement and Epistemologies of Ignorance. *Hypatia* 2006, **21**(3):1–19.

[107] Proctor RN, Schiebinger L (Eds): *Agnotology. The Making and Unmaking of Ignorance*. Stanford University Press 2008.

[108] Peels R: What Is Ignorance? *Philosophia* 2010, **38**:57–67, https://doi.org/10.1007/s11406-009-9202-8.

[109] Goldman AI, Olsso EJ: Epistemic Value. In *Reliabilism and the Value of Knowledge*, Handbook of Stuff I Care About. Edited by Pritchard D, Haddock A, Millar A, Oxford University Press 2009:19–41.

[110] DeNicola DR: *Understanding Ignorance: The Surprising Impact of What We Don't Know*. MIT Press 2017.

[111] Shea A: *Reading the OED: One Man, one Year, 21,730 pages*. Penguin 2008.

[112] Haas J, Vogt KM: Ignorance and Investigation. In *Routledge International Handbook of Ignorance Studies*. Edited by Gross M, McGoey L, Routledge 2015: 17–25.

[113] Kinsey A, Pomeroy W, Martin C: *Sexual Behavior in the Human Male*. Indiana University Press 1998, https://books.google.com/books?id=pfMKrY3VvigC.

[114] Kinsey A, for Sex Research I: *Sexual Behavior in the Human Female*. Indiana University Press 1998, https://books.google.com/books?id=9GpBB61LV14C.

[115] Masters WH, Johnson VE: *Human Sexual Response*. Little, Brown 1966.

[116] Masters W, Johnson V: *Human Sexual Inadequacy*. Little, Brown 1970, https://books.google.com/books?id=W-JrAAAAMAAJ.

[117] Descartes R: *Discourse on the Method of Rightly Conducting Reason and Seeking Truth in the Sciences*, M. Walter Dunne 1901.

[118] Ravetz JR: The Sin of Science: Ignorance of Ignorance. *Knowledge* 1993, **15**(2):157–165, https://doi.org/10.1177/107554709301500203.

[119] Burrow GN: The Body of Medical Knowledge Required Today Far Exceeds What Students Can Learn in 4 Years. *The Chronicle of Higher Education* 1990.

[120] Witte MH, Crown P, Bernas M, Garcia FA: "Ignoramics" in Medical and Premedical Education. *Journal of Investigative Medicine* 2008, **56**(7):897–901.

[121] Kerwin A: None Too Solid: Medical Ignorance. *Knowledge* 1993, **15**(2):166–185, https://doi.org/10.1177/107554709301500204.

[122] Keniston AH, Peden BF: Infusing Critical Thinking into College Courses. *Issue in Teaching and Learning* 1992, **4**:7–12.

[123] Peden BF, Keniston A: Methods for Critical Thinking. *Wisconsin Dialogue* 1991, **11**:12–34.

[124] Stocking SH: Ignorance-Based Instruction in Higher Education. *The Journalism Educator* 1992, **47**(3):43–53, https://doi.org/10.1177/107769589204700306.

[125] Witte M, Kerwin A, Witte CL: *The Curriculum on Medical Ignorance: Coursebook & Resource Manuals for Instructors and Students*. The University of Arizona College of Medicine.

[126] Schwartz MA: The Importance of Stupidity in Scientific Research. *Journal of Cell Science* 2008, **121**(11):1771–1771, http://jcs.biologists.org/content/121/11/1771.

[127] Kerwin A: Ignorance and Scientific Progress. *Lymphology* 1986, **19**:31–32.

[128] Popper K: *Logik Der Forschung. Zur Erkenntnistheorie der modernen Naturwissenschaft*. Springer-Verlag 1935.

[129] Nola R: *After Popper, Kuhn, and Feyerabend: Recent Issues in Theories of Scientific Method*. Springer Science + Business Media 2000.

[130] Feyerabend P: *Against Method*. Verso 1975.

[131] Tattersall I: What's So Special about Science? *Evolution: Education and Outreach* 2008, **1**:36–41.

[132] Weinberg S: The Methods of Science ... and Those by Which We Live. *Academic Questions* 1995, **8**(2):7–13.

[133] Sober E: *Ockham's Razor- A User's Manual*. Cambridge University Press 2015.

[134] Stuart MT, Fehige Y: Motivating the History of the Philosophy of Thought Experiments. *Hopos: The Journal of the International Society for the History of Philosophy of Science* 2021, **11**:212–221.

[135] Fehige Y: The Annus Mirabilis of 1986: Thought Experiments and Scientific Pluralism. *Hopos: The Journal of the International Society for the History of Philosophy of Science* 2021, **11**:222–240.

[136] Einstein A: How I Created the Theory of Relativity. *Physics Today* 1982, **35**:45–47.

[137] Einstein A: *Zur allgemeinen Relativitätstheorie*. Verlag der königlichen akademie der wissenschaften 1915.

[138] Dyson F, Eddington A, Davidson C: A Determination of the Deflection of Light by the Sun's Gravitational Field, from Observations Made at the Total Eclipse of May 29, 1919. *Philosophical Transactions of the Royal Society A* **220**:291–333.

[139] Abbott BP: Observation of Gravitational Waves from a Binary Black Hole Merger. *Physical Review Letters* 2016, **116**:061102.

[140] Einstein A, et al.: On the Relativity Principle and the Conclusions Drawn from It. *Jahrbuch der Radioaktivität und Elektronik* 1907, **4**:411–462.

[141] Cartlidge E: Relativity Survives Drop Test. *Science* 2017, **358**.

[142] Touboul P, et al. MICROSCOPE Mission: First Results of a Space Test of the Equivalence Principle. *Physical Review Letters* 2017, **119**:231101.

[143] Suter R: Aristotle and the Scientific Method. *The Scientific Monthly* 1939, **49**(5):468–472.

[144] Shuttleworth M: History of the Scientific Method. *Explorable* 2009, https://explorable.com/history-of-the-scientific-method.

[145] Koyré A: The Origins of Modern Science: A New Interpretation. *Diogenes* 1956, **4**(16):1–22.

[146] van Middendorp, Joost J, Sanchez GM, Burridge AL: The Edwin Smith Papyrus: A Clinical Reappraisal of the Oldest Known Document on Spinal Injuries. *European Spine Journal* 2010, **19**(11):1815–1823, www.ncbi.nlm.nih.gov/pmc/articles/PMC2989268/.

[147] Breasted JH: *The Edwin Smith Surgical Papyrus Published in Facsimile and Hieroglyphic Transliteration*. University of Chicago Press 1930.

[148] Gordetsky J, O'Brien J: Urology and the Scientific Method in Ancient Egypt. *Urology* 2009, **73**(3):476–479.

[149] Brawanski A: On the Myth of the Edwin Smith Papyrus: Is It Magic or Science? *Acta Neurochirurgica* 2012, **154**(12):2285–2291.

[150] Aristotle: *De Generatione Animalium, Volume 3* c350BC.

[151] Gomperz T: *Greek Thinkers: A History of Ancient Philosophy, Volume 4*. John Murray 1912.

[152] Shields C: Aristotle. In *Stanford Encyclopedia of Philosophy*. Edited by Zalta EN 2015, https://plato.stanford.edu/entries/aristotle/.

[153] Kattsoff LO: Ptolemy and Scientific Method: A Note on the History of an Idea. *Isis* 1947, **38**(1/2):18–22.

[154] Kuhn TS: *The Copernican Revolution: Planetary Astronomy in the Development of Western Thought, Volume 16*. Harvard University Press 1957.

[155] Hughes M: Claudius Ptolemy: 2nd Century Egyptian Astronomer. *Perth Observatory* 2016.

[156] al Khalili J: *Pathfinders: The Golden Age of Arabic Science*. Penguin Books 2010.

[157] Ahmad I: Islamic Contributions to Modern Scientific Methods. *Journal of Faith and Science Exchange* 2012:27–36, http://hdl.handle.net/2144/3968.

[158] Holmyard EJ: *Makers of Chemistry*. Clarendon Press 1931.

[159] Gorini R: Al-Haytham the Man of Experience. First Steps in the Science of Vision. *Journal of the International Society for the History of Islamic Medicine* 2003, **2**(4):53–55.

[160] Daneshfard B, Dalfardi B, Nezhad GSM: Ibn al-Haytham (965–1039 AD), the Original Portrayal of the Modern Theory of Vision. *Journal of Medical Biography* 2016, **24**(2):227–231.

[161] Bsoul LA: Classical Muslim Scholars' Development of the Experimental Scientific Method: 'Iml al-Istiqrā'/induction Approach and Methodology. *Journal of Humanities and Cultural Studies R & D* 2017, **2**(4):1–33.

[162] Scientific Method 2018, www.wiki30.com/wa?s=scientific_method.

[163] Dales RC: *Richard Grosseteste and Scientific Method, Volume 57*. University of Pennsylvania Press 1973.

[164] Crombie AC: *Grosseteste's Position in the History of Science* 1955:98–120.

[165] Clagett M: *Greek Science in Antiquity*. Abelard-Schuman 1955.

[166] Weisheipl JA: *Life and Works of St. Albert*. Pontifical Institute of Mediaeval Studies 1980 chap. 1:13–51.

[167] Lindberg DC: *The Beginnings of Western Science: The European Tradition in Philosophical, Religious, and Institutional Context, 600 BC to AD 1450*. University of Chicago Press 1992.

[168] Hume D: *A TREATISE of Human Nature: BEING An Attempt to Introduce the Experimental Method of Reasoning INTO MORAL SUBJECTS*. John Noon 1739.

[169] Wallace WA: *Albertus Magnus on Suppositional Necessity in the Natural Sciences*. Pontifical Institute of Mediaeval Studies 1980 chap. 4:103–128.

[170] Thorndike L: *A History of Magic and Experimental Science, Volume II*. Columbia University Press 1923.

[171] Easton SC: Roger Bacon and His Search for a Universal Science 1952.

[172] Hackett J: *Roger Bacon and the Sciences: Commemorative Essays 1996, Volume 57*. Brill 1997.

[173] The Center for Islamic Studies: Scientific Method. https://islamic-study.org/scientific-method/.

[174] Beale S: Four Things You Need to Know about John Duns Scotus. *Catholic Exchange* 2014, https://catholicexchange.com/four-things-need-know-john-duns-scotus.

[175] Losee J: *A Historical Introduction to the Philosophy of Science*. Oxford University Press, 4th edition 2001.

[176] Losee J: *Theories of Causality: From Antiquity to the Present*. Routledge 2017.

[177] Thorburn WM: The Myth of Occam's Razor. *Mind* 1918, **27**(107):345–353.

[178] Crombie AC: *The Significance of Medieval Discussions of Scientific Method for the Scientific Revolution*, University of Wisconsin Press 1959 chap. 3:79–101.

[179] Adams MM: *William Ockham, 2 vols, Volume 2*. University of Notre Dame Press 1987.

[180] Occam's Razor 2018, www.oed.com/view/Entry/234636?redirectedFrom=Ockham%27s+razor#eid.

[181] Pacer M, Lombrozo T: Ockham's Razor Cuts to the Root: Simplicity in Causal Explanation. *Journal of Experimental Psychology: General* 2017, **146**(12):1761–1780.

[182] Hübener W: Occam's Razor Not Mysterious. *Archiv für Begriffsgeschichte* 1983, **27**:73–92, www.jstor.org/stable/24362877.

[183] Scientific Method. In *Oxford English Dictionary*, 3rd edition, Oxford University Press 2014.

[184] Rees G, Wakely M: *The Instauratio magna Part II: Novum organum and Associated Texts*. Clarendon Press 2004.

[185] Malherbe M: *Bacon's Method of Science*. Cambridge University Press 1996 chap. 3:75–98.

[186] Snyder LJ: Renovating the Novum Organum: Bacon, Whewell and Induction. *Studies in History and Philosophy of Science* 1999, **30**(4):531–557.

[187] Einstein A: *Ideas and Opinions*. Bonanza Books 1954, https://books.google.com/books?id=FFcNAQAAMAAJ.

[188] Hawking Stephen: *A Brief History of Time*. Updated and Expanded Tenth Anniversary Edition. Bantam Books 1998, https://search.library.wisc.edu/catalog/999837008202121.

[189] Viviani V: Historical Account of the Life of Galileo Galilei. In *On the Life of Galileo*. Edited by Gattei S, Princeton University Press 2019, www.jstor.org/stable/j.ctvc772t2.

[190] Martinez AA: *Science Secrets: the Truth About Darwin's Finches, Einstein's Wife, and Other Myths*. University of Pittsburgh Press 2011.

[191] Wootton D: *Galileo: Watcher of the Skies*. Yale University Press 2010.

[192] Crombie AC: *Medieval and Early Modern Science, Volume II*. Doubleday Anchor Books 1959.

[193] Galilie G: *Il Saggiatore (The Assayer)* 1623.

[194] Crombie AC: *Robert Grosseteste and the Origins of Experimental Science 1100–1700*. Oxford University Press 1953.

[195] Galilie G: *Dialogue Concerning the Two Chief World Systems: Ptolemaic and Copernican*. Modern Library 2001.

[196] Galilie G: *Two New Sciences, Including Centers of Gravity and Force of Percussion*. University of Wisconsin Press 1974.

[197] Kuhn TS: *The Structure of Scientific Revolutions*. University of Chicago Press 1962.

[198] Noland A, Wiener PP: *Roots of Scientific Thought: A Cultural Perspective*. Basic Books 1957.

[199] Clavelin M: *The Natural Philosophy of Galileo*. MIT Press 1974.

[200] Flage DE, Bonnen CA: *Descartes and Method: A Search for a Method in Meditations*. Routledge 1999.

[201] Descartes R: *Discourse on Method and Meditations on First Philosophy. Translated by Donald A. Cress*. Hackett, 4th edition 1993.

[202] Russell B: *A History of Western Philosophy*. Simon & Shuster 1945.

[203] Browne A: Descartes's Dreams. *Journal of the Warburg and Courtauld Institutes* 1977, **40**:256–273, www.jstor.org/stable/750999.

[204] Davies R: *Descartes: Belief, Scepticism and Virtue*. Routledge 2001.

[205] Schuster J: *Whatever Should We Do with the Cartesian Method? Reclaiming Descartes for the History of Science*. Oxford University Press 1993 .

[206] Garber D: *Descartes' Metaphysical Physics*. University of Chicago Press 1992.

[207] Leibniz GW: *Philosophical Essays*. Hackett 1989.

[208] Hatfield G: Science, Certainty, and Descartes. *PSA: Proceedings of the Biennial Meeting of the Philosophy of Science Association* 1988, **1988**:249–262, www.jstor.org/stable/192888.

[209] Garber D: Descartes and Method in 1637. *PSA: Proceedings of the Biennial Meeting of the Philosophy of Science Association* 1998, **2**.

[210] Whewell W: *The Philosophy of the Inductive Sciences Founded upon Their History, Volume 2*. John W. Parker 1847.

[211] Shouls PA: Reason, Method, and Science in the Philosophy of Descartes. *Australasian Journal of Philosophy* 1972, **50**:30–39.

[212] Boyle R: *New Experiments PHYSICO-MECHANICAL Touching the Air Whereunto is added A Defense of the Author's Explication of the EXPERIMENTS, against the OBJECTIONS OF Franciscus Linus and Thomas Hobbs*. H. Hall for Thomas Robinson, 3rd edition 1662.

[213] Davidson JS: Online Annotations to Robert Boyle's Sceptical Chymist. *Journal of Chemical Education* 2003, **80**(5):487.

[214] West JB: Robert Boyle's Landmark Book of 1660 with the First Experiments on Rarified Air. *Journal of Applied Physiology* 2005, **98**:31–39.

[215] Boyle R: *Certain Physiological Essays and Other Tracts; Written at Distant Times, and on Several Occasions*. Henry Herringman, 2nd edition 1669, https://books.google.co.uk/books?id=MNVAnQAACAAJ.

[216] Cohen IB: *Franklin and Newton: An Inquiry into Speculative Newtonian Experimental Science and Franklin's Work in Electricity as an Example Thereof*. American Philosophical Society: Memoirs of the American Philosophical Society, American Philosophical Society 1956, https://books.google.com/books?id=mbUVAAAAIAAJ.

[217] Domski M: *Philosophy of Science in Newton's General Scholium* 2013 .

[218] Hall LJ, Nomura Y: Evidence for the Multiverse in the Standard Model and Beyond. *Physical Review D* 2008, **78**(3):035001–1–035001–42, https://link.aps.org/doi/10.1103/PhysRevD.78.035001.

[219] Chaudhury PJ: Newton and Hypothesis. *Philosophy and Phenomenological Research* 1962, **22**(3):344–353.

[220] Cohen IB: *The Principia, the Newtonian Style, and the Newtonian Revolution.* University of Delaware Press 2012 .

[221] Smith GE: *The Methodology of the Principia.* Cambridge University Press 2002 chap. 4:138–173.

[222] Motte A: *The Mathematical Principles of Natural Philosophy by Sir Isaac Newton, translated by Andrew Motte.* Benjamin Motte 1729.

[223] Ducheyne S: *The Main Business of Natural Philosophy.* Springer 2012.

[224] Butts RE: *William Whewell Theory or Scientific Method.* Hackett 1989.

[225] Minto W: *Logic Inductive and Deductive.* Charles Scribner's Sons 1894.

[226] Cowles HM: The Age of Methods: William Whewell, Charles Peirce, and Scientific Kinds. *Isis* 2016, **107**(4):722–737.

[227] Darwin C: *The Origin of Species.* P. F. Collier 1909.

[228] Herschel JFW: A Preliminary Discourse on the Study of Natural Philosophy. In *The Cabinet Cyclopaedia.* Edited by Lardner D. Longman, Rees, Orme, Brown, Green, and Taylor 1831.

[229] Cobb AD: Inductivism in Practice: Experiment in John Herschel's Philosophy of Science. *HOPOS: The Journal of the International Society for the History of Philosophy of Science* 2012, **2**:21–54.

[230] Butts RE: *William Whewell's Theory of Scientific Method.* University of Pittsburgh Press 1968.

[231] Todhunter I: *William Whewell, D.D., An Account of His Writings, with Selections from His Literary and Scientific Correspondence, in Two Volumes, Volume II.* Macmillan and Co. 1876.

[232] Snyder LJ: Discoverers' Induction. *Philosophy of Science* 1997, **64**(4):580–604.

[233] Yeo RR: *Scientific Method and the Rhetoric of Science in Britain, 1830–1917,* D. Reidel Publishing Company 1986:259–297.

[234] Whewell W: *History of the Inductive Sciences, from the Earliest to the Present Times, Volume 1.* John W. Parker 1837.

[235] Firestein S: *Failure: Why Science Is So Successful.* Oxford University Press 2016.

[236] Putnam H: *Mathematics, Matter, and Method, Volume 1.* Cambridge University Press, 2nd edition 1975.

[237] Macleod C: John Stuart Mill 2018, https://plato.stanford.edu/archives/fall2018/entries/mill/.

[238] Mill JS: *A system of Logic, Ratiocinative and Inductive: Being a Connected View of the Principles of Evidence and the Methods of Scientific Investigation.* Longmans, Green and Company 1884.

[239] Ryan A: *The Philosophy of John Stuart Mill.* Macmillan 1970.

[240] Anschutz RP: *The Philosophy of J. S. Mill.* Oxford University Press 1953.

[241] Mill JS: *The Earlier Letters of John Stuart Mill 1812–1848.* University of Toronto Press, *Volume 12* 1963 .

[242] Snyder LJ: *Reforming Philosophy – A Victorian Debate on Science and Society.* University of Chicago Press, Chicago Edition 2006.

[243] Mill JS: *Autobiography.* Longmans, Green, Reader, and Dyer, 3rd edition 1874.

[244] LaFollette H, Shanks N: Animal Experimentation: The Legacy of Claude Bernard. *International Studies in the Philosophy of Science* 1994, **8**(3):195–210.

[245] Roll-Hansen N: Critical Teleology: Immanuel Kant and Claude Bernard on the Limitations of Experimental Biology. *Journal of the History of Biology* 1976, **9**:59–91.

[246] Cannon WB: *The Wisdom of the Body*. W. W. Norton & Co 1932.

[247] Grmek MD: *Raisonnement Experimental et Rechereches Toxicologiques Chez Claude Bernard*. Droz 1973.

[248] Medawar P: *Pluto's Republic: Incorporating The Art of the Soluble & Induction and Intuition in Scientific Thought*. Oxford University Press 1982.

[249] Renan E: *L'Œuvre de Claude Bernard*. Bailliere 1881.

[250] Hoefer C: Causal Determinism. In *Stanford Encyclopedia of Philosophy*. Edited by Zalta EN 2016 https://plato.stanford.edu/entries/determinism-causal/#Int.

[251] Bernard C: *An Introduction to the Study of Experimental Medicine: Transl. by Henry Copley Greene*. Henry Schuman 1949.

[252] Schaffner KF: *Discovery and Explanation in Biology and Medicine*. University of Chicago Press 1993.

[253] Wikipedia: Claude Bernard 2018, https://en.wikipedia.org/wiki/Claude_Bernard.

[254] Daston L: Scientific Error and the Ethos of Belief. *Social Research* 2005, **72**:1–28, www.jstor.org/stable/40972000.

[255] Jensen MB, Janik EL, Waclawik AJ: The Early Use of Blinding in Therapeutic Clinical Research of Neurological Disorders. *Journal of Neurological Research and Therapy* 2016, **1**(2):4–16, www.ncbi.nlm.nih.gov/pubmed/27617324.

[256] Stolberg M: Inventing the Randomized Double-Blind Trial: The Nürenberg Salt Test of 1835. *Journal of the Royal Society of Medicine* 2006, **99**(12):642–643.

[257] Zabell SL: The Rule of Succession. *Erkenntnis* 1989, **31**:283–321.

[258] Samuelson PA, Koopmans TC, Stone JRN: Report of the Evaluative Committee for Econometrica. *Econometrica* 1954, **22**(2):141–146, www.jstor.org/stable/1907538.

[259] Laudan L: *Historical Essays on Scientific Methodology*, Springer Netherlands. 1981.

[260] Morgan AD: *Formal Logic or, the Calculus of Inference, Necessary and Probable*. Taylor and Walton 1847.

[261] Jevons WS: *The Principles of Science: A Treatise on Logic and Scientific Method*. Macmillan 1913.

[262] Schabas M: *A World Ruled by Number: William Stanley Jevons and the Rise of Mathematical Economics*. Princeton University Press 1990, www.jstor.org/stable/j.ctt7zv0hj.

[263] Jeffreys H: *Scientific Inference*. Cambridge University Press, 3rd edition 1973.

[264] Laudan LL: *Induction and Probability in the Nineteenth Century, Volume 74*. North Holland 1973.

[265] Boole G: *An Investigation of the Laws of Thought: On Which Are Founded the Mathematical Theories of Logic and Probabilities*. Project Gutenberg 2017, www.gutenberg.org/files/15114/15114-pdf.pdf.

[266] Mill JS: *Letter to John Elliot Cairnes*. University of Toronto Press and Routledge & Kegan Paul 1871.

[267] Niiniluoto I: Defending Abduction. *Philosophy of Science* 1999, **66**:S436–S451.

[268] Carnap R: The Two Concepts of Probability: The Problem of Probability. *Philosophy and Phenomenological Research* 1945, **5**(4):513–532, www.jstor.org/stable/2102817.

[269] Salmon WC: Preface. *Synthese* 1977, **34**:ii–2.

[270] Reichenbach H: *Experience and Prediction*. University of Chicago Press 1938.

[271] Putnam H: Reichenbach's Metaphysical Picture. *Erkenntnis* 1991, **35**:61–75.

[272] Savvatimskiy A: Measurements of the Melting Point of Graphite and the Properties of Liquid Carbon (A Review for 1963–2003). *Carbon* 2005, **43**(6):1115–1142, www.sciencedirect.com/science/article/pii/S0008622305000291.

[273] Skyrms B: *Choice and Chance: An Introduction to Inductive Logic.* Wadsworth/Thomson Learning 2000.

[274] Hempel CG: Studies in the Logic of Confirmation. *Mind* 1945, **54**(213):1–26, www.jstor.org/stable/2250886.

[275] Hempel CG, Oppenheim P: Studies in the Logic of Explanation. *Philosophy of Science* 1948, **15**(2):135–175, www.jstor.org/stable/185169.

[276] Hempel CG: Deductive-Nomological vs. Statistical Explanation. In *Scientific Explanation, Space & Time, Volume III of Minnesota Studies in the Philosophy of Science.* Edited by Feigl H, Maxwell G. University of Minnesota Press 1962:98–169.

[277] Balashov Y, Rosenberg A (Eds): *Philosophy of Science: Contemporary Readings.* Routledge 2001.

[278] Thornton R: Karl Popper 2018, https://plato.stanford.edu/archives/fall2018/entries/popper/.

[279] Frank P: *Modern Science and its Philosophy.* Harvard University Press 1950.

[280] Blumberg AE, Feigl H: Logical Positivism. *Journal of Philosophy* 1931, **28**(11):281–296.

[281] Uebel T: "Logical Positivism"-"Logical Empiricism": What's in a Name? *Perspectives on Science* 2013, **21**:58–99.

[282] Richardson A, Uebel T: *The Cambridge Companion to Logical Empiricism.* Cambridge University Press 2007.

[283] Friedman M: *Reconsidering Logical Positivism.* Cambridge University Press 1999.

[284] Stadler F: *Vienna Circle: Logical Empiricism, Volume 25.* 2nd edition. Elsevier 2015:87–94.

[285] Kant I: *Critique of Pure Reason.* Cambridge University Press 1998 .

[286] Sahu KC, et al.: Relativistic Deflection of Background Starlight Measures the Mass of a Nearby White Dwarf Star. *Science* 2017, **356**(6342):1046–1050, http://science.sciencemag.org/content/356/6342/1046.

[287] Weinberg JR: Logik der Forschung: Zur Erkenntnistheorie der Modernen Naturwissenschaft. Von KARL POPPER. *The Philosophical Review* 1936, **45**(5):511–514, www.jstor.org/stable/2180508.

[288] Popper K: *Unended Quest: An Intellectual Autobiography.* Routledge, Taylor and Francis e-Library 2005.

[289] Newton I: New Theory about Light and Colors. *Philosophical Transactions of the Royal Society* 1671:3075–3087.

[290] Feyerabend PK: Against Method: Outline of an Anarchistic Theory of Knowledge. In *Analyses of Theories and Methods of Physics and Psychology, Volume 4.* Edited by Radner M, Winokur S. University of Minnesota Press 1970:17–130.

[291] Lakatos I: *The Methodology of Scientific Research Programmes: Philosophical Papers.* Cambridge University Press 1978.

[292] Preston J, Munvar G, Lamb D: *The Worst Enemy of Science?: Essays in Memory of Paul Feyerabend.* Oxford University Press 2000.

[293] Electrodynamics. In *Oxford English Dictionary* 2019, www.oed.com/view/Entry/270111?redirectedFrom=electrodynamics&.

[294] Stadler F: Paul Feyerabend and the Forgotten "Third Vienna Circle". *Annals of the Academy of Romanian Scientists* 2014, **6**:47–66.

[295] Preston J: Paul Feyerabend. In *The Stanford Encyclopedia of Philosophy*, Winter 2016 edition. Edited by Zalta EN. Metaphysics Research Lab, Stanford University 2016.

[296] Feyerabend PK: Herbert Feigl. A Biographical Sketch. In *Mind, Matter, and Method. Essays in Philosophy and Science in Honor of Herbert Feigl.* Edited by Feyerabend P, Maxwell G, University of Minnesota Press 1996:3–13.

[297] Feyerabend P: *Killing Time*. University of Chicago Press 1995.

[298] Feyerabend P: *Science in a Free Society*. NLB 1978.

[299] Koertge N: For and against Method. *British Journal for the Philosophy of Science* 1972, **23**:274–290.

[300] Nickles T: Heuristic Appraisal: Context of Discovery or Justification? In *Revisiting Discovery and Justification: Historical and Philosophical Perspectives on the Context Distinction*, Springer Netherlands 2006:159–182.

[301] Theocharis T, Psimopoulos M: Where Science Has Gone Wrong. *Nature* 1987, **329**:595.

[302] Horgan J: Profile: Paul Karl Feyerabend: The Worst Enemy of Science. *Scientific American* 1993, **268**(5):36–37, www.jstor.org/stable/24941475.

[303] Des-Cartes R: Meditationes de prima philosophia in qva dei existentia et animæ immortalitas demonstratvr 1641.

[304] Kidd IJ: Rethinking Feyerabend: The "Worst Enemy of Science"? *PLOS Biology* 2011, **9**(10):1–3.

[305] Shaw J: Was Feyerabend an Anarchist? The Structure(s) of "Anything Goes". *Studies in History and Philosophy of Science Part A* 2017, **64**:11–21.

[306] Brown MJ, Kidd IJ: Introduction: Reappraising Paul Feyerabend. *Studies in History and Philosophy of Science* 2016, **57**:1–8.

[307] Jacobson R: 2.5 Quintillion Bytes of Data Created Every Day. How Does CPG & Retail Manage It? *IBM Consumer Products Industry Blog* 2013, www.ibm.com/blogs/insights-on-business/consumer-products/2-5-quintillion-bytes-of-data-created-every-day-how-does-cpg-retail-manage-it/.

[308] Sybrandt J, Shtutman M, Safro I: Large-Scale Validation of Hypothesis Generation Systems via Candidate Ranking. *2018 IEEE International Conference on Big Data (Big Data)* 2018:1494–1503.

[309] National Science Board: Long-Lived Digital Data Collections: Enabling Research and Education in the 21st Century. *National Science Foundation* 2005, www.nsf.gov/pubs/2005/nsb0540/start.htm.

[310] Strasser BJ: Data-Driven Sciences: From Wonder Cabinets to Electronic Databases. *Studies in History and Philosophy of Science Part C: Studies in History and Philosophy of Biological and Biomedical Sciences* 2012, **43**:85–87, www.sciencedirect.com/science/article/pii/S1369848611000859.

[311] Gray J, Szalay A: eScience – A Transformed Scientific Method. Computer Science and Technology Board of the National Research Council 2007.

[312] Bell G: Foreward. In *The Fourth Paradigm. Data-Intensive Scientific Discovery*. Edited by Hey T, Tansley S, Tolle K, Microsoft Research 2009:xi.

[313] Hey T, Tansley S, Tolle K: *The Fourth Paradigm. Data-Intensive Scientific Discovery*. Microsoft Research 2009.

[314] Duman JG, DeVries AL: Isolation, Characterization, and Physical properties of Protein Antifreezes from the Winter Flounder, Pseudopleuronectes americanus. *Comparative Biochemistry and Physiology Part B: Comparative Biochemistry* 1976, **54**(3):375–380, www.sciencedirect.com/science/article/pii/0305049176902601.

[315] Leonelli S: Introduction: Making Sense of Data-Driven Research in the Biological and Biomedical Sciences. *Studies in History and Philosophy of Science Part C: Studies in History and Philosophy of Biological and Biomedical Sciences* 2012, **43**:1–3.

[316] Anderson C: The End of Theory: The Data Deluge Makes the Scientific Method Obsolete. *Wired* 2008, **16**:07.

[317] Pigliucci M: The End of Theory in Science? *EMBO Reports* 2009, **10**(6):534–534, https://onlinelibrary.wiley.com/doi/abs/10.1038/embor.2009.111.

[318] Pyysalo S, Baker S, Ali I, Haselwimmer S, Shah T, Young A, Guo Y, Högberg J, Stenius U, Narita M, Korhonen A: LION LBD: A Literature-Based Discovery System for Cancer Biology. *Bioinformatics* 2018, **bty845**:1–9, https://dx.doi.org/10.1093/bioinformatics/bty845.

[319] Nguyen V, Bodenreider O, Minning T, Sheth A: The Knowledge Driven Exploration of Integrated Biomedical Knowledge Sources Facilitates the Generation of New Hypotheses. In *Proceedings of the First International Workshop on Linked Science (LISC 2011)* 2011.

[320] Fleming N: Computer-Calculated Compounds-Researchers Are Deploying Artificial Intelligence to Discover Drugs. *Nature* 2018, **557**:S55–S57, www.nature.com/magazine-assets/d41586-018-05267-x/d41586-018-05267-x.pdf.

[321] Bongard J, Lipson H: Automated Reverse Engineering of Nonlinear Dynamical Systems. *Proceedings of the National Academy of Sciences* 2007, **104**(24):9943–9948, www.pnas.org/content/104/24/9943.

[322] Carley KM: On Generating Hypotheses Using Computer Simulations. *Systems Engineering* 1999, **2**(2):69–77.

[323] Extance A: AI Tames the Scientific Literature. *Nature* 2018, **561**:273–274.

[324] Evans J, Rzhetsky A: Machine Science. *Science* 2010, **329**(5990):399–400.

[325] Leonelli S: *Data-Centric Biology: A Philosophical Approach*. University of Chicago Press 2106.

[326] King RD, Whelan KE, Jones FM, Reiser PGK, Bryant CH, Muggleton SH, Kell DB, Oliver SG: Functional Genomic Hypothesis Generation and Experimentation by a Robot Scientist. *Nature* 2004, **427**:247, https://doi.org/10.1038/nature02236.

[327] Gutting G: *Scientific Methodology*, John Wiley & Sons, Ltd 2017:423–432, https://onlinelibrary.wiley.com/doi/abs/10.1002/9781405164481.ch63.

[328] De Regt HW: Scientific Realism in Action: Molecular Models and Boltzmann's "Bildtheorie". *Erkenntnis* 2005, **63**(2):205–230, www.jstor.org/stable/20013358.

[329] Acosta ND, Golub SH: The New Federalism: State Policies Regarding Embryonic Stem Cell Research. *The Journal of Law, Medicine & Ethics: A Journal of the American Society of Law, Medicine & Ethics* 2016, **44**(3):419–436, www.ncbi.nlm.nih.gov/pubmed/27587447.

[330] Dresser R: Stem Cell Research as Innovation: Expanding the Ethical and Policy conversation. *The Journal of Law, Medicine & Ethics: A Journal of the American Society of Law, Medicine & Ethics* 2010, **38**(2):332–341, www.ncbi.nlm.nih.gov/pubmed/20579255.

[331] Ayala FJ: Cloning Humans? Biological, Ethical, and Social Considerations. *Proceedings of the National Academy of Sciences of the United States of America* 2015, **112**(29):8879–8886, www.ncbi.nlm.nih.gov/pubmed/26195738.

[332] Varmus H: *Embryos, Cloning, Stem Cells, and the Promise of Reprogramming*. W. W. Norton 2009 chap. 13:197–223.

[333] Cressey D: Controversial UK Research Reform Crosses Finish Line 2017, www.nature.com/news/controversial-uk-research-reform-crosses-finish-line-1.21919?utm_source=feedburner&utm_medium=feed&utm_campaign=Feed%3A+news%2Frss%2Fmost_recent+%28NatureNews+-+Most+recent+articles%29.

[334] Abbott A, Schiermeier Q: EU Science-The Next Billion. *Nature* 2019, **569**:472–475.

[335] Taylor AP: Hungarian Law Wrests Control of Research from Scientific Academy-Protestors Disapprove of Putting Scientific Institutions under the Authority of a Government-Led Committee. *The Scientist* 2019.

[336] Dunai M: Hungarian Parliament Passes Bill Tightening State Grip over Scientists. *Reuters-World News* 2019.

[337] Goldstein A: Trump Restrictions on Fetal Tissue Research Unsettle Key Studies and Scientists. *Washington Post* 2020.

[338] Worth P: Threats to Children's Health. *Catalyst* Winter 2020, **20**:8.

[339] Hawking SW: Chronology Protection Conjecture. *Physical Review D* 1992, **46**:603–611, https://link.aps.org/doi/10.1103/PhysRevD.46.603.

[340] Teplow DB, et al.: Elucidating Amyloid β-Protein Folding and Assembly: A Multidisciplinary Approach. *Accounts of Chemical Research* 2006, **39**(9):635–645, www.ncbi.nlm.nih.gov/entrez/query.fcgi?cmd=Retrieve&db=PubMed&dopt=Citation&list_uids=16981680.

[341] Merton RK: *On Social Structure and Science*. University of Chicago Press 1996, https://books.google.com/books?id=j94XiVDwAZEC.

[342] Marshall A: *Principles of Economics*. Palgrave Macmillan, 8th edition 2013.

[343] Boorstein S: *Don's Just Do Something, Sit There. A Mindfulness Retreat with Sylvia Boorstein*. Harper-Collins, 1st edition 1996.

[344] Schrödinger E: Die gegenwärtige Situation in der Quantenmechanik. *Naturwissenschaften* 1935, **23**(48):807–812, https://doi.org/10.1007/BF01491891.

[345] Maxwell JC: *Theory of Heat*. Longmans, Green and Co. 1902.

[346] Franklin A: What Makes a "Good" Experiment? *British Journal for the Philosophy of Science* 1981, **32**(4):367–379.

[347] Franklin A: *What Makes a Good Experiment?* University of Pittsburgh Press 2016.

[348] Lipton P: *Inference to the Best Explanation*. Routledge, 2nd edition 2004.

[349] Knight J: From DNA to Consciousness – Crick's Legacy. *Nature* 2004, **430**(7000):597–597, https://doi.org/10.1038/430597a.

[350] Crick F, Koch C: Towards a Neurobiological Theory of Consciousness. *Seminars in the Neurosciences* 1990, **2**:263–275.

[351] Crick F, Koch C: The Problem of Consciousness. *Scientific American* 1992, 6(2):11–17.

[352] Crick F, Koch C: A Framework for Consciousness. *Nature Neuroscience* 2003, **6**(2):119–126, https://doi.org/10.1038/nn0203-119.

[353] Urbanc B, et al.: Dynamics of Plaque Formation in Alzheimer's Disease. *Biophysical Journal* 1999, **76**(3):1330–1334.

[354] Gibney E: New Definitions of Scientific Units Are on the Horizon. Metrologists Are Poised to Change How Scientists Measure the Universe. *Nature* 2017, **550**:312–313.

[355] Tukey JW: Analyzing Data: Sanctification or Detective Work? *American Psychologist* 1969, **24**(2):83–91.

[356] Johnson DH: The Insignificance of Statistical Significance Testing. *USGS Northern Prairie Wildlife Research Center* 1999, **63**(763–772):3.

[357] Jones SR, Carley S, Harrison M: An Introduction to Power and Sample Size Estimation. *Emergency Medicine* 2003, **20**:453–458.

[358] Jørstad TS, Langaas M, Bones AM: Understanding Sample Size: What Determines the Required Number of Microarrays for an Experiment? *Trends in Plant Sciences* 2007, **12**(2):46–50.

[359] Sullivan LM, Weinberg J, Keaney JF: Common Statistical Pitfalls in Basic Science Research. *Journal of the American Heart Association* 2016, **5**:e004142.

[360] Hilbert D, Ackermann W: *Grundzüge der theoretischen logik*. Springer (Berlin) 1928.

[361] Turing AM: On Computable Numbers, with an Application to the Entscheidungsproblem. *Proceedings of the London Mathematical Society* 1937, **s2–42**:230–265, https://doi.org/10.1112/plms/s2-42.1.230.

[362] Heesen R: How Much Evidence Should One Collect? *Philosophical Studies* 2015, **172**(9):2299–2313.

[363] al-Haytham I: *The Optics of Ibn al-Haytham. Books I–III. On Direct Vision*. The Warburg Institute 1989.

[364] Whewell W: *The Philosophy of the Inductive Sciences Founded upon Their History*. John W. Parker 1837.

[365] Stegenga J, Menon T: Robustness and Independent Evidence. *Philosophy of Science* 2017, **84**(3):414–435, https://doi.org/10.1086/692141.

[366] Vosshall LB: Into the Mind of a Fly. *Nature* 2007, **450**(7167):193–197, https://doi.org/10.1038/nature06335.

[367] Gerstein MB, et al.: Comparative Analysis of the Transcriptome across Distant Species. *Nature* 2014, **512**(7515):445–448, https://doi.org/10.1038/nature13424.

[368] Wittgenstein L: *Philosophical Investigations I*. Basel Blackwell, 3rd edition 1958.

[369] Sebald WG: Don't Think but Look!. *Representations* 2010, **112**:112–139.

[370] Locke J: *The Works of John Locke, in nine volumes*. C. Baldwin, 12th edition 1824.

[371] Couch J: *Pliny's Natural History. In Thirty-Seven Books. A Translation on the Basis of That by Dr. Philemon Holland, Ed. 1601. With Critical and Explanatory Notes*. George Barclay 1847.

[372] Ptolemy C: *Ptolemy's ALMAGEST*. Duckworth 1984.

[373] Crick FH: On Protein Synthesis. *Symposia of the Society for Experimental Biology* 1958, **12**:138–63.

[374] Crick F: Central Dogma of Molecular Biology. *Nature* 1970, **227**(5258):561–563, https://doi.org/10.1038/227561a0.

[375] Raskatov JA: What Is the "Relevant" Amyloid β42 Concentration? *ChemBioChem* 2019, **20**(13):1725–1726, https://doi.org/10.1002/cbic.201900097.

[376] Douven I: Abduction. In *The Stanford Encyclopedia of Philosophy*, Summer 2017 edition. Edited by Zalta EN, Metaphysics Research Lab, Stanford University 2017.

[377] Hartshorne C, Weiss P (Eds): *The Collected Papers of Charles Sanders Peirce. 1931–1935, Volume I–VI*. Harvard University Press 1965.

[378] Bucchianico MED: Modelling High Temperature Superconductivity: A Philosophical Inquiry in Theory, Experiment and Dissent. *PhD thesis*, London School of Economics and Political Science 2009.

[379] The Peirce Edition Project (Ed): *The Essential Peirce: Selected Philosophical Writings (1893–1913)*. Indiana University Press 1998.

[380] Peirce CS: Illustrations of the Logic of Science. The Probability of Induction. *Popular Science Monthly* 1878, **12**:705–718.

[381] Anderson DR: The Evolution of Peirce's Concept of Abduction. *Transactions of the Charles S. Peirce Society* 1986, **22**(2):145–164.

[382] Plutynski A: Four Problems of Abduction: A Brief History. *HOPOS: The Journal of the International Society for the History of Philosophy of Science* 2011, **1**(2):227–248.

[383] Frankfurt HG: Peirce's Account of Inquiry. *Journal of Philosophy* 1958, **55**(14):588–592.

[384] Campos DG: On the Distinction between Peirce's Abduction and Lipton's Inference to the Best Explanation. *Synthese* 2011, **180**(3):419–442, https://doi.org/10.1007/s11229-009-9709-3.

[385] Minnameier G: Peirce-Suit of Truth: Why Inference to the Best Explanation and Abduction Ought Not to Be Confused. *Erkenntnis* 2004, **60**:75–105, www.jstor.org/stable/20013245.

[386] Peirce CS: *The Collected Papers of Charles Sanders Peirce*. Harvard University Press 1958.

[387] Quine WV: *Word and Object*. Studies in Communication, Technology Press of the Massachusetts Institute of Technology 1960.

[388] Van Fraassen BC: *Laws and Symmetry*. Oxford University Press 1989.

[389] Harman GH: The Inference to the Best Explanation. *The Philosophical Review* 1965, **74**:88–95.

[390] Mcauliffe WHB: How did Abduction Get Confused with Inference to the Best Explanation? *Transactions of the Charles S. Peirce Society* 2015, **51**(3):300–319, www.jstor.org/stable/10.2979/trancharpeirsoc.51.3.300.

[391] Burks AW: Peirce's Theory of Abduction. *Philosophy of Science* 1946, **13**(4):301–306, www.jstor.org/stable/185210.

[392] Schurz G: Patterns of Abduction. *Synthese* 2008, **164**:201–234.

[393] Mohammadian M: Abduction − The Context of Discovery + Underdetermination = Inference to the Best Explanation. *Synthese* 2019, **198**:4205–4228

[394] Hintikka J: What Is Abduction? The Fundamental Problem of Contemporary Epistemology. *Transactions of the Charles S. Peirce Society* 1998, **34**(3):503, www.jstor.org/stable/40320712.

[395] Campos DG: On the Distinction between Peirce's Abduction and Lipton's Inference to the Best Explanation. *Synthese* 2011, **180**(3):419–442, www.jstor.org/stable/41477565.

[396] Paavola S: Hansonian and Harmanian Abduction as Models of Discovery. *International Studies in the Philosophy of Science* 2006, **20**:93–108, https://doi.org/10.1080/02698590600641065.

[397] Mcauliffe WHB: How did Abduction Get Confused with Inference to the Best Explanation? *Transactions of the Charles S. Peirce Society* 2015, **51**(3):300–319, www.jstor.org/stable/10.2979/trancharpeirsoc.51.3.300.

[398] McKaughan DJ: From Ugly Duckling to Swan: C. S. Peirce, Abduction, and the Pursuit of Scientific Theories. *Transactions of the Charles S. Peirce Society* 2008, **44**(3):446–468, www.jstor.org/stable/40321321.

[399] Paavola S: Abduction through Grammar, Critic, and Methodeutic. *Transactions of the Charles S. Peirce Society* 2004, **40**(2):245–270, www.jstor.org/stable/40320991.

[400] Hempel CG: *Aspects of Scientific Explanation: And Other Essays in the Philosophy of Science*. Free Press 1965.

[401] Barnes E: Inference to the Loveliest Explanation. *Synthese* 1995, **103**(2):251–277, www.jstor.org/stable/20117399.

[402] Tversky A, Kahneman D: Judgment under Uncertainty: Heuristics and Biases. *Science* 1974, **185 4157**:1124–1131.

[403] Kahneman D, Slovic P, Tversky A (Eds): *Judgment under Uncertainty: Heuristics and Biases*. Cambridge University Press 1982.

[404] Okasha S: Van Fraassen's Critique of Inference to the Best Explanation. *Studies in History and Philosophy of Science* 2000, **31**:691–710.

[405] Harman GH: Knowledge, Reasons, and Causes. *Journal of Philosophy* 1970, **67**(21):841–855.

[406] Niiniluoto I: Social Aspects of Scientific Knowledge. *Synthese* 2018, **197**(1).

[407] Mathematica S: J. L. Heiberg 1898.

[408] Lipton P: *Inference to the Best Explanation*. Routledge 1991.

[409] Duhem P: *La théorie physique son objet et sa structure*. Chevalier et Riviere, 2nd edition 1914.

[410] Ariew R: The Duhem Thesis. *The British Journal for the Philosophy of Science* 1984, **35**(4):313–325, https://doi.org/10.1093/bjps/35.4.313.

[411] Duhem PMM: *The Aim and Structure of Physical Theory*. Princeton University Press 1954.

[412] Stanford K: Underdetermination of Scientific Theory. In *The Stanford Encyclopedia of Philosophy*, Winter 2017 edition. Edited by Zalta EN, Metaphysics Research Lab, Stanford University 2017.

[413] Quine WV: *From a Logical Point of View*. Harvard University Press 1953.

[414] Quine WV: On Empirically Equivalent Systems of the World. *Erkenntnis* 1975, **9**(3):313–328, https://doi.org/10.1007/BF00178004.

[415] Bandyopadhyay PS, Bennett JG, Higgs MD: How to Undermine Underdetermination? *Foundations of Science* 2015, **20**(2).

[416] Kyle Stanford P: *Exceeding Our Grasp*. Oxford University Press 2010.

[417] Laudan L, Leplin J: Empirical Equivalence and Underdetermination. *The Journal of Philosophy* 1991, **88**(9):449–472, www.jstor.org/stable/2026601.

[418] Kitcher P: *The Advancement of Science: Science without Legend, Objectivity without Illusions*. Springer 1996.

[419] Park S: Philosophical Responses to Underdetermination in Science. *Journal for General Philosophy of Science* 2009, **40**:115–124, https://doi.org/10.1007/s10838-009-9080-6.

[420] Stanford PK: Refusing the Devil's Bargain: What Kind of Underdetermination Should We Take Seriously? *Philosophy of Science* 2001, **68**(3):S1–S12, www.jstor.org/stable/3080930.

[421] Chamberlin TC: The Method of Multiple Working Hypotheses. *Science (Old Series)* 1890, **15**:92–96.

[422] Elliott LP, Brook BW: Revisiting Chamberlin: Multiple Working Hypotheses for the 21st Century. *BioScience* 2007, **57**(7):608–614, https://doi.org/10.1641/B570708.

[423] Yanco SW, McDevitt A, Trueman CN, Hartley LM, Wunder MB: A Modern Method of Multiple Working Hypotheses to Improve Inference in Ecology. *Royal Society Open Science* 2020, **7**:200231.

[424] Spade PV, Panaccio C: Simplicity. In *The Stanford Encyclopedia of Philosophy*, spring 2019 edition. Edited by Zalta EN, Metaphysics Research Lab, Stanford University 2019.

[425] Ball P: The Tyranny of Simple Explanations. *The Atlantic* 2016.

[426] Derkse W: *On Simplicity and Elegance-An Essay in Intellectual History*. Eburon 1992.

[427] Baker A: Simplicity. In *The Stanford Encyclopedia of Philosophy*. Edited by Zalta EN, Metaphysics Research Lab, Stanford University 2016 https://plato.stanford.edu/archives/win2016/entries/simplicity/.

[428] Sober E: What Is the Problem of Simplicity? In *Simplicity, Inference, and Modelling*. Edited by Zellner A, Keuzenkamp HA, McAleer M, Cambridge University Press 2002:13–32.

[429] Jefferys WH, Berger JO: Ockham's Razor and Bayesian Analysis. *American Scientist* 1992, **80**:64–72, www.jstor.org/stable/29774559.

[430] Barnes EC: Ockham's Razor and the Anti-Superfluity Principle. *Erkenntnis (1975–)* 2000, **53**(3):353–374, www.jstor.org/stable/20013020.

[431] Sober E: *From a Biological Point of View: Essays in Evolutionary Philosophy*. Cambridge University Press 1994, www.cambridge.org/core/books/from-a-biological-point-of-view/6C0C7E57E2F2BBBCFD2056751772E70C.

[432] Herrmann DA: PAC Learning and Occam's Razor: Probably Approximately Incorrect. *Philosophy of Science* 2020, **87**(4):685–703, https://doi.org/10.1086/709786.

[433] Valiant LG: A Theory of the Learnable. *Communications of the ACM* 1984, **27**(11):1134–1142, https://doi.org/10.1145/1968.1972.

[434] Blumer A, Ehrenfeucht A, Haussler D, Warmuth MK: Occam's Razor. *Information Processing Letters* 1987, **24**(6):377–380, www.sciencedirect.com/ science/article/pii/0020019087901141.

[435] Li M, M B Vitányi P: Inductive Reasoning and Kolmogorov Complexity. *Journal of Computer and System Sciences* 1992, **44**(2):343–384, www.sciencedirect.com/ science/article/pii/002200009290026F.

[436] Kelly K: Justification as Truth-Finding Efficiency: How Ockham's Razor Works. *Minds and Machines* 2004, **14**:485–505.

[437] Douglas H, Magnus P: State of the Field: Why Novel Prediction Matters. *Studies in History and Philosophy of Science Part A* 2013, **44**(4):580–589, www.sciencedirect.com/science/article/pii/S0039368113000198.

[438] Box GEP: Science and Statistics. *Journal of the American Statistical Association* 1976, **71**(356):791–799, www.tandfonline.com/doi/abs/10.1080/ 01621459.1976.10480949.

[439] Akaike H: Information Theory and an Extension of the Maximum Likelihood Principle. In *Proceedings of the 2nd International Symposium on Information Theory*. Edited by Petrov BN, Csaki F, Akademiai Kiado, 1973.

[440] Akaike H: A New Look at the Statistical Model Identification. *IEEE Transactions on Automatic Control* 1974, **AC-19**(6):716–723.

[441] Watanabe S: A Widely Applicable Bayesian Information Criterion. *Journal of Machine Learning Research* 2013, **14**:867–897.

[442] Spiegelhalter DJ, Best NG, Carlin BP, van der Linde A: The Deviance Information Criterion: 12 Years On. *Journal of the Royal Statistical Society: Series B (Statistical Methodology)* 2014, **76**(3):485–493, https://rss.onlinelibrary .wiley.com/doi/abs/10.1111/rssb.12062.

[443] Blankenship EE, Perkins MW, Johnson RJ: The Information-Theoretic Approach to Model Selection: Description and Case Study. *Conference on Applied Statistics in Agriculture* 2002.

[444] Burnham KP, Anderson DR: Basic Use of the Information-Theoretic Approach. In *Model Selection and Multimodel Inference: A Practical Information-Theoretic Approach*. Edited by Burnham KP, Anderson DR, Springer, 2002:98–148, https://doi.org/10.1007/978-0-387-22456-5_3.

[445] Royle JA, Dorazio RM: OCCUPANCY AND ABUNDANCE. In *Hierarchical Modeling and Inference in Ecology*. Edited by Royle JA, Dorazio RM, Academic Press 2009:127–157, www.sciencedirect.com/science/article/pii/ B9780123740977000065.

[446] Kullback S, Leibler RA: On Information and Sufficiency. *The Annals of Mathematical Statistics* 1951, **22**:79–86, www.jstor.org/stable/2236703.

[447] Burnham KP, Anderson DR: *Multimodel Inference: Understanding AIC and BIC in Model Selection*. Springer 2004, https://doi.org/10.1177/0049124104268644.

[448] Royall R: The Likelihood Paradigm for Statistical Evidence. In *The Nature of Scientific Evidence: Statistical, Philosophical, and Empirical Considerations*. Edited by Taper M, Lele S, University of Chicago Press 2010:119–152, https://books.google.com/books?id=ivwGWFM-VssC.

[449] Dijkstra EW: The Next Fifty Years 1996, www.cs.utexas.edu/users/EWD/ewd 12xx/EWD1243.PDF. [Circulated privately].

[450] Kitcher P: Explanatory Unification and the Causal Structure of the World. In *Scientific Explanation*. Edited by Kitcher P, Salmon W, University of Minnesota Press 1989:410–505 (excerpts).

[451] Kant I: *Critique of Judgement*. Hackett 1987.

[452] Hempel CG: *Philosophy of Natural Science*. Prentice-Hall 1966.

[453] Friedman M: Explanation and Scientific Understanding. *The Journal of Philosophy* 1974, **71**:5–19, www.jstor.org/stable/2024924.

[454] Salmon WC: *Scientific Explanation and the Causal Structure of the World*. Princeton University Press 1984.

[455] Railton P: Probability, Explanation, and Information. *Synthese* 1981, **48**(2):233–256, https://doi.org/10.1007/BF01063889.

[456] Hawking S, Mlodinow L: The (Elusive) Theory of Everything. *Scientific American* 2010, **303**(4):68–71, www.jstor.org/stable/26002214.

[457] de Regt HW: Wesley Salmon's Complementarity Thesis: Causalism and Unificationism Reconciled? *International Studies in the Philosophy of Science* 2006, **20**(2):129–147, https://doi.org/10.1080/02698590600814308.

[458] Marder E: Understanding Brains: Details, Intuition, and Big Data. *PLOS Biology* 2015, **13**(5):e1002147, https://doi.org/10.1371/journal.pbio.1002147.

[459] Green S (Ed): *Philosophy of Systems Biology. Perspectives from Scientists and Philosophers, Volume 20 of History, Philosophy and Theory of the Life Sciences*. Springer International Publishing 2017.

[460] Braillard PA (Ed): *Explanation in Biology. An Enquiry into the Diversity of Explanatory Patterns in the Life Sciences*. Springer 2015.

[461] Strevens M: *Depth. An Account of Scientific Explanation*. Harvard University Press 2008.

[462] Bechtel W, Richardson RC: *Discovering Complexity. Decomposition and Localization as Strategies in Scientific Research*. MIT Press, 2nd edition 2010.

[463] Brigandt I: Systems Biology and the Integration of Mechanistic Explanation and Mathematical Explanation. In *The Routledge Handbook of Mechanisms and Mechanical Philosophy*. Edited by Brigandt I, Green S, O'Malley MA, Routledge 2017 www.routledgehandbooks.com/doi/10.4324/9781315731544.ch27.

[464] Chemero A, Silberstein M: After the Philosophy of Mind: Replacing Scholasticism with Science. *Philosophy of Science* 2008, **75**:1–27, www.jstor.org/stable/10.1086/587820.

[465] Stepp N, Chemero A, Turvey MT: Philosophy for the Rest of Cognitive Science. *Topics in Cognitive Science* 2011, **3**(2):425–437.

[466] Lamb M, Chemero A: Structure and Application of Dynamical Models in Cognitive Science. *Proceedings of the Annual Meeting of the Cognitive Science Society* 2014, **36**(36):809–814.

[467] Kaplan DM, Craver CF: The Explanatory Force of Dynamical and Mathematical Models in Neuroscience: A Mechanistic Perspective. *Philosophy of Science* 2011, **78**(4):601–627, www.jstor.org/stable/10.1086/661755.

[468] Weinberg RA: The Molecules of Life. *Scientific American* 1985, **253**(4):48–57, www.jstor.org/stable/24967808.

[469] Woodward JF: *Making Things Happen: A Theory of Causal Explanation*. Oxford University Press 2004, www.loc.gov/catdir/toc/fy043/2002192596.html.

[470] Barberis M, Klipp E, Vanoni M, Alberghina L: Cell Size at S Phase Initiation: An Emergent Property of the G1/S Network. *PLOS Computational Biology* 2007, **3**(4):e64, https://doi.org/10.1371/journal.pcbi.0030064.

[471] Levy A: What Was Hodgkin and Huxley's Achievement? *British Journal for the Philosophy of Science* 2014, **65**(3):469–492.

[472] Glennan SS: Mechanisms and the Nature of Causation. *Erkenntnis* 1996, **44**:49–71, www.jstor.org/stable/20012673.

[473] Machamer P, Darden L, Craver CF: Thinking about Mechanisms. *Philosophy of Science* 2000, **67**:1–25, www.jstor.org/stable/188611.

[474] Woodward J: What Is a Mechanism? A Counterfactual Account. *Philosophy of Science* 2002, **69**(S3).

[475] Bunge M: A General Black Box Theory. *Philosophy of Science* 1963, **30**(4):346–358, www.jstor.org/stable/186066.

[476] Craver CF: *Explaining the Brain. Mechanisms and the Mosaic Unity of Neuroscience*. Clarendon Press 2007.

[477] Craver CF, Kaplan DM: Are More Details Better? On the Norms of Completeness for Mechanistic Explanations. *The British Journal for the Philosophy of Science* 2020, **71**:287–319, https://doi.org/10.1093/bjps/axy015.

[478] Batterman RW, Rice CC: Minimal Model Explanations. *Philosophy of Science* 2014, **81**(3):349–376, www.jstor.org/stable/10.1086/676677.

[479] Batterman RW: *The Devil in the Details: Asymptotic Reasoning in Explanation, Reduction, and Emergence*. Oxford University Press 2001.

[480] Seger J, Stubblefield JW: Optimization and Adaptation. In *Adaptation*, 1st edition. Edited by Rose MR, Lauder GV, Academic Press 1996:93–123.

[481] Weber E, Bouwel JV: Causation, Unification, and the Adequacy of Explanations of Facts. *THEORIA. An International Journal for Theory, History and Foundations of Science* 2009, **24**(3):301–320, http://philsci-archive.pitt.edu/10353/.

[482] Woodward J: Explanation, Invariance, and Intervention. *Philosophy of Science* 1997, **64**:S26–S41, www.jstor.org/stable/188387.

[483] Salmon WC: *Causality and Explanation*. Oxford University Press 1998.

[484] Salmon WC: *Four Decades of Scientific Explanation*. University of Pittsburgh Press 1990, www.jstor.org/stable/j.ctt5vkdm7.

[485] Emmert-Streib F, De Matos Simoes R, Mullan P, Haibe-Kains B, Dehmer M: The Gene Regulatory Network for Breast Cancer: Integrated Regulatory Landscape of Cancer Hallmarks. *Frontiers in Genetics* 2014, **5**(15):1–12, www.frontiersin.org/article/10.3389/fgene.2014.00015.

[486] Ongaro M: Explanatory Relevance. A Central Issue in the Theory of Explanation. *PhD thesis*, The London School of Economics 2017.

[487] Salmon WC: *The Foundations of Scientific Inference*. University of Pittsburgh Press 1967.

[488] Lombrozo T: Explanation and Abductive Inference. In *The Oxford Handbook of Thinking and Reasoning*, Oxford University Press 2012.

[489] Salmon MH, et al.: *Introduction to the Philosophy of Science*. Hackett Publishing Company 1992.

[490] Kyburg HE: Discussion: Salmon's Paper. *Philosophy of Science* 1965, **32**(2):147–151, https://doi.org/10.1086/288034.

[491] Levin ME, Levin MR: Flagpoles, Shadows and Deductive Explanation. *Philosophical Studies: An International Journal for Philosophy in the Analytic Tradition* 1977, **32**(3):293–299, www.jstor.org/stable/4319174.

[492] Woodward J: Scientific Explanation. In *The Stanford Encyclopedia of Philosophy (Winter edition)*. Edited by Zalta EN, Metaphysics Research Lab, Stanford University 2019, https://plato.stanford.edu/archives/win2019/entries/scientific-explanation/.

[493] Halpern JY, Pearl J: Causes and Explanations: A Structural-Model Approach. Part II: Explanations. *The British Journal for the Philosophy of Science* 2005, **56**(4):889–911, www.jstor.org/stable/3541871.

[494] Ruben DH: *Explaining Explanation*. Paradigm, 2nd edition 2012.

[495] Lewis D: Causation. *The Journal of Philosophy* 1973, **70**(17):556–567, www.jstor.org/stable/2025310.

[496] Lewis D: Causation as Influence. In *Causation and Counterfactuals*. Edited by Collins JD, Paul LA, Hall N, MIT Press 2004:75–106.

[497] Woodward J, Hitchcock C: Explanatory Generalizations, Part I: A Counterfactual Account. *Noûs* 2003, **37**:1–24, www.jstor.org/stable/3506202.

[498] Kuorikoski J: There Are No Mathematical Explanations. *Philosophy of Science* 2021, **88**(2):189–212, www.cambridge.org/core/article/there-are-no-mathematical-explanations/4FF90FDACEF2C6EA56B4859351E26E63.

[499] Nerlich G: What Can Geometry Explain? *British Journal for the Philosophy of Science* 1979, **30**:69–83.

[500] Haack S: Coherence, Consistency, Cogency, Congruity, Cohesiveness, &c.: Remain Calm! Don't Go Overboard! *New Literary History* 2004, **35**(2):167–183, www.jstor.org/stable/20057831.

[501] Thagard P: *Coherence in Thought and Action*. The MIT Press 2000.

[502] Thagard P: Explanatory Coherence. *Behavioral and Brain Sciences* 1989, **12**(3): 435–467, www.cambridge.org/core/article/explanatory-coherence/E05CB61CD64C 26138E794BC601CC9D7A.

[503] Haack S: Double-Aspect Foundherentism: A New Theory of Empirical Justification. *Philosophy and Phenomenological Research* 1993, **53**:113–128.

[504] Colombo M, Postma M, Sprenger J: Explanatory Value and Probabilistic Reasoning: An Empirical Study. *Proceedings of the Cognitive Science Society* 2016.

[505] Salmon WC: The Value of Scientific Understanding. *Philosophia* 1993, **51**:9–19.

[506] Strevens M: The Causal and Unification Approaches to Explanation Unified. *Noûs* 2004, **38**:154–176.

[507] Salmon WC: Statistical Explanation. In *The Nature and Function of Scientific Theories: Essay in Contemporary Science and Philosophy*. Edited by Colodny RG, University of Pittsburgh Press 1970:173–232, www.cambridge.org/core/article/nature-and-function-of-scientific-theories-edited-by-robert-g-colodny-pittsb urgh-university-of-pittsburgh-press-1970-pp-xv-361-price-1295/84FA935D1E1A1 C6478486F9907177801.

[508] de Regt HW: *Understanding Scientific Understanding*. Oxford University Press 2017.

[509] Millikan RA: The Electron and the Light-Quant from the Experimental Point of View. *Les Prix Nobel en 1923* 1924:1–15.

[510] Lewens T: *The Meaning of Science: An Introduction to the Philosophy of Science*. Hachette UK 2016.

[511] Will CM: The Confrontation between General Relativity and Experiment. *Living Reviews in Relativity* 2014, **17**:4.

[512] Aaltonen T, et al.: High-Precision Measurement of the W Boson Mass with the CDF II Detector. *Science* 2022, **376**(6589):170–176, https://doi.org/10.1126/science.abk1781.

[513] OPERA Collaboration and Adam, T et al.: Measurement of the Neutrino Velocity with the OPERA Detector in the CNGS Beam. *arXiv* 2011, **arXiv**:1109.4897v1 [hep-ex].

[514] Reich ES: Timing Glitches Dog Neutrino Claim. *Nature* 2012, **483**:17.

[515] Brumfiel G: Neutrinos Not Faster than Light. *Nature* 2012, https://doi.org/10.1038/nature.2012.10249.

[516] Antonello M, et al.: Measurement of the Neutrino Velocity with the ICARUS Detector at the CNGS Beam. *Physics Letters B* 2012, **713**:17–22, http://dx.doi.org/10.1016/j.physletb.2012.05.033.

[517] Bertolucci S: Neutrino Speed: A Report on the Speed Measurements of the BOREXINO, ICARUS and LVD Experiments with the CNGS Beam. *Nuclear Physics B – Proceedings Supplements* 2013, **235–236**:289–295, www.science direct.com/science/article/pii/S092056321300145X. [The XXV International Conference on Neutrino Physics and Astrophysics].

[518] OPERA Collaboration and Adam, T et al: Measurement of the Neutrino Velocity with the OPERA Detector in the CNGS Beam. *Journal of High Energy Physics* 2012, **2012**(10), http://dx.doi.org/10.1007/JHEP10(2012)093.

[519] Popper KR: *Conjectures and Refutations: The Growth of Scientific Knowledge.* Routledge 1962.

[520] Collins R: Against the Epistemic Value of Prediction over Accommodation. *Noûs* 1994, **28**(2):210–224, www.jstor.org/stable/2216049.

[521] Horwich P: *Probability and Evidence.* Cambridge University Press 1982, www.loc.gov/catdir/enhancements/fy1106/81018144-d.html.

[522] Barnes EC: Prediction versus Accommodation. In *The Stanford Encyclopedia of Philosophy*, Fall 2018 edition. Edited by Zalta EN, Metaphysics Research Lab, Stanford University 2018, https://plato.stanford.edu/archives/fall2018/entries/prediction-accommodation/.

[523] Maher P: Howson and Franklin on Prediction. *Philosophy of Science* 1993, **60**(2):329–340, https://doi.org/10.1086/289736.

[524] Maher P: Prediction, Accommodation, and the Logic of Discovery. *PSA: Proceedings of the Biennial Meeting of the Philosophy of Science Association* 1988, **1988**:273–285, https://doi.org/10.1086/psaprocbienmeetp.1988.1.192994.

[525] Howson C, Franklin A: Maher, Mendeleev and Bayesianism. *Philosophy of Science* 1991, **58**(4):574–585, www.jstor.org/stable/188481.

[526] Harker D: On the Predilections for Predictions. *The British Journal for the Philosophy of Science* 2008, **59**(3):429–453, www.jstor.org/stable/40072294.

[527] Rubin M: The Costs of HARKing. *The British Journal for the Philosophy of Science* 2020, https://doi.org/10.1093/bjps/axz050.

[528] Achinstein P: Explanation v. Prediction: Which Carries More Weight? In *PSA: Proceedings of the Biennial Meeting of the Philosophy of Science Association*, Philosophy of Science Association 1994:156–164.

[529] Keynes JM: *A Treatise on Probability.* Dover Publications 1921.

[530] Simon HA: Prediction and Hindsight as Confirmatory Evidence. *Philosophy of Science* 1955, **22**(3):227–230, https://doi.org/10.1086/287427.

[531] Worrall J: Prediction and Accommodation Revisited. *Studies in History and Philosophy of Science Part A* 2014, **45**:54–61.

[532] Wang MQ, Yan AF, Katz RV: Researcher Requests for Inappropriate Analysis and Reporting: A U.S. Survey of Consulting Biostatisticians. *Annals of Internal Medicine* 2018, **169**(8):554–558.

[533] Kerr NL: HARKing: Hypothesizing after the Results Are Known. *Personality and Social Psychology Review* 1998, **2**(3):196–217, https://doi.org/10.1207/s15327957pspr0203_4. [PMID: 15647155].

[534] Stanger-Hall K: Accommodation or Prediction? *Science* 2005, **308**(5727):1409–1412, https://science.sciencemag.org/content/308/5727/1409.3.

[535] Strevens M: *Keep Science Irrational.* Aeon Magazine 2020.

[536] Ivanova M: Poincaré's Aesthetics of Science. *Synthese* 2017, **194**(7):2581.

[537] Zeki S, Romaya J, Benincasa D, Atiyah M: The Experience of Mathematical Beauty and Its Neural Correlates. *Frontiers in Human Neuroscience* 2014, **8**, www.frontiersin.org/article/10.3389/fnhum.2014.00068.

[538] Brown S, Gao X: The Neuroscience of Beauty. *Scientific American* 2011, www.scientificamerican.com/article/the-neuroscience-of-beauty/.

[539] Murphy N: Another Look at Novel Facts. *Studies in History and Philosophy of Science Part A* 1989, **20**(3):385–388, www.sciencedirect.com/science/article/pii/0039368189900149.

[540] Lyttleton RA: The Nature of Knowledge. In *The Encyclopaedia of Ignorance: Everything You Ever Wanted to Know about the Unknown*, Pergamon Press 1977:9–17.

[541] Box JF: *R. A. Fisher, the Life of a Scientist*. New York: Wiley 1978.

[542] Box GE: Non-Normality and Tests on Variances. *Biometrika* 1953, **40**:318–335.

[543] Box G: Robustness in the Strategy of Scientific Model Building. In *Robustness in Statistics*. Edited by Launer RL, Wilkinson GN, Academic Press 1979:201–236, www.sciencedirect.com/science/article/pii/B9780124381506500182.

[544] John of Salisbury: The Metalogicon 1159.

[545] Soldner J: *Über die Ablenkung eines Lichtstrahls von seiner geradlinigen Bewegung, durch die Attraktion eines Weltkörpers, an welchem er nahe vorbei geht.* Berliner Astronomisches Jahrbuch 1921.

[546] Jaki SL: Johann Georg von Soldner and the Gravitational Bending of Light, with an English Translation of His Essay on It Published in 1801. *Foundations of Physics* 1978, **8**(11):927–950, https://doi.org/10.1007/BF00715064.

[547] Costa JT: *Wallace, Darwin, and the Origin of Species*. Harvard University Press 2014.

[548] Partridge D: Further Details Concerning the Darwin-Wallace Presentation to the Linnean Society in 1858, Including Its Submission on 1 July, Not 30 June. *Journal of Natural History* 2016, **50**(15–16):1035–1044, https://doi.org/10.1080/00222933.2015.1091102.

[549] Darwin CR, Wallace AR: On the Tendency of Species to Form Varieties; and on the Perpetuation of Varieties and Species by Natural Means of Selection. *Journal of the Proceedings of the Linnean Society of London, Zoology* 1858, **3**(9):45–62.

[550] Hamilton K: Darwin's Error: Implications for Insect Taxonomy. In *Entomological Society of America Annual Meeting* 2011.

[551] van Holstein L, Foley RA: Terrestrial Habitats Decouple the Relationship between Species and Subspecies Diversification in Mammals. *Proceedings of the Royal Society B: Biological Sciences* 2020, **287**(1923).

[552] Gee H: Science in Culture-A Year in Pangaea. *Nature* 1999, **401**(6753):530–530, https://doi.org/10.1038/44018.

[553] Cleland C: Historical Science, Experimental Science, and the Scientific Method. *Geology* 2001, **29**:987–990.

[554] Everett H: "Relative State" Formulation of Quantum Mechanics. *Reviews of Modern Physics* 1957, **29**(3):454–462, https://link.aps.org/doi/10.1103/RevModPhys.29.454.

[555] Guth AH: Inflation and Eternal Inflation. *Physics Reports* 2000, **333–334**:555–574, www.sciencedirect.com/science/article/pii/S0370157300000375.

[556] Dubray C: *The Catholic Encyclopedia*. The Encyclopedia Press 1913.

[557] Wild J: What Is Realism? *The Journal of Philosophy* 1947, **44**(6):148–158, www.jstor.org/stable/2020042.

[558] Alai M: Novel Predictions and the No Miracle Argument. *Erkenntnis* 2014, **79**(2):297–326, https://doi.org/10.1007/s10670-013-9495-7.

[559] Psillos S: The Present State of the Scientific Realism Debate. *The British Journal for the Philosophy of Science* 2000, **51**:705–728, www.jstor.org/stable/3541614.

[560] Nanay B: Singularist Semirealism. *The British Journal for the Philosophy of Science* 2013, **64**(2):371–394, www.jstor.org/stable/24563057.

[561] Peters D: What Elements of Successful Scientific Theories Are the Correct Targets for "Selective" Scientific Realism? *Philosophy of Science* 2014, **81**(3):377–397, www.jstor.org/stable/10.1086/676537.

[562] Dellsén F: Realism and the Absence of Rivals. *Synthese* 2017, **194**(7):2427–2446, https://doi.org/10.1007/s11229-016-1059-3.

[563] Morganti M: Is There a Compelling Argument for Ontic Structural Realism? *Philosophy of Science* 2011, **78**(5):1165–1176, www.jstor.org/stable/10.1086/662258.

[564] Chang H: Realism for Realistic People. *Spontaneous Generations* 2018, **9**:31–34.

[565] Fine A: *The Shaky Game: Einstein, Realism, and the Quantum Theory*. University of Chicago Press 1986.

[566] Hacking I: Experimentation and Scientific Realism. In *Representing and Intervening: Introductory Topics in the Philosophy of Natural Science*, Cambridge University Press 1983:262–275, www.cambridge.org/core/books/representing-and-intervening/experimentation-and-scientific-realism/3831C794E98605154A8B CC196F97B7B7.

[567] Thomson JJ: XL. Cathode Rays. *The London, Edinburgh, and Dublin Philosophical Magazine and Journal of Science* 1897, **44**(269):293–316, https://doi .org/10.1080/14786449708621070.

[568] Stoney GJ: LII. On the Physical Units of Nature. *The London, Edinburgh, and Dublin Philosophical Magazine and Journal of Science* 1881, **11**(69):381–390, https://doi.org/10.1080/14786448108627031.

[569] Stoney GJ: Of the "Electron," or Atom of Electricity. *Philosophical Magazine* 1894, **38**:418–420.

[570] FitzGerald GF: Dissociation of Atoms. *The Electrician* 1897, **39**:103–104.

[571] Romer A: The Experimental History of Atomic Charges, 1895–1903. *Isis* 1942, **34**(2):150–161, www.jstor.org/stable/226218.

[572] Stoney GJ: LIII. On Texture in Media, and on the Non-Existence of Density in the Elemental *Æther*. *The London, Edinburgh, and Dublin Philosophical Magazine and Journal of Science* 1890, **29**(181):467–478, https://doi.org/10 .1080/14786449008619970.

[573] Slaney K: On Empirical Realism and the Defining of Theoretical Terms. *Journal of Theoretical and Philosophical Psychology* 2001, **21**:132–152.

[574] Norris SP: The Inconsistencies at the Foundation of Construct Validation Theory. *New Directions for Program Evaluation* 1983:53–74, https://doi.org/10 .1002/ev.1344.

[575] Musgrave A: The "No Miracle" Argument for Scientific Realism. *The Rutherford Journal* 2006–2007, **2**.

[576] Hacking I: *Representing and Intervening: Introductory Topics in the Philosophy of Natural Science*. Cambridge University Press 1983, https://books.google.com/ books?id=4hIQ5fGf-_oC.

[577] Cartwright N: *How the Laws of Physics Lie*. Oxford University Press 1983.

[578] Giere RN: *Science without Laws*. University of Chicago Press 1999.

[579] Hacking I: Extragalactic Reality: The Case of Gravitational Lensing. *Philosophy of Science* 1989, **56**(4):555–581, www.jstor.org/stable/187781.

[580] McMullin E: Explanatory Success and the Truth of Theory. In *Scientific Inquiry in Philosophical Perspective*. Edited by Rescher N, University Press of America 1987.

[581] Chakravartty A, Van Fraassen BC: What Is Scientific Realism? *Spontaneous Generations: A Journal for the History and Philosophy of Science* 2018, **9**:12–25.

[582] Shaw JCO: Why the Realism Debate Matters for Science Policy: The Case of the Human Brain Project. *Spontaneous Generations: A Journal for the History and Philosophy of Science* 2018, **9**:82–98.

[583] Poincaré H: *Science and Hypothesis*. The Walter Scott Publishing Co., Ltd 1905.

[584] Worrall J: Structural Realism: The Best of Both Worlds? *Dialectica* 1989, **43**(1/2):99–124, www.jstor.org/stable/42970613.

[585] de Senarmont H, Émile Verdet, Fresnel L (Eds): *Oeuvres complétes d'Augustin Fresnel*. Imprimer Impériale 1866.

[586] Brewster D: IX On the Laws Which Regulate the Polarisation of Light by Reflexion from Transparent Bodies. *Philosophical Transactions of the Royal Society* 1815, **105**:125–159.

[587] Maxwell JC: *A Dynamical Theory of the Electromagnetic Force*. The Royal Society 1865.

[588] Hågenvik HO, Skaar J: Magnetic Permeability in Fresnel's Equation. *Journal of the Optical Society of America B* 2019, **36**(5):1386–1395, http://josab .osa.org/abstract.cfm?URI=josab-36-5-1386.

[589] Wright AS: Fresnel's Laws, *ceteris paribus*. *Studies in History and Philosophy of Science Part A* 2017, **64**:38–52, www.sciencedirect.com/ science/article/pii/S0039368116301182.

[590] Chakravartty A: Semirealism. *Studies in History and Philosophy of Science Part A* 1998, **29**(3):391–408.

[591] Ralón L: Interview with Anjan Chakravartty. *Figure/Ground* 2020, http:// figureground.org/interview-with-anjan-chakravartty/.

[592] Ghins M: Defending Scientific Realism without Relying on Inference to the Best Explanation. *Axiomathes* 2017, **27**(6):635–651, https://doi.org/10.1007/s10516-017-9356-0.

[593] Ghins M: Can Common Sense Realism Be Extended to Theoretical Physics? *Logic Journal of the IGPL* 2005, **13**:95–111, https://doi.org/10.1093/jigpal/jzi006.

[594] Pratchett T, Briggs S: *Lords and Ladies*. Bloomsbury Publishing 2021, https:// books.google.com/books?id=AeslEAAAQBAJ.

[595] Alai M: Scientific Realism, Metaphysical Antirealism and the No Miracle Arguments. *Foundations of Science* 2020, **28**(2):1–24, https://doi.org/10.1007/ s10699-020-09691-z.

[596] Putnam H: *Reason, Truth and History*. Cambridge University Press 1981, www.cambridge.org/core/books/reason-truth-and-history/17C4C420E3BFE409FD 6673C262BF1446.

[597] Putnam H: *Philosophy in an Age of Science*. Harvard University Press 2012.

[598] O'Hear A: *An Introduction to the Philosophy of Science*. Clarendon Press 1989.

[599] Chakraborty S: *Understanding Meaning and World: A Relook on Semantic Externalism*. Cambridge Scholars Publishing 2016.

[600] Massimi M: 2017 Wilkins–Bernal–Medawar Lecture: Why Philosophy of Science Matters to Science. *Notes and Records of the Royal Society* 2018, **73**(3):1–15.

[601] Falk D: Cosmos, Quantum and Consciousness: Is Science Doomed to Leave Some Questions Unanswered? *Scientific American* 2019.

[602] Saatsi J: Historical Inductions, Old and New. *Synthese* 2019, **196**(10):3979–3993.

[603] Niiniluoto I: *Truth-Seeking by Abduction*. Springer 2018.

[604] Niiniluoto I: Optimistic Realism about Scientific Progress. *Synthese* 2017, **194**(9):3291–3309, https://doi.org/10.1007/s11229-015-0974-z.

[605] Park S: On the Evolutionary Defense of Scientific Antirealism. *Axiomathes* 2014, **24**(2):263–273, https://doi.org/10.1007/s10516-013-9225-4.

[606] Sankey H: Scientific Realism: An Elaboration and a Defence. *Theoria: A Journal of Social and Political Theory* 2001, **98**:35–54, www.jstor.org/stable/41802172.

[607] Laudan L: A Confutation of Convergent Realism. *Philosophy of Science* 1981, **48**:19–49, www.jstor.org/stable/187066.

[608] Putnam H: *Meaning and the Moral Sciences, Volume 29.* Routledge and Kegan Paul 1978.

[609] Hardy JA, Higgins GA: Alzheimer's Disease: The Amyloid Cascade Hypothesis. *Science* 1992, **256**:184–185.

[610] Ono K, Condron MM, Teplow DB: Structure-neurotoxicity Relationships of Amyloid β-protein Oligomers. *Proceedings of the National Academy of Sciences USA* 2009, **106**(35):14745–14750, www.ncbi.nlm.nih.gov/entrez/query.fcgi? cmd=Retrieve&db=PubMed&dopt=Citation&list_uids=19706468.

[611] Saint Anselm: *Du Cacu Diaboli (The Fall of the Devil).* The Arthur J. Banning Press 2000.

[612] Mayo D: P-Values on Trial: Selective Reporting of (Best Practice Guides Against) Selective Reporting. *Harvard Data Science Review* 2020, **2**. [Https://hdsr.mitpress.mit.edu/pub/bd5k4gzf].

[613] Gettier EL: Is Justified True Belief Knowledge? *Analysis* 1963, **23**(6):121–123.

[614] Clay M: *Teaching Theory of Knowledge.* Council for Philosophical Studies 1986.

[615] Zagzebski L: The Inescapability of Gettier Problems. *The Philosophical Quarterly* 1994, **44**(174):65–73, https://doi.org/10.2307/2220147.

[616] Zagzebski LT: *Epistemic Values: Collected Papers in Epistemology.* Oxford University Press 2020.

[617] Edgeworth FY: Methods of Statistics. *Journal of the Statistical Society of London* 1885:181–217, www.jstor.org/stable/25163974.

[618] Boring EG: Mathematical vs. Scientific Significance. *Psychological Bulletin* 1919, **16**:335–338, https://doi.org/10.1037/h0074554.

[619] Hald A: *A History of Probability and Statistics and Their Applications before 1750.* John Wiley & Sons 2003.

[620] Student: The Probable Error of a Mean. *Biometrika* 1908, **6**:1–25, www.jstor.org/stable/2331554.

[621] Fisher RA: Correlation Coefficients in Meteorology. *Nature* 1935, **136**, https://doi.org/10.1038/136474b0.

[622] Fisher RA: *Statistical Methods for Research Workers.* Oliver and Boyd, 5th edition 1934.

[623] McShane BB, Gal D: Statistical Significance and the Dichotomization of Evidence. *Journal of the American Statistical Association* 2017, **112**(519):885–895, https://doi.org/10.1080/01621459.2017.1289846.

[624] Hurlbert SH, Levine RA, Utts J: Coup de Grâce for a Tough Old Bull: "Statistically Significant" Expires. *The American Statistician* 2019, **73**(sup1):352–357, https://doi.org/10.1080/00031305.2018.1543616.

[625] Wasserstein RL, Schirm AL, Lazar NA: Moving to a World Beyond "p < 0.05". *The American Statistician* 2019, **73**(sup1):1–19, https://doi.org/10.1080/00031305.2019.1583913.

[626] Amrhein V, Trafimow D, Greenland S: Inferential Statistics as Descriptive Statistics: There Is No Replication Crisis if We Don't Expect Replication. *The American Statistician* 2019, **73**(sup1):262–270, https://doi.org/10.1080/00031305.2018.1543137.

[627] Goodman SN: How Sure Are You of Your Result? Put a Number on It. *Nature* 2018, **564**:7.

[628] Amrhein V, Greenland S, McShane B: Retire Statistical Significance-Supplement. *Nature* 2019, **567**, https://media.nature.com/original/magazine-assets/d41586-019-00857-9/16614318.

[629] Wasserstein RL, Lazar NA: The ASA Statement on p-Values: Context, Process, and Purpose. *The American Statistician* 2016, **70**(2):129–133, https://doi.org/10.1080/00031305.2016.1154108.

[630] Comment: Five Ways to Fix Statistics. *Nature* 2017, **551**:557–559.

[631] This Week: Significant Debate. *Nature* 2019, **567**:283.

[632] Leek J, Peng RD: *P* Values Are Just the Tip of the Iceberg. *Nature* 2015, **520**:612.

[633] Gelman A: Online Discussion of the ASA Statement on Statistical Significance and P-Values. *The American Statistician* 2016, **70**.

[634] Mayo DG: P-Value Thresholds: Forfeit at Your Peril. *European Journal of Clinical Investigation* 2019, **49**(10):e13170.

[635] Mayo D: *Statistical Inference as Severe Testing: How to Get beyond the Statistics Wars*. University Printing House 2018.

[636] Ioannidis JPA: The Importance of Predefined Rules and Prespecified Statistical Analyses: Do Not Abandon Significance. *Journal of the American Medical Association* 2019, **321**(21):2067–2068, https://doi.org/10.1001/jama.2019.4582.

[637] Szucs D, Ioannidis JPA: When Null Hypothesis Significance Testing Is Unsuitable for Research: A Reassessment. *Frontiers in Human Neuroscience* 2017, **11**:390, www.frontiersin.org/article/10.3389/fnhum.2017.00390.

[638] Baker M: 1,500 Scientists Lift the Lid on Reproducibility. *Nature* 2016, **533**(7604):452–454, https://doi.org/10.1038/533452a.

[639] Localio AR, Wong JB, Cornell JE, Griswold ME, Goodman SN: Inappropriate Statistical Analysis and Reporting in Medical Research: Perverse Incentives and Institutional Solutions. *Annals of Internal Medicine* 2018, **169**:577–578.

[640] Forscher BK: Chaos in the Brickyard. *Science* 1963, **142**(3590):339.

[641] Alberts B, Kirschner MW, Tilghman S, Varmus H: Rescuing US Biomedical Research from Its Systemic Flaws. *Proceedings of the National Academy of Sciences* 2014, **111**(16):5773–5777, www.pnas.org/content/111/16/5773.abstract.

[642] Pashler H, Wagenmakers EJ: Editors' Introduction to the Special Section on Replicability in Psychological Science: A Crisis of Confidence? *Perspectives on Psychological Science* 2012, **7**(6):528–530, https://doi.org/10.1177/1745691612465253.

[643] Makel MC, Plucker JA, Hegarty B: Replications in Psychology Research: How Often Do They Really Occur? *Perspectives on Psychological Science* 2012, **7**(6):537–542, https://doi.org/10.1177/1745691612460688.

[644] Heesen R: Why the Reward Structure of Science Makes Reproducibility Problems Inevitable. *The Journal of Philosophy* 2018, **115**(12):661–674.

[645] Fletcher SC: How (Not) to Measure Replication. *European Journal for Philosophy of Science* 2021, **11**(2):1–27.

[646] Randall D, Welser C: The Irreproducibility Crisis of Modern Science. *National Association of Scholars* 2018.

[647] Freedman LP: On Rigor and Replication. *Science* 2017, **356**(6333):34.

[648] Begley CG, Ellis LM: Raise Standards for Preclinical Cancer Research. *Nature* 2012, **483**(7391):531–533, https://doi.org/10.1038/483531a.

[649] Harris R: *How Sloppy Science Creates Worthless Cures, Crushes Hope, and Wastes Billions*. Basic Books 2017.

[650] Stroebe W, Postmes T, Spears R: Scientific Misconduct and the Myth of Self-Correction in Science. *Perspectives on Psychological Science* 2012, **7**(6):670–88.

[651] Stapel DA, Lindenberg S: Coping with Chaos: How Disordered Contexts Promote Stereotyping and Discrimination. *Science* 2011, **332**(6026):251–253.

[652] Fiedler S: Bad Apples and Dirty Barrels: Outliers and Systematic Institutional Failures. https://ethics.harvard.edu/blog/bad-apples-and-dirty-barrels 2013, https://ethics.harvard.edu/blog/bad-apples-and-dirty-barrels.

[653] Nosek B: An Open, Large-Scale, Collaborative Effort to Estimate the Reproducibility of Psychological Science. *Perspectives on Psychological Science* 2012, **7**(6):657–660.

[654] Yong E: Replication Studies: Bad Copy. *Nature* 2012, **485**(7398):298–300, https://doi.org/10.1038/485298a.

[655] Ioannidis J: Why Science Is Not Necessarily Self-Correcting. *Perspectives on Psychological Science* 2012, **7**:645–654.

[656] Bem D: Feeling the Future: Experimental Evidence for Anomalous Retroactive Influences on Cognition and Affect. *Journal of Personality and Social Psychology* 2011, **100**(3):407–425.

[657] Rouder JN, Morey RD: A Bayes Factor Meta-analysis of Bem's ESP Claim. *Psychonomic Bulletin & Review* 2011, **18**(4):682–689, https://doi.org/10.3758/s13423-011-0088-7.

[658] Simmons JP, Nelson LD, Simonsohn U: False-Positive Psychology: Undisclosed Flexibility in Data Collection and Analysis Allows Presenting Anything as Significant. *Psychological Science* 2011, **22**(11):1359–1366.

[659] Wagenmakers EJ, Wetzels R, Borsboom D, van der Maas HLJ: Why Psychologists Must Change the Way They Analyze Their Data: The Case of psi: Comment on Bem (2011). *Journal of Personality and Social Psychology* 2011, **100**(3):426–432.

[660] Diaconis P: [Replication and Meta-Analysis in Parapsychology]: Comment. In *Statistical Science, Volume 6*, Institute of Mathematical Statistics 1991:386–386, www.jstor.org/stable/2245731.

[661] Jeffreys H: *Theory of Probability*. Oxford University Press, 3rd edition 1961.

[662] Silberzahn R, et al.: Many Analysts, One Data Set: Making Transparent How Variations in Analytic Choices Affect Results. *Advances in Methods and Practices in Psychological Science* 2018, **1**(3):337–356, https://doi.org/10.1177/2515245917747646.

[663] Botvinik-Nezer, et al.: Variability in the Analysis of a Single Neuroimaging Dataset by Many Teams. *Nature* 2020, **582**(7810):84–90.

[664] Begley S: Fevered Debate over Alzheimer's Origins Causes Deep Divisions. *The Wall Street Journal* 2004.

[665] Yong E: Psychology's Replication Crisis Is Running Out of Excuses. *The Atlantic* 2018.

[666] Stupple A, Singerman D, Celi LA: The Reproducibility Crisis in the Age of Digital Medicine. *NPJ Digital Medicine* 2019, **2**:2.

[667] Errington TM, et al.: Investigating the Replicability of Preclinical Cancer Biology. *eLife* 2021, **10**:e71601, https://doi.org/10.7554/eLife.71601.

[668] Aker M, et al.: Direct Neutrino-Mass Measurement with Subelectronvolt Sensitivity. *Nature Physics* 2022, **18**:160–166.

[669] Stanford Linear Accelerator Center Hybrid Facility Photon Collaboration, et al.: Charm Photoproduction Cross Section at 20 GeV. *Physical Review Letters* 1983, **51**(3):156–159, https://link.aps.org/doi/10.1103/PhysRevLett.51.156.

[670] Hardwig J: The Role of Trust in Knowledge. *The Journal of Philosophy* 1991, **88**(12):693–708, www.jstor.org/stable/2027007.

[671] Medawar P: Is the Scientific Paper a Fraud? *From a BBC Talk* 1964.

[672] Ioannidis JPA: Why Most Published Research Findings Are False. *PLOS Medicine* 2005, **2**(8):e124, https://doi.org/10.1371/journal.pmed.0020124.

[673] Chang H: Ontological Principles and the Intelligibility of Epistemic Activities. In *Scientific Understanding: Philosophical Perspectives*. Edited by De Regt H, Leonelli S, Eigner K, University of Pittsburgh Press 2009.

[674] Friedman M: Objectivity and History: Reviewed Work(s): The Advancement of Science: Science without Legend, Objectivity without Illusions by Philip Kitcher. *Erkenntnis* 1996, **44**(3):379–395, www.jstor.org/stable/20012698.

[675] Merton RK: Science and the Social Order. *Philosophy of Science* 1938, **5**(3), www.jstor.org/stable/184832.

[676] Parr D: This Is Why We Stand against GM Crops 2014, www.new scientist.com/article/dn26445-greenpeace-this-is-why-we-stand-against-gm-crops/.

[677] Project TNG: About Non-GMO Project 2017, www.nongmoproject.org/about/mission/. [Online; Stand 19. December 2012].

[678] Finocchiaro MA: *The Galileo Affair: A Documentary History*. No. 1 in California Studies in the History of Science, University of California Press 1989.

[679] Shaw D: Medical Miracles of Misguided Media? *The Los Angeles Times* 2000, http://articles.latimes.com/2000/feb/13/news/mn-63989.

[680] Institute of Medicine (US) Immunization Safety Review C: Immunization Safety Review: Vaccines and Autism. Tech. rep., Institute of Medicine of the National Academies 2004.

[681] DeStefano F, Price CS, Weintraub ES: Increasing Exposure to Antibody-Stimulating Proteins and Polysaccharides in Vaccines Is Not Associated with Risk of Autism. *The journal of pediatrics* 2013, **163**(2):561–567.

[682] Irzik G, Kurtulmus F: What Is Epistemic Public Trust in Science? *British Journal for the Philosophy of Science* 2019, **70**(4):1145–1166.

[683] Offit PA: *Deadly Choices: How the Anti-Vaccine Movement Threatens Us All*. Basic Books 2011.

[684] Mnookin S: *The Panic Virus: A True Story of Medicine, Science, and Fear*. Simon & Shuster 2011.

[685] Hou CY: The Number of Measles Cases This Year Is Already More than All of the Cases Reported in 2018. *The Scientist* 2019.

[686] Akst J: Measles Epidemic Rocks Madagascar. *The Scientist* 2019.

[687] Wadman M: Measles Cases Have Tripled in Europe, Fueled by Ukrainian Outbreak. *Science* 2019.

[688] Yeager A: US Measles Cases Continue to Climb toward Record High. *The Scientist* 2019.

[689] Sheaffer R: Uncritical Publicity for Supposed "Independent UFO Investigation" Demonstrates Media Gullibility. *Skeptical Inquirer* 1998, **22**, www.csicop.org/si/show/massive_uncritical_publicity.

[690] Lemonick MD: Science on the Fringe: ESP, UFOs and Reincarnation Are Treated with Respect at the World's Most Bizarre Scientific Conference. *Time* 2005, **165**(22):53.

[691] Avaaz: Facebook: From Election to Insurrection. How Facebook Failed Voters and Nearly Set Democracy Aflame March 28, 2021, https://secure.avaaz.org/campaign/en/facebook_election_insurrection/.

[692] Avaaz: How Facebook Can Flatten the Curve of the Coronavirus Infodemic April 15, 2020, https://secure.avaaz.org/campaign/en/facebook_corona virus_misinformation/.

[693] Loomba S, de Figueiredo A, Piatek SJ, de Graaf K, Larson HJ: Measuring the Impact of COVID-19 Vaccine Misinformation on Vaccination Intent in the UK and USA. *Nature Human Behaviour* 2021, **5**(3):337–348, https://doi.org/10.1038/s41562-021-01056-1.

[694] Millar B: Misinformation and the Limits of Individual Responsibility. *Social Epistemology Review and Reply Collective* 2021, **10**(12):8–21.

[695] Sharma A, Shoukry S, Espinosa P: The Window for Climate Action Has Not Yet Closed. *United Nations Intergovernmental Panel on Climate Change (IPCC)* 2017, https://unfccc.int/files/essential_background/background_publications_htmlpdf/application/pdf/conveng.pdf.

[696] Ogien A: Doubt, Ignorance and Trust: On the Unwarranted Fears Raised by the Doubt-Mongers. In *Routledge International Handbook of Ignorance Studies*. Edited by Gross M, McGoey L, Routledge 2015:199–205.

[697] Levi SC, Sela R: *Islamic Central Asia: An Anthology of Historical Sources*. Indiana University Press 2010.

[698] Goldsmith HS: A New Approach to the Treatment of Alzheimer's Disease: The Need for a Controlled Study. *Journal of Alzheimer's disease: JAD* 2011, **25**:209–212.

[699] Tobinick E, Gross H, Weinberger A, Cohen H: TNF-alpha Modulation for Treatment of Alzheimer's Disease: A 6-Month Pilot Study. *MedGenMed: Medscape General Medicine* 2006, **8**:25.

[700] Novella S: Enbrel for Stroke and Alzheimer's. *Science-Based Medicine* 2013, https://sciencebasedmedicine.org/enbrel-for-stroke-and-alzheimers/.

[701] Leshner AI, Landis S, Stroud C, Downey A: *Preventing Cognitive Decline and Dementia: A Way Forward*. The National Academies Press 2017.

[702] Fowler JG, Reisenwitz TH, Carlson L: Deception in Cosmetic Advertising Examining Cosmetic Advertising: Examining Cosmetics Advertising Claims in Fashion Magazine Ads. *Journal of Global Fashion Marketing* 2015, **6**:194–206.

[703] Highfield R: Baffled by the Beauty Adverts? So Is a Nobel Prizewinner. *The Telegraph* 2005.

[704] World Health Organization: What Are Genetically Modified (GM) Organisms and GM Foods? 2014, www.who.int/foodsafety/areas_work/food-technology/faq-genetically-modified-food/en/.

[705] Melo EO, Canavessi AMO, Franco MM, Rumpf R: Animal Transgenesis: State of the Art and Applications. *Journal of Applied Genetics* 2007, **48**:47–61.

[706] Zhang C, Wohlhueter R, Zhang H: Genetically Modified Foods: A Critical Review of Their Promise and Problems. *Food Science and Human Wellness* 2016, **5**(3):116–123, www.sciencedirect.com/science/article/pii/S2213453016300295.

[707] Ammann K: Genomic Misconception: A Fresh Look at the Biosafety of Transgenic and Conventional Crops. A Plea for a Process Agnostic Regulation. *New Biotechnology* 2014, **31**:1–17, www.sciencedirect.com/science/article/pii/S1871678413000605.

[708] Ames BN, Profet M, Gold LS: Dietary Pesticides (99.99% All Natural). *Proceedings of the National Academy of Sciences* 1990, **87**(19):7777–7781, www.pnas.org/content/87/19/7777.abstract.

[709] Gould F: *Genetically Engineered Crops Experiences and Prospects*. The National Academies Press 2016.

[710] Nicolia A, Manzo A, Veronesi F, Rosellini D: An Overview of the Last 10 Years of Genetically Engineered Crop Safety Research. *Critical Reviews in Biotechnology* 2014, **34**:77–88.

[711] Blancke S, Van Breusegem F, De Jaeger G, Braeckman J, Van Montagu M: Fatal Attraction: The Intuitive Appeal of GMO Opposition. *Trends in Plant Science* 2015, **20**:414–418.

[712] Weise E: Academies of Science Find GMOs Not Harmful to Human Health. *USA Today* 2016.

[713] Haack S: Federal Philosophy of Science: A Deconstruction- and a Reconstruction. *5 N.Y.U. J.L. & Liberty* 2010, **5**(2):394–435.

[714] Foster K, Huber P: *Judging Science. Scientific Knowledge and the Federal Courts.* MIT Press 1999.

[715] Abboud A: Daubert v. Merrell Dow Pharmaceuticals, Inc. (1993). *The Embryo Project Encyclopedia* May 29, 2017, https://embryo.asu.edu/pages/daubert-v-merrell-dow-pharmaceuticals-inc-1993.

[716] Sunstein CR: Beyond the Precautionary Principle. *University of Pennsylvania Law Review* 2003, **151**(3):1003–1058, www.jstor.org/stable/3312884.

[717] Scheidel W, Meeks E: Orbis: The Stanford Geospatial Network Model of the Roman World. https://orbis.stanford.edu 2021.

[718] Zuckerman H, Merton RK: Patterns of Evaluation In Science: Institutionalisation, Structure and Functions of the Referee System. *Minerva* 1971, **9**:66–100, https://doi.org/10.1007/BF01553188.

[719] Garfield E: Introducing *Citation Classics*. The Human Side of Scientific Reports. *Current Comments* 1977, **1**:5–7.

[720] Beaver D, Rosen R: Studies in Scientific Collaboration. *Scientometrics* 1978, **1**: 65–84.

[721] Heesen R: Communism and the Incentive to Share in Science. *Philosophy of Science* 2017, **84**(4):698–716.

[722] Samuelson PA: Economists and the History of Ideas. *The American Economic Review* 1962, **52**:1–18, www.jstor.org/stable/1823476.

[723] Merton RK: The Matthew Effect in Science. *Science* 1968, **159**(3810):56–63, https://doi.org/10.1126/science.159.3810.56.

[724] Merton RK: The Matthew Effect in Science. 2. Cumulative Advantage and the Symbolism of Intellectual Property. *Isis* 1988, **79**(299):606–623.

[725] Bitan G, Lomakin A, Teplow DB: Amyloid β-Protein Oligomerization – Prenucleation Interactions Revealed by Photo-Induced Cross-Linking of Unmodified Proteins. *Journal of Biological Chemistry* 2001, **276**(37):35176–35184.

[726] Le Châtelier HL: Sur un énoncé général des lois des équilibres chimiques. *Comptes Rendus* 1884, **99**:786–789.

[727] Lomakin A, Teplow DB, Kirschner DA, Benedek GB: Kinetic Theory of Fibrillogenesis of Amyloid b-Protein. *Proceedings of the National Academy of Sciences USA* 1997, **94**(15):7942–7947, www.ncbi.nlm.nih.gov/pubmed/9223292.

[728] Einstein A: Über die von der molekularkinetischen Theorie der Wärme geforderte Bewegung von in ruhenden Flüssigkeiten suspendierten Teilchen. *Annalen Der Physik* 1905, **17**(8):549–560.

[729] Maddox B: The Double Helix and the "Wronged Heroine". *Nature* 2003, **421**(6921):407–408, https://doi.org/10.1038/nature01399.

[730] Zuckerman HA: Nobel Laureates in the United States: A Sociological Study of Scientific Collaboration. *PhD thesis*, Columbia University 1965.

[731] Zuckerman H: Nobel Laureates in Science: Patterns of Productivity, Collaboration, and Authorship. *American Sociological Review* 1967, **32**:391–403.

[732] Zuckerman H: *Scientific Elite: Nobel Laureates in the United States.* Free Press 1977.

[733] Gieryn T, Oberlin K: Science, Sociology of. *International Encyclopedia of the Social & Behavioral Sciences* 2015, **21**:261–267.

[734] Shapin S: Here and Everywhere: Sociology of Scientific Knowledge. *Annual Review of Sociology* 1995, **21**:289–321, https://doi.org/10.1146/annurev.so.21.080195.001445.

[735] Longino H: The Social Dimensions of Scientific Knowledge. In *The Stanford Encyclopedia of Philosophy*. Edited by Zalta EN, Metaphysics Research Lab, Stanford University 2019.

[736] Hardwig J: Epistemic Dependence. *The Journal of Philosophy* 1985, **82**(7):335–349, www.jstor.org/stable/2026523.

[737] Turnbull HW, Hall AR, Tilling L, Scott JF (Eds): *The Correspondence of Isaac Newton, Volume 1*. Cambridge University Press for the Royal Society 1959.

[738] Merton RK: *On the Shoulders of Giants: A Shandean Postscript*. Harcourt 1985.

[739] Crick F: *What Mad pursuit – A Personal View of Scientific Discovery*. Basic Books 1989.

[740] Goldman AI: Experts: Which Ones Should You Trust? *Philosophy and Phenomenological Research* 2001, **63**:85–110, www.jstor.org/stable/3071090.

[741] Burge T: Content Preservation. *The Philosophical Review* 1993, **102**(4):457–488, www.jstor.org/stable/2185680.

[742] Foley R: Egoism in Epistemology. In *Socializing Epistemology: The Social Dimensions of Knowledge*. Edited by Schmitt FF, Rowman & Littlefield 1994.

[743] Sankey H: Kuhn's Changing Concept of Incommensurability. *The British Journal for the Philosophy of Science* 1993, **44**(4):759–774, www.jstor.org/stable/688043.

[744] Feyerabend P: Explanation, Reduction and Empiricism. In *Scientific Explanation, Space, and Time (Minnesota Studies in the Philosophy of Science), Volume III*. Edited by Feigl H, Maxwell G, University of Minneapolis Press 1962:28–97.

[745] Oberheim E, Hoyningen-Huene P: The Incommensurability of Scientific Theories. In *Stanford Encyclopedia of Philosophy*. Edited by Zalta EN, Metaphysics Research Lab, Stanford University Fall 2018.

[746] Walker TC: The Perils of Paradigm Mentalities: Revisiting Kuhn, Lakatos, and Popper. *Perspectives on Politics* 2010, **8**(2):433–451, www.jstor.org/stable/25698611.

[747] Gholson B, Barker P: Kuhn, Lakatos, and Laudan. Applications in the History of Physics and Psychology. *American Psychologist* 1985, **40**:755–769.

[748] Reese H, Overton W: Models of Development and Theories of Development. In *Life-Span Developmental Psychology: Research and Theory*. Edited by Goulet LR, Baltes PB, Academic Press 1970:115–145.

[749] Palermo D: Is a Scientific Revolution Taking Place in Psychology? *Social Studies of Science* 1971, **1**:135–155.

[750] Lakatos I, Musgrave A: *Criticism and the Growth of Knowledge*. Proceedings of the 1965 International Colloquium in the Philosophy of Science Cambridge: University Press 1970, www.loc.gov/catdir/enhancements/fy0642/78105496-d.html.

[751] Duhem PMM: *ΣΩZEIN TA ΦAINOMENA: Essai sur la notion de théorie physique de Platon àGalilée*. A. Hermann et Fils 1908.

[752] Duhem PMM: *To Save the Phenomena. An Essay on the Idea of Physical Theory from Plato to Galileo*. University of Chicago Press 1969.

[753] Lakatos I, Feyerabend PK, Motterlini M: *For and against Method: Including Lakatos's Lectures on Scientific Method and the Lakatos-Feyerabend Correspondence*. University of Chicago Press 1999.

[754] Rescorla M: Convention. In *Stanford Encyclopedia of Philosophy*. Edited by Zalta EN, The Metaphysics Research Lab Center for the Study of Language and Information, Stanford University Summer 2019.

[755] Lilienfeld S: *Great Readings in Clinical Science: Essential Selections for Mental Health Professionals*. Pearson 2012.

[756] Laudan L: From Theories to Research Traditions. In *Readings in the Philosophy of Science*. Edited by Brody BA, Grandy RE, Prentice Hall 1989.

[757] Aspect A: From Huygens' Waves to Einstein's Photons: Weird Light. *Comptes Rendus Physique* 2017, **18**(9):498–503, www.sciencedirect.com/science/article/pii/S1631070517301032.

[758] Laudan LL: *Progress and Its Problems: Toward a Theory of Scientific Growth.* University of California Press 1977, https://doi.org/10.1007/BF02074137.

[759] Cole J: Wittgenstein's Neurophenomenology. *Medical Humanities* 2007, **33**:59–64, https://mh.bmj.com/content/33/1/59.

[760] Holt J: Suicide Squad. *The New York Times* 2009, www.nytimes.com/2009/03/01/books/review/Holt-t.html?_r=1&ref=review.

[761] Lackey DP: What Are the Modern Classics? The Baruch Poll of Great Philosophy in the Twentieth Century. *The Philosophical Forum* 1999, **30**(4):329–346.

[762] Beaney MA: What Is Analytic Philosophy? In *The Oxford Handbook of the History of Analytic Philosophy.* Edited by Beaney MA, Oxford University Press 2013:3–29.

[763] Harton, Jr MC: *A Critical Examination of the Volitional Theory of Action.* McMaster University 1976.

[764] Frith CD, Haggard P: Volition and the Brain – Revisiting a Classic Experimental Study. *Trends in Neurosciences* 2018, **41**(7):405–407, www.ncbi.nlm.nih.gov/pubmed/29933770.

[765] Haggard P, Libet B: Conscious Intention and Brain Activity. *Journal of Consciousness Studies* 2001, **8**(11):47–63.

[766] Libet B, Gleason CA, Wright EW, Pearl DK Time of Conscious Intention to Act in Relation to Onset of Cerebral Activity (Readiness-Potential): The Unconscious Initiation of a Freely Voluntary Act. *Brain* 1983, **106**(3):623–642, https://doi.org/10.1093/brain/106.3.623.

[767] Trevena J, Miller J: Brain Preparation before a Voluntary Action: Evidence against Unconscious Movement Initiation. *Consciousness and Cognition* 2010, **19**:447–456, www.sciencedirect.com/science/article/pii/S1053810009001135.

[768] Pockett S, Purdy SC: Are Voluntary Movements Initiated Preconsciously? The Relationships between Readiness Potentials, Urges, and Decisions. In *Conscious Will and Responsibility*, 1st edition. Edited by Sinnott-Armstrong W, Nadel L, Oxford University Press 2011:34–46.

[769] Libet B: Do We Have Free Will? In *Conscious Will and Responsibility*, 1st edition. Edited by Sinnott-Armstrong W, Nadel L, Oxford University Press 2011:34–46.

[770] Niven PR, Vasshus TI: *Business in Cartoons* Corporater 2021.

Index

Printed in the United States
by Baker & Taylor Publisher Services